GREENWICH LIBRARY

ADHESIVE PARTICLE FLOW

A Discrete-Element Approach

Adhesive Particle Flow: A Discrete-Element Approach offers a comprehensive treatment of adhesive particle flows at the particle level. This book adopts a particle-level approach oriented toward directly simulating the various fluid, electric field, collision, and adhesion forces and torques acting on the particles, within the framework of a discrete-element model.

It is ideal for professionals and graduate students working in engineering and atmospheric and condensed matter physics, materials science, environmental science, and other disciplines in which particulate flows have a significant role. The presentation is applicable to a wide range of flow fields, including aerosols, colloids, fluidized beds, and granular flows.

It describes both physical models of the various forces and torques on the particles as well as practical aspects necessary for efficient implementation of these models in a computational framework.

Jeffrey S. Marshall is a Professor in the School of Engineering at the University of Vermont. He is a Fellow of the American Society of Mechanical Engineers. He obtained a Ph.D. in Mechanical Engineering from the University of California, Berkeley. Dr. Marshall taught at the University of Iowa from 1993 to 2006 and was Chair of the Mechanical and Industrial Engineering Department from 2001 to 2005. He is a recipient of the ASME Henry Hess Award and the U.S. Army Research Office Young Investigator Award. He has authored more than 95 journal articles and book chapters and one textbook, *Inviscid Incompressible Flow* (2001).

Shuiqing Li is a Professor in the Department of Thermal Engineering at Tsinghua University. He obtained a Ph.D. in Engineering Thermophysics from Zhejiang University. He was a visiting scholar at the University of Leeds in 2004–2005, at the University of Iowa in 2006, and at Princeton University in 2010–2011. Dr. Li is a recipient of the National Award for New Century Excellent Talents (2009) and the Tsinghua University Award for Young Talents on Fundamental Studies (2011). He shared a Chinese National Teaching Award on Combustion Theory. He has been awarded five fundamental grants from the Natural Science Foundation of China and has authored more than 40 journal articles.

Adhesive Particle Flow

A DISCRETE-ELEMENT APPROACH

Jeffrey S. Marshall
University of Vermont

Shuiqing Li
Tsinghua University

CAMBRIDGE
UNIVERSITY PRESS

CAMBRIDGE
UNIVERSITY PRESS

32 Avenue of the Americas, New York, NY 10013-2473, USA

Cambridge University Press is part of the University of Cambridge.

It furthers the University's mission by disseminating knowledge in the pursuit of
education, learning, and research at the highest international levels of excellence.

www.cambridge.org
Information on this title: www.cambridge.org/9781107032071

© Jeffrey S. Marshall and Shuiqing Li 2014

This publication is in copyright. Subject to statutory exception
and to the provisions of relevant collective licensing agreements,
no reproduction of any part may take place without the written
permission of Cambridge University Press.

First published 2014

Printed in the United States of America

A catalog record for this publication is available from the British Library.

Library of Congress Cataloging in Publication Data
Marshall, Jeffrey S. (Jeffrey Scott), 1961–
Adhesive particle flow : a discrete-element approach / Jeffrey S. Marshall,
Shuiqing Li, University of Vermont, Tsinghua University.
    pages   cm
Includes bibliographical references.
ISBN 978-1-107-03207-1 (hardback)
1. Granular flow.   2. Adhesion.   3. Discrete element method.   I. Li, Shuiqing,
1975–   II. Title.
TA357.5.G47M37   2014
620.1′06–dc23   2013040678

ISBN 978-1-107-03207-1 Hardback

Cambridge University Press has no responsibility for the persistence or accuracy of
URLs for external or third-party Internet Web sites referred to in this publication
and does not guarantee that any content on such Web sites is, or will remain,
accurate or appropriate.

To Marilyn and Yun,
and to Jodie, Eric, Emily, Paul, Jonathan, and Zelin

Contents

Preface

There has been a rapid increase in the number of research papers over the past decade concerning flow of adhesive particles. Interest is driven in part by a focus on particulate flow problems with small particle sizes, for which adhesive force becomes increasingly important compared with particle inertia or gravity. Literature on flow of adhesive particles is found both in the standard particulate flow and fluid mechanics journals and in more specialized journals dealing with applications in areas such as ash filtration, aerosol and cloud modeling, dust mitigation, nanoparticle deposition, ceramics manufacturing, fouling of MEMS devices, food science, bioengineering, microfluidics, sediment transport, and production of biofuels. Unlike previous research involving adhesive particles, which employed a population-balance method, this recent work has adopted a mesoscale particle-level approach that simulates the various fluid, electric field, collision, and adhesion forces and torques acting on individual particles, enabling study of the collaborative dynamics governing the interaction of groups of many agglomerates consisting of large numbers of particles. Particle-level modeling of such problems is made possible both by improved physical models of the various forces and torques acting on the particles, and by improved computational algorithms for handling systems with a wide range of time scales.

Adhesive particle flows arise in many applications in industry, nature, and the life sciences. In the field of manufacturing, applications include dust fouling of electronic equipment, 3D printing, manufacturing and surface treatment of ceramic materials, and electrospray processes. A variety of new microscale and nanoscale devices have been designed whose manufacturing requires the precise placement of nanoparticles and nanotubes onto a substrate using some type of dispersion process. Microfluidic processes used for biological assay ("lab-on-a-chip") rely on the ability to manipulate and sort particles and biological cells, which can be treated as particles. Algae biofuel production requires the ability to process and optimize flows with suspended algae cells, which respond to near-surface turbulent flow fields and light shading by other algae cells. Blood flow involves not only interaction of red and white blood cells, but also interaction of blood cells with platelets and other particles (e.g., liposomes or cancer cells) that might be transported in the blood. Particulate pollution problems are of great concern in many parts of the world due to ash from combustion processes that needs to be captured before it escapes into the atmosphere. In many of these examples and in a wide variety of others, adhesion of particles (via a variety of mechanisms) plays a critical role. The physics of these processes are often controlled by agglomerate formation and breakup processes at the particle scale, and in cases such as nanoparticle dispersion it is desirable to precisely control the motion of individual particles in the presence of complex flow geometry.

The objective of this book is to provide a comprehensive account of modern particle-level approaches for analyzing and simulating particulate flows at the mesoscale, with particular focus on flows involving adhesive particles. Although several different modeling approaches are described, the book focuses specifically on the soft-sphere discrete-element method (DEM), which is useful for a wide range of particulate flow problems. DEM shares a similar Lagrangian computational methodology with molecular dynamics methods, but at the same time it makes use of extensive modeling for fluid-induced forces on particles and for interparticle interaction via collision, adhesion, and electric field effects. The book is structured according to different types of approximations and computational models used for flow simulation. The first chapter discusses various applications of flows with adhesive particles and the associated dimensionless parameters governing them. This chapter also introduces various approximations that are commonly made when analyzing particulate fluids. The second chapter compares different modeling approaches for adhesive particle flows as a function of length and time scales of the problem, and examines different types of multiscale modeling approaches. This chapter introduces the discrete-element model and compares it to other mesoscale and macroscale models for particulate flows, such as molecular dynamics, Brownian dynamics, dissipative particle dynamics, the discrete parcel method, and the population balance method. The third chapter summarizes forces and torques that occur during particle collision for cases with no adhesion forces, including elastic and dissipative normal forces as well as resistance to sliding, rolling, and twisting motions. The effect of adhesion on collision forces is discussed in the fourth chapter, including van der Waals forces, electrical double-layer repulsion, protein binding forces often found in cell interaction problems, liquid bridging, and sintering forces. Different fluid-induced forces on particles are discussed in the fifth chapter, including a scaling analysis to assess when different fluid forces can be neglected. This chapter also discusses particle interaction with acoustic radiation. Turbulent dispersion models are discussed in the sixth chapter, with an emphasis on accurate modeling of the particle collision rate in turbulent flows and its relationship to small-scale concentration field heterogeneity. Chapter 7 extends the discrete-element method to nonspherical particles, which are common in applications such as blood flow, biofuel combustion, and food processing. Particle interactions with electric and magnetic fields are discussed in the eighth chapter. These forces are important for many particle adhesion problems as well as for control of particulate flows in many applications. Chapter 9 examines the differences between flows with micron-sized particles and those with nanoscale particles. These differences arise from noncontinuum effects due to the fact that nanoscale particles are often of a similar size as the mean-free-path of the surrounding fluid. Chapter 10 discusses issues that arise during computer implementation of discrete-element methods, including numerical stiffness, numerical instability, and challenges of computing particulate flows in complex domains. This chapter also discusses various measures used to characterize particle agglomerates. Chapter 11 describes select applications of discrete-element modeling of particulate flow problems, which are selected to illustrate interesting physical phenomena exhibited by particles interacting with fluids and electric fields.

Acknowledgments

We would like to thank the many students and former students with whom we have collaborated on studies of particulate flows, several of whose work played a significant role in shaping this book. Particular thanks are extended to Jennifer Chesnutt, Guanqing Liu, Kyle Sala, John Mousel, Greg Hewitt, Auston Maynard, Mengmeng Yang, and Yiyang Zhang, each of whose work is featured in different sections of the book. Comments on this research were provided by many colleagues, and we particularly acknowledge valuable discussions with Charley Wu, Colin Thornton, Norman Chigier, Chung K. Law, Aibing Yu, Jun-ru Wu, H.S. Udaykumar, Louis Rossi, Stephen Tse, Albert Ratner, Yulong Ding, Pratim Biswas, Jonathan Seville, Eric Loth, Stefan Luding, and Cetin Cetinkaya. Professors V.C. Patel and Qiang Yao and Dr. John R. Grant are particularly acknowledged for their invaluable mentoring and friendship throughout our careers.

Funding to support the work of JSM on particulate flow from NASA (NNX12AI15A, NNX13AD40A, NNX08AZ07A), the U.S. National Science Foundation (DGE-1144388, CBET-1332472), the U.S. Department of Energy (DE-FG36 – 08G088182), the Caterpillar Corporation, and the University of Iowa Facilities Management Group is greatly appreciated. SL particularly acknowledges support from the National Science Foundation of China (No. 50306012, 50776054, 50976058, and 51176094) in his early career, and from the National Key Basic Research and Development Program (No. 2013CB228506) to work across disciplines.

Assistance in production of the book was provided by Runru Zhu, Wenwei Liu, Melissa Faletra, and Yihua Ren in producing some of the figures; by Bing Chen for code assistance; and by the students enrolled in SL's *Introduction to Particle Transport* class at Tsinghua. Special thanks are extended to Emily Marshall for handling all of the permissions, and to our editor Peter Gordon at Cambridge University Press and our project manager Adrian Pereira at Aptara, Inc., for their enthusiasm, encouragement, and professionalism throughout the writing and production of the book.

1

Introduction

1.1. Adhesive Particle Flow

A particulate flow is one in which a moving fluid interacts with a large number of discrete solid particles. The category is extraordinarily broad, encompassing everything from suspended dust carried by atmospheric winds to avalanches of debris or snow rolling down a hillside. Widely varying industrial, biological, and environmental processes can be interpreted as particulate flows, encompassing areas of study such as sediment transport by stream and coastal flows, aerosol dynamics, colloidal suspensions, fluidized bed reactors, granular flows, slurries, and nanoparticle dispersions. There are also many situations where a suspension of biological cells can be interpreted as a particulate fluid, which extends the notion of particulate flow to problems such as blood flow and algal suspensions. Finally, there are many aspects of the methods used to analyze and model particulate flows that can be either directly applied or applied with small modifications to other types of multiphase flows, including droplet dispersions and bubbly flows, assuming that the deformation of the droplets and bubbles is minimal.

Despite the many different forms in which we encounter them, there are a number of characteristics that are shared by most particulate flows. Some of these characteristics arise from the interaction of the individual particles with the surrounding fluid. For instance, a particulate flow past a blunt body tends to exert a higher drag force than the body would experience in a fluid with no particles. One characteristic of particulate fluids associated with this increased drag is the fact that the effective viscosity of the fluid increases due to the presence of particles, even if the particles move with the same velocity as the surrounding fluid (Einstein, 1906). A second characteristic of particulate fluids is the high degree of dissipation that occurs due to the interfacial force caused by particle motion relative to the fluid velocity field, for instance, as the fluid accelerates around a body. This relative particle-fluid motion also causes oscillatory flows associated with sound waves and turbulent fluctuations to attenuate rapidly in particulate fluids (Epstein and Carhart, 1953). A third characteristic of particulate fluids is that they can be highly abrasive due to collision of particles with a surface, for instance, in a sand blasting application. This abrasive quality of particulate fluids arises from the fact that heavy particles with more inertia than the surrounding fluid will drift relative to fluid streamlines as they curve around a body, leading particles to instead collide with the body surface, imparting their momentum to the body. Of course, fluid flows in general, and multiphase flows in particular, are highly nonlinear, and so none of these characteristics hold invariably. For instance, Lashkov (1992) showed that at small concentrations the presence of particles can serve to accelerate occurrence of drag crisis in flow past a bluff body, thus moving the point of boundary layer separation backward on the body surface. Although the local skin friction is increased by the presence of particles, the pressure drag on the body is sufficiently decreased by the backward motion of the separation point that the net drag on the body decreases due to the presence of particles in the flow in a select range of Reynolds numbers.

Other characteristics of particulate flows are associated with the interaction of the particles with each other. A common simplification in the fluid flow literature is that of *dilute particulate flow*, in which case the particles are assumed to be so few in number and so small as to never interact with each other. While from a modeling point of view this is often a convenient assumption, in many applications particle collisions play a vital role. Particle collisions occur due to several different processes. Inertia- and diffusion-induced collisions occur due to the drift of particles relative to the surrounding fluid that is caused by the particle inertia and by Brownian motion, respectively. The direction and degree of particle drift varies across the flow field, and these differences give rise to particles originating from different parts of the flow colliding with each other. Another cause of particle collisions is fluid shear stress. Shear-induced collisions occur due to the finite particle size as particles in a faster-moving layer of the shear flow move past particles in a slower-moving layer. Collisions may also occur when particles exert forces on each other, as might be caused by the long-range Coulomb force when particles are electrically charged (Zhang et al., 2011).

Particle collisions have a number of important consequences. For instance, collisions affect the particle transport by modifying the direction of particle velocity, thereby increasing the rate of particle diffusion in the fluid. Collisions can cause transfer of electrons between the particles, leading to accumulation of electrical charge on the particles, a process known as *tribocharging*. This process is responsible, for instance, for charging of the dust particles on Mars and associated radio and electrical discharges emitted by the dust storms (Renno et al., 2003). Violent collisions can cause fracture and breaking of the particles, or more moderate collisions can over time cause a gradual smoothening of the particle surfaces. Most importantly, in the presence of an adhesive force, collisions can cause particles to stick together.

Particle adhesion typical occurs for suspensions of small particles, with diameter of about 10 microns or smaller, for which the adhesive force can overcome the gravitational force and the elastic rebound force acting on the particles. Colliding particles held together by relatively weak adhesive forces (such as van der Waals attraction or capillary force) are generally called *agglomerates*, whereas *aggregates* occur when particles adhere together by strong forces, for example by covalent, ionic, or sintering bonds. This terminology, first established by Nichols et al. (2002), is not universally accepted in the particulate flow literature, and we often find the terms switched or referred to instead as soft agglomerate and hard agglomerate, respectively.

There are a wide variety of different adhesive forces that act between particles during and after particle collisions. Kinloch (1987) proposed six distinct categories of particle adhesion: *dispersive adhesion* due to van der Waals attraction of molecules in the contact region between the particles; *electrostatic adhesion* due to electrical attraction of charged or polarized particles; *diffusive adhesion* due to diffusion of atoms or molecules from one particle to the next, such as occurs with sintering of ceramic or metal powders; *chemical adhesion* due to swapping or sharing of atoms of the two particles, as occurs with ionic and covalent bonding processes, respectively; *hydrogen-bond adhesion* due to formation of a weak bond when atoms such as oxygen, nitrogen, or fluorine share a hydrogen nucleus; and finally *mechanical adhesion*, which occurs when the void or pore space between the particles is filled with a third medium that mechanically holds the particles together.

A common example of mechanical adhesion occurs for slightly wet granular particles in air, in which each particle is surrounded by a thin liquid film. The liquid films on two colliding particles connect to form a *liquid bridge*, which fills the space between the particles and holds the particles together by capillary force, modified by a viscous effect within the liquid as the particles are moved toward or away from each other (Ennis et al., 1990). Dry dust particles usually adhere to each other or to the surfaces of a surrounding solid body by a combination of electrostatic and van der Waals forces. Electrostatic forces are relatively weak, but since they act over a relatively long range, electrostatic forces are effective at transporting dust particles together or moving them toward the surface of a body immersed in the flow. Once collision has occurred, the stronger (but relatively shorter-range) van der Waals force typically takes over and is primarily responsible for holding the dust particles together (Feng and Hays, 2003; Gady et al., 1996). Biological cells adhere to one another by a type of chemical adhesion called *ligand-receptor* binding (Bell, 1978), in which ligand proteins on the cell surface connect to specific receptor proteins on the surface of the opposing cell to form a molecular link between the cells. This cellular bonding is responsible for adhesion of white blood cells to endothelial cells on a blood vessel wall prior to transport of the white blood cells into surrounding tissue, among other things. These and other examples of flows with adhesive particles are discussed in the final section of this chapter.

There exists extensive literature on particulate flows, including many excellent books and review articles. Much of this literature, however, deals with either dilute particulate flow or with granular flow of nonadhesive particles. Most existing literature on adhesive particles takes what is called a *population-balance* approach, in which analytical models are used to cluster agglomerates of particles into larger "effective particles" without taking into account the dynamics of the individual particles that make up the agglomerate (Friedlander, 2000). The population-balance approach can be very useful for large-scale systems provided that the conditions under which it is derived are satisfied. This requires, for instance, that the particle concentration is sufficiently small, that particles are spherical and have negligible inertia, that the particle distribution is homogeneous, that collisions are binary, and that the flow length scales are much larger than the particle agglomerates, so that we can treat agglomerates as existing in a simple shear flow (Reinhold and Briesen, 2012).

The rise in prominence of microscale and nanoscale engineering in recent years, including microscale description of biological systems, has resulted in a great deal of increased focus on processes that are outside the scope of traditional population-balance models. Such problems include processes where particles interact with microscale structures that have a length scale comparable to that of the particle agglomerates, processes with high particle concentrations or with nonspherical particles, processes involving strong gradients in particle concentration, or processes where particle dispersion relative to the flow streamlines plays an important role. Such problems require a fundamentally different approach for modeling and simulation, based on tracking the dynamics of individual particles both while they are traveling alone through the fluid and while they are colliding with and adhering to other particles. At the same time, the modeling approach should allow a sufficiently large number of particles in the simulation to enable examination of the collective dynamics of adhesive particles as large agglomerates interact with each other and with fluid flows in complex domains. Such collective dynamics often require simulations ranging from tens of thousands to millions of particles. This requirement puts such problems out of the scope of microscale approaches, which simulate the flow around each individual particle.

Within the past few decades, a variety of computational methods have been developed that model particle collective dynamics at an intermediate (or mesoscale) level. These mesoscale methods, which we broadly classify as *Lagrangian particle methods*, attempt to strike a middle ground in which the motions of individual particles are resolved, but sufficiently coarse modeling is used to describe the particle interactions that large numbers of particles can be accommodated. Lagrangian particle methods have developed rapidly in application areas such as atomic-level simulation of molecular processes, colloidal fluids and aerosols, granular flows, and polymeric fluids. A number of different variations of these methods have arisen, examples being molecular dynamics (MD), Brownian dynamics (BD), dissipative particle dynamics (DPD), and the discrete-element method (DEM). Lagrangian particle methods share the feature that the computational elements (e.g., atoms, molecules, particles) are allowed to interact for a period of time under prescribed interaction laws.

These different methods have important differences in both approach and applicability. MD deals with individual atoms with dimensions of roughly 0.1 nm by using a computational time step on the order of 1 femtosecond (10^{-15}s), while the duration of the overall computation typically ranges from picoseconds to nanoseconds. At present, MD is too time consuming to apply for particles with sizes larger than about 10 nm. At a larger scale, DPD simulates the dynamic and rheological properties of simple and complex fluids by using a computational element that represents a group of fluid molecules, rather than individual molecules or atoms. Brownian dynamics (BD), or the more general Langevin dynamics (LD), introduces the notion of a fluid continuum surrounding a set of molecules or small particles, which is used to represent effects of a solvent or gas phase without simulating the individual molecular dynamics of this surrounding fluid. The surrounding fluid continuum exerts both a drag force and a random (Brownian) force on the computational elements. Use of the fluid continuum approximation, together with neglect of particle inertia in BD, enables the BD method to employ much higher time steps than the MD or DPD methods.

This book focuses on the *discrete element method*, in which the motion, collision, and adhesion of individual particles are resolved in time and space, but analytical models are used to approximate the interaction of the particles with the surrounding fluid. The discrete element method was originally developed for nonadhesive granular flows, for which it is used extensively to simulate collisions of individual granules with each other (Bertrand et al., 2005; Cundall and Strack, 1979). Although for macroscopic granular particles the dynamics are governed mostly by gravity and collisional and frictional forces, for adhesive microparticles the dominant interactions include electrostatic (Coulomb) and van der Waals forces (Aranson and Tsimring, 2006). There has recently been rapid progress on understanding the physics related to the intermolecular and surface forces at the microscale (Israelachvili, 2011), which enable us to develop more rational adhesive contact models. In this book, we examine the discrete element method in a way that is applicable for a wide range of particulate flows, from granular flows to nanoparticle flows. The book places particular emphasis on *adhesive particle flows*, which consist of suspensions of adhesive particles immersed in a fluid flow. The modeling approach for such flows must deal with collision and adhesion of the particles to each other and to surrounding substrates, with interaction of the particles with the surrounding fluid, with complex flow domains, and with the effect of external fields, such as electrostatic or acoustic fields, on the particles.

1.2. Dimensionless Parameters and Related Simplifications

Dimensionless parameters are used in fluid mechanics to delineate regimes of motion and to establish limitations of various assumptions and approximations used in obtaining a problem solution. A large number of dimensionless parameters are used in describing particulate flows; however, among these, a few play a particularly important role in characterizing the flow type and the various approximations used in flow analysis. While various other dimensionless parameters will be introduced throughout the book, it is useful to discuss a few particular parameters early on so as to introduce and characterize the different assumptions typically used in analyzing particle flows.

1.2.1. Stokes Number

A particulate flow in which particles are strongly influenced by the surrounding fluid is called a *suspension*, examples being colloids and aerosols. At the opposite extreme, particulate flows where there is little influence between the particles and the fluid are called *granular flows*. Where a specific particulate flow falls on this spectrum from suspension to granular flow is governed largely by the *Stokes number*, denoted by St. The Stokes number is defined as the ratio of the characteristic response time of a particle to the characteristic fluid time scale. By way of motivation in developing these time scales, we consider a simplified form of the momentum equation for a single particle in which the particle momentum changes only due to the drag force between the particle and the surrounding fluid. It is further assumed that the particle is very small and slow-moving relative to the fluid and that it is spherical in shape, so that the classic Stokes drag solution gives the drag force on the particle as

$$\mathbf{F}_d = -3\pi \mu d (\mathbf{v} - \mathbf{u}). \tag{1.2.1}$$

In this equation, d is the particle diameter, μ is the fluid viscosity, and $\mathbf{v}$ and $\mathbf{u}$ are the particle velocity and the fluid velocity evaluated at the particle centroid (in the absence of the particle), respectively. With these approximations, the simplified particle momentum equation becomes

$$m\frac{d\mathbf{v}}{dt} = -3\pi\mu d(\mathbf{v} - \mathbf{u}), \tag{1.2.2}$$

where m is the mass of a single particle and d/dt denotes the rate of change in time at a point following the particle. If the fluid flow in which the particle is immersed has a characteristic velocity scale U and length scale L, dimensionless velocity and time variables can be defined as

$$\mathbf{v}' = \mathbf{v}/U, \quad \mathbf{u}' = \mathbf{u}/U, \quad t' = tU/L. \tag{1.2.3}$$

Substituting these dimensionless variables into (1.2.2) yields

$$\frac{mU}{3\pi\mu dL}\frac{d\mathbf{v}'}{dt'} = -(\mathbf{v}' - \mathbf{u}'). \tag{1.2.4}$$

The fluid advection time scale is $\tau_f = L/U$ and the particle response time is $\tau_p = m/3\pi d\mu$. From these time scales, the Stokes number can be defined by

$$\text{St} \equiv \frac{\tau_p}{\tau_f} = \frac{mU}{3\pi\mu dL}, \tag{1.2.5}$$

which is equal to the coefficient multiplying the dimensionless particle acceleration in (1.2.4). Writing particle mass $m = \pi\rho_p d^3/6$ in terms of particle density ρ_p gives an alternative expression for Stokes number as

$$\text{St} = \frac{\rho_p d^2 U}{18\mu L}. \tag{1.2.6}$$

When the Stokes number is much smaller than unity, the particles nearly follow the fluid streamlines and the magnitude of the particle drift velocity relative to the fluid is small compared to the fluid velocity scale U. When the Stokes number is much larger than unity, the particles respond only very slowly to the fluid force. Some of the most interesting flows occur when the Stokes number is near unity, for which the fluid flow has a strong effect on the particle transport but the particles also disperse significantly relative to the fluid flow. An example illustrating particle transport in the vortex street wake behind a square cylinder is given in Figure 1.1 from Jafari et al. (2010), using a two-dimensional lattice-Boltzmann method. In this figure, the particles are initiated upstream in a random configuration and transported past the cylinder by the fluid flow. Computations are shown with two values of the particle Stokes number. In Figure 1.1a, with $\text{St} = 0.001$, the particles are advected with the fluid flow and exhibit random positions independent of the fluid vortices. In Figure 1.1b, with $\text{St} = 1$, the particles are expelled from the vortex cores by the action of centrifugal force, forming thin particle sheets surrounding the vortex structures in the wake.

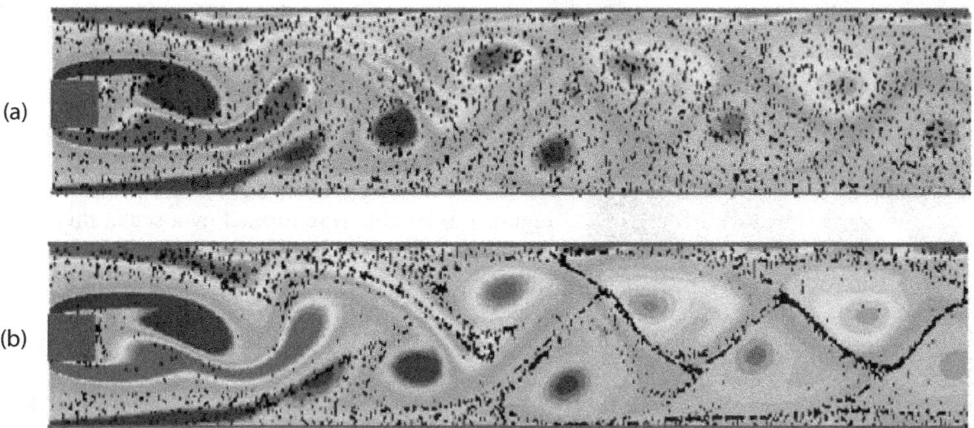

Figure 1.1. Instantaneous particle dispersion patterns from a numerical simulation of a plane wake behind a square cylinder at (a) St = 0.001 and (b) St = 1.0. [Reprinted with permission from Jafari et al. (2010).]

1.2.2. Density Ratio

The ratio of the fluid density to the particle density, $\chi \equiv \rho_f/\rho_p$, plays an important role in determining how particles respond to curvature in the fluid streamlines. To see this, we consider the problem of a particle placed in the flow field of a line vortex of strength Γ. In addition to the drag force given by (1.2.1), the particle is subject to two other forces that act in the radial direction relative to the vortex axis. The first of these forces is the centrifugal force, which acts to expel the particle from the vortex. In an inertial reference frame, the centrifugal force is included as part of the $d\mathbf{v}/dt$ term in (1.2.2), so to better highlight this effect we examine the particle in a frame of motion that rotates at a rate $\Omega_F = u_\theta/r = \Gamma/2\pi r^2$, where r is the radial position of the particle centroid and $u_\theta = \Gamma/2\pi r$ is the fluid azimuthal velocity at radius r. In this frame, the centrifugal force is

$$\mathbf{F}_c = m\Omega_F^2\mathbf{r}, \tag{1.2.7}$$

where $\mathbf{r} = r\mathbf{e}_r$ is the particle radial position times the unit vector $\mathbf{e}_r$ in the radial direction. The second force that acts on the particle in the radial direction arises from the fact that the pressure has a minimum value at the vortex axis and increases away from the axis, such that there exists a radial pressure gradient surrounding the line vortex given by

$$\nabla p = \frac{\rho_f \Gamma^2}{4\pi^2 r^3}\mathbf{e}_r. \tag{1.2.8}$$

This pressure gradient induces a force on the particle, similar to the buoyancy force that acts on a beach ball held under water in the presence of a hydrostatic pressure gradient. The value of the resulting pressure-gradient force acting on the particle is (Marshall, 2001)

$$\mathbf{F}_p = -V_p\nabla p = -\frac{\rho_f V_p \Gamma^2}{4\pi^2 r^3}\mathbf{e}_r, \tag{1.2.9}$$

Figure 1.2. Bubble ring formed by a scuba diver by bubble entrainment into a vortex ring core.

where V_p is the particle volume. The centrifugal force acts to throw the particle out of the vortex and the pressure-gradient force acts to draw the particle inward toward the vortex axis.

The ultimate fate of the particle is determined by the sum of these two radial forces, given by

$$\mathbf{F}_c + \mathbf{F}_p = \frac{(\rho_p - \rho_f)V_p\Gamma^2}{4\pi^2 r^3}\mathbf{e}_r. \tag{1.2.10}$$

If the particle density is greater than the fluid density, the net force is oriented in the positive radial direction and the particle drifts outward, away from the vortex axis. In this case, the particles are referred to as "heavy" and exhibit patterns similar to that shown in Figure 1.1b in which particles collect in high-concentration sheets that surround the vortex structures. This effect leads to development of a high degree of particle concentration intermittency in certain turbulent flows, and it also accounts for the formation of streaky patterns in the near-wall region of turbulent boundary layers where the particles are centrifuged out of the cores of the quasi-streamwise vortices that form in the near-wall turbulent flow (Kaftori et al., 1995). If the particle density is less than the fluid density, the net force in (1.2.10) is negative and the particle is pulled toward the vortex axis. For example, cavitation bubbles formed along a propeller surface are pulled into the trailing vortex behind the propeller blade, leading to the formation of the distinctive gas-filled propeller trailing vortex. Similarly, air bubbles blown by a scuba diver can be entrained into the core of a vortex ring, forming the so-called "bubble-ring" seen in Figure 1.2. Whales and dolphins have been observed spending a great deal of time forming and playing with bubble rings.

1.2.3. Length Scale Ratios

The *dimensionless particle diameter* is defined as the ratio of particle diameter d and fluid length scale L. We use the term "bulk flow" to refer to the flow field minus the perturbations caused by the particles. In particulate fluids, it is usual to assume that the bulk flow seen by an individual particle is relatively simple, often consisting of just a uniform flow and a shear flow, so that simple analytical models can be constructed for the fluid forces on the particles. This condition is fulfilled provided that the value of d/L is sufficiently small compared to unity. If the bulk flow seen by an individual particle is complicated (i.e., if d/L is not small compared to unity), then it is usually necessary to numerically model the complete flow field around each particle, which limits the number of particles that can be considered in the computation.

To give a specific example, consider the flow of red blood cells (the "particles") of "effective" diameter d in a blood vessel of diameter L. The value of the dimensionless particle diameter, d/L, depends on which section of the cardiovascular system is under consideration. In the large arteries $d/L \approx 0.001$, so the changes in the bulk flow occur on a much larger length scale than the particle diameter. In small arteries and arterioles, d/L has values between 0.02 and about 0.08, which are still sufficiently small that the particles can be regarded as seeing relatively simple bulk flows. On the other hand, in the capillaries, d/L has values ranging from 0.5 to 1, for which the simple drag and lift expressions used in particulate flow models break down and it is necessary to numerically simulate the entire flow field around each red blood cell.

The dimensionless particle diameter is also an important parameter for studies of turbulent flow of particulate fluids. For such problems, the fluid length scale L can often be replaced by the turbulence integral length scale. For small values of d/L, the particles are found to attenuate the turbulence. This attenuation arises from a combination of increased mixture inertia due to addition of particles, particle drag that occurs from relative motion of the particle through the fluid (including radial drift of the particles in response to the turbulent eddies), and enhanced effective viscosity of the fluid (Balachandar and Eaton, 2010). By contrast, for high values of d/L the particles are found to accentuate the turbulence due to injection of new turbulent eddy structures in the particle wakes. In this case, each particle acts to transition kinetic energy from the bulk flow into turbulent kinetic energy within the particle wake. Buoyancy-induced instabilities that arise from density variations in the fluid as the particles cluster in certain preferential locations can also enhance the turbulence intensity (Elghobashi and Truesdell, 1993). The transition from attenuation of turbulence to enhancement of turbulence was examined by a number of investigators (Gore and Crowe, 1989; Hetsroni, 1989). Based on the plot reprinted in Figure 1.3, Gore and Crowe (1989) proposed that turbulence is attenuated by the presence of particles when $d/L < 0.1$ and that particles enhance turbulence when $d/L \geq 0.1$.

Another important length scale ratio for particle transport is the *Knudsen number*, defined as the ratio of mean free path λ of the molecules in the fluid to the particle diameter, or

$$\mathrm{Kn} \equiv \frac{\lambda}{d}. \tag{1.2.11}$$

The flow past a particle can be treated using the standard continuum assumption if $\mathrm{Kn} < 10^{-3}$ (Crowe et al., 2012). For particles in gases, the mean free path of the fluid can often be large enough that this criterion is exceeded. For instance, air is primarily nitrogen, which has a mean free path of $\lambda = 10^{-7}$m. Dust particles have a typical diameter of about 10 microns, or $d = 10^{-5}$m. The Knudsen number for a dust particle in air is therefore approximately $\mathrm{Kn} \cong 0.01$, which is about an order of

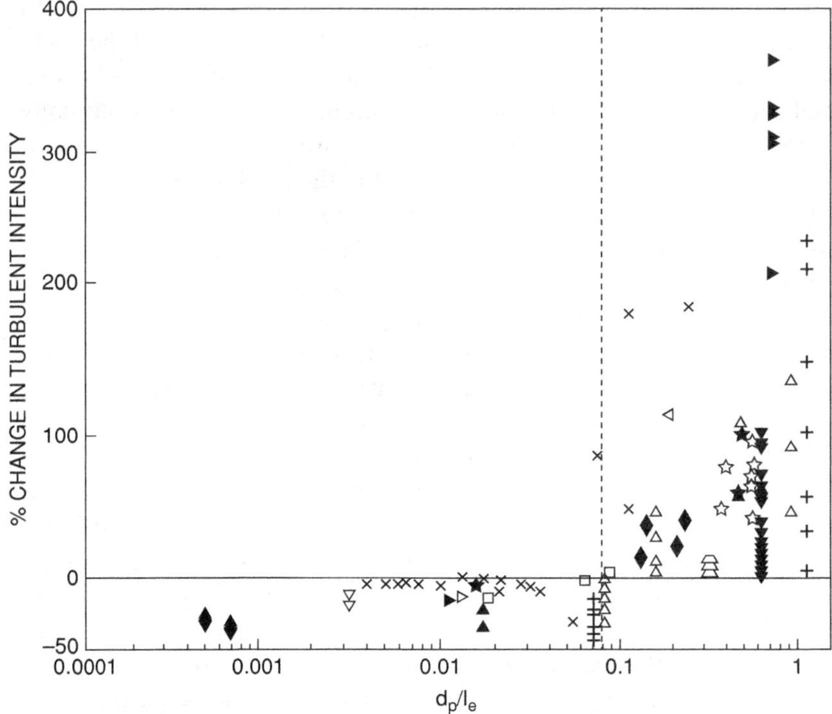

Figure 1.3. Attenuation and enhancement of turbulence by particles as a function of the ratio of dimensionless particle diameter to the turbulence integral length scale. The symbols represent experimental data collected from many different researchers, as listed by Gore and Crowe (1989). [Reprinted with permission from Gore and Crowe (1989).]

magnitude larger than the criterion given earlier for application of the continuum approximation. The most common effect of violation of this criterion for particulate flows is the presence of nonzero slip of the fluid past the particle, which modifies the Stokes drag expression (1.2.1), as discussed in Section 5.1.

1.2.4. Particle Reynolds Number

The *Reynolds number* is a ratio of inertial to viscous force. The instantaneous particle Reynolds number Re_p is defined in terms of the particle instantaneous velocity relative to the fluid as

$$\mathrm{Re}_p \equiv \frac{\rho_f d\,|\mathbf{v} - \mathbf{u}|}{\mu}, \tag{1.2.12}$$

where ρ_f is the fluid density. The difference $\mathbf{v}_d \equiv \mathbf{v} - \mathbf{u}$ is called the particle *slip velocity* or *drift velocity*, depending on the context. Because $|\mathbf{v} - \mathbf{u}|$ varies in time as a particle moves about, Re_p is also a function of time. However, it follows from (1.2.4) and the scaling relationship (1.2.3) that $\mathbf{v} - \mathbf{u} = O(\mathrm{St}\,U)$ for small values of the Stokes number, so a characteristic particle Reynolds number $\overline{\mathrm{Re}}_p$ can be defined which is a constant for the flow field as follows:

$$\overline{\mathrm{Re}}_p \equiv \frac{\rho_f d(\mathrm{St}\,U)}{\mu} = \left(\frac{d}{L}\right)\mathrm{St}\,\mathrm{Re}_f, \tag{1.2.13}$$

where $\mathrm{Re}_f \equiv \rho_f L U/\mu$ is the flow Reynolds number.

For small values of the particle Reynolds number, the simple Stokes drag force (1.2.1) can be used for the particle drag, assuming that other particles are sufficiently far away. For sufficiently small particles immersed in a gas, fluid slip at the particle surface becomes important and other forces, such as Brownian motion, also play an important role. Many particulate flows are characterized by small particle Reynolds numbers, including most colloidal and aerosol suspensions. More complicated drag expressions exist for cases where the particle Reynolds number is not small, which is often the case in particulate flows that arise in applications such as biofuel combustion, agricultural grain processing, bed-load sediment transport, and pneumatic coal feeders, as well as in many granular flows in general. The range of particle sizes for which adhesion forces play a significant role tend to have small particle Reynolds number, although problems involving wet granular media might pose an exception.

The flow Reynolds number, Stokes number, density ratio, and dimensionless particle diameter are not independent of each other, but can be related to each other in a variety of ways. For instance, the Stokes number can be written in terms of the flow Reynolds number as

$$\mathrm{St} = \frac{1}{18\chi} \left(\frac{d}{L}\right)^2 \mathrm{Re}_f. \tag{1.2.14}$$

Substituting (1.2.14) into (1.2.13) gives the averaged particle Reynolds number as

$$\overline{\mathrm{Re}}_p = \frac{1}{18\chi} \left(\frac{d}{L}\right)^3 \mathrm{Re}_f^2. \tag{1.2.15}$$

Combining (1.2.15) and (1.2.14) gives the Stokes number in terms of the averaged particle Reynolds number as

$$\mathrm{St}^2 = \frac{1}{18\chi} \left(\frac{d}{L}\right) \overline{\mathrm{Re}}_p. \tag{1.2.16}$$

1.2.5. Particle Concentration and Mass Loading

The particle *volume concentration*, or simply particle concentration, ϕ is defined as the ratio of the volume $V_{S,p}$ occupied by all of the particles in a system divided by the total system volume $V_{S,f} + V_{S,p}$, or

$$\phi \equiv \frac{V_{S,p}}{V_{S,f} + V_{S,p}}, \tag{1.2.17}$$

where $V_{S,f}$ is the part of the system volume occupied by the fluid. This definition yields a single concentration value for the entire flow field. It is also possible to define a local concentration field that varies as a function of position and time within a flow, which is discussed in detail in Chapter 10.

A useful length scale ratio can be derived from the concentration value. Let us assume that a system consists of spherical particles of diameter d that are arranged in a uniform array with *particle spacing distance* ℓ between each pair of adjacent particles. Each cubic cell in this array of volume $V = \ell^3$ includes a particle volume equal to the volume of a single particle, or $V_p = \pi d^3/6$. The particle concentration is therefore given by

$$\phi = \frac{V_p}{V} = \frac{\pi}{6} \left(\frac{d}{\ell}\right)^3. \tag{1.2.18}$$

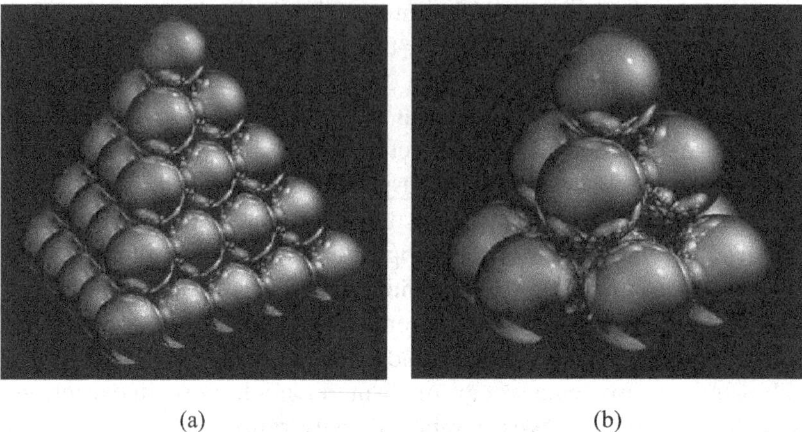

(a)   (b)

Figure 1.4. Two popular close-packed arrangements of spheres: (a) the face-centered cubic (fcc) and (b) hexagonal-centered cubic (hcc).

Solving (1.2.18) for the particle spacing ratio ℓ/d gives

$$\frac{\ell}{d} = \left(\frac{\pi}{6\phi}\right)^{1/3}. \tag{1.2.19}$$

This result is useful for determining the significance of particle-particle collisions in a particulate flow. If $\ell/d \gg 1$, the particles will rarely collide with each other and the particulate flow can be considered to be *dilute*. As ℓ/d approaches unity, the rate of particle collisions increases rapidly, and the particulate flow is said to be *dense*. Of course, if the particles are adhesive then even infrequent collisions will over sufficient time lead to formation of agglomerates, so even in the limit of large ℓ/d particle collisions cannot be neglected in adhesive particle flows.

The volume concentration has an upper bound, $\phi_{\max}$, when the particles are in the maximum packing arrangement. The value of $\phi_{\max}$ depends on the particle shape and arrangement in the maximum packing configuration. For spherical particles maximum packing occurs in a close-packed arrangement such as those shown in Figure 1.4, in which each sphere is in contact with exactly 12 other spheres. Among the popular close-packed arrangements are the face-centered cubic (fcc) and hexagonal-centered cubic (hcc) arrangements. All close-packed arrangements yield the same theoretical value of maximum concentration, given by

$$\phi_{\max} = \frac{\pi}{\sqrt{18}} \cong 0.7405. \tag{1.2.20}$$

In practice, particulate flows cannot reach this high value of volume concentration because the particles are not arranged in a regular packing, but are instead randomly carried along by the flow. The volume concentration for a randomly packed particle bed varies from about 0.52 to 0.74, depending on the method used to deposit the particles (Dullien, 1992).

The *mass concentration*, ϕ_M, is defined as the ratio of the mass of the particles in a system to the total mass of the system, or

$$\phi_M \equiv \frac{\rho_p V_{S,p}}{\rho_f V_{S,f} + \rho_p V_{S,p}}. \tag{1.2.21}$$

In a dilute flow, the volume concentration is small, such that $V_{S,p} \ll V_{S,f}$ and $\phi \cong V_{S,p}/V_{S,f} \ll 1$. In this case, the volume and mass concentrations can be related as

$$\phi_M \cong \frac{\phi}{\chi + \phi}. \tag{1.2.22}$$

We do not neglect the ϕ term in the denominator of (1.2.22), because the density ratio χ can also be very small. Some authors define mass concentration differently as the ratio of particle mass to fluid mass (e.g., Crowe et al., 2012), so that the volume and mass concentration in a dilute flow will be related simply as $\phi_M \cong \phi/\chi$. In the present book we use the definition in (1.2.21), which is more in line with the traditional interpretation of concentration.

The *mass loading* (or sometimes simply the *loading*), Z, is defined as the ratio of mass flow rate $\dot{m}_p$ of the particles to the mass flow rate $\dot{m}_f$ of the fluid, or

$$Z \equiv \frac{\dot{m}_p}{\dot{m}_f}. \tag{1.2.23}$$

If the fluid and particle velocities are approximately the same (e.g., for low Stokes numbers), then the mass loading reduces to a ratio of the particle mass to the fluid mass in the system, or

$$Z \cong \frac{\rho_p V_{S,p}}{\rho_f V_{S,f}}. \tag{1.2.24}$$

For dilute flows, the volume concentration can be approximated as $\phi \cong V_{S,p}/V_{S,f}$, so (1.2.24) reduces to

$$Z \cong \frac{\phi}{\chi}. \tag{1.2.25}$$

The mass loading plays a key role in determining the effect of the particles on the fluid flow. In a solution of particulate flows with small concentration values, it is often useful to make the assumption of *one-way coupling*, which assumes that the particles do not affect the fluid flow. This assumption greatly simplifies the flow computation because we can solve for the fluid flow independently of the particles, and then solve for the particle motion in a known fluid flow field. In flows where this assumption is not valid we must utilize full *two-way coupling*, in which the particle and fluid motion are fully coupled to each other, often requiring an iterative solution approach.

Validity of the one-way coupling assumption requires that the drag exerted on the fluid by the particles is much less than the momentum flux of the particulate flow. For a dilute flow system ($\phi \ll 1$) with N particles, in which the particles are assumed to have a sufficiently small Reynolds number so that the Stokes drag law (1.2.1) is valid, a *momentum coupling parameter* Π can be defined as the ratio of net particle drag force exerted on the fluid to the fluid momentum flux, giving

$$\Pi = \frac{3\pi \mu d N v_d}{\rho_f U^2 L^2}, \tag{1.2.26}$$

where v_d denotes the magnitude of the characteristic particle slip velocity $\mathbf{v}_d \equiv \mathbf{v} - \mathbf{u}$. The one-way coupling approximation can be considered to be valid when the value of Π is sufficiently small, typically less than about 10%.

The number of particles N in (1.2.26) can be written in terms of the volume concentration as

$$N = \frac{6\phi L^3}{\pi d^3}. \tag{1.2.27}$$

Because the slip velocity has order of magnitude $v_d = O(\text{St}\, U)$ for a flow at small Stokes number, substituting these estimates for N and v_d into (1.2.26) and using the definition (1.2.6) for Stokes number gives the result that the momentum coupling parameter at small Stokes numbers is approximately equal to the mass loading, or

$$\Pi = \frac{\phi}{\chi} \cong Z. \tag{1.2.28}$$

In this limit, the fluid and particle velocities are equal to each other to first order in the Stokes number, so the approximation (1.2.25) applies. An extension of this estimate valid for finite Stokes numbers was derived by Crowe et al. (2012) as

$$\Pi = \frac{1}{1 + \text{St}} \frac{\phi}{\chi}. \tag{1.2.29}$$

These results indicate that the assumption of one-way coupling is valid for small mass loadings, and that this assumption becomes even more accurate as the Stokes number increases with a fixed mass loading.

1.2.6. Bagnold Number

The *Bagnold number*, Ba, is used to measure the importance of particle collisions on stress transmission in a particulate flow, and can loosely be interpreted as a ratio of the order of magnitude of particle collision stress to that of viscous fluid stresses in the flow field. The Bagnold number is defined mathematically by (Bagnold, 1954)

$$\text{Ba} = \frac{\rho_p d^2 \lambda_B^{1/2} \dot{\gamma}}{\mu}, \tag{1.2.30}$$

where $\dot{\gamma}$ is the fluid shear rate and λ_B is called the linear concentration, defined in terms of the particle volume concentration ϕ and the maximum packing concentration ϕ_{max} by

$$\lambda_B = \frac{1}{(\phi_{\text{max}}/\phi)^{1/3} - 1}. \tag{1.2.31}$$

Writing the shear rate in terms of a fluid velocity scale U and length scale L as $\dot{\gamma} = U/L$, the Bagnold number can be written as a function of the Stokes number and the particle concentration as

$$\text{Ba} = 18\, \text{St}\, \lambda_B^{1/2}. \tag{1.2.32}$$

Based on a series of experiments for the dynamics of large particles in shear flows, Bagnold separated the flow dynamics into a "viscous" regime in which the flow is dominated by viscous fluid stresses when Ba is sufficiently small (typically, Ba < 40) and a "grain-inertia" regime in which the stress is primarily transmitted by particle collisions when Ba is sufficiently large (typically Ba > 400 to 450). In the grain-inertia regime, the particle system behaves as a granular flow and the effect of the interstitial fluid is relatively small.

1.2.7. Adhesion Parameter

In flows of adhesive particles, the set of dimensionless parameters that govern the particle transport must also include some parameter associated with the particle adhesive force. This can be done in a variety of ways, but several recent studies have defined an *adhesion parameter* as a ratio of the adhesive force and the particle inertia (e.g., Li and Marshall, 2007). Transport of particles in the micrometer size range is dominated by a balance between particle inertia and drag force, so use of either viscous drag or particle inertia in the denominator of this parameter yields an essentially equivalent parameter. In this book, the adhesion parameter, Ad, is defined by

$$\mathrm{Ad} = \frac{2\gamma}{\rho_p U^2 d}, \tag{1.2.33}$$

where γ is the adhesive surface energy, which is equal to half the work required to separate two surfaces that are adhesively bound per unit surface area. This parameter has an obvious relationship to the Weber number, $\mathrm{We} = \rho_f U^2 d/2\sigma$, which is a measure of the relative importance of the fluid inertia compared with its surface tension σ. For large values of the adhesion parameter, particles tend to stick together upon collision, forming particle agglomerates whose size is determined primarily by the bending and shear forces exerted by the fluid drag force on the agglomerated particles. For small values of the adhesion parameter, colliding particles tend to separate from each other, and the primary role of the adhesive force is to reduce the particle rebound velocity.

1.3. Applications

There are a large number of applications in nearly all areas of engineering, biology, agriculture, and environmental science involving flow of adhesive particles. In this section, five application areas are briefly described that illustrate different aspects and interesting physics occurring in adhesive particle flows.

1.3.1. Fibrous Filtration Processes

Filtration is one of the primary processes used in many chemical and materials purification applications. Filtration also plays a critical role in environmental processes used for production of clean air and drinking water. Filtration is vital for clean room operations used in electronics manufacturing, as well as in microfabrication and nanofabrication industries. Filtration processes involve a mechanical separation of particles from a fluid stream by collision and adhesion of particles onto the filtration media. There are many types of filtration media, varying from a granular bed (such as soil) to paper or cloth. All filtration media share the characteristic that the fluid flow is forced to meander back and forth in a tortuous path as it passes through the medium. Filtration differs from sieving in that a sieving process involves capture of particles that are typically larger than the sieve hole size, whereas in filtration the particles are much smaller than the channel or gap size in the filtration medium. The example discussed in the current section deals with *fibrous filters*, which consist of a

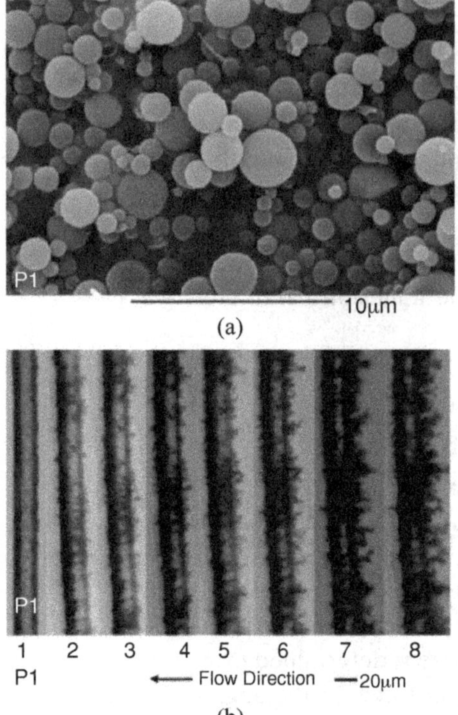

Figure 1.5. Photographs showing (a) fly ash particles recovered from the exhaust stream of a combustion process and (b) a time series showing adhesion of the fly ash particles onto a fiber, with flow orthogonal to the fiber. [Reprinted with permission from Huang et al. (2006).]

mesh of thin cylindrical structures (fibers) joined together to form a structure similar to a cloth or porous paper.

A study of the micromechanics of fibrous filtration processes was reported by Huang et al. (2006), with a subsequent computational study by Li and Marshall (2007). Huang et al. (2006) used fly ash particles recovered from a combustion process exhaust stream (Figure 1.5a), which were forced to flow orthogonally to a single fiber. A time series of images showing capture of the particles by the fiber is given in Figure 1.5b, starting with a relatively clean fiber (on the left) and ending with a fiber that is nearly saturated with captured particles (on the right).

Particles entrained in the flow collide with the fiber by a combination of two mechanisms, illustrated in Figure 1.6. The first mechanism involves the inertia of the particles. Particles in a gas flow are much denser than the surrounding gas, and tend to have Stokes number values near or above unity, thus allowing significant inertial drift of the particles. As the fluid streamlines bend near the stagnation point at the front of the fiber, the greater inertia of the particles carries them forward and causes the particles to collide with the front surface of the fiber (Figure 1.6a). The second mechanism involves the finite size of the particles in a shear flow, and can allow for particle collision with the fiber even for cases of neutrally buoyant particles in a liquid. Shear-induced collision becomes important as the distance between streamlines gradually narrows as the fluid accelerates to move around the fiber (Figure 1.6b). Particles with centroids positioned on streamlines lying very close to the fiber surface can collide with the surface due to the finite particle size as the fluid carries the particles close to the fiber surface. This type of collision tends to deposit particles along the sides of the fiber. Particles can also collide with other particles

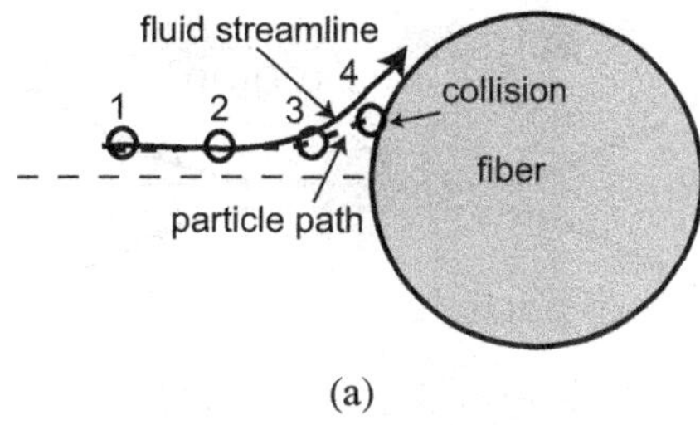

Figure 1.6. Schematic illustrating (a) inertia-induced collision and (b) shear-induced collision of a particle onto a cylindrical fiber.

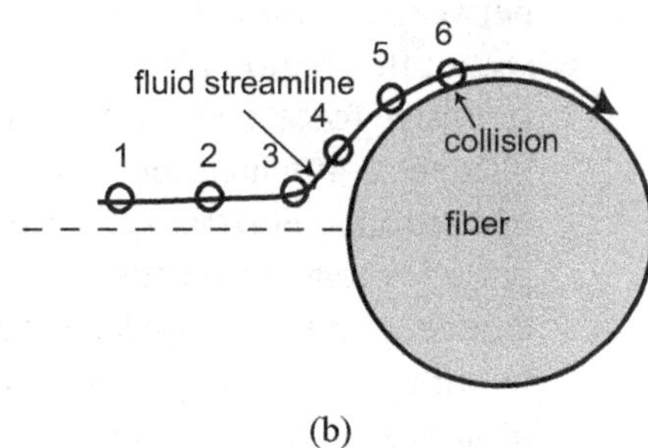

that are already attached to the fiber, forming dendritic structures that reach out into the flow field, as seen in the time series in Figure 1.5b.

The primary adhesion force involved in fibrous filters is the van der Waals attraction force. The van der Waals force can be very strong, but it decays quickly with distance away from the particle surface. For instance, the adhesive force between two parallel planar surfaces separated by a distance h can be written as

$$F = -\frac{A}{6\pi h^3} + \frac{B}{h^9}, \qquad (1.3.1)$$

where A and B are positive constants. The first term in this expression is the attractive van der Waals force, and the second term is the short-range repulsive force. Typically, the van der Waals force between two surfaces is small if the separation distance between the surfaces is more than about $\delta_{vdw} \cong 10$ nm. Particles used for filtration processes, such as smoke or dust particles, have typical diameters d of 1–100 µm, or between 2 and 4 orders of magnitude larger than the length scale of the adhesive force. This observation has a number of consequences in modeling of the particle adhesion process. First, it should be obvious that van der Waals adhesion is significant only for colliding particles, as particles that do not collide are typically too far apart for the van der Waals force to be significant. Second, the fact that δ_{vdw}/d is so small implies that the particle adhesive force is highly sensitive to small amounts of particle deformation. To illustrate this second point, consider a case where two colliding particles are idealized as perfect spheres of equal diameter d that are touching at a single point (the *contact point*). Elementary geometry indicates that the separation distance between the particle surfaces will exceed δ_{vdw} within a distance from the contact point equal to approximately $\sqrt{\delta_{vdw}d}$, so only within this small region near the contact point is the adhesive force significant. For instance, for a 10 µm diameter particle, the value of $\sqrt{\delta_{vdw}d}$ is about 0.3 µm, assuming $\delta_{vdw} \sim 10$ nm. Within this small region around the contact point, the particle surfaces are pulled strongly toward each other by the van der Waals force, whereas outside of this region there is little force

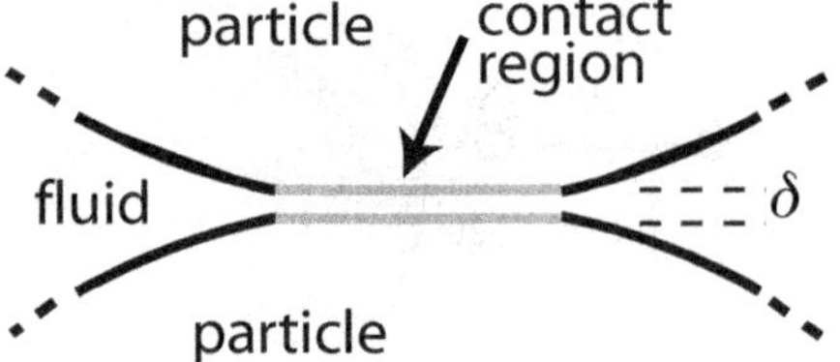

Figure 1.7. Flattened contact region that forms near the contact point of two otherwise spherical particles, showing a separation distance δ between the particle surfaces within the contact region.

between the particle surfaces. The attractive pull of the surfaces toward each other within this region is balanced at very small separation distances by the short-range repulsive force, so that the particle surfaces tend to adopt an equilibrium separation distance δ at which these two forces are in balance. As a result of this combination of a strong short-range force acting on the surfaces of much larger particles, the particles tend to elastically deform within a small *contact region* surrounding the contact point, forming a flattened region in which the particle surfaces are separated by an approximately constant distance δ (Figure 1.7). In some models for contact of adhesive particles, the van der Waals force is considered to act only within this contact region.

Particles captured by a fiber attach onto other particles captured by the fiber to form chain-like dendritic structures that project outward into the flow field (Tien et al. 1977). As these fibers grow, they experience increased bending force due to the increased fluid force acting on the particles within the dendrite. Eventually, the dendrites become so long that the adhesion force between the particles can no longer sustain the required bending force of the structure, and the dendritic structures either break or bend backwards onto the fiber (Figure 1.8). The latter process often disturbs other dendritic structures, leading to breakage of structures and release of particles downstream in a domino effect. These breaking and bending processes lead over time to an equilibrium state, in which the rate of new particles captured by the fiber is equal to the rate at which particles are lost from the fiber by breakage of the dendritic structures.

1.3.2. Extraterrestrial Dust Fouling

One of the most severe challenges of performing manned or robotic operations on the Moon or on planets such as Mars is the presence of fine dust. Neither the Moon nor

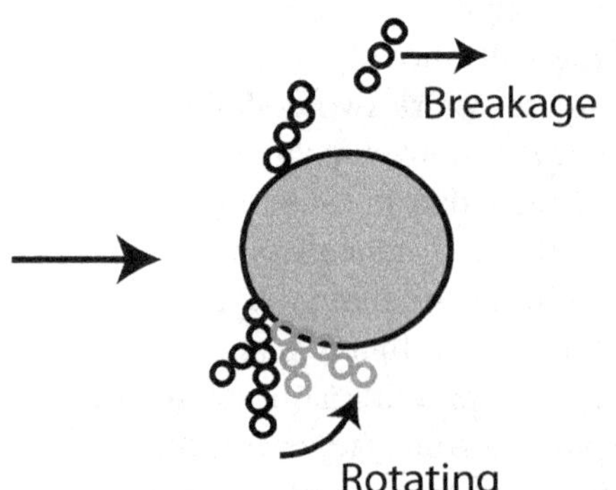

Figure 1.8. Schematic showing attachment of particle chains onto a fiber, and two mechanisms for chain failure leading to an equilibrium state of captured particles.

(a) (b)

Figure 1.9. Photographs showing (a) an astronaut in a dusty space suit during a moon walk and (b) dust on astronaut Eugene Cernan's suit inside the lunar capsule following a moon walk during the Apollo 17 mission.

Mars has liquid water, so instead of forming clays, fine dust particles remain loose in the soil. On the Moon, these dust particles become electrically charged by the action of the solar wind, whereas on Mars the dust particles become electrically charged primarily through triboelectric effects that result from the frequent planetary dust storms. When viewed under a microscope, the dust particles are found to be very sharp and irregular in shape.

The presence of very fine, charged dust particles leads to a wide range of problems in extraterrestrial exploration missions. For instance, during the Apollo missions to the moon in the 1960s and early 1970s, dust was found to quickly adhere onto the astronauts' space suits during moon walks, and was then carried back into the astronauts' living space when they returned to the lunar module (Figure 1.9). Once tracked into the habitat, the fine dust particles formed an aerosol that produced respiratory problems when breathed by the astronauts, even after only a few days' residence. Being charged, the dust particles cling to everything within the habitat, from electrical instrumentation to air recycling units. For longer space missions, the dust fouling would, over time, be expected to cause instrument failure and severe health problems for the astronauts unless adequate precautions are taken. An indication of the significance of dust fouling was given during astronaut Eugene Carnan's technical briefing following the Apollo 17 mission, in which he states: "I think dust is probably one of our greatest inhibitors to a nominal operation on the Moon. I think we can overcome other physiological or physical or mechanical problems except dust."

Dust fouling has also been a critical factor limiting operation of the rover missions to Mars. In this case the primary issue has been decreased efficiency of the solar panels that provide energy to the rover due to covering of the panels by a dust coating (Tanabe, 2008). Dust storms have been a particular threat to the rovers, leading to significant deposition of dust layers on the solar panels. On the other hand, dust devils and high wind events associated with dust storms have also helped

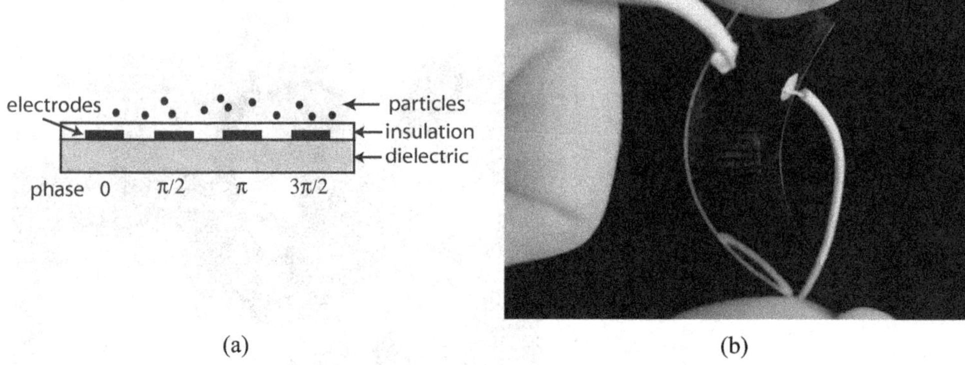

(a)　　　　　　　　　　　　　　　(b)

Figure 1.10. (a) Cross-section showing a 4-phase electric curtain design with plate-like electrodes and (b) a sample of the transparent "dust shield" under development at NASA Kennedy Space Center for dust mitigation on solar panels.

to remove dust from the rovers' panels, resulting in large variations in system power over time.

The characteristic feature of lunar and Martian dust that makes it so difficult to manage is the fact that the dust particles are electrically charged. Applications with charged dust particles in flowing fluids are also common on earth, for instance, in electrostatic precipitators used to remove ash from the exhaust streams of coal power plants and in control of particle mixing, separation, and transport in microfluidic systems (Markarian et al., 2003). Electrostatic interactions influence particle transport both by the dipole field induced on the particle by an external electric field, leading to dielectrophoretic (DEP) force on the particle, and by the monopole field associated with charged particles, leading to a Coulomb force on the particle in the presence of an external electric field. The Coulomb force between two charged particles decays in proportion to distance squared, which is slow compared to the decay rate of the van der Waals force. For this reason, charged particles have significant interaction via their mutual electric fields over distances that are large compared to the particle radius.

A wide variety of approaches have been proposed for dust mitigation in planetary exploration on the Moon and Mars. One of the more interesting and potentially effective approaches is the use of an electric curtain device (Calle et al., 2009). An electric curtain consists of a series of many parallel electrodes embedded on a dielectric surface and covered by a thin layer of insulation (Figure 1.10). Each electrode has a high-voltage oscillating potential, such that the phase of the potential oscillation differs by an amount $2\pi/n$ between neighboring electrodes, where n is the number of electrodes in each group before the phase repeats. A standing wave is obtained with $n = 2$, and all larger values of n correspond to traveling electrostatic waves on the curtain. The traveling electrostatic waves can carry particles with them in a number of different modes, including a mode where particles are levitated above the surface, a mode where particles intermittently hop along the surface, and a mode where particles roll back and forth on the surface. Particle interaction and adhesion to the dielectric surface modifies these modes, leading to emergent behavior of

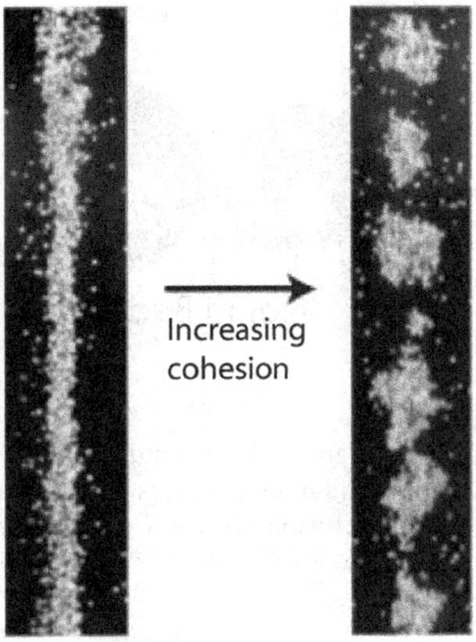

Increasing cohesion

Figure 1.11. Photographs showing streams of 130 μm copper spheres both in a dry state (left) and mixed with a small amount of oil (right) emitted from a small opening in the bottom of a hopper. [Reprinted with permission from Royer et al. (2009).]

the particles in the system that can be significantly different than the behavior of individual particles (Liu and Marshall, 2010a,b).

1.3.3. Wet Granular Material

Granular materials such as soil, rocks, snow, and volcanic debris are of obvious importance in a large number of environmental and geophysical processes, as well as in biological processes involving motion of organisms that live in the soil. Granular mechanics also plays an important role in engineering processes in the pharmaceutical, chemical, and materials processing industries. Agricultural grain sorting and materials handling are dominated by granular flow mechanics. Although most of the literature on mechanics of granular materials involves cohesionless particles, in many practical applications the presence of moisture within the granular matrix leads to particle adhesive force that plays an important role in the overall system mechanics.

It was found that the addition of small amounts of a liquid, even with a volume fraction of $10^{-4} \sim 10^{-2}$, into a dry granular system can lead to strong cohesion force between particles with significant changes of the macroscopic system behavior (Fiscina et al., 2010; Samadani and Kudrolli, 2000). For instance, Figure 1.11 shows a stream of dry copper grains of diameter $d = 130 \pm 30$ μm falling out of a nozzle with a diameter of 4 mm (left). As a small amount ($\sim 3 \times 10^{-4}$ by volume) of mineral oil is mixed with the copper grains, the uniform stream of dry granular material breaks up into particle clusters. The break-up process appears to be qualitatively similar to the capillary break-up of a liquid jet into droplets (Royer et al., 2009). Interparticle cohesive forces were measured in this study by recording force-displacement curves of individual grains brought into contact and pulled apart using an atomic force microscope. The results indicate significantly larger pull-off forces for the oil-wetted copper particles than for the dry particles.

The larger pull-off force of wet particles compared with dry particles is caused by the interaction of thin liquid films that form around each particle in the presence

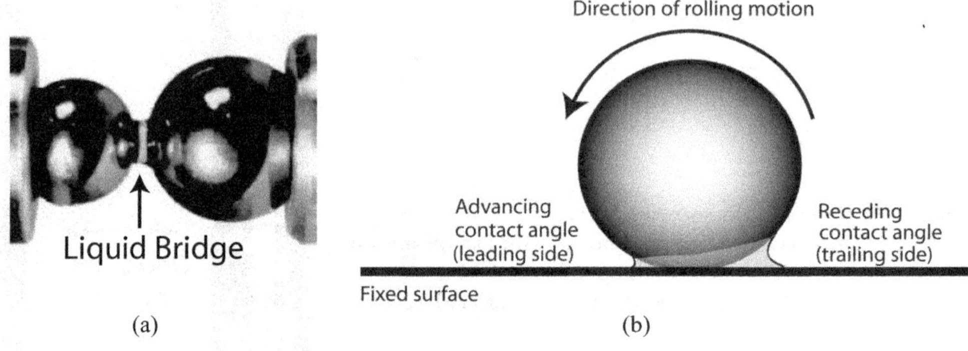

Figure 1.12. Liquid bridge formed between two spherical particles: (a) photograph of liquid bridge on particles pulled apart in normal direction and (b) schematic of liquid film capillary deformation on a rolling sphere. [Figure (a) reprinted with permission from Soulié et al. (2006). Figure (b) reprinted with permission from Schade and Marshall (2011).]

of a small amount of moisture. When particles move sufficiently close to each other for a collision to occur, the liquid films of the colliding particles touch and merge into one another. As the two particles are pulled away from each other following collision, the liquid films of the two particles form a connection between the particles that is known as a *liquid bridge*. An example of a liquid bridge is shown in Figure 1.12a for two particles of different sizes pulled apart along a line connecting the particle centroids. A liquid bridge exerts a force between the particles due to both the interfacial surface tension at the liquid-gas-solid contact line and the pressure decrease within the liquid that develops from interfacial curvature of the liquid bridge. When the particles move relative to each other, viscous effects due to liquid motion within the film introduce an additional viscous force that must be added to the attractive capillary force between the particles (Ennis et al., 1990). The result of this liquid bridging force is to cause particles to adhere to container walls and to each other, forming particle agglomerates that resist gravitational and other forces, such as fluid shear, that try to tear them apart.

A less-studied consequence of the presence of liquid films around particles in a wet granular flow is the effect of film asymmetry on the rolling motion of particles along container walls or during particle collisions with each other. Rolling is a critical form of particle interaction, which is typically far more likely to occur for systems of small particles than are sliding or twisting motions due to the lower threshold for onset of rolling than for these other motions. It is well known that van der Waals adhesion causes a resistance to particle rolling (Dominik and Tielens, 1995), which can have important consequences on the motion of particulate systems. However, it was not realized until a recent paper by Bico et al. (2009) that liquid films can also cause a resistance to rolling motion of particles in a wet granular flow. As shown in a study by Schade and Marshall (2011), rolling of a particle causes the liquid bridge connecting the particle to a flat surface to become anisotropic, such that the liquid-gas interface contact line moves inward on one side and outward on the other side of the particle (Figure 1.12b). The net result of the contact line movement is to induce a capillary torque on the particle in such a direction as to resist the rolling motion.

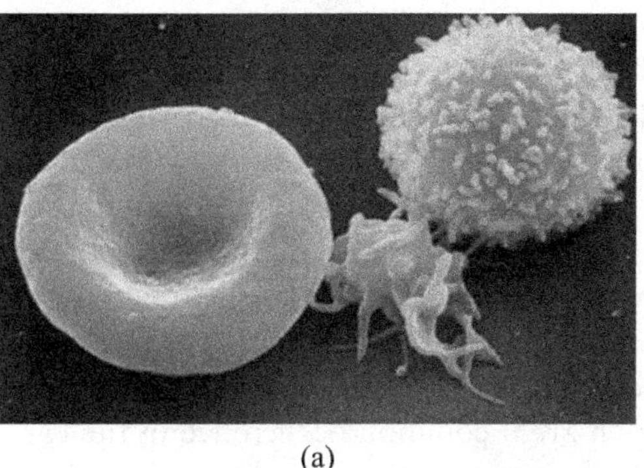
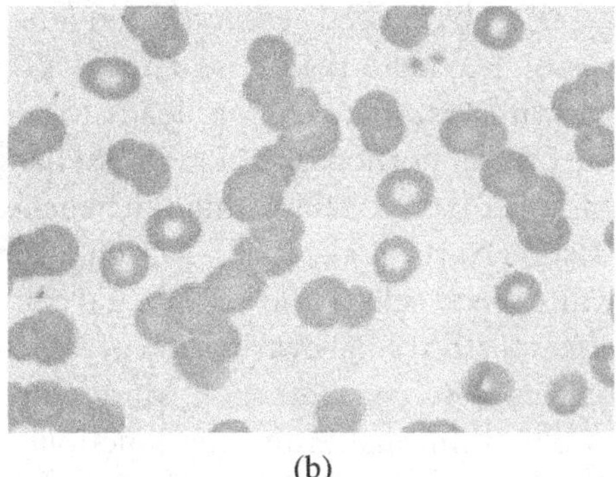

(a)   (b)

Figure 1.13. (a) Scanning electron microscope image showing a red blood cell (left), an activated platelet (middle), and a white blood cell (T-lymphocyte) (right). (b) Rouleaux formed of red blood cells at low shear stress levels. [Figure (b) reprinted with permission of 2009 Rector and Visitors of the University of Virginia, Charles E. Hess, M.D., and Lindsey Krstic, B.A.]

1.3.4. Blood Flow

Fluid flows consisting of suspensions of biological cells can often be considered to behave as if they were particulate flows. This idealization is useful for modeling environmental flows with microorganisms, such as to determine the response of microalgae cell suspensions to turbulent mixing (Marshall and Sala, 2011) or to study instabilities that can occur in a fluid with a suspension of self-propelled microorganisms (Pahlavan and Saintillan, 2011). In such applications, it is necessary to endow the particles with additional properties to account for processes such as cell growth, cell division, microorganism competition, and bacterial locomotion.

An area in which particulate flow models are particularly useful to represent suspensions of biological cells occurs in flows within the human cardiovascular system (Chesnutt and Marshall, 2009), in which the particulate matter consists of a combination of red blood cells (RBCs, or erythrocytes), white blood cells (leukocytes), and platelets immersed in a fluid called the plasma (Figure 1.13a). Together, blood cells constitute about 45% of total blood volume. The most numerous type of blood cells are red blood cells, which have the form of biconcave disks with thickness of about 2 µm and a maximum diameter between 6 and 8 µm. RBCs number between 4 and 6 million cells per cubic millimeter of blood. A mature RBC has no nucleus, but instead has the form of a capsule that is filled with hemoglobin, which holds the oxygen that is carried through the body by the cardiovascular flow. In the very small capillary vessels, measuring only 8–10 µm in diameter, the RBCs must deform by large amounts to pass through narrow constrictions. On the other hand, in the larger arterioles (approximately 80 µm diameter), and throughout the different small and large arteries (300 µm to 2 cm diameter), very little RBC deformation is observed. At low values of shear stress, RBCs adhere to each other to form chain-like agglomerates called rouleaux (Figure 1.13b). Formation of rouleaux is not common in normal physiological blood flow conditions, but they can form under various disease states such as diabetes, infections, and cancer when blood protein levels are high, as well as in low-shear wake regions behind obstructions in the circulatory system.

Red blood cells exhibit a tendency to collect near the central part of a blood vessel, leaving a region near the walls with low RBC concentration. This process, known as *marginalization*, leads to a state where the RBCs flow primarily within the central portion of a blood vessel and other blood cells (such as white blood cells and platelets) collect within the near-wall ("cell-free") regions with low numbers of RBCs (Aarts et al., 1988). Marginalization is important for regulating transport of white blood cells to the endothelial cells along the vessel walls, and it has a significant effect on the radial distribution of effective blood viscosity, which varies as a function of local RBC concentration. Marginalization also plays an important role on the flow of blood through branching junctions, which are a common occurrence in the cardiovascular system. When the flow rate through one of the outlet branches of the junction is substantially less than that through the other outlet branch, it is observed that the branch with lower flow rate has significantly lower RBC concentration than the outlet with the higher flow rate. This phenomenon, called *plasma skimming*, arises because the lower flow rate outlet primarily draws fluid from the near-wall cell-free layer whereas the high flow rate outlet draws fluid primarily from the RBC-rich central region of the inlet vessel. Significant differences in RBC concentration ("hematocrit") develop within the body as a result of plasma skimming, and the resulting variation in blood viscosity is known to provide a number of important physiological benefits (Jonsson et al., 1992).

White blood cells (leukocytes) are produced in the bone marrow and are a key element of the immune system, responsible for protecting the body from infectious diseases. There are several different types of white blood cells with different functions and different sizes, although typical white blood cells are approximately spherical in shape with diameter ranging from 7 to 15 μm. The number of white blood cells can vary greatly with disease state, but under normal conditions there are approximately 7,000 white blood cells per cubic millimeter of blood, or about 1 white blood cell for every 600–700 red blood cells. White blood cells can adhere to endothelial cells along the blood vessel wall, which is sometimes followed by passage of the white blood cell into the surrounding tissue. This adhesion process is influenced by rolling of the white blood cells along the wall and by collision of white blood cells with passing red blood cells during the adhesion process (Melder et al., 2000).

Platelets make up the final category of major types of blood cells. Platelets are irregularly shaped cell fragments which are approximately 2–3 μm in diameter, and number about 1 platelet for every 20 red blood cells. Unactivated platelets travel in the blood waiting for a signal to become activated. Once activated, platelets change form and become highly adhesive, gathering at the site of an injury to the blood vessel and acting to initiate and regularize the formation of a blood clot to seal the vessel. Excessively high platelet levels are associated with development of strokes and heart attacks. The platelet activation process occurs through a complex series of chemical reactions, which can be triggered by a number of factors. Among these triggering factors is exposure to high shear stress, which can occur both naturally and during closing operations of mechanical heart valves. The high fluid shear levels observed on mechanical heart valves can lead to thrombus formation, which limits useful life span of the valves (Bluestein et al., 2004).

1.3.5. Aerosol Reaction Engineering

In recent years, there has been a surge in development of tailored nanoparticles (including nanowires, nanotubes, and nanofilms) because of their wide range of industrial applications, such as selective catalysts, chemical sensors, optical films/fibers, additives for nanofluids, and fuel additives for energetic chemical propulsion. As tailored nanoparticles come into increased industrial usage, it is necessary to place greater emphasis on the problem of how to synthesize these particles on an industrial scale with high throughput under carefully controlled conditions. *Aerosol reactors* are technological systems used for the large-scale synthesis of nanoparticles of designed properties (Friedlander, 2000). Within aerosol reactors, precursors are converted to monomers by processes such as oxidation and pyrolysis, producing a supersaturated vapor that undergoes nucleation to form primary particles. These particles then grow via the collision-sintering mechanism to form aggregate nanoparticles. Aerosol reactors employ a vapor-phase approach that is different from conventional wet-chemistry methods for nanoparticle production, and offers a number of distinct advantages. According to the mode of heat supply to the process, aerosol reactors can be classified as various types, such as flame, laser, plasma, hot-wall, and electric-heating. Key governing factors are aerosol precursor properties and reactor process conditions (e.g., temperature).

More recently, another widely used vapor-phase method, vapor deposition (either chemical or physical), has received increased attention for production of thin films, in which materials in a vapor state are directly condensed to form solid-phase materials on a temperature-controlled substrate (Chen and Mao, 2007). By combining advantages of both vapor-phase flame aerosol reactors and deposition techniques, a one-step stagnation flame aerosol synthesis was developed, in which high-quality nanostructured TiO_2 films can be produced at large growth rates in a single step (Zhang et al., 2012). The synthesis process can be classified into two zones, consisting of a predeposition gaseous zone and a postdeposition zone. In the former, the sintering occurs between two colliding particles induced by either the inertial or the shear collision processes. In the latter, the sintering happens between two quasistatic contacting particles on the substrate.

The roles of these two distinct sintering mechanisms in determining the morphologies of TiO_2 films, which can be inferred from the specific surface area (SSA) of the films, are illustrated in Figure 1.14. When the substrate temperature is low enough, the on-substrate sintering is quite weak due to the low local temperature, with particles "frozen" upon deposition. The SSA dramatically decreases as the precursor concentration increases. The SSA of produced films is mainly dependent on the in-flame collision-sintering process prior to deposition, which we term the "in-flame-collision-sintering controlled" regime. In contrast, when the substrate temperature is high, SSA dependence on precursor concentration is much weaker, with SSA varying in a narrow range from 91.4 m^2/g to 129.7 m^2/g. Subplots A and B of Figure 1.14 display TEM images of TiO_2 films for precursor concentrations of 116.4 ppm and 291.0 ppm, respectively, with the substrate at 763 K. The size of primary particles in subplot B (291.0 ppm) is only slightly larger than that in subplot A (116.4 ppm), although precursor concentration is about 2.5 times larger. It

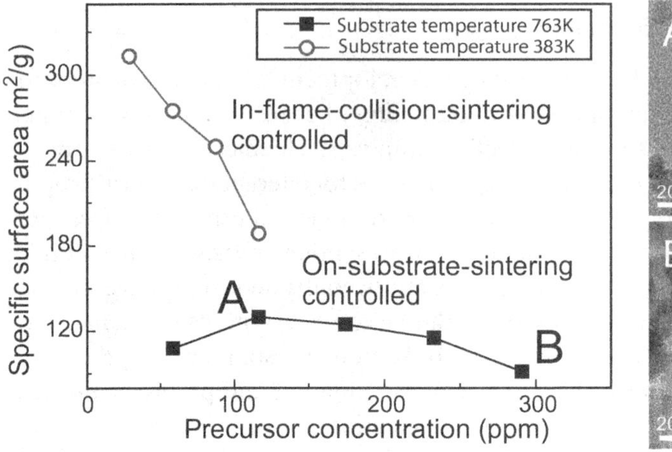
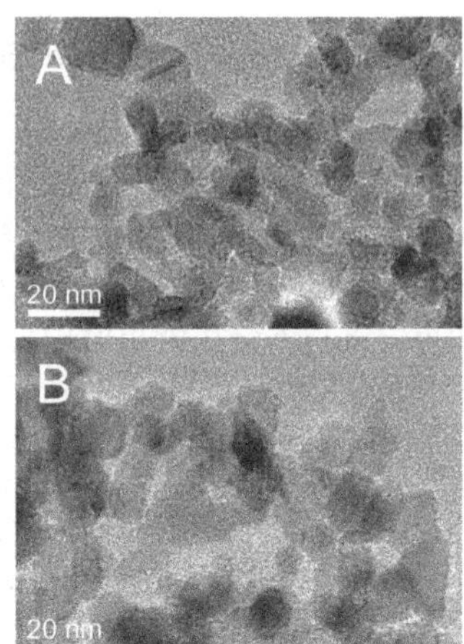

Figure 1.14. Specific surface area (SSA) of TiO_2 nanofilms for different precursor concentrations using a stagnation flame aerosol reactor. Open circles represent the data for a substrate temperature of 383 K, while the filled squares correspond to a substrate temperature of 763 K. The TEM images on the right-hand side show the morphologies of the nanofilms at (A) 116.4 ppm and (B) 291.0 ppm, for a substrate temperature of 763 K.

is concluded that on-substrate sintering of primary particles after deposition greatly reconstructs the TiO_2 film morphology.

REFERENCES

Aarts PAMM, van den Broek SAT, Prins GW, Kuiken GDC, Sixma JJ, Heethaar RM. Blood platelets are concentrated near the wall and red blood cells, in the center in flowing blood. *Arterioscler* **8**, 819–824 (1988).

Aranson IS, Tsimring LS. Patterns and collective behavior in granular media: theoretical concepts. *Review of Modern Physics* **78**, 614–692 (2006).

Bagnold RA. Experiments on a gravity-free dispersion of large solid spheres in a Newtonian fluid under shear. *Proceedings of the Royal Society of London A* **225**(1160), 49–63 (1954).

Balachandar S, Eaton JK. Turbulent dispersed multiphase flow. *Annual Review of Fluid Mechanics* **42**, 111–133 (2010).

Bell GI. Models for the specific adhesion of cells to cells. *Science* **200**, 618–627 (1978).

Bertrand F, Leclaire LA, Levecque G. DEM-based models for the mixing of granular materials. *Chemical Engineering Science* **60**, 2517–2531 (2005).

Bico J, Ashmore-Chakrabarty J, McKinley GH, Stone HA. Rolling stones: The motion of a sphere down an inclined plane coated with a thin liquid film. *Physics of Fluids* **21**, 082103 (2009).

Bluestein D, Yin W, Affeld K, Jesty J. Flow-induced platelet activation in mechanical heart valves. *Journal of Heart Valve Disease* **13**(3), 5001–5008 (2004).

Calle CI, Buhler CR, McFall JL, Snyder SJ. Particle removal by electrostatic and dielectrophoretic forces for dust control during lunar exploration missions. *Journal of Electrostatics* **67**, 89–92 (2009).

Chen XB, Mao SS. Titanium dioxide nanomaterials: Synthesis, properties, modifications, and applications. *Chemical Reviews* **107**, 2891–2959 (2007).

Chesnutt JKW, Marshall JS. Blood cell transport and aggregation using discrete ellipsoidal particles. *Computers & Fluids* **38**, 1782–1794 (2009).

Crowe CT, Schwarzkopf JD, Sommerfeld M, Tsuji Y. *Multiphase Flows with Droplets and Particles*, 2nd ed., CRC Press, Boca Raton, Florida (2012).

Cundall PA, Strack ODL. A discrete numerical model for granular assemblies. *Geotechnique* **29**, 47–65 (1979).

Dominik C, Tielens AGGM. Resistance to rolling in the adhesive contact of two elastic spheres. *Philosophical Magazine A* **72**(3), 783–803 (1995).

Dullien FAL. *Porous Media: Fluid Transport and Pore Structure*, 2nd ed., Academic Press, San Diego (1992).

Einstein A. A new determination of the molecular dimensions. *Annalen der Physik* **19**, 289–306 (1906).

Elghobashi S, Truesdell GC. On the two-way interaction between homogeneous turbulence and dispersed solid particles. I. Turbulence modification. *Physics of Fluids A* **5**, 1790–1801 (1993).

Ennis BJ, Li J, Tardos GI, Pfeffer R. The influence of viscosity on the strength of an axially strained pendular liquid bridge. *Chemical Engineering Science* **45**(10), 3071–3088 (1990).

Epstein PS, Carhart RR. The absorption of sound in suspensions and emulsions. *Journal of the Acoustical Society of America* **25**(3), 553–565 (1953).

Feng JQ, Hays DA. Relative importance of electrostatic forces on powder particles. *Powder Technology* **135**, 65–75 (2003).

Fiscina JE, Lumay G, Ludewig F, Vandewalle N. Compaction dynamics of wet granular assemblies. *Physical Review Letters* **105**, 048001 (2010).

Friedlander SK. *Smoke, Dust and Haze: Fundamentals of Aerosol Dynamics*, Oxford University Press, New York (2000).

Gady B, Schleef D, Reifenberger R. Identification of electrostatic and van der Waals interaction forces between a micrometer-size sphere and a flat substrate. *Physical Review B* **53**, 8065–8070 (1996).

Gore RA, Crowe CT. Effect of particle size on modulating turbulence intensity. *International Journal of Multiphase Flow* **15**, 279–285 (1989).

Hetsroni G. Particles-turbulence interaction. *International Journal of Multiphase Flow* **15**, 735–746 (1989).

Huang B, Yao Q, Li SQ, Zhao HL, Song Q, You CF. Experimental investigation on the particle capture by a single fiber using microscopic image technique. *Powder Technology* **163**, 125–133 (2006).

Israelachvili JN. *Intermolecular and Surface Forces*. 3rd ed., Academic Press, Boston (2011).

Jafari S, Salmanzadeh M, Rahnama M, Ahmadi G. Investigation of particle dispersion and deposition in a channel with a square cylinder using the lattice Boltzmann method. *Journal of Aerosol Science* **41**, 198–206 (2010).

Jonsson V, Bock JE, Nielsen JB. Significance of plasma skimming and plasma volume expansion. *Journal of Applied Physiology* **72**(6), 2047–2051 (1992).

Kaftori D, Hetsroni G, Banerjee S. Particle behavior in the turbulent boundary layer. 1. Motion, deposition, and entrainment. *Physics of Fluids* **7**(5), 1095–1106 (1995).

Kinloch AJ. *Adhesion and Adhesives*, Chapman and Hall, London (1987).

Lashkov VA. Drag of a cylinder in a two-phase flow. *Fluid Dynamics* **27**(1), 93–97 (1992).

Li SQ, Marshall JS. Discrete-element simulation of micro-particle deposition on a cylindrical fiber in an array. *Journal of Aerosol Science* **38**, 1031–1046 (2007).

Li SQ, Marshall JS, Liu G, Yao Q. Adhesive particulate flow: The discrete element method and its application in energy and environmental engineering. *Progress in Energy and Combustion Science* **37**, 633–668 (2011).

Liu G, Marshall JS. Effect of particle adhesion and interactions on motion by traveling waves on an electric curtain. *Journal of Electrostatics* **68**, 179–189 (2010a).

Liu G, Marshall JS. Particle transport by standing waves on an electric curtain. *Journal of Electrostatics* **68**, 289–298 (2010b).

Markarian JAP, Markarian N, Yeksel M, Khusid B, Farmer KR, Acrivos A. Particle motions and segregation in dielectrophoretic microfluidics. *Journal of Applied Physics* **15**, 4160–4169 (2003).

Marshall JS. *Inviscid Incompressible Flow*. John Wiley and Sons, New York (2001).

Marshall JS, Chesnutt JKW, Udaykumar HS. Mesoscale analysis of blood flow. In *Image-based Computational Modeling of the Human Circulatory and Pulmonary Systems*, KB Chandran, HS Udaykumar, J Reinhardt, editors, Springer Publications, New York, Chapter 6, pp. 235–266 (2011).

Marshall JS, Sala K. A stochastic Lagrangian approach for simulating the effect of turbulent mixing on algae growth rate in photobioreactors. *Chemical Engineering Science* **66**, 384–392 (2011).

Melder RJ, Yuan J, Munn LL, Jain RK. Erythrocytes enhance lymphocyte rolling and arrest in vivo. *Microvascular Research* **59**, 316–322 (2000).

Nichols G, Byard S, Bloxham MJ, Botterill J, Dawson NJ, Dennis A, Diart V, North NC, Sherwood JD. A review of the terms agglomerate and aggregate with a recommendation for nomenclature used in powder and particle characterization. *Journal of Pharmaceutical Sciences* **91**(10), 2103–2109 (2002).

Pahlavan AA, Saintillan D. Instability regimes in flowing suspensions of swimming microorganisms. *Physics of Fluids* **23**, 011901 (2011).

Reinhold A, Briesen H. Numerical behavior of a multiscale aggregation model – coupling population balances and discrete element models. *Chemical Engineering Science* **70**, 165–175 (2012).

Renno NO, Wong A-S, Altreya SK. Electrical discharges and broadband radio emission by Martian dust devils and dust storms. *Geophysical Research Letters* **30**(22), 2140 (2003).

Royer JR, Evans DJ, Oyarte L, Guo Q, Kapit E, Mobius ME, Waitukaitis SR, Jaeger HM. High-speed tracking of rupture and clustering in freely falling granular streams. *Nature* **459**, 1110–1113 (2009).

Samadani A, Kudrolli A. Segregation transitions in wet granular matter. *Physical Review Letters* **85**, 5102–5105 (2000).

Schade P, Marshall JS. Capillary effects on a particle rolling on a plane surface in the presence of a thin liquid film. *Experiments in Fluids* **51**(6), 1645–1655 (2011).

Soulié F, Cherblanc F, El Youssoufi MS, Saix C. Influence of liquid bridges on the mechanical behavior of polydisperse granular materials. *International Journal for Numerical and Analytical Methods in Geomechanics* **30**, 213–228 (2006).

Tanabe K. Modeling of airborne dust accumulation on solar cells at the Martian surface. *Acta Astronautica* **62**, 683–685 (2008).

Tien C, Wang, CS, Barot DT. Chainlike formation of particle deposits in fluid-particle separation. *Science* **196**, 983–985 (1977).

Zhang YY, Li SQ, Yan W, Yao Q, Tse SD. Role of dipole-dipole interaction on enhancing Brownian coagulation of charge-neutral nanoparticles in the free molecular regime. *Journal of Chemical Physics* **134**, 084501(2011).

Zhang YY, Li SQ, Yan W, Yao Q, Tse SD. Direct synthesis of nanostructured TiO_2 films with controlled morphologies by stagnation swirl flames. *Journal of Aerosol Science* **44**, 71–82 (2012).

Modeling Viewpoints and Approaches

2.1. A Question of Scale

Particulate flow problems are inherently multiscale. The need for a multiscale approach is evident simply by estimating particle numbers in most applications, such as in the adhesive particle applications described in Chapter 1. For instance, a dust layer with a thickness of only one 10 μm diameter particle requires on the order of 10^{10} particles to cover a solar panel with surface area of 1 m^2. As seen in Figure 1.5a, fly ash particles measure about 2 μm in diameter. A coating one fly ash particle thick covering a single 20 μm diameter fiber measuring 10 cm long would contain about 1.5 million particles. A single cross-knit filter sheet measuring 10 cm on each side, with a 50% void ratio, would contain approximately 7.5×10^9 particles. There are on average about 5 million red blood cells per cubic millimeter of whole blood. The human body contains about 5.6 liters of blood, or about 2.8×10^{13} red blood cells. There are fewer platelets and white blood cells in the body, numbering about 1.4×10^{12} and 4×10^{10}, respectively. The number of particles involved in wet granular flows varies widely depending on the specific application. However, the common comparison of a value that is so large as to be essentially uncountable as comparable to the "number of grains of sand in all the beaches of the world" gives an indication that even wet granular flows often involve very large numbers of particles. Somewhat surprisingly, a number of people have tried to determine even this uncountable value, perhaps the most widely cited being a recent estimate by a researcher at the University of Hawaii that there are 7.5×10^{18} grains of sand on all the world's beaches.[1] On a more human scale, it was recently estimated that there are approximately 5×10^{11} grains of sand on a regulation beach volleyball court.[2]

A full simulation of a multiphase flow would entail not only tracking such huge numbers of particles, but also solving for the fluid flow field in the interparticle region surrounding each particle. For particles that collide with each other, or that reside in clusters or agglomerates, the fluid flow about even a single particle can itself require a multiscale approach, particularly when one accounts for the large difference in scale between the particle diameter and the small gap size within the contact region separating colliding particles. Because of this large disparity in scales

---

[1] http://www.hawaii.edu/suremath/jsand.html.
[2] http://www.6footsix.com/my_weblog/how-many-grains-of-sand-i.html.

and the large number of particles involved in most multiphase flow problems, a wide range of different modeling approaches have been developed for simulation of particulate fluids. Each of these modeling approaches seeks to describe a specific scale of particulate flow, although there is a great deal of overlap between the capabilities of the different approaches.

In this chapter, we simplify the various approaches for simulation of adhesive particle flows as dealing with one of three basic scales of modeling. At the *macroscale*, the simulation model does not follow individual particles, but instead invokes some type of approximation to represent groups of particles by a single computational element. There are various different approximations that are used for this purpose, which are described in the second section of this chapter. At the *mesoscale*, the individual particles are followed by the simulation method, but various analytical and numerical approximations are invoked to accelerate the computation. The most common of these approximations is to introduce analytical models for the different fluid forces acting on the particles. Again, a wide range of computational approaches fit into this general class of mesoscale models, a summary of which is given in the third section of this chapter.

At the *microscale*, we attempt to fully model both the motion of individual particles and the fluid flow about each particle. This involves not merely placing grid points in the region between the particles, but properly demonstrating that the solutions are independent of grid resolution and time step size used. Interparticle flow fields in general are very difficult to accurately simulate, due to the moving particle boundaries and the interaction of particles with the wakes of other particles. Interparticle flows for adhesive particles are especially difficult to simulate due to the wide variation in length scale between that associated with the average particle spacing and the very small length scale associated with the gap separating the contact surfaces of colliding particles. Adhesive particles are continuously in contact with each other, with contacts forming and breaking in an ongoing process. Microscale simulations are often used for describing flows in microscale geometries, such as a red blood cell in a capillary vessel or a bubble in a microchannel, for which other modeling approaches are not appropriate. Microscale simulations are also used to understand particle interactions within small regions of a larger fluid flow, for instance, in order to develop and validate models for particle interaction to be used in mesoscale or macroscale simulation approaches.

2.2. Macroscale Particle Methods

2.2.1. Discrete Parcel Method

A common approach for modeling particulate flows with large numbers of particles is to approximate groups, or parcels, of particles by a single representative particle with some average velocity $\mathbf{v}$ (Figure 2.1). This method has been known by a variety of names in the literature, including Particle-Source-in-Cell (Crowe et al., 1977) and Multiphase-Particle-in-Cell (Andrews and O'Rouke, 1996). In models of plasmas these particle parcels are sometimes referred to as super-particles (Dawson, 1983). Methods in this general category were termed *discrete parcel methods* (DPMs) by Crowe et al. (2012), and we have retained this terminology. In a discrete parcel

Figure 2.1. Schematic diagram illustrating the discrete parcel method, which replaces a group of particles by a single representative particle with average velocity **v**, together with a cloud of virtual particles having group radius R. 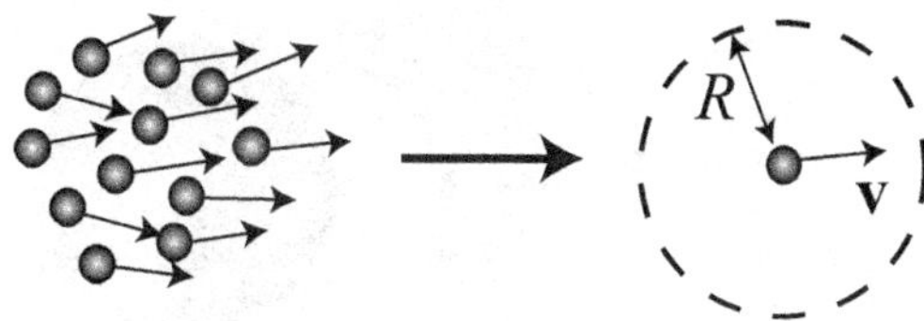

method, each representative particle carries along with it a group of N_C virtual particles, where the characteristic length scale of the group is set equal to the group radius R. The representative particles are transported in the same manner as any individual particle, by solution of the particle momentum equation balancing the rate of particle momentum with fluid forces acting on the particle. Because the number of particles per parcel can have any value, the discrete parcel method is capable of simulating transport of practically unlimited numbers of particles, subject only to resolution limitations. This is a common simulation approach for flows of dilute particles, in which it is assumed that the particles are far enough apart so as not to interact with each other.

A variety of extensions of the discrete parcel method have been proposed. For instance, to represent the effect of turbulence on particle dispersion, Zhou and Yao (1992) proposed increasing the parcel radius R as a function of time. Sommerfeld (2001) proposed a stochastic method to account for effects of particle collisions on the motion of the representative particles in the discrete parcel method, in which a random fluctuation velocity that mimics the effects of collisions with other (nonresolved, or fictitious) particles is added to the velocity of each representative particle at each time step. In dense particle flows, multiple particle collisions can result in transmission of stress through the two-phase medium. To account for this effect within a discrete parcel method, Andrews and O'Rourke (1996) inserted an additional term in the particle momentum equation that is proportional to the gradient of the "solids stress," given by $-\nabla\tau/\rho_p\phi$, where $\tau = P_s\phi^\beta/(\phi_{cp} - \phi)$ is a function of the particle concentration ϕ, ϕ_{cp} is the concentration in a close-packed state, and P_s and β are empirical constants (see also Harris and Crighton, 1994). Typical values proposed by Snider (2007) for these coefficients are $P_s = 100\,$pa, $\beta = 2$, and $\phi_{cp} = 0.7$.

A second type of discrete parcel method for dense particle flows was proposed by Patankar and Joseph (2001). This paper treated parcels as if they are large particles, so that parcel-parcel collisions are modeled similarly to particle-particle collisions in a discrete element model, with the difference that the parcels are larger and may have different elastic and dissipation parameters than the constituent particles. Validation studies for discrete parcel methods were reported by Snider (2007) and Benyahia and Galvin (2010). Snider reported good agreement between the Andrews-O'Rourke DPM model and experimental results for several different granular flow problems. Benyahia and Galvin similarly reported good agreement for the concentration field for both the Andrews-O'Rourke and the Patankar-Joseph DPM formulations for several different validation problems in comparison to results from full discrete element simulations. However, this paper noted significant differences in other flow quantities, such as granular temperature and particle mean velocity fields. Some of these differences can be mitigated by assigning parcels different properties than the

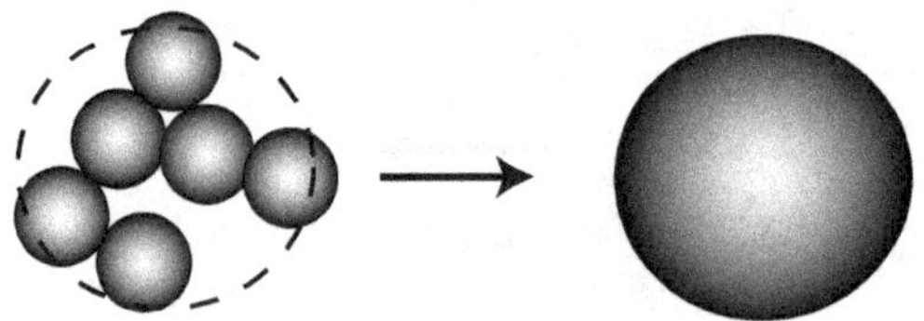

Figure 2.2. Schematic diagram illustrating the population balance method, which replaces each particle agglomerate by a single representative particle with diameter larger than that of the actual particles.

constituent particles, for instance, for restitution coefficient in the Patankar-Joseph approach. A significant amount of research needs to be performed on validation and extension of DPM under different flow conditions. DPM has not been frequently used for flows with adhesive particles, due in part to the difficulty of accounting for adhesion of particles belonging to different (and possibly overlapping) parcels.

2.2.2. Population Balance Method

The population balance method (PBM) is based on the idea of replacing an agglomerate of particles by a single effective particle with volume set equal to the sum of the volumes of the constituent particles making up the agglomerate, as shown schematically in Figure 2.2. Collision and adhesion of two agglomerates results in the elimination of the effective particles representing the original two agglomerates and the formation of a new effective particle with volume equal to the sum of that of the original two agglomerates. The elimination and formation of effective particles in a population balance approach may also depend on processes such as particle growth, nucleation, and agglomerate break-up. If we consider the particle collision and agglomeration process alone, the governing equation for the rate of change of concentration n_i of a particle with volume V_i due to coagulation with other particles is given by a discrete form of the Smoluchowski equation (Smoluchowski, 1917) as

$$\left[\frac{dn_k}{dt}\right]_{coag} = \frac{1}{2}\sum_{i+j=k}\beta(V_i,V_j)n_i n_j - n_k\sum_{i=1}^{\infty}\beta(V_i,V_k)n_i, \quad k = 1, 2, \ldots, \qquad (2.2.1)$$

where $\beta(V_i, V_j)$ represents a measure of collision/agglomeration frequency called the *coalescence kernel*. The first term on the right-hand side in (2.2.1) represents the increase of particles of size k due to collision and agglomeration of smaller particles whose volumes sum up to V_k. The factor of $1/2$ is used to avoid double-counting when the same particle pair is considered with i and j interchanged. The second term on the right-hand side of (2.2.1) represents the decrease in concentration of particles of size k due to collision of these particles with other particles.

Summaries of various coalescence kernels in free-molecular, continuum, and turbulent regimes, as well as under the effects of various external forces, were given by Friedlander (2000) and Cameron et al. (2005). Extensions to the population balance equation to account for agglomerate breakage were reviewed by Ramkrishna (2000). Particle growth in aerosol suspensions by gas-to-particle conversion was considered in the context of population balance theory by Friedlander (2000) and McCoy (2002), which is important for crystallization, granulation, and aerosol-reaction processes.

The effect of convective and diffusive processes on the rate of change of n_k is governed by the advection-diffusion equation, given by

$$\frac{\partial n_k}{\partial t} + \nabla \cdot (n_k \mathbf{v}) = \nabla \cdot (D \nabla n_k) + \left[\frac{dn_k}{dt}\right]_{coag} - \nabla \cdot (n_k \mathbf{v}_t). \qquad (2.2.2)$$

In this equation, $\mathbf{v}$ is the particle velocity, D is a diffusion coefficient, and $\mathbf{v}_t$ is a particle terminal velocity resulting from an external force field (e.g., gravity, electric field, thermophoresis, etc.). All of these terms depend on the PBM particle size. The term in brackets on the right-hand side of (2.2.2) represents the rate of change of n_k due to particle agglomeration, which is given by (2.2.1). If mass transfer processes are present that allow the particles to grow or evaporate with time, an additional term must be added to the right-hand side of (2.2.1) to incorporate these effects. The equation (2.2.2) is often called the *general dynamic equation* (GDE) for aerosol flows.

A variety of different approaches have been used in the literature for solving the GDE. Perhaps the most efficient approach is to a priori assume a particle size distribution (e.g., log-normal) which is characterized by certain coefficients, and then to use (2.2.2) to estimate the values of these coefficients as functions of time and position. Alternatively, the size distribution might be characterized by moments of the particle number density of the form

$$\mu_k \equiv \int_0^\infty r^k g(r) dr, \qquad (2.2.3)$$

where r is the particle radius and $g(r)$ is the size distribution function. The GDE (2.2.2) is used to solve for the time evolution of these moments (Frenklach and Harris, 1987). The most accurate method to solve the GDE is the sectional approach, in which the particle size distribution is divided up into sections, or bins, and the GDE is used to solve for the number of particles in each bin.

The population balance method is very popular for applications with adhesive particles because it provides for rapid simulation of systems with large numbers of agglomerated particles. However, it should be emphasized that accuracy of the method requires satisfaction of a number of restrictions. In particular, the traditional analytic models used for the coalescence kernel require low particle mass fraction, spherical particles, binary collisions, single-size component particles, and homogeneous agglomerate size distribution. Significant errors in the collision model assumed for the standard coalescence kernel used in population balance models can arise at small particle volumetric concentrations. For instance, as shown in Figure 2.3, computations reported by Heine and Pratsinis (2007) and Trzeciak et al. (2006) using the Langevin dynamics method indicate that the dilute-flow approximation to the particle collision rate is in error by between 10% to 400% as the particle concentration increases from 0.001 to 0.1, respectively. A number of researchers have recently utilized variations of the discrete element method in efforts to motivate new coalescence kernel expressions that can be used in situations with larger particle concentrations and with heterogeneous flows (Freireich et al., 2011; Reinhold and Briesen, 2012; Cameron et al., 2005; Gantt et al., 2006).

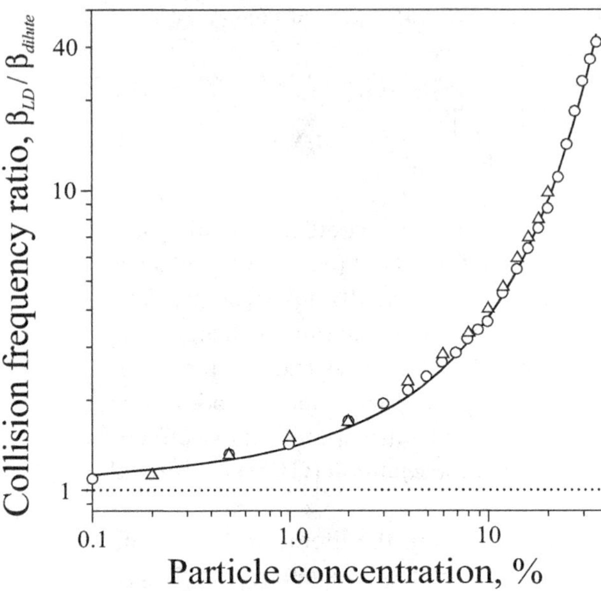

Figure 2.3. Plot showing the ratio of the computed collision frequency to the analytical estimate for dilute suspensions typically used in population balance models as a function of the particle concentration. Circles and the solid line represent data from a Langevin dynamics calculation and a best fit curve by Heine and Pratsinis (2007), and triangles represent numerical data by Trzeciak et al. (2006). [Reprinted with permission from Heine and Pratsinis (2007).]

2.3. Mesoscale Particle Methods

Mesoscale methods for particulate flow range from methods designed for simulation of molecular interactions to methods developed specifically for particulate systems. The various methods invoke different approximations to speed up the computation, which are valid for different ranges of particle sizes. In the current section, we review four mesoscale models, each of which invokes approximations that are significantly different from the others. The molecular dynamics (MD) method was developed for simulating interactions of individual atoms having prescribed potential fields with each other. In particulate flows, MD is often used for problems such as simulating growth of nanoparticles or interaction of nanoparticles with the surrounding solvent (Tian, 2008). Although MD is more properly considered to be a microscale method from the point of view of particulate systems, we have included it in this section on mesoscale modeling because the basic MD formulation is the starting point used in many mesoscale modeling approaches. The dissipative particle dynamics (DPD) method was developed to simulate groups of molecules, in a manner similar in spirit to how the discrete parcel method simulates groups of particles (Section 2.2.1). Both due to this coarse-graining approach and due to the fact that DPD employs soft potentials, instead of the sharp potentials typical of MD, DPD can be employed for significantly larger time and spatial scales than MD. The Brownian dynamics (BD) method replaces individual interactions between small colloidal particles and the molecules of the surrounding solvent with a stochastic forcing term, achieving considerable speed-up over methods such as MD and DPD that attempt to explicitly resolve particle-molecular interactions. Similar to MD, Brownian dynamics approximates adhesive and collision forces between particles by an interaction potential.

The MD, DPD, and BD methods were all designed either for molecules or for particles with diameters in the nanoscale range, such as for a colloidal particle whose size is on the same order of magnitude (3–10 nm) as the length scales of the

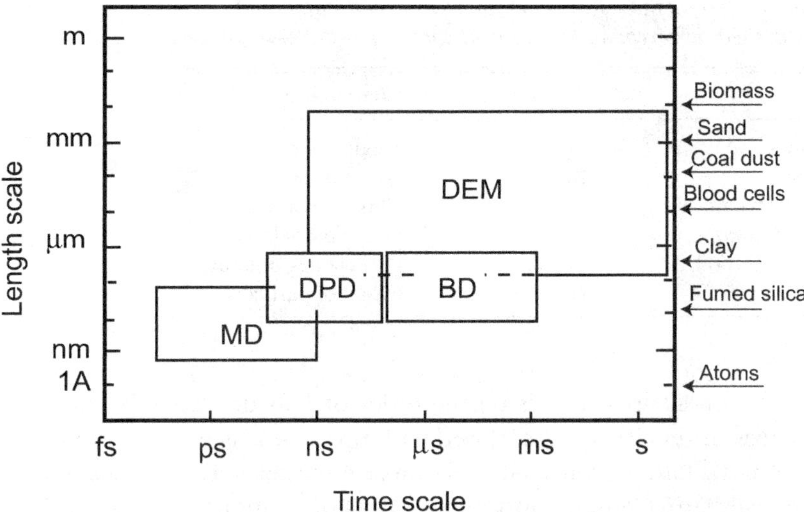

Figure 2.4. Diagram showing the approximate time and length scales for particle flow simulation methods, including molecular dynamics (MD), dissipative particle dynamics (DPD), Brownian dynamics (BD), and discrete-element method (DEM).

electrical double-layer and van der Waals forces that govern its adhesion processes. By contrast, the discrete element method (DEM) was developed for particles with diameters significantly larger than the length scales of adhesive forces, typically applying to cases with particle diameters above about 100 nm. The key change in DEM compared to these other methods is that particles in this size range can no longer be approximated as spheres, because the small elastic deformation of the particle surface that occurs when two particles collide can be of a similar order of magnitude to the length scale of the adhesive force. In such cases, the particle deformation greatly modifies the area on the particle surface over which significant adhesive interaction occurs.

A schematic diagram illustrating the particle length and time scales for which each of these models is generally used is given in Figure 2.4. We note, however, that there is a great deal of variation in the literature on the names used for different methods and the types of approximations imposed. Regardless of the specific type of computational element examined, MD, DPD, BD, and DEM are all computer simulation approaches in which the elements are allowed to interact for a period of time in a Lagrangian framework under various imposed laws of motion. A listing of some of the models that will be discussed in this chapter along with their computational elements is given in Table 2.1.

MD was developed for atoms or molecular groupings with sizes from 0.1 nm to more than 10 nm, with the computational time interval below about 1 ns. The time and length scales of the coarse-graining DPD approach are about one to two orders of magnitude larger than the respective scales for traditional MD. Larger time scales (milliseconds) can be used for BD simulations with particles such as rigid protein cells as compared with MD simulations of macromolecules (typically nanoseconds), even for identical length scales. The DEM approach can treat particles ranging from about 100 nm to several millimeters. The common computational time step of

Table 2.1. *Listing of different mesoscale and macroscale computational methods, along with their acronym and the typical computational element used*

Method	Acronym	Computational element
Molecular Dynamics	MD	Single atom
Brownian Dynamics	BD	Atom or particle
Dissipative Particle Dynamics	DPD	Cluster of atoms
Discrete Element Method	DEM	Single particle
Population Balance Method	PBM	Particle agglomerate
Discrete Parcel Method	DPM	Cloud of particles

DEM with micrometer-size particles is on the order of 1 μs or below. Because of the large differences in time scales associated with processes such as collision and fluid transport in particulate systems with micron-size particles, DEM calculations tend to be numerically stiff. Consequently, many DEM simulations utilize a multiple time scale computational structure to separately resolve processes that occur at very different time scales (Marshall, 2009).

2.3.1. Molecular Dynamics

In traditional molecular dynamics simulations, each computational element represents an individual atom. Interaction of the atoms is governed by an assumed potential field representing nonbonded forces between the atoms, including electrostatic (Coulomb) force, attractive (van der Waals) force, and repulsive (steric) forces, together with expressions for the bonded forces that occur between atoms in a molecule. Electrostatic force is important for ions, where the number of electrons and protons are unequal, giving the atom a net charge. Van der Waals force arises from either permanent or induced atomic polarization, giving rise to electric dipole-dipole interaction between nearby molecules. The steric repulsion acts on a scale on the order of the atomic radius, and is caused by interaction between the electron clouds of two interacting atoms.

A popular potential for nonbonded interactions is the Lennard-Jones potential, in which the interaction energy $U(r)$ is given by

$$U(r) = 4\varepsilon_{LJ}\left[\left(\frac{r_p}{r}\right)^{12} - \left(\frac{r_p}{r}\right)^{6}\right], \tag{2.3.1}$$

where r is the distance between two atoms, ε_{LJ} is the interaction energy parameter, and $r_p \equiv d/2$ is the atomic "radius." The first term in (2.3.1) represents the short-range repulsive force and the second term is the attractive van der Waals force. The electron dynamics can be ignored when developing these force potentials based on the Born-Oppenheimer approximation, which states that electron dynamics is sufficiently rapid that electrons can be assumed to respond instantaneously relative to the time scale for atomic motion. The interaction force acting between the atoms is $\mathbf{F} = -\nabla U$, which is oriented along the line passing through the atom centers. The Lennard-Jones potential is called a pair-wise potential because it deals with only two atoms at a time. There are a wide range of other potentials designed for modeling atomic interactions in various situations, which account for effects such as electrostatic interaction and multiple-atom interactions. With the development of quantum mechanical simulation tools, such as the *ab initio* method, it has become increasingly common to utilize smaller-scale quantum mechanics simulations to motivate the potential field used in subsequent MD simulations.

The time and length scale restrictions of classical MD severely limit the method. For instance, in order to capture the thermal motion of the atoms as well as the stiffness imposed by the forces representing molecular bonds, the time step for MD must be very small, on the order of 1 femtosecond (10^{-15} s). With current generation computer systems, MD computations are limited to computations with total time duration of about 1 ns and length scales of about 10 nm. Some investigators utilize MD for larger-scale systems using a coarse-graining approach, in which the computational particles represent an entire molecule, a group of molecules, or even a nanoparticle (Tian, 2008). Of course, accuracy of the potential function in such approaches becomes an increasingly important issue as the MD particle structure becomes more complex.

2.3.2. Brownian Dynamics

Brownian motion refers to the seemingly random motion of a small particle observed when it is suspended in a fluid. Among the earliest written accounts of this motion is that of Roman philosopher Lucretius (c. 60 B.C.), who described the random, tumbling motion of dust particles in air as proof of the existence of atoms. Botanist Robert Brown provided a detailed account of the random motion of pollen grains when viewed under the microscope in 1827 while studying the plant life of the South Seas (Brown, 1828). Einstein (1905) and Smoluchowski (1906) provided among the first mathematical theories for Brownian motion, the former using the measured mean value of the second moment of particle displacement $< x^2 >$ as a means of deducing Avogadro's number. A stochastic differential equation governing the velocity $\mathbf{v}$ of a small particle in the presence of Brownian motion, with no mean fluid velocity field, was proposed by Langevin (1908) in the form

$$m\frac{d\mathbf{v}}{dt} = -\zeta\mathbf{v} + \mathbf{F}^R(t), \tag{2.3.2}$$

where $\zeta = 3\pi\mu d$ for the Stokes drag law and $\mathbf{F}^R(t)$ is a rapidly fluctuating force that represents the force imposed on the particle from impact of molecules of the surrounding fluid. Instead of resolving each individual molecular impact, the forcing term $\mathbf{F}^R(t)$ in the Langevin equation is chosen to be a random variable with Gaussian probability distribution having zero mean and variance $2\zeta k_b T$, where k_B is Boltzmann's constant (the ratio of the ideal gas constant to Avogadro's number) and T is the absolute temperature. For this choice of random forcing function, it follows that the autocorrelation of the random forcing is given by

$$\langle F^R(t)F^R(t')\rangle = 2\zeta k_B T\delta(t - t'), \tag{2.3.3}$$

where $\delta(\cdot)$ is the Dirac delta. Over long time periods, it can be shown from the system of equations presented here that the mean value of the second moment of particle displacement increases linearly in time, or

$$< x^2 > = 2k_B T t / \zeta. \tag{2.3.4}$$

Brownian dynamics (BD) was introduced as a mesoscale computational method by Ermak and McCammon (1978) as a simplification of the more general Langevin dynamics (LD) method. In LD, it is typical to add a particle interaction term to the Langevin equation (2.3.2), giving

$$m\frac{d\mathbf{v}}{dt} = -\zeta\,\mathbf{v} + \sum_j \mathbf{F}_j^C(t) + \mathbf{F}^R(t). \tag{2.3.5}$$

The particle interaction force $\mathbf{F}_j^C$ is determined using a conservative potential, such as the Lennard-Jones potential (2.3.1), and the sum in (2.3.5) is over particles that are nearby to the given particle. Langevin dynamics is often applied for applications where the computational elements represent small colloidal particles, sections of a long-chain molecule, or large-size atoms. The small solvent atoms are not explicitly resolved in this method, but instead the effect of these solvent atoms on the computational particle is modeled by the sum of the Stokes drag and random forcing terms in (2.3.5). This simplification yields a considerable speed-up for computations using LD compared with those using MD, for which every atom of both the solvent and the particles are resolved.

Brownian dynamics (BD) invokes the further simplification that the particle inertia in (2.3.5) is negligible. In the absence of particle collisions, BD thus assumes that the Brownian motion of each particle is balanced by hydrodynamic drag. Use of the fluid continuum approximation, together with neglect of particle inertia, enables BD to use significantly larger time steps than MD, as indicated in Figure 2.4. More information on Brownian dynamics can be obtained from the review article by Chen and Kim (2004).

2.3.3. Dissipative Particle Dynamics

Dissipative particle dynamics (DPD) was introduced by Hoogerbrugge and Koelman (1992) as a type of coarse-grained MD, in which each DPD computational particle represents a collection of many molecules. The DPD approach was later extended to colloidal fluids (Dzwinel and Yuen, 2002), complex fluids such as blood flow (Dzwinel et al., 2003; Filipovic et al., 2008; Pivkin et al., 2009), and polymeric fluids (Symeonidis et al., 2005; Wijmans et al., 2001). Español and Warren (1995) derived the Fokker-Planck equation for the DPD method and the corresponding fluctuation-dissipation theorem, demonstrating that the DPD model yields collective behavior consistent with classical hydrodynamics. A review of the DPD approach was given by Groot and Warren (1997).

Similar to the Langevin dynamics approach illustrated by (2.3.5), the DPD model imposes three distinct types of interaction forces between any pair of computational particles i and j, representing a dissipative force $\mathbf{F}_{ij}^D$, a conservative force $\mathbf{F}_{ij}^C$, and a random force $\mathbf{F}_{ij}^R$. However, there are considerable differences with LD regarding the expressions used for these forces. In particular, in DPD all forces act between the computational particles, and there is no notion of a surrounding continuum fluid representing a solvent or gaseous interstitial region. Galilean invariance requires that these forces depend only on the vector separating their centroids $\mathbf{r} = \mathbf{r}_i - \mathbf{r}_j$ and the relative centroid velocity $\mathbf{v}_R = \mathbf{v}_i - \mathbf{v}_j$, rather than the particle centroid positions and velocities independently. It is also useful to define the unit vector $\mathbf{n}$ pointing from the centroid of particle i to the centroid of particle j by $\mathbf{n} \equiv (\mathbf{r}_j - \mathbf{r}_i)/r$, where $r = |\mathbf{r}_j - \mathbf{r}_i|$ is the distance between the particle centroids. Expressions for the dissipative and random forces were proposed by Español and Warren (1995) for $r < r_C$, where r_C is a prescribed cut-off distance, as

$$\mathbf{F}_{ij}^D = -\gamma_{dpd}\omega_D(r)(\mathbf{n} \cdot \mathbf{v}_R)\mathbf{n}, \quad \mathbf{F}_{ij}^R = \sigma_{dpd}\omega_R(r)\zeta_{ij}\mathbf{n}. \tag{2.3.6}$$

In these equations, ζ_{ij} is a Gaussian random variable with unit variance, and γ_{dpd} and σ_{dpd} are the friction coefficient and the amplitude of the random forcing, respectively. The functions $\omega_D(r)$ and $\omega_R(r)$ govern how these forces decay as the distance between the particles increases. The balance between the dissipative and random terms serves as a thermostat for the DPD simulation, such that the dissipative force tends to cool the system and the random force tends to heat the system. The fluctuation-dissipation theorem for DPD requires that the coefficients of these two terms are related to the absolute temperature T of the medium by

$$\omega_R^2(r) = \omega_D(r), \quad \sigma_{dpd}^2 = 2k_B T \gamma_{dpd}, \tag{2.3.7}$$

where k_B is Boltzmann's constant.

The conservative force $\mathbf{F}_{ij}^C$ in DPD has the form of a soft potential, and thus avoids the numerical stiffness associated with rapid variation of $\mathbf{F}_{ij}^C$ with r in potentials such as the Lennard-Jones potential (2.3.1). A simple linear form is often assumed for both the conservative force and for the spatial decay function $\omega_R(r)$ of the form

$$\mathbf{F}_{ij}^C = \begin{cases} a[1 - r/r_C]\mathbf{n} & \text{for } r < r_C \\ 0 & \text{for } r > r_C \end{cases}, \quad \omega_D(r) = \omega_R^2(r) = \begin{cases} 1 - r/r_C & \text{for } r < r_C \\ 0 & \text{for } r > r_C \end{cases} \tag{2.3.8}$$

where a is an adjustable parameter related to the material compressibility. The resulting momentum equation for a DPD particle i thus has the form

$$m\frac{d\mathbf{v}_i}{dt} = \sum_{\substack{j=1 \\ j \neq i}}^{N} \left(\mathbf{F}_{ij}^C + \mathbf{F}_{ij}^D + \mathbf{F}_{ij}^R\right). \tag{2.3.9}$$

Because DPD particles represent a group of molecules, instead of individual atoms, and because of the use of soft potentials and lack of covalent bonds between DPD particles, the DPD method can be used for problems involving length and time scales that are one to two orders of magnitude greater than can be used for MD. A time step limitation for DPD was presented by Español and Warren (1995) as $\Delta t/t_c \ll 1$, where $t_C = r_C/v_{rms}$ and $v_{rms} = (3k_B T/m)^{1/2}$ is the root-mean square particle velocity.

2.3.4. Discrete Element Method

The discrete element method (DEM) is similar to BD in that it solves for motion of individual particles using a continuum approximation of the surrounding fluid. The most significant difference between DEM and BD concerns the model used for interparticle contact forces. Whereas it is typical in BD simulations to employ simple conservative potentials, such as the Lennard-Jones potential, to describe interactions of very small particles (on the nanometer scale), such potentials do not adequately represent the interparticle forces present for collision of the larger (micron-size) particles used for DEM simulations. Instead, DEM is founded on the assumption that the particle size is substantially larger than the characteristic length scale of van der Waals force or other adhesion force, which for van der Waals interaction is on the order of a nanometer. One consequence of this assumption is that in DEM, particle adhesion force has no effect until two particles collide. This situation is unlike the MD, DPD, or BD models, which are intended for smaller particles for which adhesion forces act over length scales on the order of the particle diameter. These larger particles also have more inertia than the nanometer scale particles treated with the MD, DPD, and BD schemes, so it is typically necessary to account for resistance to sliding, rolling, and twisting motions of the particles in addition to the interaction force directed along the particle normal vector $\mathbf{n}$. Since the various modes of frictional resistance depend on the relative tangential motion between the particle surfaces at the contact point, which in turn depends on both the particle rotation rate and centroid velocity, in DEM simulations it is necessary to solve governing equations for both particle linear and angular momentum.

Two very different variations of DEM models exist, called the *hard-sphere* and *soft-sphere* models. In the soft-sphere model, the particle momentum and angular momentum equations, given by

$$m\frac{d\mathbf{v}}{dt} = \mathbf{F}_F + \mathbf{F}_A + \mathbf{F}_E, \qquad I\frac{d\mathbf{\Omega}}{dt} = \mathbf{M}_F + \mathbf{M}_A + \mathbf{M}_E, \qquad (2.3.10)$$

are solved during the entire period in which two colliding particles are touching each other. In these equations, m and I are the particle mass and moment of inertia, $\mathbf{v}$ and $\mathbf{\Omega}$ are the particle velocity and rotation rate, $\mathbf{F}_F$ and $\mathbf{M}_F$ are the force and torque exerted by the surrounding fluid, $\mathbf{F}_A$ and $\mathbf{M}_A$ are the force and torque due to the collision and adhesion forces, and $\mathbf{F}_E$ and $\mathbf{M}_E$ are the force and torque induced by interaction with the surrounding electric field. The time derivative in (2.3.10) is evaluated on a moving particle. The soft-sphere model requires that the computational time step be sufficiently small to resolve the period during which the collision occurs. Because particle elastic collisions tend to occur over very small time scales, soft-sphere DEM computations tend to be numerically stiff. This difficulty is handled in practice by use of multiple time-step algorithms, and also in some cases by use of elastic moduli for the particles that are smaller than the actual elastic moduli. By contrast, the hard-sphere model assumes that collisions between particles are pair-wise additive and occur instantaneously. The particle collisions are thus accounted for using particle impulse equations. The hard-sphere model removes the stiffness associated with the need to resolve the time interval during which particles collide, allowing this method to use much larger time steps than the soft-sphere model. However, the hard-sphere model cannot handle problems with collisions of three or more particles or problems in which particles remain in contact for a prolonged time period. Consequently, most studies of adhesive particle flows utilize the soft-sphere model. Those studies that do use the hard-sphere model for adhesive particles typically assume that particles that adhere together are "frozen" upon collision with each other. However, this assumption ignores deformation of particle agglomerates by fluid-induced forces and by collisions with other agglomerates, which have been found to be a common and important process governing the physics of the agglomerated particles.

As discussed in Section 1.3.1, adhesive particles in the micron size range cannot be treated as spheres when they collide, because the length scale associated with particle adhesive force is typically on the same order of magnitude, or smaller, than that associated with particle elastic deformation. A common type of soft-sphere DEM model makes a very different assumption – that the particle adhesive forces only act within the flattened part of the colliding particle surfaces called the contact region (Figure 1.7). For this model, the adhesive interaction between two micrometer-sized particles is approximated by the adhesion force between two parallel flat plates, rather than between two spheres. This observation is the basis of the JKR theory (Johnson, Kendell, and Roberts, 1971) of elastic adhesion. Since the particle deformation is caused by the force acting between the particles, which is subsequently influenced by the extent of particle deformation, the interaction of adhesive micron-size particles is a highly nonlinear process. As a consequence, adhesion and collision forces between the particles cannot simply be added together, as they can for two spherical nanoparticles, but instead a new model must be formulated for the particle contact forces that includes the combination of elastic and adhesive effects.

There has recently been quite a lot of activity with DEM for adhesive particles, both in terms of applying the method to a wide variety of problems and in extending the method to deal with particle interactions with different field effects, such as acoustic and electric fields, and to handle different types of adhesive forces. These extensions of DEM are the focus of most of the remainder of this book, so more detailed discussion of these developments is deferred until later chapters.

2.4. Microscale Dynamics of Elastohydrodynamic Particle Collisions

Microscale methods simulate the detailed interaction of particles with the surrounding fluid flow, or with other fields such as electric, acoustic, or thermal fields. There might be some confusion as to what exactly constitutes a microscale simulation given the broad range of scales covered by the mesoscale methods discussed in the previous section. In general, a mesoscale method employs a relatively simple approximate model for the interaction of the computational particles, whereas a microscale method directly simulates the surrounding field distortions that give rise to the interaction. For example, molecular dynamics is a mesoscale method for simulating interaction of atoms that uses a simple potential model (e.g., the Lennard-Jones potential) to approximate the interaction of atoms. By contrast, a direct simulation of the quantum field around the atoms using the ab initio method would represent a microscale simulation of atomic interaction. For particulate fluids, Brownian dynamics or the discrete element method represent mesoscale methods that employ simple models for particle interactions, whereas a microscale simulation would directly simulate the flow around each particle and the particle deformation during collisions.

There is a broad literature devoted to simulation of detailed flow fields about groups of spherical particles, droplets, and bubbles, as well as the various computational methods used for such problems. A thorough review of this literature is outside the scope of what we wish to cover in this book. Instead, in the current section we consider a specific example of a microscale simulation by examining the problem of particle elastohydrodynamic interaction, which deals with the combined elastic deformation and the squeeze-film fluid dynamics that occurs in the contact region between two colliding particles. The fluid dynamics in this region is of great importance in constructing physical scaling parameters for DEM models, since this thin region of the flow field governs the extent to which the two particle surfaces come together in a dynamic collision process.

2.4.1. Microscale Simulations of Elastohydrodynamic Interactions

A classic series of papers on the elastohydrodynamics of the head-on collision of two particles immersed in a viscous fluid was published in the 1980s by Robert Davis and his students at the University of Colorado, along with E.J. Hinch from Cambridge University. In this work, the particle elastic deformation was computed using a boundary-element formulation, and the fluid flow within the thin gap region separating the particles was computed using a lubrication equation of the form

$$\frac{\partial h}{\partial t} = \frac{1}{12\mu r} \frac{\partial}{\partial r}\left[rh^3 \frac{\partial p}{\partial r}\right]. \tag{2.4.1}$$

The gap thickness $h(r, t)$ and the averaged pressure $p(r, t)$ over the gap are both functions of the radial coordinate r and time t. The lubrication theory assumes that the thickness of the gap between the particles is much smaller than the contact region radius and that the flow within this squeeze-film region has a low Reynolds number, evaluated based on the gap thickness.

The boundary-element theory assumes small elastic displacement of the particle. An effective radius R and a parameter θ are defined by

$$R \equiv \left(r_1^{-1} + r_2^{-1}\right)^{-1}, \quad \theta \equiv \frac{1 - \nu_1^2}{\pi E_1} + \frac{1 - \nu_2^2}{\pi E_2}, \tag{2.4.2}$$

where r_1 and r_2 are the radii, ν_1 and ν_2 are the Poisson ratios, and E_1 and E_2 are the elastic moduli of the two colliding particles. The extent of elastic deformation can be characterized based on the dimensionless elasticity parameter ε_E, which is defined as

$$\varepsilon_E \equiv \frac{4\theta\mu v_0 R^{3/2}}{x_0^{5/2}}, \tag{2.4.3}$$

where μ is the fluid viscosity and the particle impact velocity v_0 is evaluated when the minimum gap thickness has a value x_0. The parameter x_0 is a measure of the gap thickness between the particles, but its value is somewhat arbitrary subject to the restrictions $x_0/R \ll 1$, $\varepsilon_E \ll 1$, and $\mathrm{Re}\, x_0/R \ll 1$, where $\mathrm{Re} = \rho v_0 R/\mu$ is the particle Reynolds number. The computations presented by Davis et al. (1986) select

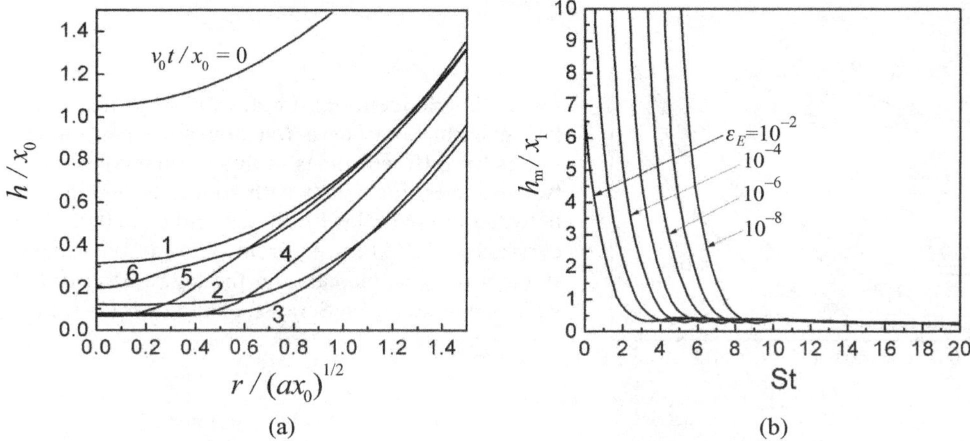

Figure 2.5. Plots showing details of particle collision with no adhesion: (a) sequence of profiles of the dimensionless gap thickness as a function of dimensionless radius for different times for St = 5 and $\varepsilon_E = 0.01$ and (b) plot of minimum gap half-thickness divided by the axial length scale x_1 as a function of the Stokes number for different values of the elasticity parameter. [Reprinted with permission from Davis et al. (1986).]

$x_0 = 0.01\,R$. Using this value for x_0, typical values of ε_E for collisions of 1–100 μm diameter particles in air and water lie in the range 10^{-7}–10^{-5}. An alternative length scale characterizing the gap thickness is motivated by noting that the numerator of (2.4.3) has dimensions of length to the 5/2 power, so that a length scale x_1 can be defined by

$$x_1 \equiv (4\theta\mu R^{3/2}v_0)^{2/5}. \tag{2.4.4}$$

Numerical results from Davis et al. (1986) are plotted in Figure 2.5. Figure 2.5a plots the ratio of the particle gap thickness $h(r, t)$ to the axial length scale x_0 as a function of the ratio of radius r to the radial length scale $(Rx_0)^{1/2}$. Curves are drawn at different times during the particle collision. It is of note that the separation distance between the particles decreases to some minimum value, but for times after this minimum value is reached the particle deformation forms a flattened contact region between the two particles. Motion of the particles toward each other at times after this minimum separation is achieved results in growth of the contact region, but not in decrease of the particle minimum separation distance. Figure 2.5b plots the ratio of the minimum gap thickness achieved during the collision, h_m, to the alternative axial length scale x_1 as a function of Stokes number for different values of the elasticity parameter ε_E. This plot demonstrates that at a sufficiently high Stokes number, the ratio of the minimum gap thickness to the length scale x_1 approaches a value that is independent of the elasticity parameter. This result provides strong evidence that the parameter x_1 gives a correct scaling for the minimum gap thickness for collision of smooth particles in the absence of adhesive force.

In later work, Barnocky and Davis (1989) examined effects of density and viscosity variation that occurs due to high pressures within the contact region. Serayssol and Davis (1986) examined the effect of adhesive forces on the elastohydrodynamics of particle collisions. An interesting result of the latter paper is shown in Figure 2.6, which plots h/x_0 as a function of $r/(Rx_0)^{1/2}$ for a case with adhesion force acting

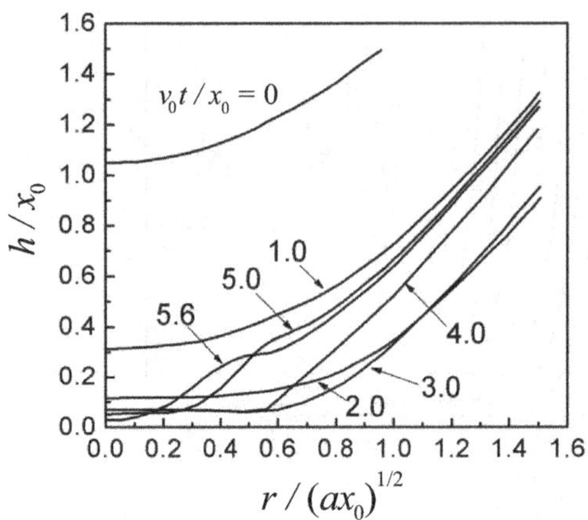

Figure 2.6. Sequence of profiles showing dimensionless gap thickness as a function of dimensionless radius for different times during normal collision of two particles, for a case with adhesive force acting between the particles for St = 5 and $\varepsilon_E = 0.01$. The curves are plotted for different values of the dimensionless time, as indicated in the figure. [Reprinted with permission from Serayssol and Davis (1986).]

between the particle surfaces. Similar to the case without adhesion shown in Figure 2.5a, the particle surfaces form a flattened contact region as the particles approach each other. However, for the adhesion case, it is observed that as the particles begin moving apart during the rebound phase of the collision, an interesting necking phenomenon occurs (at times 5.0–5.6 for the example shown in Figure 2.6) during which the surface of each particle is pulled outward toward the opposing particle by the adhesion force. This adhesion-induced necking behavior plays an important role in the contact dynamics of adhesive particles.

2.4.2. Experimental Results for Two-Particle Collisions

Extensive experimental studies have been performed to examine the effect of interstitial fluid on particle-particle and particle-wall interactions. In the simplest experiments, in which two particles collide with each other with velocities oriented along the normal direction **n** relative to the other particle, the experiments measure the relative velocity of the particles before and after collision, given by

$$v_I \equiv |v_2(t_C - t_0) - v_1(t_C - t_0)|, \quad v_R \equiv |v_2(t_C + t_0) - v_1(t_C + t_0)|, \quad (2.4.5)$$

where t_C is the time at which the collision occurs and t_0 is a sufficiently large time interval that for $t < t_C - t_0$ the particles have negligible effect on each other. The ratio of the relative normal particle speed after collision to that before collision is called the restitution coefficient, e, and is given by

$$e \equiv v_R/v_I. \quad (2.4.6)$$

In general, restitution coefficient is a function of all parameters associated with processes that lead to energy dissipation during the collision process. This includes both energy loss due to viscous dissipation within the fluid as well as energy loss due to inelastic deformation of the solid particles. In granular flows, which typically involve relatively large particles in air, the majority of the energy dissipation occurs due to solid phase energy loss, and the fluid loss is often neglected. On the other hand, for the small particles that are typically of concern for adhesive particle flows,

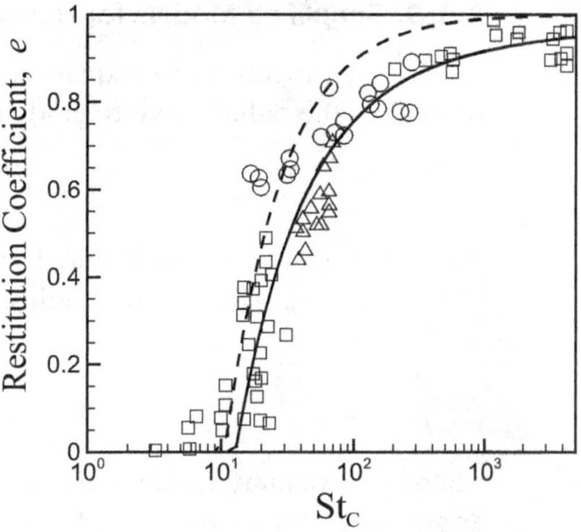

Figure 2.7. Comparison of predicted total restitution coefficient e_{tot} (solid line) and spherical particle restitution coefficient e_{sph} (dashed line) with the experimental data of Yang and Hunt (2006) for collision of identical spherical particles, made from steel (squares), glass (circles), and Delrin (triangles). The predictions are performed with parameter values $\varepsilon_E = 10^{-7}$, $\delta/r_p = 5 \times 10^{-5}$, and $h_{0,init}/r_p \cong 0.02$. [Reprinted with permission from Marshall (2011).]

the collision velocity is relatively small and the fluid losses often dominate over the solid losses.

Estimates for the fluid losses are usually expressed as functions of the Stokes number, defined in (1.2.5). It was demonstrated by Yang and Hunt (2006) that restitution coefficients for collisions between particles of two different radii, r_1 and r_2, with two different masses, m_1 and m_2, can be approximately collapsed onto a single curve using the effective radius R defined in (2.4.2) and an effective mass M defined by

$$M \equiv \left(m_1^{-1} + m_2^{-1}\right)^{-1}. \tag{2.4.7}$$

The Stokes number for collision processes is defined with particle mass replaced by M, the length scale L set equal to R, and the velocity scale U set equal to the impact velocity v_I, such that (1.2.5) becomes

$$St_C = \frac{Mv_I}{6\pi\mu R^2}. \tag{2.4.8}$$

The data collapse presented by Yang and Hunt (2006) includes data for two equal-size colliding spheres made of different materials and with different masses, unequal-size spheres, and sphere collision with a wall, where the wall is treated as a sphere of infinite radius. Yang and Hunt (2006) even found that this data collapse can be applied to oblique collisions if the restitution coefficient is defined only in terms of the normal component of the particle velocities before and after collision. The full plot is rather cluttered, so we present a reduced plot with data only for two equal-size spheres in Figure 2.7. A particularly interesting feature of this plot is that the restitution coefficient is essential equal to zero if the collision Stokes number is less than 10. This case applies to many applications involving adhesive particles, which tend to involve particles of a sufficiently small size that the adhesive forces are of a comparable magnitude to other forces that might exist, such as the gravitational and fluid drag forces and the particle inertia.

2.4.3. Simplified Models for Restitution Coefficient in a Viscous Fluid

Integrating the lubrication equation (2.4.1) in the radial direction over the interval $(0, r)$ gives the radial pressure gradient within the squeeze-film as

$$\frac{\partial p}{\partial r} = \frac{6\mu r}{h^3} \frac{\partial h}{\partial t}. \tag{2.4.9}$$

Integrating (2.4.9) a second time in the radial direction, this time over the interval (r, r_p), where r_p is the particle radius, gives the squeeze-film pressure distribution as

$$p(r, t) - p_0 = -6\mu \int_r^{r_p} \frac{\xi}{h^3} \frac{\partial h}{\partial t} d\xi, \tag{2.4.10}$$

where ξ is a dummy variable of integration and the pressure is assumed to approach an ambient value p_0 as $r \to r_p$. The normal damping force F_{nd} exerted by the squeeze-film fluid along the axial direction $\mathbf{n}$ is obtained by integrating the pressure field from (2.4.10) over the circular area covered by the film, which after using integration by parts yields

$$F_{nd} = -12\pi \mu \int_0^{r_p} \left[\int_r^{r_p} \frac{\xi}{h^3} \frac{\partial h}{\partial t} d\xi \right] r dr = -6\pi \mu \int_0^{r_p} \frac{r^3}{h^3} \frac{\partial h}{\partial t} dr. \tag{2.4.11}$$

The damping force between two spherical particles that are approaching each other can be obtained from (2.4.11) using

$$\frac{\partial h}{\partial t} = \frac{dh_0}{dt} = f(t), \quad h - h_0 \cong \frac{r^2}{r_p}, \tag{2.4.12}$$

where $h_0(t) \equiv h(0, t)$ is the gap thickness along the line separating the particle centroids and we assume $h/r_p \ll 1$. Defining $\eta \equiv r/\sqrt{r_p h_0}$ and using (2.4.12), the integral in (2.4.11) becomes

$$F_{nd} = -\frac{6\pi \mu r_p^2}{h_0} \frac{dh_0}{dt} \int_0^{(h_0/r_p)^{-1/2}} \frac{\eta^3}{(1 + \eta^2)^3} d\eta. \tag{2.4.13}$$

Since $h_0/r_p \ll 1$, the upper bound for the integration in (2.4.13) can be approximated by infinity, for which case the integral is equal to $1/4$. The resulting expression for the normal damping force on the spherical particle becomes

$$F_{nd}(t) = -\frac{3\pi \mu r_p^2}{2h_0} \frac{dh_0}{dt}. \tag{2.4.14}$$

The normal force expression (2.4.14) applies only prior to and following collision of two particles. During the collision itself, the viscous damping force is dominated by the fluid flow near the outer edge of the contact region as this region grows and contracts with motion of the particle centroids toward or away from each other. An expression for normal damping force during collision was derived by Marshall (2011) using lubrication theory as

$$F_{nd} = \frac{3\pi \mu r_p^2}{2\delta} \frac{d\delta_N}{dt} \left(1 + \frac{\delta_N}{\delta}\right), \tag{2.4.15}$$

where $\delta_N(t)$ is the "normal overlap" of the particles and δ is the gap thickness between the particle surfaces within the contact region. The normal overlap for collision of two equal spheres is given by

$$\delta_N = 2r_p - |\mathbf{x}_i - \mathbf{x}_j|, \tag{2.4.16}$$

where $|\mathbf{x}_i - \mathbf{x}_j|$ denotes the distance between the sphere centroids. At the instant of particle collision, the variables in (2.4.15) have the values $h_0(t) = \delta$, $\delta_N(t) = 0$, and $\dot{\delta}_N(t) = -\dot{h}_0(t)$. Substituting these values into (2.4.15) yields the equation (2.4.14) for two spherical particles, thus guaranteeing that the normal force expressions (2.4.14) and (2.4.15) are continuous the throughout the collision process.

Head-on collision of two identical, nonadhesive particles of radius r_p and equal mass m, with centroid positions $\pm[r_p + x(t)]$ and velocities $\mp v(t)$, is governed by the equations

$$\frac{dx}{dt} = v, \quad m\frac{dv}{dt} = F_{nd} + F_{ne}, \tag{2.4.17}$$

where F_{nd} is the normal damping force, given by (2.4.14) before and after the particles collide and by (2.4.15) during the collision, and F_{ne} is the elastic rebound force that acts to push the particles apart during the collision. The two particles are assumed to have initial centroid positions $\pm x_0$ and to travel with initial velocity $\mp v_0$ toward each other.

It is helpful to define dimensionless variables $v' = v/v_0$, $x' = x/x_0$, and $t' = tv_0/x_0$. Prior to the collision, substituting (2.4.14) in (2.4.17) with $h_0 = 2x$ and $dh_0/dt = -2v$, the governing equation can be written in dimensionless form as

$$v'\frac{dv'}{dx'} = -\frac{1}{\mathrm{St}_C}\frac{v'}{x'}, \tag{2.4.18}$$

where St_C is the collision Stokes number, with $M = m/2$, $R = r_p/2$, and $v_I = 2v_0$ for two equal colliding spheres. The left-hand side of (2.4.18) is obtained using chain rule by writing $v' = v'(x')$ and using (2.4.17)$_1$. Dividing (2.4.18) by v' and integrating over x' yields

$$\frac{v(x)}{v_0} = 1 - \frac{1}{\mathrm{St}_C}\ln(x_0/x). \tag{2.4.19}$$

This equation describes the decrease in velocity of the spheres as they approach each other. If we assume that the spheres collide with no loss in energy during the collision, then the resulting restitution coefficient e_{sph} is obtained from (2.4.19) as

$$e_{sph} = 1 - \frac{2}{\mathrm{St}_C}\ln(x_0/\delta), \tag{2.4.20}$$

where δ is the equilibrium gap thickness between the particle surfaces within the contact region. The factor of 2 in the last term in (2.4.20) accounts for the motion of the two spheres both toward and away from each other. On the other hand, energy might also be lost during the collision process itself, either within the solid phase due to deformation of the particles or within the fluid phase due to viscous flow associated with change in the size of the contact region, with an expression for the latter damping force given by (2.4.15). If the restitution coefficient associated with the collision itself, not including the motion of the two spheres toward each other prior to the collision or away from each other following the collision, is denoted by e_{col}, then the equation for the total restitution coefficient becomes

$$e_{tot} = e_{col} - \frac{1 + e_{col}}{\mathrm{St}_C} \ln(x_0/\delta). \tag{2.4.21}$$

The predicted values of e_{sph} from (2.4.20) and of e_{tot} from (2.4.21) as functions of the Stokes number are plotted in Figure 2.7 as dashed and solid lines, respectively, in comparison with the experimental data of Yang and Hunt (2006) for collision of two equal-size spheres. The value of e_{col} in (2.4.21) was obtained by a numerical integration of the equations of motion using the expression (2.4.15) for the normal damping force and the classical Hertz law for the elastic force. Although both expressions agree reasonably well with the data for small Stokes numbers, the total restitution coefficient given by (2.4.21) fits the data considerably better as the Stokes number increases.

REFERENCES

Andrews MJ, O'Rourke PJ. The multiphase particle-in-cell method for dense particulate flows. *International Journal of Multiphase Flow* **22**, 379–402 (1996).

Barnocky G, Davis RH. The influence of pressure-dependent density and viscosity on the elastohydrodynamic collision and rebound of two spheres. *Journal of Fluid Mechanics* **209**, 501–519 (1989).

Benyahia S, Galvin JE. Estimation of numerical errors related to some basic assumptions in discrete particle methods. *Industrial & Engineering Chemistry Research* **49**, 10588–10605 (2010).

Brown R. A brief account of microscopical observations made in the months of June, July and August, 1827, on the particles contained in the pollen of plants; and on the general existence of active molecules in organic and inorganic bodies. *Philosophical Magazine* **4**, 161–173 (1828).

Cameron IT, Wang FY, Immanuel CD, Stepanek F. Process systems modelling and applications in granulation, A review. *Chemical Engineering Science* **60**(14), 3723–3750 (2005).

Chen JC, Kim AS. Brownian dynamics, molecular dynamics, and Monte Carlo modeling of colloidal system. *Advances in Colloid and Interface Science* **112**, 159–173 (2004).

Crowe CT, Sharma MP, Stock DE. The particle-source-in-cell method for gas droplet flow. *Journal of Fluids Engineering* **99**, 325–332 (1977).

Crowe CT, Schwarzkopf JD, Sommerfeld M, Tsuji Y. *Multiphase Flows with Droplets and Particles*, CRC Press, Boca Raton, Florida (2012).

Davis RH, Serayssol J-M, Hinch EJ. The elastohydrodynamic collision of two spheres. *Journal of Fluid Mechanics* **163**, 479–497 (1986).

Dawson JM. Particle simulation of plasmas. *Reviews of Modern Physics* **55**(2), 403–447 (1983).

Dzwinel W, Yuen DA. Mesoscopic dispersion of colloidal agglomerate in a complex fluid modelled by a hybrid fluid-particle model. *Journal of Colloid and Interface Science* **247**, 463–480 (2002).

Dzwinel W, Boryczko K, Yuen DA. A discrete-particle model of blood dynamics in capillary vessels. *Journal of Colloid and Interface Science* **258**, 163–173 (2003).

Einstein A. On the movement of small particles suspended in stationary liquids required by the molecular-kinetic theory of heat. *Annalen der Physik* **17**, 549–560 (1905).

Ermak DL, McCammon JA. Brownian dynamics with hydrodynamic interactions. *Journal of Chemical Physics* **69**(4), 1352–1360 (1978).

Español P, Warren P. Statistical mechanics of dissipative particle dynamics. *Europhysics Letters* **30**(4), 191–196 (1995).

Filipovic N, Kojic M, Tsuda A. Modelling thrombosis using dissipative particle dynamics method. *Philosophical Transactions of the Royal Society of London A* **366**, 3265–3279 (2008).

Freireich B, Li J, Litster J, Wassgren C. Incorporating particle-flow information from discrete-element simulations in population balance models of mixer-coaters. *Chemical Engineering Science* **66**, 3592–3604 (2011).

Frenklach M, Harris SJ. Aerosol dynamics modeling using the method of moments. *Journal of Colloidal and Interface Science* **118**(1), 252–261 (1987).

Friedlander AK. *Smoke, Dust and Haze: Fundamentals of Aerosol Dynamics*, 2nd ed., Oxford University Press, U.S.A. (2000).

Gantt JA, Cameron IT, Litster JD, Gatzke EP. Determination of coalescence kernels for high-shear granulation using DEM simulations. *Powder Technology* **170**(2), 53–63 (2006).

Groot RD, Warren PB. Dissipative particle dynamics: bridging the gap between atomistic and mesoscopic simulation. *Journal of Chemical Physics* **107**(11), 4423–4435 (1997).

Harris SE, Crighton DG. Solitons, solitary waves, and voidage disturbances in gas-fluidized beds. *Journal of Fluid Mechanics* **266**, 243–276 (1994).

Heine MC, Pratsinis SE. Brownian coagulation at high concentration. *Langmuir* **23**, 9882–9890 (2007).

Hoogerbrugge PJ, Koelman JMVA. Simulating microscopic hydrodynamic phenomena with dissipative particle dynamics. *Europhysics Letters* **19**(3), 155–160 (1992).

Johnson KL, Kendall K, Roberts AD. Surface energy and the contact of elastic solids. *Proceedings of the Royal Society of London A* **324**, 301–313 (1971).

Langevin P. On the theory of brownian motion 1908. *C. R. Acad. Sci. (Paris)* **146**, 530–533 (1908).

Li SQ, Marshall JS, Liu G, Yao Q. Adhesive particulate flow: The discrete element method and its application in energy and environmental engineering. *Progress in Energy and Combustion Science* **37**(6), 633–668 (2011).

Marshall JS. Discrete-element modeling of particulate aerosol flows. *Journal of Computational Physics* **228**, 1541–1561 (2009).

Marshall JS. Viscous damping force during head-on collision of two spherical particles. *Physics of Fluids* **23**(1), 013305 (2011).

McCoy BJ. A population balance framework for nucleation, growth, and aggregation. *Chemical Engineering Science* **57**, 2279–2285 (2002).

Patankar NA, Joseph DD. Lagrangian numerical simulation of particulate flows. *International Journal of Multiphase Flow* **27**, 1685–1706 (2001).

Pivkin IV, Richardson PD, Karniadakis GE. Effect of red blood cells on platelet aggregation. *IEEE Engineering in Medicine and Biology Magizine* **28**(2), 32–37 (2009).

Ramkrishna D. *Population Balances – Theory and Applications to Particulate Systems in Engineering*. Academic Press, San Diego (2000).

Reinhold A, Briesen H. Numerical behavior of a multiscale aggregation model: Coupling population balances and discrete element models. *Chemical Engineering Science* **70**, 165–175 (2012).

Serayssol J-M, Davis RH. The influence of surface interactions on the elastohydrodynamic collision of two spheres. *Journal of Colloid and Interface Science* **114**(1), 54–66 (1986).

Smoluchowski M von. Zur kinetischen Theorie der Brownschen Molekularbewe-gung und der Suspensionen. *Annalen der Physik* **21**, 756–780 (1906).

Smoluchowski M von. Versuch einer mathematischen Theorie der Koagulationkinetik kollider lösungen. *Z. Phys. Chem.* **92**, 129–168 (1917).

Snider DM. Three fundamental granular flow experiments and CPFD predictions. *Powder Technology* **176**, 36–46 (2007).

Sommerfeld M. Validation of stochastic Lagrangian modeling approach for inter-particle collisions in homogeneous turbulence. *International Journal of Multiphase Flow* **27**, 1829–1858 (2001).

Symeonidis V, Karniadakis GE, Caswell B. Dissipative particle dynamics simulations of polymer chains: scaling laws and shearing response compared to DNA experiments. *Physical Review Letters* **95**, 076001 (2005).

Tian P. Molecular dynamics simulations of nanoparticles. *Annual Reports on the Progress of Chemistry, Section C Physical Chemistry* **104**, 142–164 (2008).

Trzeciak TM, Podgórski A, Marijnissen JCM. Stochastic calculation of collision kernels: Brownian coagulation in concentrated systems. In: 5th *World Congress on Particle Technology (WCPT5)*, Orlando, Florida, p. 202d (2006).

Wijmans CM, Smit B, Groot RD. Phase behavior of monomeric mixtures and polymer solutions with soft interactions potentials. *Journal of Chemical Physics* **114**(17), 7644–7654 (2001).

Yang F-L, Hunt ML. Dynamics of particle-particle collisions in a viscous liquid. *Physics of Fluids* **18**, 121506 (2006).

Zhou Q, Yao SC. Group modeling of impacting spray dynamics. *International Journal of Heat and Mass Transfer* **35**, 121–129 (1992).

Contact Mechanics without Adhesion

Contact mechanics refers to the solid phase stresses and deformation that occur when two bodies collide, resulting in the formation of a contact force and torque exerted on the bodies. As discussed in Chapter 1, for micrometer-sized particles the contact forces are changed in a nonlinear manner by the presence of adhesive forces. For this reason, the contact force and the adhesive force are not additive, but instead collision with adhesion must be examined as a combined mechanical theory. Although the theory of particle contact in the presence of adhesion reduces to that with no adhesion as a limiting case, it is nevertheless of value to consider the problem of contact with no adhesion by itself both as an introduction to the broader theory and because adhesionless contact has a large number of applications for problems involving granular materials.

3.1. Basic Concepts

Two spherical particles with radii r_i and r_j, elastic moduli E_i and E_j, Poisson ratios v_i and v_j, and shear moduli $G_i = E_i/2(1 + v_i)$ and $G_j = E_j/2(1 + v_j)$ are in contact with each other. Typical values of elastic modulus and Poisson ratio for different materials are shown in Table 3.1. An effective particle radius R and effective elastic and shear moduli, E and G, are defined by

$$\frac{1}{R} \equiv \frac{1}{r_i} + \frac{1}{r_j}, \qquad \frac{1}{E} \equiv \frac{1 - v_i^2}{E_i} + \frac{1 - v_j^2}{E_j}, \qquad \frac{1}{G} \equiv \frac{2 - v_i}{G_i} + \frac{2 - v_j}{G_j}. \qquad (3.1.1)$$

If no force is applied between the two particles, the contact occurs only at a point. In the presence of a force F_n along the direction of the vector $-\mathbf{n}$ normal to the particle surface, each particle is deformed in the region near the contact point such that the particles contact each other over a finite region of circular shape called the *contact region*. The point at the center of the contact region, which is usually the first point to make contact as two spherical particles collide, is called the *contact point*. The radius a of the contact region is called the *contact radius*. An important parameter characterizing the particle deformation within the contact region is the *normal overlap* δ_N, defined as the distance between the contact points on the particles if the particles had remained undeformed (Figure 3.1). If $\mathbf{x}_i$ and $\mathbf{x}_j$ denote

Table 3.1. *Typical values for Young's modulus and Poisson's ratio of representative materials. [Extracted from data in Engineering Tool Box (http://www.engineeringtoolbox.com).]*

Materials	Young's modulus, E_i (GPa)	Poisson's ratio, v_i
Metals		
Aluminum	69–120	0.33
Cast Iron	200	0.21–0.26
Copper	120	0.33
Gold	75	0.42
Magnesium	45	0.35
Stainless Steel	180	0.30–0.31
Steel	200	0.27–0.30
Titanium	110	0.34
Soil		
Clay	5–50	0.30–0.45
Concrete	30	0.2
Sand	10–80	0.20–0.45
Other Materials		
Glass	50–90	0.18–0.3
Foam	0.01–0.15	0.10–0.40
Rubber	0.01–0.1	~0.50

the centroid positions of the two spherical particles, such that $\mathbf{n} = (\mathbf{x}_j - \mathbf{x}_i)/|\mathbf{x}_j - \mathbf{x}_i|$ is a unit vector pointing from the centroid of particle i to the centroid of particle j, then the normal overlap is given by

$$\delta_N = r_i + r_j - |\mathbf{x}_i - \mathbf{x}_j|. \tag{3.1.2}$$

If the contact region radius a is small compared to the particle radius, the contact region radius and the normal overlap can be related geometrically as

$$a = \sqrt{R\delta_N}. \tag{3.1.3}$$

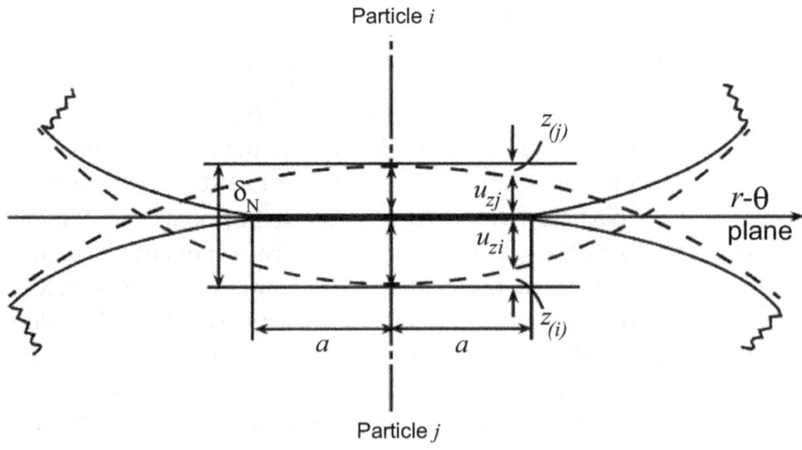

Figure 3.1. Geometry of two spherical particles in contact, illustrating the normal overlap and the contact radius. The dashed lines denote what the particle surfaces would have been if they had remained spherical, and the solid lines denote the deformed particle surfaces.

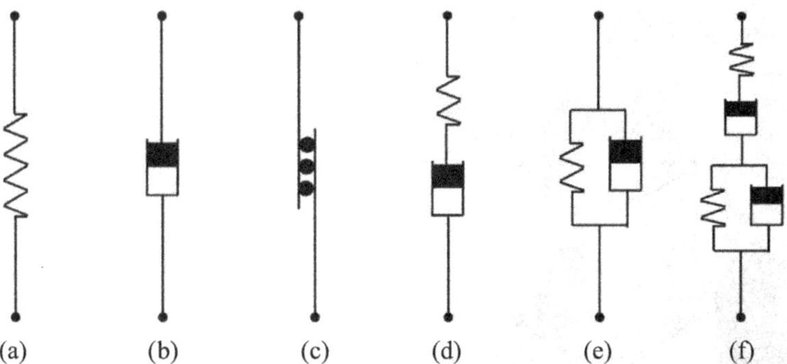

Figure 3.2. Schematics of classic rheological models in continuum mechanics. (a) Elastic; (b) viscous; (c) slider; (d) Maxwell; (e) Kelvin; (f) Burger's.

For two spherical particles, the contact region forms a circle centered at the contact point. At any point on the contact surface with radial distance r from the circle center, the corresponding elastic displacements of the two bodies in the direction normal to the contact surface ($\pm z$), u_{zi} and u_{zj}, must satisfy the kinematic relationship

$$u_{zi} + u_{zj} = \delta_N - \frac{r^2}{2R}. \tag{3.1.4}$$

Even for two particles that are considered to be in contact, there still exists an equilibrium gap between the particle surfaces with thickness δ, which has a value ranging between 0.16 and 0.40 nm for smooth surfaces. For rough surfaces, the gap size is on the order of the roughness length scale. Due to the small size of this gap, it is generally negligible for the purpose of modeling elastic deformation for nonadhesive surfaces, but it plays a major role in the contact mechanics of adhesive surfaces discussed in Chapter 4.

There are many different rheological models relating force and displacement in modeling deformation of solids, which in the current context relate the normal contact force F_n and the normal overlap δ_N. Several of these models are shown schematically in Figure 3.2. The simplest such relationship is that of a purely elastic response, as is typical of Hooke's law for elastic deformation, which is represented in Figure 3.2a by a spring. The elastic model does not dissipate energy or lead to plastic material deformation, enabling the corresponding $F_{ne} - \delta_N$ curve to be the same for loading and unloading processes. The purely viscous response, as is typical of a Newtonian viscous fluid, is represented in Figure 3.2b by a damping dashpot, such that the force is a function of the rate of deformation, for example, $F_n \propto d\delta_N/dt$. The rigid–perfectly plastic response, represented as a slider in Figure 3.2c, indicates that there is no deformation ($\delta_N = 0$) until a yield point is reached, and after that the deformation continues at a constant loading force. The Maxwell response model in Figure 3.2d consists of a spring and a dashpot arranged in series, which implies that the elastic and viscous elements experience the same force but exhibit different displacements. In the Maxwell model, the total displacement is the sum of elastic and viscous displacements. The Kelvin response model shown in Figure 3.2e comprises a spring and dashpot acting in parallel. In the Kelvin model, the elastic and viscous forces are different but both the spring and dashpot experience the same displacement. Finally, Burger's model, shown in Figure 3.2f, is a combination of the Kelvin and Maxwell models.

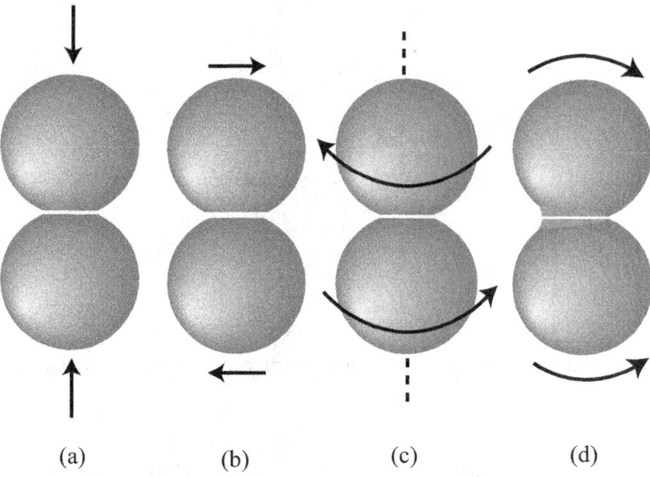

Figure 3.3. Modes of particle interaction: (a) normal impact; (b) sliding; (c) twisting; and (d) rolling.

The Kelvin model is widely used for describing the normal contact force between particles in contact. In this model, the total normal force F_n is the sum of the elastic normal force F_{ne} and the dissipative normal force F_{nd}. In addition to the normal forces, solids in contact can also exhibit resistance to other motions associated with relative tangential displacement of their surfaces. Among these motions we include sliding, twisting, and rolling, shown schematically in Figure 3.3. The resistance to normal displacement consists of a force acting in the $-\mathbf{n}$ direction passing through the particle centroids, so that for spherical particles this force exerts no torque on the particles. The particle resistance to other types of motions might have the form of a force, a torque, or even a combination of a force and a torque. The sliding resistance exerts a force F_s acting in a direction $\mathbf{t}_S$, corresponding to the direction of relative motion of the particle surfaces at the contact point projected onto a plane orthogonal to $\mathbf{n}$. The sliding resistance also imposes a torque on the particle in the $\mathbf{n} \times \mathbf{t}_S$ direction with magnitude $r_i F_s$, where r_i is the radius of particle i. The twisting resistance exerts a torque M_t on the particle in the $\mathbf{n}$ direction, but no force. Rolling resistance exerts a torque M_r on the particle in the $\mathbf{t}_R \times \mathbf{n}$ direction, where $\mathbf{t}_R$ is the direction of the "rolling velocity" $\mathbf{v}_L$, which is defined in Section 3.6. The total collision force $\mathbf{F}_A$ and torque $\mathbf{M}_A$ acting on particle i can be written as the sum of these various components as

$$\mathbf{F}_A = -F_n \mathbf{n} + F_s \mathbf{t}_S, \quad \mathbf{M}_A = r_i F_s (\mathbf{n} \times \mathbf{t}_S) + M_r (\mathbf{t}_R \times \mathbf{n}) + M_t \mathbf{n}. \tag{3.1.5}$$

The negative sign before F_n in (3.1.5) is used in order to conform with the convention that a positive value of F_n corresponds to a repulsive force and a negative value corresponds to an attractive force. The contact mechanics of nonspherical particles is somewhat different from that for spheres. This matter is addressed in Chapter 7.

3.2. Hertz Theory: Normal Elastic Force

The foundational theory of nonadhesive elastic contact of two elastic bodies was developed over a century ago by Hertz (1882). Interestingly, his initial research was driven not by an effort to develop a theory for colliding particles, or even colliding elastic bodies. He was instead trying to determine how the optical properties of stacked glass lenses are modified due to slight deformation of their surfaces by elastic deformation. Despite various extensions, the Hertz theory has stood the test of time relatively unchanged and has found application in a wide variety of engineering problems.

3.2.1. Derivation

The basic approximation introduced by Hertz to make an analytical solution of the contact problem tractable is to treat each of the colliding bodies as an elastic half-space loaded within the small contact region in such a manner as to produce displacements satisfying the kinematic relationship (3.1.4). This simplification allows us to use the well-developed methods available for solving elasticity problems in an elastic half-space for contact problems (Love, 1952), without having to worry about satisfaction of boundary conditions on the particle surface outside of the contact region. In order for this simplification to be valid, it is necessary that the contact area radius is small compared to the particle radius. This restriction is usually well satisfied by particles made of metals, hard plastic, glass, or ceramics operating within their elastic limits, but caution must be taken when applying contact mechanics models to particles formed of soft materials, such as biological cells. A second approximation made in the Hertz theory is that the contact surfaces are frictionless, so that only normal stress is transmitted between them. This approximation is discussed in more detail in Sections 3.3 and 3.4.

We consider a cylindrical polar coordinate system with coordinates (r, θ, z), such that $z = 0$ corresponds to the top boundary of an elastic half-space that fills the region $z < 0$. It is known from the kinematic result (3.1.4) that the displacement u_z in the normal (z) direction within a circle of radius a on the surface $z = 0$ must vary in proportion to r^2. Hertz observed that a displacement of this form can be obtained by imposing a surface pressure distribution of the form

$$p = p_0[1 - (r/a)^2]^{1/2}, \qquad (3.2.1)$$

where the coefficient p_0 is related to the total elastic normal force F_{ne} on the particle by

$$F_{ne} = 2\pi \int_0^a p(r)r\,dr = \frac{2}{3}p_0\pi a^2. \qquad (3.2.2)$$

A solution for the problem of a point force P oriented in the normal direction on the surface of an elastic half-space was given by Timoshenko and Goodier (1970, 402), based on the original solution by Boussinesq (1885). This solution gives the surface displacement as

$$u_z\big|_{z=0} = \frac{P(1 - \nu_i^2)}{\pi E_i r}, \qquad (3.2.3)$$

where ν_i is the Poisson's ratio and E_i is the elastic modulus of the material. If we now consider a distributed load with pressure $p(r)$ over a circle of radius a on the surface of the elastic half-space, the load P in (3.2.3) can be set equal to an infinitesimal load $p\,da$ applied at a position $\mathbf{x}'$ on the half-space surface. By superposition, the resulting displacement at any point $\mathbf{x}$ on the surface is given by

$$u_z(\mathbf{x}) = \frac{1 - \nu_i^2}{\pi E_i} \int_S \frac{p(\mathbf{x}')}{|\mathbf{x}' - \mathbf{x}|} da', \tag{3.2.4}$$

where S denotes the interior of the circle of radius a on the surface of the half-space. Substituting (3.2.1) into (3.2.4) and performing the integration yields a solution for normal displacement within the contact region of particle i of the form (Johnson, 1985, 92)

$$u_{zi} = \frac{1 - \nu_i^2}{E_i} \frac{\pi p_0}{4a} (2a^2 - r^2). \tag{3.2.5}$$

Writing a similar expression for the displacement of the surface of particle j and substituting into the kinematic requirement (3.1.4) yields

$$u_{zi} + u_{zj} = \frac{\pi p_0}{4aE} (2a^2 - r^2) = \delta_N - \frac{r^2}{2R}. \tag{3.2.6}$$

From (3.2.6), we can write

$$a = \pi R p_0 / 2E, \qquad \delta_N = \pi a p_0 / 2E. \tag{3.2.7}$$

Solving for p_0 in (3.2.7) and substituting into (3.2.2) gives

$$F_{ne} = \frac{4Ea^3}{3R} = K\delta_N^{3/2}, \tag{3.2.8}$$

where the stiffness coefficient K is given by

$$K = \tfrac{4}{3}E\sqrt{R}. \tag{3.2.9}$$

The equation (3.2.8) is the classical Hertz force expression for elastic contact of colliding bodies. A useful alternative form for this expression is given by

$$F_{ne} = k_N \delta_N, \tag{3.2.10}$$

where the stiffness k_N is written in terms of contact region radius a as

$$k_N = \tfrac{4}{3}Ea. \tag{3.2.11}$$

It should be noted that (3.2.10) does not yield a linear dependence between elastic force and the normal overlap, because the contact region radius a in (3.2.11) varies in proportion to $\sqrt{\delta_N}$ from (3.1.3).

3.2.2. Two-Particle Collision

To illustrate the results of the normal elastic force model derived earlier, the problem of head-on elastic collision of two particles, labeled Particle 1 and Particle 2, traveling in a vacuum in the horizontal (x) direction is considered. The particle centroid positions and velocities at the initial time satisfy $x_1 < x_2$ and $v_1 > v_2$. The faster-moving particle on the left catches up with the slower-moving particle on the right, leading to a collision between the two particles. During the collision, the displacement of each particle is described by the particle momentum equation

$$m_1 \frac{d^2 x_1}{dt^2} = -F_{ne}, \quad m_2 \frac{d^2 x_2}{dt^2} = F_{ne}. \tag{3.2.12}$$

Substituting (3.2.12) into the time derivative of the definition (3.1.2) of the normal overlap, $\delta_N = r_1 + r_2 - (x_2 - x_1)$, gives

$$\frac{d^2 \delta_N}{dt^2} = \frac{d^2(x_1 - x_2)}{dt^2} = -\frac{F_{ne}}{M}, \tag{3.2.13}$$

where $M = (m_1^{-1} + m_2^{-1})^{-1}$ is the effective mass of the particles.

Substituting the Hertz expression for normal elastic force (3.2.8) into (3.2.13) yields a differential equation for the normal overlap as

$$\frac{d^2 \delta_N}{dt^2} + \frac{K}{M} \delta_N^{3/2} = 0. \tag{3.2.14}$$

Multiplying (3.2.14) by $d\delta_N/dt$, we can write the result as

$$\frac{1}{2} \frac{d}{dt}\left(\frac{d\delta_N}{dt}\right)^2 + \frac{2K}{5M} \frac{d}{dt}(\delta_N^{5/2}) = 0. \tag{3.2.15}$$

Integrating (3.2.15) over time gives

$$\frac{1}{2}\left(\frac{d\delta_N}{dt}\right)^2 + \frac{2K}{5M} \delta_N^{5/2} = C, \tag{3.2.16}$$

where C is a constant of integration. If v_{10} and v_{20} are the velocities of Particle 1 and 2, respectively, at the instant of time $t = 0$ just prior to collision and we define an approach velocity $v_0 \equiv v_{10} - v_{20}$, then the constant of integration is given by

$$C = \tfrac{1}{2} v_0^2. \tag{3.2.17}$$

The coefficient C is equal to the kinetic energy of the two-particle system divided by the effective mass M when measured in a coordinate system in which the origin coincides with the center of inertia.

Both the maximum overlap and the maximum repulsive elastic force are attained when the particle approach velocity vanishes, such that $d\delta_N/dt = 0$. Solving for the maximum value of δ_N from (3.2.16) and (3.2.17) yields

$$\delta_{\max} = \left(\frac{5Mv_0^2}{4K}\right)^{2/5} = \left(\frac{15Mv_0^2}{16ER^{1/2}}\right)^{2/5}, \tag{3.2.18}$$

where the last expression is obtained using (3.2.9). Substituting (3.2.18) into (3.1.3) gives the maximum contact region radius as

$$a_{\max} = \sqrt{R\delta_{\max}} = \left(\frac{5Mv_0^2}{4K}\right)^{1/5} R^{1/2} = \left(\frac{15Mv_0^2 R^2}{16E}\right)^{1/5}. \tag{3.2.19}$$

Equation (3.2.16) is of a separable form, and can be rearranged as

$$dt = \frac{d\delta_N}{\sqrt{v_0^2 - \frac{16}{15}(ER^{1/2}\delta_N^{5/2}/M)}}. \tag{3.2.20}$$

Defining $y \equiv \delta_N/\delta_{max}$ and integrating (3.2.20) between the time $t = 0$ of initial contact and the time $\tau_C/2$ of maximum overlap, the total contact time τ_C is obtained as

$$\tau_C = \frac{2\delta_{max}}{v_0} \int_0^1 \frac{dy}{\sqrt{1 - y^{5/2}}} = \frac{4\sqrt{\pi}}{5} \frac{\delta_{max}}{v_0} \frac{\Gamma(2/5)}{\Gamma(2/5 + 1/2)} \cong 2.943 \frac{\delta_{max}}{v_0}, \qquad (3.2.21)$$

where $\Gamma(\cdot)$ is the gamma function. Substituting the result (3.2.18) for the maximum overlap gives the contact time as

$$\tau_C \cong 2.868 \left(\frac{M^2}{E^2 R v_0} \right)^{1/5}. \qquad (3.2.22)$$

The prediction that contact time decreases with increase in impact velocity v_0, as predicted by (3.2.22), is verified by experimental measurements (e.g., Goldsmith, 1960; Stevens and Hrenya, 2005).

As a specific example, we consider the impact of 1 mm diameter glass spheres. It is assumed that the approach velocity $v_0 = 1$ m/s, and the elastic modulus, Poisson ratio, and density for glass are $E_1 = E_2 = 55$ GPa, $v_1 = v_2 = 0.3$, and $\rho_p = 2500$ kg/m^3. For identical spheres, $R = r_1/2$, $M = m_1/2 = (2/3)\pi r_1^3 \rho_p = (16/3)\pi R^3 \rho_p$, and $E = E_1/2(1 - v_1^2)$. Substituting these values into (3.2.18), (3.2.19) and (3.2.21) gives $\delta_{max} = 2.22$ μm, $a_{max} = 33.3$ μm, and $\tau_C = 6.53$ μs. The extremely short contact time obtained in this example illustrates some of the challenges involved in computing flows with colliding particles, for which the time scale associated with particle collisions is often six or more orders of magnitude smaller than the time scale associated with the fluid flow in which the particles are entrained.

3.3. Normal Dissipation Force

In an elastic contact model, the force-displacement response is the same during loading and unloading, implying that the net kinetic energy of the particles is conserved during the collision. In the practice, there is a loss of kinetic energy to heat during collision processes that can be characterized by the *restitution coefficient e* defined in (2.4.6). A completely elastic collision, such as one governed by the Hertz expression in the momentum equation (3.2.13), results in a restitution coefficient of unity. A completely inelastic collision, for which there is no particle rebound, corresponds to a restitution coefficient of zero. All real collisions occur with restitution coefficients between these two extremes.

3.3.1. Physical Mechanisms

The kinetic energy loss is in general due to a combination of dissipation within the particle (solid-phase dissipation) and dissipation within the fluid surrounding the particle (fluid-phase dissipation). The solid-phase dissipation can be attributed to a wide range of factors. For instance, in high-velocity impacts, plastic deformation of the material or localized cracking near the contact region may occur. Because the current study is concerned with motion of adhesive particles immersed in fluid

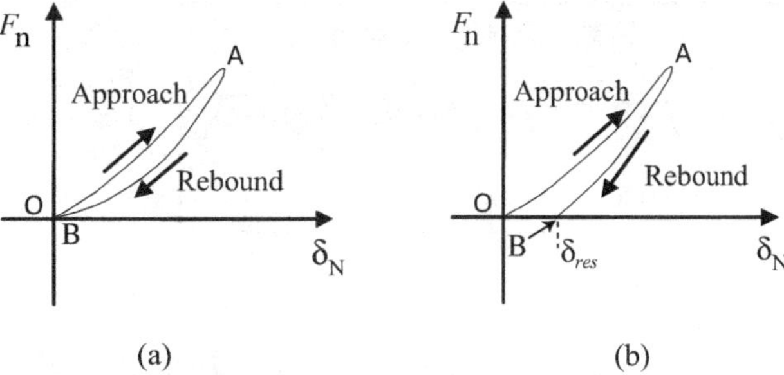

Figure 3.4. Hysteretic behavior of the load-displacement curve for (a) viscoelastic and (b) plastic material behavior.

flows, high-velocity impacts are generally outside the scope of this text. Interfacial slip within the contact region during normal impact can lead to kinetic energy loss; however, as noted by Johnson (1985, 119), such slip only occurs in the normal collision problem if the elastic moduli of the two colliding particles are different.

Solid-phase dissipation occurring in low-velocity collision of two identical particles is generally associated with what is called *viscoelastic response*. Classical elasticity theory is developed utilizing an assumption that the material deformations change so slowly that the material is approximately static. Although this quasi-static approximation yields an expression for the normal elastic force of two colliding particles that is in reasonable agreement with experimental findings, in reality the collisions occur over relatively short time intervals. The rapid change in particle deformation during collisions gives rise to a viscous response caused by internal friction within the solid, which is proportional to the rate of change of the material deformation. The sum of this elastic response and the viscous response characterizes a viscoelastic material response, which can be visually identified by the presence of hysteresis in the force–displacement curve (i.e., the plot of F_n versus δ_N). Hysteresis causes the $F_n - \delta_N$ relationship to be different as the two particles approach each other compared with when they move apart from each other, as illustrated in Figure 3.4. Differences in viscoelastic and plastic material behavior are apparent in this curve by the fact that the deformation returns to zero as the load vanishes in a viscoelastic material, whereas it does not in a plastically deforming material.

The Hertz theory of normal elastic collision also assumes that the particle surfaces are perfectly smooth. Of course, all materials are formed of molecules, which cluster into crystals, grains, and so on, so no surface can be perfectly smooth. Instead, the notion of smoothness implies that the length scale associated with the roughness elements on the surface is much smaller than some comparable length scale in the problem. For the problem of collision of two particles, a natural comparison length scale is the maximum particle overlap δ_{max}, given for Hertz theory by (3.2.18). It has previously been discussed that δ_{max} is typically much smaller than the particle diameter. For instance, in the example given in Section 3.2 for collision of two 1 mm diameter glass particles, we found that $\delta_{\mathrm{max}} \cong 2.2\ \mu$m. Because of the small size of δ_{max}, the contact mechanics of surfaces can be substantially influenced by roughness even for surfaces with very small roughness elements. For instance, energy dissipation during a collision process can be caused by sliding friction and localized plastic deformation as the highest roughness elements on each surface contact each other at

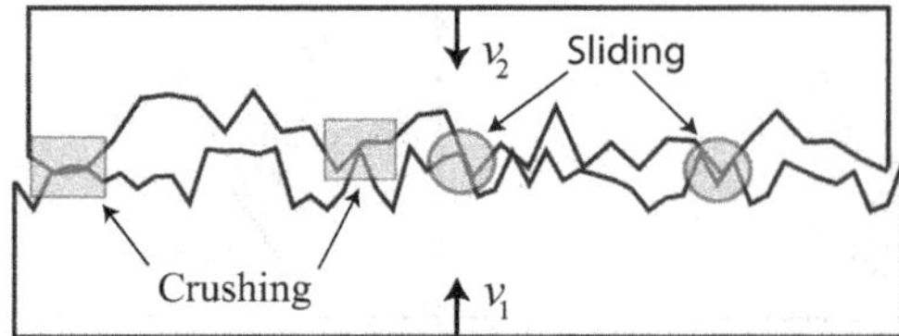

Figure 3.5. Schematic diagram showing the roughness elements on the surfaces of two colliding particles interacting with each other.

various angles and are either crushed by high local stresses or slide relative to each other as the surfaces move closer together (Figure 3.5).

Another important mechanism for reduction of the restitution coefficient is the presence of adhesion forces between the particle surfaces. This mechanism becomes important for low-velocity collisions of relatively small particles. Even for cases where the adhesion forces are far smaller than is necessary to stop the particle rebound completely, adhesion still exerts a force on particles which decreases the restitution coefficient. There are many types of adhesive forces that can act between particles, which are discussed in detail in Chapter 4. Since in models of adhesive particle flows the adhesion force is treated explicitly in a combined elastic-adhesion force model (see Chapter 4), we do not in this section include effects of adhesion forces in modeling F_{nd}. Nevertheless, adhesion forces are present in nature and may influence experimental data used for validation of normal dissipative force expressions.

In addition to the solid-phase dissipation mechanisms discussed earlier, fluid-phase friction is an important source of dissipation for the low-velocity impacts of small particles that are typically of concern in adhesive particle flows. The fluid-phase dissipation is typically dominated by the high shear stress motions within the lubrication film separating the two particles as they approach or move away from each other, or else from the fluid motion near the outer edges of the contact region as the contact region area changes during the collision. Fluid dissipative forces are discussed in detail in Section 2.4, including forces prior to, during, and following collision of the particles. The fluid-phase dissipation often dominates effects of the solid-phase dissipation for low-velocity impacts of small particles, particularly in cases where the collision Stokes number St_C (defined in (2.4.8)) is small. As a consequence, it is often possible to predict the restitution coefficient quite well in such cases using the fluid dissipation force alone, as shown in Figure 2.7.

Figure 3.6 presents a compilation of experimental data for restitution coefficient for particle collision studies with different materials. These experiments were conducted in air with relatively high impact velocities and sufficiently large particles so that the fluid-phase dissipation force is small compared with the solid-phase dissipation force. The relevant mechanical and geometrical properties used for the data in Figure 3.6 are listed in Table 3.2, along with sources of the data. The data exhibit a general trend for the restitution coefficient to decrease with increase in the particle approach velocity v_0. It is noted from (2.4.8) that the collision Stokes number St_C is proportional to particle approach velocity, and so that the expression (2.4.21) for restitution coefficient under purely fluid dissipation force is consistent with this trend. However, as shown in Section 3.3.2, this trend is also consistent with certain models used for describing the solid-phase dissipation force.

Table 3.2. *Material properties used for measuring the coefficient of restitution in the previous literature (corresponding to data in Figure 3.6)*

	Material	E_i (GPa)	σ_i	ρ_p (kg/m³)	r_i (mm)[a]	Investigator
1	Stainless Steel (Grade 316)	193	0.35	7830	12.7	Stevens & Hrenya (2005)
2	High Carbon Steel	211	0.295	8184	16.5	Kuwabara Kono & (1987)
3	Brass(Cu70/Zn30)	110	0.331	8418	15.0	
5	Al_2O_3/Al^a	360/70	0.23/0.35	3900/6500	5.00	Gorham & Kharaz (2000)
4	Brass	96	0.36	8522	20.0	
6	Aluminum	68.5	0.36	2707	20.0	Goldsmith (1960)
7	Lead	16.2	0.45	11373	20.0	
8	Marble	71.0	0.30	2900	20.0	
9	Cork	0.05	0	1654	16.6 25.4	Kuwabara & Kono (1987)
10	Nylon	0.35	0.3	1140	12.7 6.35	Labous et al. (1997)

[a] Case of Al_2O_3 sphere impacting with Al alloy plate, instead of two-sphere impact.

3.3.2. Models for Solid-Phase Dissipation Force

We consider two colliding particles in the viscoelastic regime. The normal force on the particles is modeled by the Kelvin rheological model shown schematically in Figure 3.2, consisting of a spring and a dashpot acting in parallel, such that the spring and dashpot experience different forces but an equal deformation. The total normal force on each particle is the sum of the Hertzian elastic force F_{ne} and the normal dissipative force F_{nd}. The dissipation force F_{nd} for a dashpot is proportional to the approach velocity of the particles, and so is given by

$$F_{nd} = \eta_N \mathbf{v}_R \cdot \mathbf{n}. \tag{3.3.1}$$

In this equation, $\mathbf{v}_R$ is the relative particle surface velocity at the contact point and η_N is the normal dissipation coefficient. The surface velocity of particle i at the contact point is given for a spherical particle by $\mathbf{v}_{C,i} = \mathbf{v}_i + \mathbf{\Omega}_i \times \mathbf{r}_i$, where $\mathbf{r}_i = r_i \mathbf{n}$ is

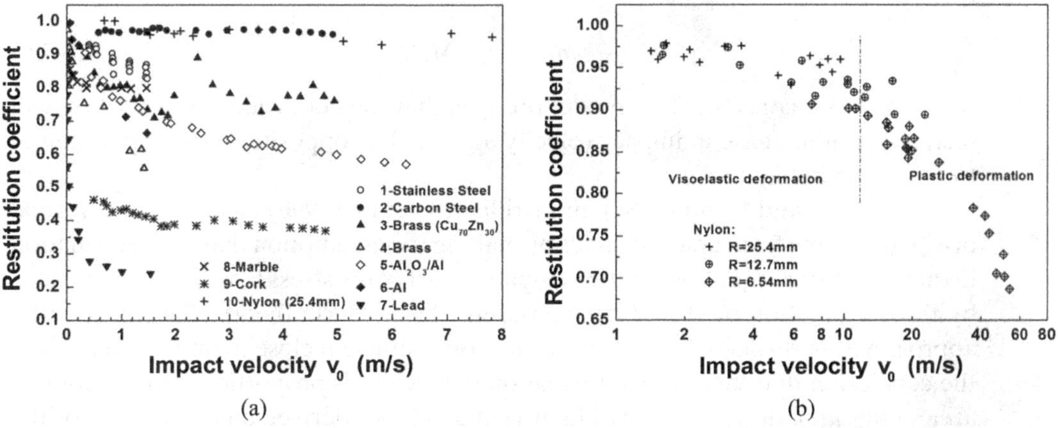

Figure 3.6. Experimental data for restitution coefficient as a function of the initial normal approach velocity, from sources listed in Table 3.2. (a) Viscoelastic regime; (b) viscoelastic and plastic deformation regimes.

the vector from the particle center to the contact point, $\mathbf{v}_i = d\mathbf{x}_i/dt$ is the particle centroid velocity, and $\boldsymbol{\Omega}_i$ is the particle angular velocity. For the opposing particle j, we can write $\mathbf{r}_j = -r_j\mathbf{n}$. The relative particle surface velocity at the contact point is then given by

$$\mathbf{v}_R = \mathbf{v}_{C,i} - \mathbf{v}_{C,j}. \tag{3.3.2}$$

The normal component of $\mathbf{v}_R$ can also be written in terms of the normal overlap δ_N as

$$\mathbf{v}_R \cdot \mathbf{n} = \frac{d\delta_N}{dt}, \tag{3.3.3}$$

such that $\mathbf{v}_R \cdot \mathbf{n}$ is positive for particles that are approaching each other and negative for particles that are moving away from each other. Using the result (3.3.3), it is noted that the expressions for fluid dissipation force both before collision (2.4.14) and during collision (2.4.15) can be written in the form (3.3.1).

Using both the Hertz theory for the elastic normal force and the expression (3.3.1) for the dissipative normal force, the governing equation for the normal overlap becomes

$$\frac{d^2\delta_N}{dt^2} + \frac{\eta_N}{M}\frac{d\delta_N}{dt} + \frac{K}{M}\delta_N^{3/2} = 0. \tag{3.3.4}$$

The restitution coefficient can be written in terms of the time derivative $\dot{\delta}_N$ of the normal overlap at time $t = 0$, just after onset of collision, and at time $t = \tau_C$, just before the end of the collision, as

$$e = -\frac{\dot{\delta}_N(\tau_C)}{\dot{\delta}_N(0)}. \tag{3.3.5}$$

A wide variety of approaches have been taken in the literature to determine expressions for the normal dissipation coefficient η_N. The existence of many different approaches for modeling normal dissipation is not surprising given the different physical mechanisms that give rise to the solid-phase dissipation. Lee and Herrmann (1993) assumed that η_N is constant and proportional to the reduced mass M, such that

$$\eta_N = \alpha_{LH}M, \tag{3.3.6}$$

where α_{LH} is a constant. This model predicts that the coefficient of restitution will increase with increase in impact velocity $\dot{\delta}_N$, which is opposite to the experimental trend shown in Figure 3.6.

Kuwabara and Kono (1987) and Brilliantov et al. (1996) examined the normal dissipation force from first principles by making the assumption that the deformation in the constitutive equation for the dissipative part of the stress tensor can be replaced by the deformation predicted by the purely elastic Hertz theory. This quasi-static approximation is valid for values of restitution coefficient close to unity, and leads to the conclusion that the normal dissipation coefficient is proportional to the contact area radius $a(t)$, or $\eta_N \propto \delta_N^{1/2}$. Brilliantov et al. (1996) derived an expression for the normal dissipation coefficient as

$$\eta_N = \alpha_{KKB}K\delta_N^{1/2}, \tag{3.3.7}$$

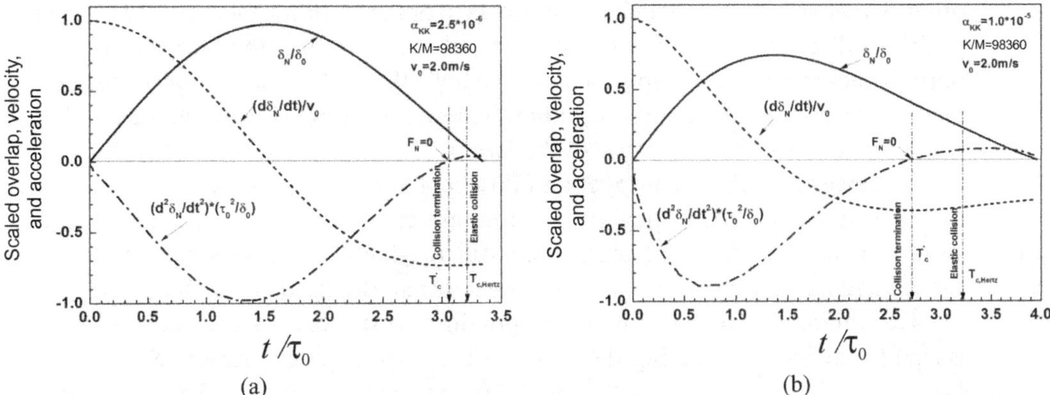

Figure 3.7. Time variation of the scaled overlap, velocity, and acceleration for two colliding particles, computed using the Kuwabara-Kono-Brilliantov model for the normal dissipation coefficient. (a) Weak damping; (b) strong damping.

where K is the Hertz stiffness coefficient given by (3.2.9) and α_{KKB} is a material constant with units of time. Substituting (3.3.7) into (3.3.4) gives

$$\frac{d^2\delta_N}{dt^2} + \alpha_{KKB}\frac{K}{M}\delta_N^{1/2}\frac{d\delta_N}{dt} + \frac{K}{M}\delta_N^{3/2} = 0. \tag{3.3.8}$$

For initial conditions given by $\delta_N(0) = 0$ and $\dot{\delta}_N(0) = v_0$, the Hertz theory for two purely elastic spheres yields a length scale δ_0 and a time scale τ_0 with order of magnitude

$$\delta_0 = \left(\frac{K}{M}\right)^{-2/5} v_0^{4/5}, \quad \tau_0 = \left(\frac{K}{M}\right)^{-2/5} v_0^{-1/5}. \tag{3.3.9}$$

A plot showing the time variation of δ_N, $\dot{\delta}_N$, and $\ddot{\delta}_N$ during the collision process, made dimensionless using the length and time scales in (3.3.9), is given in Figure 3.7. The computations are performed using material constants for Al_2O_3/Al, which are given in Table 3.2, and particles with diameter 10 mm, impact velocity $v_0 = 2$ m/s, and $K/M = 98{,}360$ m$^{-1/2}$s^{-2}. Results are given in Figure 3.7 for two cases, one with fairly weak dissipation ($\alpha_{KKB} = 2.5 \times 10^{-6}$ s) and restitution coefficient $e = 0.73$, and another with strong dissipation ($\alpha_{KKB} = 1.0 \times 10^{-5}$ s) and a much smaller restitution coefficient of 0.29. The time required for the collision is somewhat delayed by the dissipative term compared with what it would have been in a purely elastic case, which is indicated by a vertical dashed line in Figure 3.7.

These restitution coefficients are typically obtained using the assumption that the beginning and end of the collision correspond to times at which $\delta_N = 0$. However, it is observed in Figure 3.7 that the second derivative term $\ddot{\delta}_N$ becomes positive near the end of the collision. The total normal force F_n is proportional to $-\ddot{\delta}_N$, so a change from negative to positive values of $\ddot{\delta}_N$ implies a change in sign of F_n from positive to negative. It is recalled that a positive value of F_n acts to push the particles apart, whereas a negative value of F_n acts to pull the particles together. This change in sign of F_n occurs when the magnitude of the normal dissipation force term F_{nd} exceeds that of the normal elastic force term F_{ne}. As the particles move toward each other both F_{ne} and F_{nd} have a positive sign, resisting the particle motion toward each other. However, as the particles move away from each other the sign of F_{ne} remains positive whereas that of F_{nd} becomes negative, such that the two normal forces oppose each other. As δ_N approaches zero near the end of the collision process, the magnitude of the elastic force F_{ne} approaches zero more quickly than does that of the dissipative force F_{nd} because of the larger 3/2 power of δ_N in (3.2.8) compared to the 1/2 power in (3.3.7). Schwager and Pöschel (2007) argued that this state in which the total normal force is negative is nonphysical, and that in reality the particle surfaces would separate with the particle still in a deformed state, so that the collision should end at the time at which $F_n = 0$ rather than the time at which $\delta_N = 0$.

Determination of the time corresponding to the end of the collision process is significant for determining the value of the restitution coefficient. Schwager and Pöschel (2008) derived an approximate analytical solution of (3.3.8) by expanding the equation in powers of $\sqrt{t/\tau_0}$, where τ_0 is the Hertzian contact time scale given in (3.3.9). This expansion yields a series approximation for restitution coefficient e, based on the usual assumption that the collision terminates at a time for which $\delta_N = 0$, as

$$e = 1 + \sum_{n=0}^{\infty} c_n [\alpha_{KKB}(K/M)^{2/5}]^n v_0^{n/5}. \tag{3.3.10}$$

An alternative expression for e was derived using the observation that the particles will actually lose physical contact while still in a deformed state (such that $\delta_N > 0$) at a time for which $F_n = 0$, which yields the restitution coefficient as

$$e = 1 + \sum_{n=0}^{\infty} h_n [\alpha_{KKB}(K/M)^{2/5}]^{n/2} v_0^{n/10}. \tag{3.3.11}$$

The values of the coefficients c_n and h_n are listed in Table 3.3. A plot comparing restitution coefficient results obtained from these analytical solutions with results from a numerical solution of (3.3.8) and from experimental data is given in Figure 3.8. The KKB model qualitatively predicts the trend of decreasing e with increasing impact velocity. However, the model significantly underpredicts the slope of the experimental data. Stevens and Hrenya (2005) observed a similar trend for predictions of the KKB model for impact of stainless steel spheres.

Numerous other expressions for η_N have been proposed, but of the available alternative expressions that proposed by Tsuji et al. (1992) is of particular interest because it leads to a value for the restitution coefficient that is independent of impact velocity. In this expression, an assumption is made that $\eta_N \propto (Mk_N)^{1/2}$, where k_N is the normal stiffness coefficient, such that

$$\eta_N = \alpha_{TTI}\sqrt{MK}\delta_N^{1/4}, \tag{3.3.12}$$

where α_{TTI} is a dimensionless constant. Substituting this expression into (3.3.4) gives

$$\frac{d^2\delta_N}{dt^2} + \alpha_{TTI}\sqrt{\frac{K}{M}}\delta_N^{1/4}\frac{d\delta_N}{dt} + \frac{K}{M}\delta_N^{3/2} = 0. \tag{3.3.13}$$

Setting $\hat{t} = t/\tau_0$ and $\hat{\delta} = \delta_N/\delta_0$ gives

$$\frac{d^2\hat{\delta}}{d\hat{t}^2} + \alpha_{TTI}\hat{\delta}^{1/4}\frac{d\hat{\delta}}{d\hat{t}} + \hat{\delta}^{3/2} = 0. \tag{3.3.14}$$

Table 3.3. *Table of the first ten values of c_n and the first twenty values of h_n for the Schwager and Pöschel (2008) analytic solution of the restitution coefficient from the KKM model*

n	c_n	h_n
1	−1.153 448 854	0
2	0.798 266 555 3	−1.153 448 856
3	−0.522 882 560 9	0
4	0.348 742 667 8	0.798 266 558 1
5	−0.233 098 126 0	0.266 666 666 7
6	0.156 682 147 7	−0.522 882 565 7
7	−0.105 818 782 8	−0.461 379 542 4
8	0.071 765 282 42	0.348 742 673 7
9	−0.048 857 172 37	0.452 351 049 6
10	0.033 373 471 94	−0.146 431 464 4
11		−0.367 728 299 2
12		−0.043 248 9833
13		0.281 804 232 5
14		0.147 852 5872
15		−0.179 442 059 0
16		−0.178 466 032 6
17		0.065 933 588 82
18		0.171 358 617 8
19		0.025 249 822 3
20		−0.137 923 498 6

The restitution coefficient e is therefore affected only by the coefficient α_{TTI} in (3.3.14), as all other coefficients in both the governing equation and the initial condition have been removed by scaling. The TTI model is particularly suitable for materials in which the restitution coefficient exhibits weak dependence on impact velocity over the range of interest in the problem, for which it is simplest to prescribe a fixed value for the restitution coefficient.

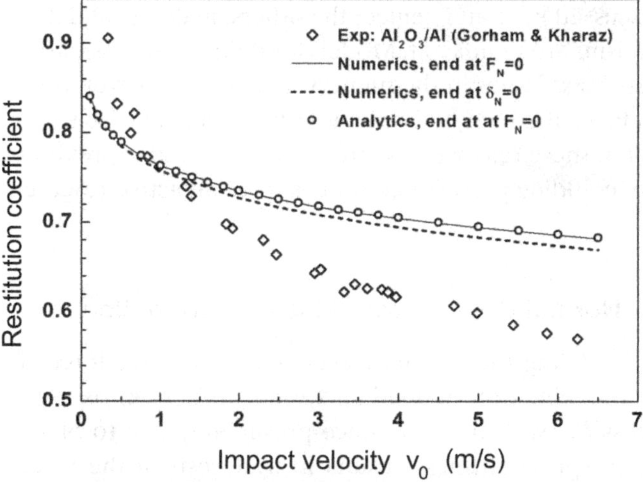

Figure 3.8. Comparison of predictions of the KKB model for normal dissipation from Equation (3.3.8) (solid line) with experimental data of Gorham and Kharaz (2000) (symbols).

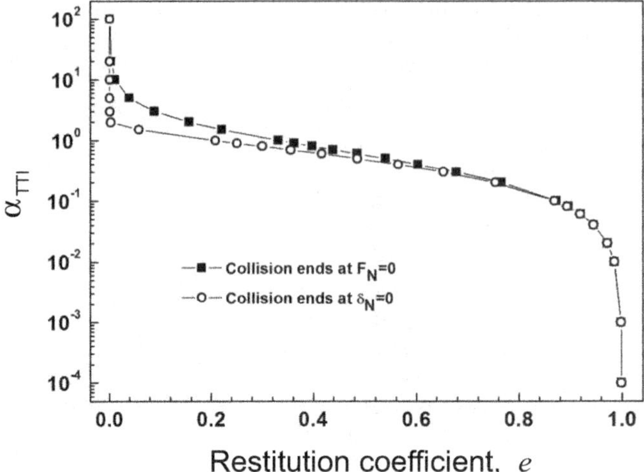

Figure 3.9. The restitution coefficient as a function of the damping coefficient α_{TTI} for two-particle collision predicted by the TTI model.

Equation (3.3.14) was solved numerically using a fourth-order Runge-Kutta method. In each case, the overlap increases to some maximum value $\hat{\delta} = \hat{\delta}_{max}$ at a time $\hat{t} = \hat{t}_{max}$, and then decreases again to zero as the two particles rebound from each other. In Figure 3.9, the restitution coefficient e is plotted as a function of α_{TTI} both for cases where the collision is assumed to end at $\hat{\delta} = 0$ and for cases where it is assumed to end at $F_n = 0$.

Despite these various developments, many issues remain in selecting an accurate, simple model for the normal dissipation force that includes all of the various mechanisms that may be significant for a given problem. Part of the challenge is that models for solid-phase dissipation force are typically developed for large particles colliding at high velocities, as might be typical of granular flows, whereas those for fluid-phase dissipation force are often suited more for low-velocity collisions. Similarly, models developed for the normal dissipation force often account for only one of the various mechanisms discussed here and neglect the others, making validation difficult when multiple mechanisms are significant. Models for fluid-phase dissipation force can give rise to severe stiffness issues in the numerical analysis, so even if the models are accurate they can be challenging to implement in practical models with large particle numbers. For all of these reasons, research into improved expressions for normal dissipation force of colliding particles continues to be an active research area.

3.4. Hysteretic Models for Normal Contact with Plastic Deformation

An alternative approach for modeling the normal elastic and dissipative forces is the so-called *hysteretic model*, in which the unloading spring stiffness k_U is much larger than the loading stiffness k_L, with the difference presumably due to plastic deformation. This difference in spring stiffness results in hysteresis in the force-deformation diagram, leading to energy loss during the collision. The important contributions to this class of model were pioneered by Walton and Braun (1986),

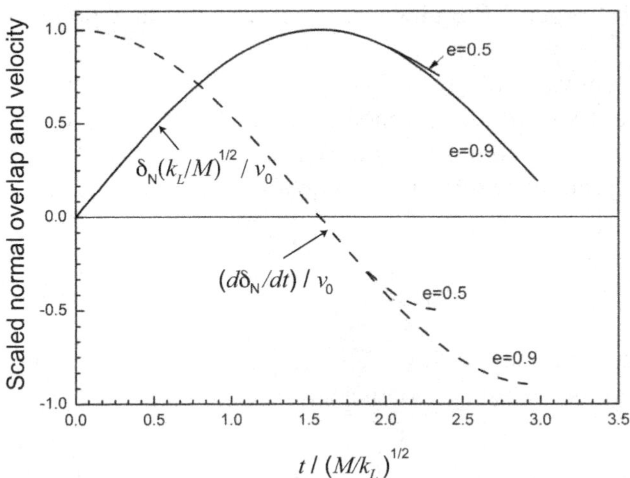

Figure 3.10. Time variation of the scaled overlap and velocity for two colliding particles, computed using the Walton-Braun (1986) model with coefficient of restitution e.

followed by the work by Sadd and Tai (1993), Thornton (1997), and Vu-Quoc and Zhang (1999). Both linear and nonlinear (i.e., Hertzian) stiffness coefficients have been used in the literature.

The simple linear hysteretic model, first proposed by Walton and Braun (1986), is illustrated in Figure 3.4b. The normal contact force is assumed in this model to have the form

$$F_n = \begin{cases} k_L \delta_N & \text{Loading} \\ k_U(\delta_N - \delta_{res}) & \text{Unloading} \end{cases} \tag{3.4.1}$$

where δ_{res} is the residual overlap at the end of the unloading period, known as the final resulting plastic deformation between two colliding particles. Substituting this linear hysteretic model into the momentum equation (3.2.13) for two-particle collision, an analytic solution can be obtained. Results of this solution for the scaled normal overlap and impact velocity are shown in Figure 3.10. As two particles initially collide and move from the origin O to point A in Figure 3.4b under the stiffness k_L, this solution gives the maximum overlap and normal force as $\delta_{max} = v_0(M/k_L)^{1/2}$ and $F_{n,max} = v_0(Mk_L)^{1/2}$, respectively. As the particles move apart, going from point A to point B in Figure 3.4b under a stiffness k_U, the residual normal overlap is obtained as $\delta_{res} = (1 - k_L/k_U)\delta_{max} = v_0(1 - k_L/k_U)(M/k_L)^{1/2}$. The linear hysteretic model leads to a coefficient of restitution given by

$$e = \sqrt{k_L/k_U}, \tag{3.4.2}$$

such that the duration of contact is

$$\tau_0 = \frac{\pi}{2}M^{1/2}\left(k_L^{-1/2} + k_U^{-1/2}\right). \tag{3.4.3}$$

Because it is not practicable to keep track of the effect of plastic deformation on the shape of each particle over time, the model assumes that particles start off as spherical at each new contact. Put another way, as F_n approaches zero during the unloading process, the model "forgets" the plastic deformation, so that point B approaches the origin O.

This model leads to a constant value of e regardless of the value of the approach velocity v_0, which is inconsistent with the experimental trend of the $e - v_0$ curve shown in Figure 3.6b in the plastic deformation regime. To address this issue, Walton and Braun (1986) modified their model by relating the unloading stiffness k_U to both k_L and $F_{n,\max}$ in a linear manner,

$$k_U = k_L + sF_{n,\max}, \tag{3.4.4}$$

where s is an empirical parameter with units of m^{-1}. Using $F_{n,\max} = v_0(Mk_L)^{1/2}$, this assumption yields an expression for restitution coefficient as

$$e = \sqrt{k_L/k_U} = \frac{1}{\sqrt{1 + sv_0(M/k_L)^{1/2}}}. \tag{3.4.5}$$

Stevens and Hrenya (2005) showed that the Walton-Braun model with variable e exhibits reasonably good predictions for restitution coefficient when compared with experiment data for impact of stainless steel spheres, but that this model underpredicts the influence of the impact velocity on the contact time scale. The experimental data reported in this paper for restitution coefficient generally lie between the predictions of this model and those of the KKB model discussed in Section 3.3. Of course, it should be kept in mind that these comparisons were performed with two empirical fitting coefficients, k_L and s.

The existing experimental data for particle collisions (Stevens and Hrenya, 2005; Kruggel-Emden et al., 2007) indicate that although linear models can be tuned to provide reasonable predictions for restitution coefficients, a full nonlinear (Hertzian) model is necessary to also obtain accurate predictions of contact duration time. Thornton (1997) proposed a nonlinear hysteretic model for the contact of two elastic, perfectly plastic spheres. In this model, the collision of two approaching particles is decomposed into three parts. In the first part (elastic loading), the particle loading behaves like an elastic material as governed by the Hertz theory until the material yield point is reached. In the second part, the yield point is reached and the material behaves in a perfectly plastic manner as the loading is continued. The third part of the process involves unloading as the particles move away from each other. The normal force on the particle is assumed to have the following form in each of these three parts of the collision process:

$$F_n = \begin{cases} K_L\delta_N^{3/2} & \delta_N < \delta_Y & \text{for } v_0 \geq 0 \\ K_L\delta_Y^{3/2} + \pi p_y R(\delta_N - \delta_Y). & \delta_N \geq \delta_Y & \text{for } v_0 \geq 0 \\ K_U(\delta_N - \delta_Y)^{3/2} & & \text{for } v_0 < 0 \end{cases} \tag{3.4.6}$$

where K_L and K_U are the coefficients of stiffness during loading and unloading (given by (3.2.9)), p_y is the material yield point (a material property), and δ_Y is the normal overlap at the yield point, which can be calculated by $\delta_Y = R(\pi p_y/2E)^2$. Starting from Hertzian contact theory, the Thornton model provides very good estimates for contact time but it is reported to overpredict the dependency of e on v_0 compared with the experimental data. An expression for the restitution coefficient can be obtained from the model as

$$e = 1.442 \left[1 - \frac{(v_y/v_0)^2}{6} \right]^{1/2} \left[\frac{v_y/v_0}{v_y/v_0 + 2\sqrt{[6 - (v_y/v_0)^2]/5}} \right]^{1/4}. \qquad (3.4.7)$$

Because it does not include viscoelastic effects, the Thornton model predicts that $e = 1$ for $v_0 < v_y$, where v_y refers to the yield velocity for plastic deformation estimated by $v_y = 3.194 p_y^{5/2} R^{3/2}/E^2 M^{1/2}$ (e.g., for stainless steel v_y is about 0.3 m/s). This prediction is at odds with experimental data, such as those shown in Figure 3.6b, which exhibit substantial reduction in restitution coefficient by viscoelastic effects prior to onset of the plastic deformation regime.

3.5. Sliding and Twisting Resistance

Sliding occurs when the surface of one particle slips relative to that of another particle during contact, due to relative tangential motion of the particle surfaces within the contact region. Dissipation of kinetic energy is associated with this slip, where the rate of energy dissipation is equal to the slip velocity times the slip resistance force F_s. The slip force acts in the direction of the slip velocity $\mathbf{v}_S(t)$, which is the tangent projection of the relative surface velocity $\mathbf{v}_R$ to the particle surface at the contact point, or

$$\mathbf{v}_S = \mathbf{v}_R - (\mathbf{v}_R \cdot \mathbf{n})\mathbf{n}, \qquad (3.5.1)$$

where $\mathbf{v}_R$ is defined in (3.3.2). The slip direction is given by the unit vector $\mathbf{t}_S = \mathbf{v}_S/|\mathbf{v}_S|$. Because the sliding force acts at the contact point in a direction that does not pass through the particle centroid, there is an associated sliding torque, given for particle i by $r_i F_s(\mathbf{n} \times \mathbf{t}_S)$.

Twisting is closely related to sliding, with the difference that in a pure twisting motion the slip velocity at the contact point vanishes, but there is slip at surrounding points within the contact region due to difference in the rotation rate of the particles in a direction along the normal vector $\mathbf{n}$. For rigid-body motions, in sliding the slip velocity within the contact region is equal to that at the contact point, whereas in a twisting motion the slip velocity increases linearly within the contact region with distance from the contact point. The resulting particle response to the twist motion is a torque M_t oriented parallel to the unit normal vector $\mathbf{n}$. As discussed in the next section, elastic deformation of the material modifies this observation, allowing slip between the two surfaces in only part of the contact region.

3.5.1. Physical Mechanisms of Sliding and Twisting Resistance

Sliding resistance arises from energy dissipation in both the fluid and the solid phases. In the solid phase, sliding is resisted by impact of the roughness elements on the two surfaces as they move relative to each other. The presence of a fluid within the gap between the sliding surfaces introduces additional friction within the fluid phase. For sliding with perfectly smooth surfaces, the sliding resistance would be entirely due to the fluid friction, the magnitude of which increases with decrease in gap thickness

Table 3.4. *Typical values of static friction coefficient for collision of two bodies made of the same material. [Extracted from data in Engineering Tool Box (http://www. engineeringtoolbox.com).]*

	Static friction coefficient	
Materials	Dry	Lubricated
Metals		
Aluminum	1.05–1.35	0.3
Iron	1.0	0.15–0.2
Copper	1.0	0.08
Graphite	0.1	0.1
Magnesium	0.6	0.08
Platinum	1.2	0.25
Silver	1.4	0.55
Steel	0.8	0.16
Zinc	0.6	0.04
Other Materials		
Glass	0.9–1.0	0.1–0.6
Nylon	0.15–0.25	
Polystyrene	0.5	0.5
Teflon	0.04	0.04

and approaches infinity as the gap thickness approaches zero. However, for a finite sliding velocity the fluid pressure within the contact region also increases as the gap thickness decreases. As a result, the presence of the fluid tends to increase the gap thickness between the surfaces, thereby decreasing the sliding resistance and providing a lubricating effect.

The problem of solid-phase sliding resistance between two bodies with uniform pressure distribution between the body surfaces is a classical problem of mechanics, which was largely solved in the eighteenth century by scientists such as Amontons, Euler, Belidor, and Coulomb. As taught in elementary physics books, two bodies in contact will not slide relative to each other until the external tangential force F_{tang} between the two bodies exceeds a critical value F_{crit}, equal to the product of the normal force F_n and a static friction coefficient μ_S. The value of the friction coefficient depends on the materials under consideration, the nature of the surface (e.g., whether it is polished or rough), the gap thickness between the surfaces, and the properties of any fluid contained within this gap. Since it depends on so many different factors, determination of friction coefficient can often be quite uncertain. A listing of typical friction coefficients is given in Table 3.4, but the values listed in this table are based only on the type of material and do not include important features such as surface preparation and fluid lubrication.

The case where the external tangential force between the bodies, F_{tang}, is less than the critical force $F_{crit} = \mu_S F_n$ is called *static friction*, because in this state the two surfaces are at rest relative to each other. In this state, the response force F_s acting on each body is equal to the imposed tangential force F_{tang} and the surfaces are in equilibrium. If F_{tang} exceeds this critical force, then the surfaces begin to move relative to each other and the surfaces are in a state called *dynamic friction*, or sometimes kinetic friction. The response force during dynamic friction can be

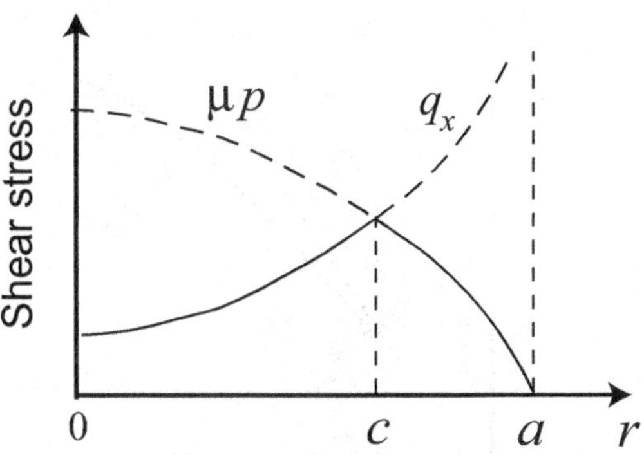

Figure 3.11. Plot showing the variation of shear stress for the case with no slip (q_x) and the critical shear stress (μp) as functions of radius within a contact region of radius a, for particle collision subject to a tangential force F_{tang}. The shear stress for the case with partial slip is indicated by a solid line, where slip starts at radius $r = c$.

written as $F_s = \mu_f F_n$, where the dynamic friction coefficient μ_f is generally smaller than the static friction coefficient μ_S for a given problem. The dynamic friction coefficient is often taken as a constant, particularly for cases in which the range of the sliding velocity is fairly narrow. However, in general, dynamic friction coefficient is a function of surface roughness, temperature, and slip velocity (Sang et al., 2008).

It was pointed out by Mindlin (1949) and Mindlin and Deresiewicz (1953) that because the Hertz pressure field (3.2.1) is not uniform over the contact region, different regions of the contact surface begin to slip at different times. A given point on the contact surface will start slipping when the local shear stress in the direction of tangential loading, q_x, exceeds the product of the static friction coefficient and the local pressure, or $\mu_S p$. According to the Hertz pressure distribution (3.2.1), the lowest pressure value occurs at the outer edge of the contact region and the pressure increases as one progresses inward, achieving a maximum at the contact point. If a tangential force difference with magnitude F_{tang} is imposed on two colliding particles with no slipping, a shear stress q_x forms within the contact region with the radial distribution

$$q_x = q_0(1 - r^2/a^2)^{-1/2}, \qquad (3.5.2)$$

where $q_0 = F_{tang}/2\pi a^2$. Thus, the largest shear stress occurs on the outer part of the contact region. As a consequence, slip starts within a ring near the outer part of the contact region and progresses inward, with decreasing radius c as the tangential force is increased. The ratio of the slip ring radius c to the contact region radius a is given by (Johnson, 1985, 218)

$$\frac{c}{a} = \left(1 - F_{tang}/\mu_S F_n\right)^{1/3}. \qquad (3.5.3)$$

As shown in Figure 3.11, in the partial slip case the shear stress increases with radius from the contact point and the pressure decreases with radius until the slip ring is reached, at which point $q_x = \mu_S p$. For radii greater than this value, the shear stress remains proportional to the pressure and hence decreases with radius. As the value of F_{tang} is increased, the radius of the slip ring grows smaller, until at $F_{tang} = \mu_S F_n$ the slip ring radius decreases to zero and the two surfaces are free to translate relative to each other in a state of dynamic friction. A plot of the tangential displacement obtained from the solution with partial slip is given in Figure 3.12, in comparison to the linear displacement variation that would be obtained with no slip. For small values of tangential force the tangential displacement predicted by the two approaches are similar, but as the tangential force increases to its maximum value

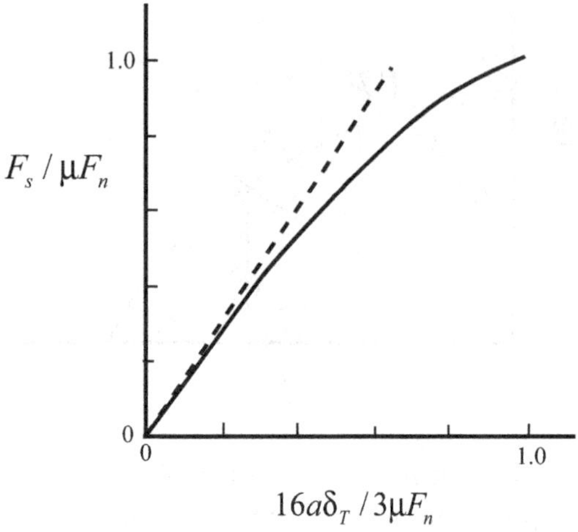

Figure 3.12. Plot of the tangential displacement δ_T of two spheres in static contact as a function of the ratio of tangential force to critical tangential force for cases with no slip (dashed line) and partial slip (solid line).

the displacement for the partial slip case increases to nearly twice the value for the no-slip case.

3.5.2. Sliding Resistance Model

In computations with the discrete element method, the tangential force on two colliding particles is not generally known. It is therefore necessary to introduce some type of model that relates the tangential force to known quantities, such as the particle centroid positions or velocity. A variety of options have been proposed in the literature, including the assumption that the tangential force is proportional to the normal force, or $F_s = -\mu_f |F_n|$ (Haff and Werner, 1986; Schäfer and Wolf, 1995), and the assumption that the tangential force is proportional to the sliding velocity, or $F_s = -\eta_T \mathbf{v}_S \cdot \mathbf{t}_S$ (Taguchi, 1992; Gallas et al., 1992; Luding et al., 1994). However, as discussed in the review by Schäfer et al. (1996), both of these choices lead to qualitative inconsistencies with experimental data for sliding. One of the most successful models for sliding resistance is the spring-dashpot-slider model proposed by Cundall and Strack (1979) and subsequently used in a large number of studies. In this model, shown schematically in Figure 3.13, the sliding resistance in the static friction state is absorbed by a combination of a spring and a dashpot. The force due to the spring is proportional to the particle tangential displacement δ_T, given by

$$\delta_T = \int_{t_0}^{t} \mathbf{v}_S(\xi) \cdot \mathbf{t}_S d\xi, \tag{3.5.4}$$

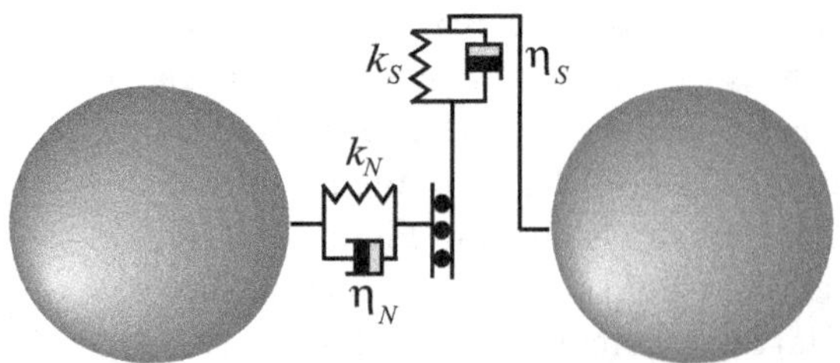

Figure 3.13. Schematic diagram of the spring-dashpot-slider model proposed by Cundall and Strack (1979) for the sliding resistance of two colliding particles.

where t_0 is the time of initial particle impact. The dashpot force is linearly proportional to the sliding velocity $\mathbf{v}_S$. The complete Cundall-Strack model thus has the form

$$F_s = -k_T \delta_T \cdot \mathbf{t}_S - \eta_T \mathbf{v}_S \cdot \mathbf{t}_S. \tag{3.5.5}$$

An expression for the tangential stiffness coefficient k_T was derived by Mindlin (1949) and can be written in terms of the effective shear modulus G, defined in (3.1.1), and the contact region radius $a(t)$ as

$$k_T = 8Ga(t). \tag{3.5.6}$$

Many investigators omit the dashpot term in (3.5.5), or else assign η_T a constant value. Other authors (e.g., Tsuji et al., 1992; Cleary et al., 1998) assume that η_T is approximately equal to the normal viscous damping coefficient η_N.

When the magnitude of the tangential force reaches a critical value $F_{crit} = \mu_S|F_n|$, the surfaces of the two particles start to slide relative to each other and the system enters into a condition of dynamic friction. This value for the critical force is the same as the value of F_{tang} for which the radius c of the slip ring, given in (3.5.3), goes to zero. If $|F_{tang}| > F_{crit}$, the particle surfaces slip relative to each other, which is represented by the slider in Figure 3.11. In this dynamic friction state, the sliding response force is given by

$$F_s = -\mu_f |F_n|. \tag{3.5.7}$$

3.5.3. Twisting Resistance Model

Twisting occurs when the two colliding particles have different rotation rate along the normal direction $\mathbf{n}$, as illustrated in Figure 3.3c. A relative twisting rate Ω_T is defined by

$$\Omega_T = (\mathbf{\Omega}_i - \mathbf{\Omega}_j) \cdot \mathbf{n}. \tag{3.5.8}$$

A rotational spring-dashpot-slider model for twisting resistance has a form analogous to (3.5.5), giving the twisting resistance torque in the normal direction as

$$M_t = -k_Q \int_{t_0}^{t} \Omega_T(\tau)d\tau - \eta_Q \Omega_T. \tag{3.5.9}$$

The time integral in this expression is equal to the angular displacement of the particle since the time of initial particle impact (t_0). The torsional stiffness and dissipation coefficients can be related to the linear stiffness and dissipation coefficients for sliding as

$$k_Q = k_T a^2/2, \quad \eta_Q = \eta_T a^2/2. \tag{3.5.10}$$

Similar to the case for tangential force, slip within the contact region first occurs within a radial ring starting at the outer edge of the contact region and progressing inward as the imposed twisting torque is increased. The radius of the slip ring shrinks to zero when the magnitude of the imposed twisting moment equals a critical value $M_{t,crit}$, whose value is given by (Johnson, 1985, 233)

$$M_{t,crit} = \frac{3\pi}{16} a F_{crit},\tag{3.5.11}$$

where $F_{crit} = \mu_S |F_n|$ is the critical linear tangential force for sliding of the particle surfaces. For imposed twisting moments with magnitude greater than $M_{t,crit}$, the torsional resistance is given by

$$M_t = -\frac{3\pi}{16} a \mu_f |F_n| \Omega_T / |\Omega_T|.\tag{3.5.12}$$

3.6. Rolling Resistance

Rolling occurs when particle rotation rates and centroid translation velocities are set such that the particles can move tangentially relative to each other without slipping. Rolling motion is known to be a primary mechanism by which assemblages of particles deform under a shearing load, particularly in cases with densely packed particles such as in granular materials or agglomerates of adhesive particles (Oda et al., 1982; Bardet, 1994; Iwashita and Oda, 1998). Although the rolling motion of a sphere generally has lower energy dissipation than an equivalent sliding motion, there is nevertheless a resistance to rolling that takes the form of a torque imposed on the particle in such a direction as to decrease the rolling velocity. As a consequence, a spherical particle set into a rolling motion on a horizontal planar substrate will not continue to roll forever, but will gradually slow down and eventually come to rest. There are two separate challenges associated with incorporation of the rolling resistance in discrete element models. The first is the challenge of clearly defining the rolling velocity for arbitrary motions of a particle assemblage. Although it is a simple thing to identify rolling of a particle on a plane, it is no simple matter to clearly distinguish between rolling, sliding, and rigid body rotation of a group of colliding particles moving in different directions. Once the rolling velocity is clearly defined, the second challenge is to understand the physical mechanism that gives rise to rolling resistance and to introduce a simple model that accurately represents this mechanism. This also is not a simple matter, as multiple mechanisms play a role in determining the rolling resistance.

3.6.1. Rolling Velocity

It is desired to define a velocity that characterizes rolling motion during collision of two spherical particles, labeled Particle 1 and Particle 2, with radii r_1 and r_2, respectively. To simplify the problem, we first consider the case of two-dimensional motion in which all motion lies in the x-y plane. The positions of the two particles are sketched in Figure 3.14 at times t and $t + \Delta t$. The centroid position vectors are given by $\mathbf{x}_1$ and $\mathbf{x}_2$ at time t and by $\mathbf{x}_1'$ and $\mathbf{x}_2'$ at time $t + \Delta t$, so the displacement vectors for each particle over this time interval are given by

$$d\mathbf{u}_1 = \mathbf{x}_1' - \mathbf{x}_1, \quad d\mathbf{u}_2 = \mathbf{x}_2' - \mathbf{x}_2.\tag{3.6.1}$$

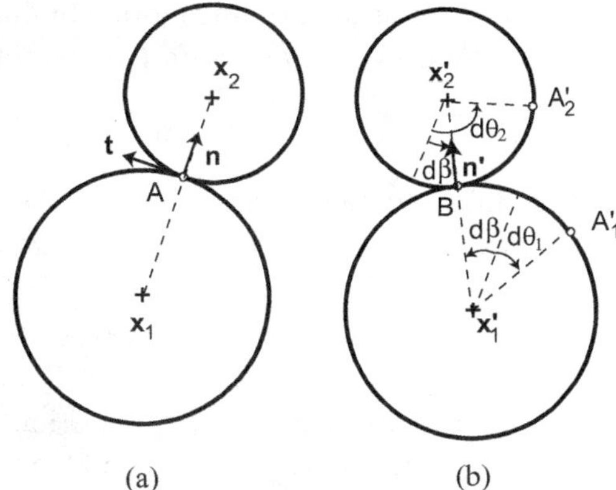

Figure 3.14. Schematic diagram showing rolling rotation of two particles with radii r_1 and r_2 at times (a) t and (b) $t + \Delta t$.

The unit normal vector pointing from the centroid of Particle 1 to that of Particle 2 is denoted by $\mathbf{n}$ at time t and by $\mathbf{n}'$ at time $t + \Delta t$. The angle between vector $\mathbf{n}$ and vector $\mathbf{n}'$ is denoted by $d\beta$. A unit tangent vector $\mathbf{t}$ is defined in the x-y plane in a direction tangent to the particle surfaces at the contact point, and a third unit vector $\mathbf{b} = \mathbf{n} \times \mathbf{t}$ is defined in a direction orthogonal to the plane of motion. Vectors $\mathbf{r}_1$ and $\mathbf{r}_2$ extending from the particle centroids to the contact point are defined by

$$\mathbf{r}_1 = r_1\mathbf{n}, \quad \mathbf{r}_2 = -r_2\mathbf{n}, \tag{3.6.2}$$

so $d\beta$ can be written as

$$d\beta = \frac{(d\mathbf{u}_2 - d\mathbf{u}_1) \cdot \mathbf{t}}{r_1 + r_2} = \frac{(d\mathbf{u}_2 - d\mathbf{u}_1) \cdot \mathbf{t}}{(\mathbf{r}_1 - \mathbf{r}_2) \cdot \mathbf{n}}. \tag{3.6.3}$$

The numerator in (3.6.3) denotes the relative tangential displacement of the particle centroids during the time interval of length Δt.

In addition to translation of the centroids, each particle rotates about its centroid by an angle $d\theta_1$ and $d\theta_2$, respectively, during the time interval, where a positive angle indicates rotation in the counterclockwise direction. In Figure 3.14a, the contact point at time t is denoted by A. The material point on each particle moves during the time interval to occupy positions denoted by A_1' and A_2' on the two particles at time $t + \Delta t$. The new contact point at time $t + \Delta t$ is denoted by B in Figure 3.14b. The arc length along the surface of Particle 1 from point B to point A_1' at time $t + \Delta t$ is denoted by da_1. Similarly, the arc length from B to point A_2' along the surface of Particle 2 is denoted by da_2. We take the arc length as being positive if the corresponding angle is in the counterclockwise direction, and negative otherwise. Referring to Figure 3.14b, the arc lengths can be written as

$$da_1 = r_1(d\theta_1 - d\beta), \quad da_2 = r_2(d\theta_2 - d\beta). \tag{3.6.4}$$

Pure rolling motion occurs at the contact point if $da_1 = -da_2$, whereas pure sliding motion occurs when $da_1 = da_2$. In a rigid-body rotation, $da_1 = da_2 = 0$. Following Iwashita and Oda (1998), we define a *rolling displacement* $d\mathbf{u}_L$ by

$$d\mathbf{u}_L = \tfrac{1}{2}(da_2 - da_1)\mathbf{t}. \tag{3.6.5}$$

For either pure sliding motion or pure rigid body rotation $d\mathbf{u}_L = 0$, whereas for pure rolling motion $d\mathbf{u}_L = da_2\mathbf{t}$. Substituting (3.6.4) into (3.6.5) gives

$$d\mathbf{u}_L = \tfrac{1}{2}[r_2 d\theta_2 - r_1 d\theta_1 - (r_2 - r_1)d\beta]\mathbf{t}. \qquad (3.6.6)$$

Substituting the expression (3.6.3) for $d\beta$ into (3.6.6) and using the notation (3.6.2), we can write the rolling displacement as

$$d\mathbf{u}_L = -\frac{1}{2}\left\{ d\theta_1 \times \mathbf{r}_1 + d\theta_2 \times \mathbf{r}_2 - \frac{(\mathbf{r}_1 + \mathbf{r}_2) \cdot \mathbf{n}}{(\mathbf{r}_1 - \mathbf{r}_2) \cdot \mathbf{n}}[(d\mathbf{u}_2 - d\mathbf{u}_1) \cdot \mathbf{t}]\mathbf{t}\right\}. \qquad (3.6.7)$$

The vectors $d\theta_1$ and $d\theta_2$ are oriented in the $\mathbf{b}$ direction, so that the cross products $d\theta_1 \times \mathbf{r}_1$ and $d\theta_2 \times \mathbf{r}_2$ in (3.6.7) are oriented in the $\mathbf{t}$ direction.

The problem of defining rolling displacement for arbitrary motion of two colliding particles of arbitrary shape was examined by Bagi and Kuhn (2004). They found that for three-dimensional motion of two colliding spherical particles, the result (3.6.7) applies provided that the product $[(d\mathbf{u}_2 - d\mathbf{u}_1) \cdot \mathbf{t}]\mathbf{t}$ in the last term is replaced by the projection of the difference $d\mathbf{u}_2 - d\mathbf{u}_1$ in the plane normal to $\mathbf{n}$, which is given by $[\mathbf{n} \times (d\mathbf{u}_2 - d\mathbf{u}_1)] \times \mathbf{n}$. The rolling displacement for general three-dimensional collisions of spherical particles is then given by

$$d\mathbf{u}_L = -\frac{1}{2}\left\{ d\theta_1 \times \mathbf{r}_1 + d\theta_2 \times \mathbf{r}_2 - \frac{(\mathbf{r}_1 + \mathbf{r}_2) \cdot \mathbf{n}}{(\mathbf{r}_1 - \mathbf{r}_2) \cdot \mathbf{n}}[\mathbf{n} \times (d\mathbf{u}_2 - d\mathbf{u}_1)] \times \mathbf{n}\right\}. \qquad (3.6.8)$$

The *rolling velocity* $\mathbf{v}_L$ can be obtained by dividing $d\mathbf{u}_L$ by the time increment Δt and taking the limit $\Delta t \to 0$, which yields

$$\mathbf{v}_L = -\frac{1}{2}\left\{ \mathbf{\Omega}_1 \times \mathbf{r}_1 + \mathbf{\Omega}_2 \times \mathbf{r}_2 - \frac{(\mathbf{r}_1 + \mathbf{r}_2) \cdot \mathbf{n}}{(\mathbf{r}_1 - \mathbf{r}_2) \cdot \mathbf{n}}[\mathbf{n} \times (\mathbf{v}_2 - \mathbf{v}_1)] \times \mathbf{n}\right\}. \qquad (3.6.9)$$

An alternative expression for the rolling velocity in terms of the slip velocity $\mathbf{v}_S$, defined in (3.5.1), can be obtained by writing

$$\mathbf{v}_S = \mathbf{v}_R - (\mathbf{v}_R \cdot \mathbf{n})\mathbf{n} = (\mathbf{n} \times \mathbf{v}_R) \times \mathbf{n}. \qquad (3.6.10)$$

Recalling that the relative particle surface velocity $\mathbf{v}_R = \mathbf{v}_{C,i} - \mathbf{v}_{C,j}$, where $\mathbf{v}_{C,i} = \mathbf{v}_i + \mathbf{\Omega}_i \times \mathbf{r}_i$ is the surface velocity of particle i at the contact point, we can write for the two-particle collision problem

$$\mathbf{v}_R = \mathbf{v}_{C,1} - \mathbf{v}_{C,2} = \mathbf{v}_1 - \mathbf{v}_2 + \mathbf{\Omega}_1 \times \mathbf{r}_1 - \mathbf{\Omega}_2 \times \mathbf{r}_2. \qquad (3.6.11)$$

Because the cross-product terms in (3.6.11) are already tangent to the unit normal vector $\mathbf{n}$, substituting (3.6.11) into (3.6.10) gives

$$\mathbf{v}_S = [\mathbf{n} \times (\mathbf{v}_1 - \mathbf{v}_2)] \times \mathbf{n} + \mathbf{\Omega}_1 \times \mathbf{r}_1 - \mathbf{\Omega}_2 \times \mathbf{r}_2. \qquad (3.6.12)$$

Solving for $[\mathbf{n} \times (\mathbf{v}_2 - \mathbf{v}_1)] \times \mathbf{n} = -\mathbf{v}_S + \mathbf{\Omega}_1 \times \mathbf{r}_1 - \mathbf{\Omega}_2 \times \mathbf{r}_2$ from (3.6.12) and substituting into (3.6.9) yields an expression for the rolling velocity as

$$\mathbf{v}_L = R(\mathbf{\Omega}_2 - \mathbf{\Omega}_1) \times \mathbf{n} + \frac{1}{2}\frac{r_2 - r_1}{r_1 + r_2}\mathbf{v}_S, \qquad (3.6.13)$$

where R is the effective radius defined in (3.1.1).

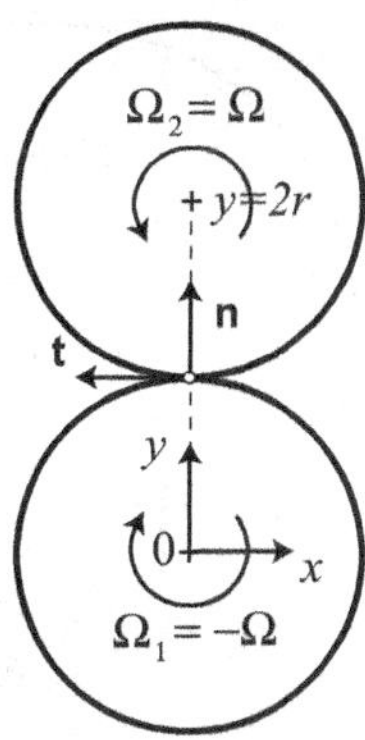

Figure 3.15. Sketch of the collision of two equal-size particles, as used for the example in Section 3.6.1.

For the special case of two equal-size spheres, (3.6.13) reduces to the simple expression

$$\mathbf{v}_L = R(\mathbf{\Omega}_2 - \mathbf{\Omega}_1) \times \mathbf{n}, \qquad (3.6.14)$$

indicating that in this case rolling velocity is proportional to the difference in rotation rate of the two colliding particles. As an example, we consider a case with two equal-size particles with radius $r_p = 2R$ in a state of pure rolling with the centroid of particle 1 at the origin of a Cartesian coordinate system and that of particle 2 at $(x, y, z) = (0, 2r_p, 0)$, as sketched in Figure 3.15. The rotation rates of the particles are given by $\mathbf{\Omega}_1 = -\Omega\mathbf{e}_z$ and $\mathbf{\Omega}_2 = \Omega\mathbf{e}_z$, thus satisfying the condition $r_1\Omega_1 + r_2\Omega_2 = 0$ for pure rolling in two dimensions. Then, $\mathbf{n} = \mathbf{e}_y$ and $(\mathbf{\Omega}_2 - \mathbf{\Omega}_1) \times \mathbf{n} = -2\Omega\mathbf{e}_x$, so (3.6.14) gives $\mathbf{v}_L = -r_p\Omega\mathbf{e}_x$. Letting $\mathbf{b} = \mathbf{e}_z$ gives $\mathbf{t} = \mathbf{b} \times \mathbf{n} = -\mathbf{e}_x$ for this example, so we can write $\mathbf{v}_L = r_p\Omega\mathbf{t}$. The rolling velocity in this pure-rolling example is therefore equal to the difference between the particle centroid velocity and the particle material velocity at the contact point.

3.6.2. Physical Mechanism of Rolling Resistance

The primary physical mechanisms underlying rolling resistance in low-velocity collisions of nonadhesive particles are the solid-phase viscoelastic resistance and the viscous fluid resistance. Adhesive force introduces an important third mechanism for rolling resistance, which is discussed in Section 4.2. A detailed examination of the solid-phase rolling resistance was given by Tabor (1955), who concluded that although slip does occur during rolling motion in the outer part of the contact region, the magnitude of the resistance torque that would be caused by slip is insufficient to account for the observed rolling resistance. The relationship between rolling resistance and the normal dissipative force that occurs during particle impact was discussed by Brilliantov and Pöschel (1998, 1999). A schematic diagram showing the different solid stresses acting on a particle rolling in the counterclockwise direction along a flat plane is given in Figure 3.16. Both elastic and dissipative stresses act in the direction normal to the planar surface of the contact region, as discussed in Sections 3.2 and 3.3. The elastic stress depends only on the displacement of the particle surface and it is therefore the same on both sides of the rolling particles. Consequently, the elastic stress produces a normal force in the vertical direction in Figure 3.16, but no net moment on the particle. As discussed in Section 3.3, the dissipative stress

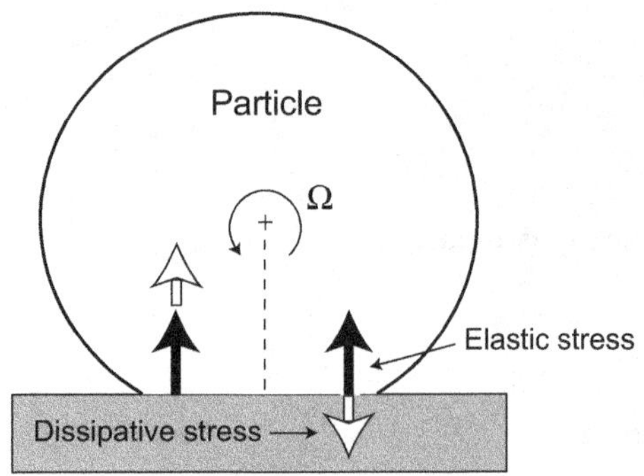

Figure 3.16. Schematic diagram showing a particle rolling to the left (with rotation in the counterclockwise direction) on a plane surface. The direction of the elastic stress on each side of the particle is indicated by black arrows, and the direction of the dissipative stress is indicated by white arrows.

acts in a direction so as to resist the relative motion of the particle and the plate. For the rolling particle shown in Figure 3.16, the left-hand side of the particle is moving downward into the plate, so that the dissipative stress is positive upward on the left-hand side. The right-hand side of the particle is moving upward away from the planar surface, and so the dissipative stress on the right-hand side is downward. The dissipative stresses on the two sides of the particle cancel out one another, and therefore produce zero net vertical force. However, the dissipative stresses do produce a torque on the particle acting in the clockwise direction, resisting the rolling motion of the particle.

For very small particles immersed in a fluid medium, energy dissipation due to the fluid viscous shear also makes a significant contribution to the overall rolling resistance. For the problem of a sphere in contact with a plane wall, application of the incompressible Navier-Stokes equations together with the usual no-slip boundary conditions to the fluid surrounding a perfectly smooth sphere results in the conclusion that an infinite force on the particle is required in order for the particle to roll! This singularity is resolved in practice by small deformation of the particle and microscopic roughness of the particle surface, both of which lead to formation of a small gap between the particle surface and the plane wall within the contact region. A detailed examination of the effect of a wall on the forces and torque acting on a nearby particle, including the case of a particle rolling along a wall, is given in Section 5.6.

3.6.3. Model for Rolling Resistance

Iwashita and Oda (1998) proposed a rotational spring-dashpot-slider model for rolling resistance similar to the Cundell-Strack model for sliding resistance. In this model, the rolling torque M_r is set equal to the sum

$$M_r = -k_R \xi_R - \eta_R \mathbf{v}_L \cdot \mathbf{t}_R, \qquad (3.6.15)$$

where $\xi_R = \int_{t_0}^{t} \mathbf{v}_L(\tau) \cdot \mathbf{t}_R d\tau$ is the rolling displacement of the particle and $\mathbf{t}_R = \mathbf{v}_L / |\mathbf{v}_L|$ is the direction of rolling. The first term in (3.6.15) represents a spring force and the second term accounts for the effect of damping. The rolling torque is given by (3.6.15) if $|M_r|$ is less than a critical value $M_{r,crit}$, beyond which $M_r = -M_{r,crit}$.

This basic framework of the model given by (3.6.15) has been used by a large number of investigators, for cases with no adhesion as well as for cases with adhesion. However, not all investigators use all of the terms in this model, and there are a number of differences in how the coefficients are set. Iwashita and Oda (1998) assumed that the rolling stiffness and the sliding stiffness are of the same order of magnitude, and thus set $k_R = r_i k_T$. As discussed in Section 4.2, both the spring and the slider mechanisms in (3.6.15) play a particularly important role when accounting for the effect of adhesion on the rolling resistance. Focusing on the nonadhesive case, Brilliantov and Pöschel (1998) omitted the spring term in (3.6.15) to derive an expression for the damping coefficient η_R using a quasi-static approximation in which the purely elastic (Hertz) solution $\mathbf{u}_{el}$ is used for the displacement $\mathbf{u}$ and the rate of change of displacement $\dot{\mathbf{u}}$ is determined from

$$\dot{\mathbf{u}}(\mathbf{r}) = \mathbf{\Omega} \cdot (\mathbf{r} \times \nabla)\mathbf{u}_{el}(\mathbf{r}). \tag{3.6.16}$$

These expressions for $\mathbf{u}$ and $\dot{\mathbf{u}}$ are substituted into the constitutive equation for the dissipative stress, which is then integrated to obtain the rolling resistance torque. The resulting expression has the form of the damping term in (3.6.15) with the coefficient η_R given by

$$\eta_R = \mu_R |F_{ne}|. \tag{3.6.17}$$

Using the analogy with the normal dissipation force F_{nd}, Brilliantov and Pöschel (1999) showed analytically that the coefficient μ_R can be related to the same constant coefficient α_{KKB} that was introduced in the expression (3.3.7) for the F_{nd}, giving

$$\mu_R = \tfrac{2}{3}\alpha_{KKB}. \tag{3.6.18}$$

REFERENCES

Bagi K, Kuhn M. A definition of particle rolling in a granular assembly in terms of particle translations and rotations. *Journal of Applied Mechanics* **71**, 493–501 (2004).

Bardet JP. Observations on the effects of particle rotations on the failure of idealized granular materials. *Mechanics of Materials* **18**, 159–182 (1994).

Boussinesq J. *Application des Potentiels a l'Etude de l'Equilibre et du Mouvement des Solides Elastiques*, Gauthier-Villars, Paris (1885).

Brilliantov NV, Pöschel T. Rolling friction of a viscous sphere on a hard plane. *Europhysics Letters* **42**(5), 511–516 (1998).

Brilliantov NV, Pöschel T. Rolling as a "continuing collision." *The European Physical Journal B* **12**, 299–301 (1999).

Brilliantov NV, Spahn F, Hertzsch J-M, Pöschel T. Model for collisions in granular gases. *Physical Review E* **53**(5), 5382–5392 (1996).

Cleary PW, Metcalfe G, Liffman K. How well do discrete element granular flow models capture the essentials of mixing processes? *Applied Mathematical Modeling* **22**, 995–1008 (1998).

Cundall PA, Strack ODL. A discrete numerical model for granular assembles. *Geotechnique* **29**(1), 47–65 (1979).

Gallas JAC, Hermann HJ, Sokolowski S. Molecular dynamics simulation of powder fluidization in two dimensions. *Physica A* **189**, 437–446 (1992).

Gorham DA, Kharaz AH. The measurement of particle rebound characteristics. *Powder Technology* **112**, 193–202 (2000).

Goldsmith, W. *Impact: The Theory and Physical Behaviour of Colliding Solids*, Arnold, London (1960).

Haff PK, Werner BT. Computer simulation of the mechanical sorting of grains. *Powder Technology* **48**(3), 239–245 (1986).

Hertz H. Über die Berührung fester elastischer Körper. *J. reine und angewandte Mathematik* **92**, 156–171 (1882).

Iwashita K, Oda M. Rolling resistance at contacts in simulation of shear band development by DEM. *Journal of Engineering Mechanics* **124**(3), 285–292 (1998).

Johnson, KL. *Contact Mechanics*, Cambridge University Press (1985).

Kruggel-Emden H, Simsek E, Rickelt S, Wirtz S, Scherer V. Review and extension of normal force models for the discrete element method. *Powder Technology* **171**, 157–173 (2007).

Kuwabara G, Kono K. Restitution coefficient in a collision between two spheres. *Japanese Journal of Applied Physics* **26**(8), 1230–1233 (1987).

Labous L, Rosato AD, Dave RN. Measurements of collisional properties of spheres using high-speed video analysis. *Physical Review E* **56**(5), 5717–5725 (1997).

Lee J, Herrmann HJ. Angle of repose and angle of marginal stability: Molecular dynamics of granular particles. *J. Phys. A: Math. Gen.* **26**, 373–383 (1993).

Love AEH. *A Treatise on the Mathematical Theory of Elasticity*, 4th ed., Cambridge University Press, Cambridge (1952).

Luding S, Clement E, Blumen A, Rajchenbach J, Duran J. Onset of convection in molecular dynamics simulations of grains. *Physical Review E* **50**(3), R1762–R1765 (1994).

Mindlin RD. Compliance of elastic bodies in contact. *Journal of Applied Mechanics* **16**, 259–268 (1949).

Mindlin RD, Deresiewicz H. Elastic spheres in contact under varying oblique forces. *Journal of Applied Mechanics* **20**, 327–344 (1953).

Oda M, Konishi J, Nemat-Nasser S. Experimental micromechanical evaluation of strength of granular materials: Effects of particle rolling. *Mechanics of Materials* **1**, 269–283 (1982).

Sadd MH, Tai QM. A contact law effects on wave-propagation in particulate materials using distinct element modeling. *International Journal of Non-Linear Mechanics* **28**(2), 251–265 (1993).

Sang Y, Dubé M, Grant M. Dependence of friction on roughness, velocity, and temperature. *Physical Review E* **77**, 036123 (2008).

Schäfer J, Wolf DE. Bistability in simulated granular flow along corrugated walls. *Physical Review E* **51**(6), 6154–6157 (1995).

Schäfer J, Dippel S, Wolf DE. Force schemes in simulations of granular materials. *J. Phys. I France* **6**, 5–20 (1996).

Schwager T, Pöschel T. Coefficient of restitution and linear-dashpot model revisited. *Granular Matter* **9**, 465–469 (2007).

Schwager T, Pöschel T. Coefficient of restitution for viscoelastic spheres: The effect of delayed recovery. *Physical Review E* **78**, 051304 (2008).

Stevens AB, Hrenya CM. Comparison of soft-sphere models to measurements of collision properties during normal impacts. *Powder Technology* **154**, 99–109 (2005).

Tabor D. The mechanism of rolling friction. 2. The elastic range. *Proceedings of the Royal Society of London A* **229**(1177), 198–220 (1955).

Taguchi R-H. New origin on a convective motion: Elastically induced convection in granular materials. *Physical Review Letters* **69**(9), 1367–1370 (1992).

Thornton C. Coefficient of restitution for collinear collisions of elastic perfectly plastic spheres. *Journal of Applied Mechanics* **64**, 383–386 (1997).

Timoshenko, SP, Goodier, JN, *Theory of Elasticity*, 3rd Ed., McGraw-Hill Publishers, New York (1970).

Tsuji Y, Tanaka T, Ishida T. Lagrangian numerical simulation of plug flow of cohesionless particles in a horizontal pipe. *Powder Technology* **71**, 239–250 (1992).

Vu-Quoc L, Zhang X. An elastoplastic force-displacement model in the normal direction: displacement-driven version. *Proceedings of the Royal Society of London. Series A, Mathematical and Physical Sciences* **455**, 4013–4044 (1999).

Walton OR, Braun RL. Viscosity, granular temperature, and stress calculations for shearing assemblies of inelastic, frictional disks. *Journal of Rheology* **30**, 949–980 (1986).

4 Contact Mechanics with Adhesion Forces

As particles become sufficiently small, adhesion phenomena that are insignificant at larger scales begin to become important. The specific particle length scale at which adhesion becomes significant depends on the type of adhesive force. For instance, van der Waals adhesion generally becomes important for particles with diameter on the order of magnitude of a micrometer, whereas liquid bridging can be important for particles as large as a millimeter or more (Li et al., 2011). The presence of field forces that bring particles toward each other, such as electric or magnetic fields, can enhance the effect of adhesive forces such as the van der Waals force, causing them to be significant even for particle sizes for which they would not otherwise have been important. The presence of significant adhesive forces can be established experimentally by the observation that the area of the contact region is larger than predicted by the Hertz theory and the observation that the contact region does not vanish when the load is removed. The significance of adhesion also depends on the magnitude of the applied load and on the roughness and cleanliness of the surface. Within the context of DEM, we typically assume that adhesive forces between particles act on length scales that are much smaller than the particle size, such that in most cases the adhesive forces have negligible effect until two or more particles collide. An exception occurs for the case of capillary forces between particles, which typically act on a scale comparable to the particle diameter.

This chapter discusses various forces that give rise to adhesion between particles, including van der Waals force, electric double-layer repulsion force, ligand-receptor binding of biological cells, liquid bridging forces of particles surrounded by a thin liquid film in a gaseous environment, and sintering forces at relatively higher temperature. Several of these adhesion forces arise from relatively weak bonds, leading to formation of particle agglomerates, whereas the sintering forces give rise to much stronger bonds leading to formation of particle aggregates (also sometimes called "hard agglomerates"). These five different types of adhesive forces are by no means exhaustive, but they are representative of the modeling involved in adhesive DEM and are involved in numerous problems of interest in engineering applications.

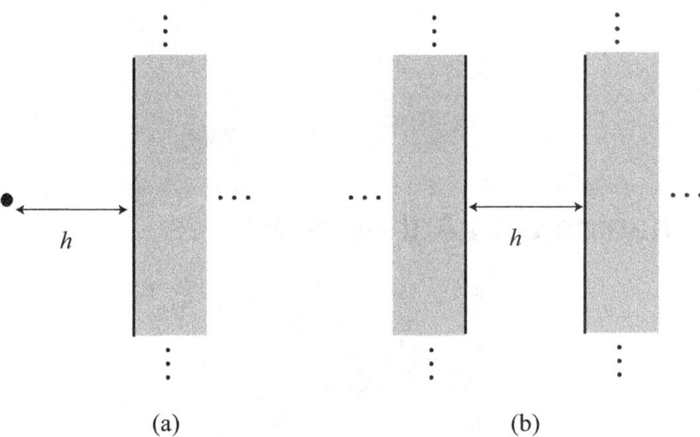

Figure 4.1. Diagrams illustrating (a) a single molecule or atom at a distance h from the surface of a semi-infinite body and (b) two semi-infinite bodies with surfaces separated by a distance h.

4.1. Basic Concepts and the Surface Energy Density

We consider two spherical particles approaching each other with minimum separation distance $h_0(t)$ between the particle surfaces. As the particles collide with each other they deform within a small contact region of radius $a(t)$, where it is assumed that the separation distance between the particle surfaces within the contact region has a constant value δ. The gap thickness is assumed to remain uniform throughout the contact region and to remain constant in time, so that movement of the particle centroids toward or away from each other results only in a change of the contact region radius $a(t)$.

The interaction potential between two microscopic elements, such as atoms or molecules, separated by a distance r is assumed to have a power law form, given by

$$w_{m-m}(r) = -C/r^n. \tag{4.1.1}$$

For instance, $n = 6$ corresponds to the potential associated with the van der Waals attraction force, in which the constant C includes the sum of dispersion, induction, and orientation forces between molecules. The corresponding force between the microscopic elements is given by $F(r) = -dw_{m-m}(r)/dr$, where a positive value of F corresponds to a repulsive force. Using the assumption of additivity, the net adhesive force between two bodies is equal to the sum of the adhesive forces caused by the interaction potential of all microscopic elements within the two bodies. The net interaction energy between a "free" molecule located a distance h away from the surface of a semi-infinite solid (Figure 4.1a) with molecule number density ρ within the solid can be derived by summing all pair interactions (4.1.1) between the free molecule and the molecules within the solid, giving the equation (Israelachvili, 1991, 156)

$$w_{m-p}(h) = -2\pi C\rho \int_h^\infty \int_0^\infty \frac{r}{(r^2 + z^2)^{n/2}} \, dr \, dz$$

$$= -2\pi C\rho/(n-2)(n-3)h^{n-3} \quad \text{for } n > 3. \tag{4.1.2}$$

The interaction energy $dw_{p-p}(z)$ between a thin sheet of material of width dz having unit area in the x-y plane and molecule number density ρ_1 located a distance z from a semi-infinite solid with molecule number density ρ_2 can be obtained from (4.1.2) as $dw_{p-p}(z) = -2\pi C\rho_2\rho_1 dz/(n-2)(n-3)z^{n-3}$. Integrating this expression from a distance h to infinity gives the interaction energy per unit area between two solids with planar surfaces separated by a distance h (Figure 4.1b) as

$$w_{p-p}(h) = -\frac{2\pi C\rho_2}{(n-2)(n-3)} \int_h^\infty \frac{\rho_1 dz}{z^{n-3}} = -\frac{2\pi C\rho_1\rho_2}{(n-2)(n-3)(n-4)h^{n-4}}. \qquad (4.1.3)$$

The normal force per unit area (i.e., the pressure) between the two materials is related to the derivative of the interaction energy by

$$p(h) = -dw_{p-p}/dh. \qquad (4.1.4)$$

In the case of the long-range van der Waals force (with $n = 6$), (4.1.3) reduces to

$$w_{p-p}(h) = -\pi C\rho_1\rho_2/12h^2. \qquad (4.1.5)$$

The Hamaker coefficient A_{12} for medium 1 interacting with medium 2 across a vacuum is defined by $A_{12} = \pi^2 C\rho_1\rho_2$, so that (4.1.5) becomes

$$w_{p-p}(h) = -A_{12}/12\pi h^2. \qquad (4.1.6)$$

The Hamaker coefficient A_{12} can be approximately written in terms of the Hamaker coefficient A_{11} for medium 1 interacting with itself and the respective coefficient A_{22} for medium 2 interacting with itself as

$$A_{12} \cong \sqrt{A_{11}A_{22}}. \qquad (4.1.7)$$

If media 1 and 2 interact across a space filled by a third medium 3, the resulting Hamaker coefficient can be approximated by

$$A_{123} \cong \sqrt{A_{131}A_{232}}, \qquad (4.1.8)$$

where A_{131} can be written in terms of A_{11} and A_{33} as

$$A_{131} \cong A_{11} + A_{33} - 2A_{13} \cong \left(\sqrt{A_{11}} - \sqrt{A_{33}}\right)^2. \qquad (4.1.9)$$

The final expression in (4.1.9) is obtained with use of (4.1.7) to write $\sqrt{A_{11}A_{33}} \cong A_{13}$. Using (4.1.9) for both A_{131} and A_{232} and substituting into (4.1.8) yields the useful approximation

$$A_{123} \cong \left(\sqrt{A_{11}} - \sqrt{A_{33}}\right)\left(\sqrt{A_{22}} - \sqrt{A_{33}}\right). \qquad (4.1.10)$$

Expressions (4.1.7), (4.1.8), (4.1.9), and (4.1.10) are examples of what is called *combining relations*. Such approximations are reasonably accurate, to within 15% or so, provided that dispersion forces dominate the van der Waals force and that the dielectric constants are not especially high. More accurate equations for calculation of Hamaker coefficients were given by Israelachvili (1991). The value of the Hamaker coefficient of most condensed phases, either solid or liquid, is typically in the range of 0.4×10^{-19} J to 4×10^{-19} J in a vacuum. Table 4.1 summarizes typical values of Hamaker constant, A_{131}, for two identical macroscopic bodies (identified as medium 1) interacting in air or across water (identified as medium 3).

Table 4.1. *Hamaker coefficients for two identical macroscopic bodies (media 1 and 2) interacting across a third medium*

Interacting media (symmetry $A_{131} = A_{313}$)		Hamaker constant A (unit: 10^{-20} J)
Medium 1	Medium 3	
Water/Water	Air	3.7–5.5
Pentane/Pentane	Air	3.75
Pentane/Pentane	Water	0.34
Dodecane/Dodecane	Air	5.0
Dodecane/Dodecane	Water	0.4–0.5
SiO_2		
Silica/Silica	Air	6.5
Silica/Silica	Dodecane	0.10–0.15
Fused quartz/quartz	Water	0.5–1.0
Fused quartz/quartz	Octane	–
Mica/Mica	Air	7–10
Mica/Mica	Water	1.–2.9
α-Alumina/α-Alumina	Air	15
α-Alumina/α-Alumina	Water	2.7–5.2
Metals		
Ag/Ag, Au/Au or Cu/Cu	Air	20–50
Ag/Ag, Au/Au or Cu/Cu	Water	10–40
Polymers		
Polystyrene/Polystyrene	Air	6.6–7.9
Polystyrene/Polystyrene	Water	0.95–1.3
Teflon/Teflon	Air	3.8
Teflon/Teflon	Water	0.33

Forces between atoms and molecules often have a long-range attractive (van der Waals) force and a short-range repulsive force. This behavior is apparent, for instance, in the Lennard-Jones potential given in (2.3.1), in which the exponent in (4.1.1) is equal to 6 for the attractive force and 12 for the repulsive force. The resulting variation of interaction potential between two parallel surfaces can be computed using (4.1.3) for each of these terms individually and then adding to obtain the total value as a function of the distance h between the surfaces. Using (4.1.4), this procedure results in a pressure distribution of the form

$$p(h) = -Ah^{-n} + Bh^{-m}, \tag{4.1.11}$$

where A and B are positive constants and $m > n$, yielding a pressure variation curve similar to that shown Figure 4.2. For instance, for the case where the molecules of the two parallel surfaces interact via the Lennard-Jones potential, the resulting exponents in (4.1.11) are obtained from (4.1.3) and (4.1.4) as $n = 3$ and $m = 9$.

The equilibrium gap thickness δ for two particles colliding in a vacuum is defined as the value of separation distance h at which the attractive and repulsive (steric) forces between the surfaces are equal to each other, corresponding to the point $p(\delta) = 0$ in Figure 4.2. Aside from setting the value of δ, the steric repulsive force is not usually included in contact mechanics calculations. The reason for this is that

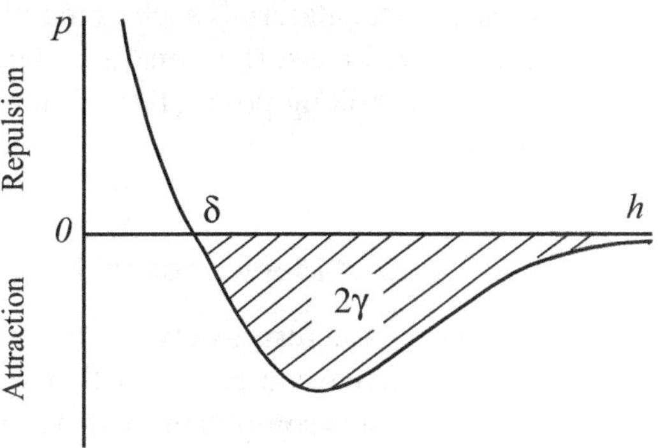

Figure 4.2. Plot showing the variation of pressure p as a function of distance h between two surfaces caused by the sum of the repulsive and attractive van der Waals forces. The equilibrium gap thickness δ is equal to the value of h at which p vanishes, and the cross-hatched area above the curve to the right of $h = \delta$ is equal to twice the surface energy density γ.

the steric force decays so quickly that if the separation distance h is even slightly larger than δ the steric force will be very small, whereas the longer range attractive force may have changed very little. The surface energy density γ is defined as half the work required to separate two surfaces with initial separation distance h per unit surface area. In Figure 4.2, the surface energy density corresponds to half the area under the curve of p versus h to the right of the equilibrium point $h = \delta$, as indicated by the cross-hatched region. For a liquid, the surface energy density is the same as the surface tension. The surface energy density of a solid reflects the nature of the bonds, either weak or strong, between the atoms which constitute it. For strong bonds between atoms or molecules, typical values of surface energy density in solids range between 100 and 500 mJ/m^2 for ionic crystals to 1,000–3,000 mJ/m^2 for many metals. For the weak van der Waals interactions between two surfaces separated by a vacuum-filled gap with separation distance δ, surface energy density can be obtained from (4.1.6) as

$$\gamma = \tfrac{1}{2}[w_{p-p}(\infty) - w_{p-p}(\delta)] = A/24\pi\delta^2, \tag{4.1.12}$$

where A is the Hamaker coefficient for the given material. For problems involving interaction of solids made of two different materials over a vacuum, the surface energy density can be approximated using the relationship (4.1.7) for the Hamaker coefficient as $\gamma \cong \sqrt{\gamma_1\gamma_2}$, where γ_1 and γ_2 are the surface energy densities of the individual materials.

The minimum separation distance δ for two macroscopic surfaces in contact should be substantially smaller than the intermolecular (or interatomic) center-to-center distance. For most materials, experimental studies have recommended values for δ ranging between 0.15 and 0.40 nm. However, use of the value $\delta = 0.165$ nm in (4.1.12) predicts the surface energy density in good agreement with measured values for many different solids and liquids. For example, for polystyrene and Teflon, using the Hamaker coefficients listed in Table 4.1 and $\delta = 0.165$ nm, the surface energy density is calculated as 32.1 and 18.5 mJ/m^2, respectively, and the corresponding experimental values are 33.0 and 18.3 mJ/m^2. This simple approximation works well for most materials, except strongly polar or hydrogen-bonding materials.

Interestingly, the surface energy of a material is dependent on the same intermolecular forces that determine the material's latent heat and boiling point. As a consequence, materials such as metals with high boiling points ($T_B > 2{,}000°C$) commonly have high surface energies, larger than about 1,000 mJ/m^2; meanwhile, water with a low boiling point (100°C) has a much lower surface energy density value of 73 mJ/m^2.

4.2. Contact Mechanics with van der Waals Force

The interaction forces between two noncharged contacting surfaces includes a long-range attractive component with a length scale of approximately 10 nm and a short-range repulsive component with a length scale of about 0.1 nm. The long-range (van der Waals) force is associated with a variety of effects, typically referred to as induction, orientation, and dispersion force, all of which are associated with molecular or atomic polarization effects. The dominant long-range force is the dispersion force, which is associated with the fact that the instantaneous positions of the electrons surrounding an atomic nucleus give rise to a polarized electric dipole field at any instant of time. This electric dipole field induces surrounding atoms to also become polarized, which leads these other atoms also to be surrounded by electric dipole fields. The dispersion force is associated with the electrostatic interaction of these induced dipoles with each other, leading to a dielectrophoretic attraction between the dipoles. The short-range repulsive force, often called steric or exchange repulsion, is associated with overlap of the electron clouds of two different atoms, which leads to strong electrostatic repulsion. This force is nearly zero outside of a distance approximately equal to the atomic radius.

4.2.1. Models for Normal Contact Force

A variety of simplified models exist to account for the effect of van der Waals adhesion on the combined adhesion-elastic force during collision of two particles. The appropriateness of these models depends primarily on the size of the length scale associated with elastic deformation of the particle relative to that associated with the adhesive force. From the Hertz solution, Equations (3.2.8) and (3.2.9), the normal overlap δ_N can be written in terms of the normal elastic force F_{ne} in the absence of adhesion as

$$\delta_N = \left(\frac{9}{16} \frac{F_{ne}^2}{E^2 R} \right)^{1/3}, \tag{4.2.1}$$

where E and R are the effective elastic modulus and radius defined in (3.1.1). For purposes of scaling analysis, the elastic normal force F_{ne} in (4.2.1) is taken as being of the same order of magnitude as the critical pull-off force F_C, which is shown in the following to be proportional to the product of R and the surface energy density γ. Setting $F_{ne} \propto R\gamma$ in (4.2.1) and ignoring the constants (because this is a scaling analysis) gives an estimate for the order of magnitude of the normal overlap caused by adhesive force between two particles as

$$\delta_N \propto \left(\frac{R\gamma^2}{E^2} \right)^{1/3}. \tag{4.2.2}$$

Figure 4.3. Schematic diagrams showing regions with strong adhesion forces acting between surfaces of colliding particles: (a) particles remain spherical and adhesion force acts along the close parts of the spherical surface (DMT approximation); (b) particles develop flattened contact region and adhesion forces act only within this region (JKR approximation).

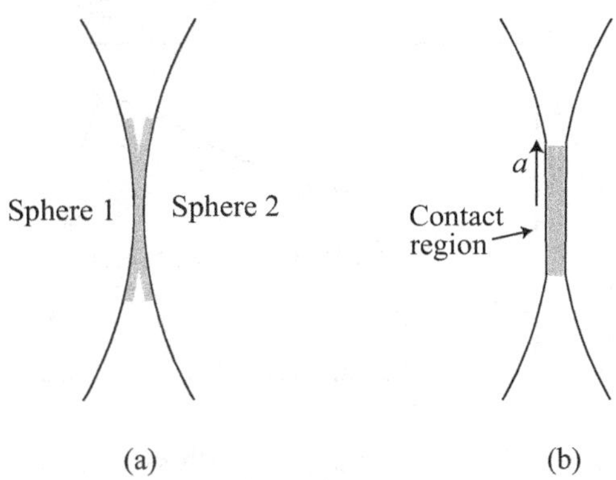

The result (4.2.2) yields a scaling for the elastic deformation of the interface of a spherical particle in contact with a second particle due to the adhesive force between the particles. A scaling for the characteristic length scale of the van der Waals adhesion force is given by the equilibrium gap thickness δ defined by (4.1.12). The ratio of these two length scales yields a dimensionless parameter called the *Tabor parameter* (Tabor, 1977), defined as

$$\lambda_T \equiv \left(\frac{4R\gamma^2}{E^2\delta^3} \right)^{1/3}. \tag{4.2.3}$$

The Tabor parameter is important for determining the extent to which elastic deformation of the colliding particles influences the adhesive force. Specifically, if $\lambda_T \ll 1$ it can be assumed that there is no effect of elastic deformation on the adhesive force, and so the Hertz elastic rebound force and the adhesive force for two spherical particles in contact can simply be added together to obtain the total force on the particle (Figure 4.3a). In the other extreme, if $\lambda_T \gg 1$ then the length scale of the adhesive force will be much smaller than that of the particle elastic deformation. In this case, the particles cannot be regarded as spheres when determining the adhesive force, but instead it can be assumed that the adhesive force only acts within the flattened contact region, as shown in Figure 4.3b. Within this contact region, the particle surfaces can be approximated as two infinite flat planes for purposes of determining the adhesive force. These two extremes are the basis of two leading theories of adhesive particle contact dynamics, the theory for which was developed respectively by Derjaguin, Muller, and Toporov (1975) for small λ_T (the DMT model) and by Johnson, Kendall, and Roberts (1971) for large λ_T (the JKR model). A model appropriate for intermediate values of λ_T was developed by Maugis (1992) based on an approximation to the adhesive potential similar to that suggested by Dugdale. The resulting Maugis-Dugdale (M-D) model includes the DMT model as a special case. In the extreme case of very small particles where the particle size approaches the length scale of the adhesive force, given by the equilibrium gap thickness δ, all of the models break down. In this case, particles can be assumed to be rigid (so that there is no elastic rebound force) and the full Lennard-Jones potential can be applied to each spherical particle as a function of distance away from the opposing particle. This extreme, called the Bradley model, violates our previous assumption that adhesion forces are not important until after particle collision. The

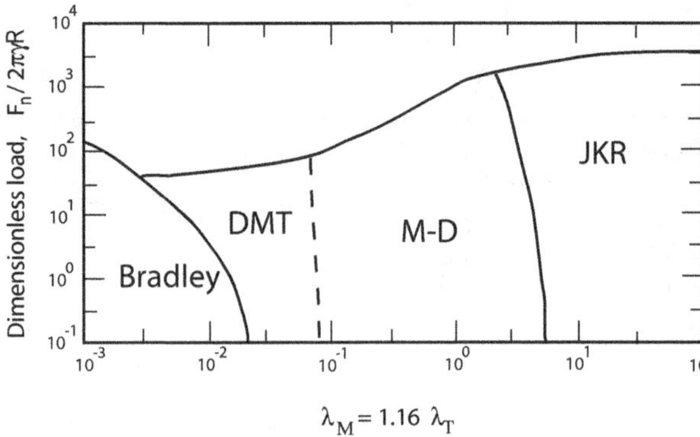

Figure 4.4. Adhesion map showing regimes of different adhesive contact models as a function of the dimensionless normal force and the Maugis parameter λ_M, which can be related to the Tabor parameter defined in (4.2.3) by $\lambda_M = 1.16\lambda_T$. [Based on data of Johnson and Greenwood (1997). Reprinted with permission.]

conditions required for the Bradley model to be valid are difficult to satisfy, even for very small particles in the nanoscale size range.

A map showing the regimes in which each of these adhesive contact models apply was developed by Johnson and Greenwood (1997), and a version of this map is shown in Figure 4.4. The regimes are organized in terms of the normal load F_n, normalized by the characteristic pull-off adhesive force $2\pi R\gamma$, and a parameter $\lambda_M = \sigma_0(9R/4\pi\gamma E^2)^{1/3}$, where σ_0 is an upper limit on the stress. Following Maugis (1992), Johnson and Greenwood (1997) chose the maximum stress as $\sigma_0 = 1.03(2\gamma/\delta)$, where δ is the equilibrium gap thickness obtained from the Lennard-Jones potential. This choice results in a relationship between λ_M and the Tabor parameter λ_T as $\lambda_M = 1.16\lambda_T$. The Hertz zone, in which adhesion is ignored, is defined on this map as the region where the adhesive force is less than 5% of the maximum elastic rebound force. The JKR region is taken to be that region where the material displacement δ_a caused by adhesive forces is large compared to the equilibrium gap thickness δ, where for the map in Figure 4.4 it was assumed that $\delta_a/\delta > 20$ in the JKR region. Similarly, the DMT region was characterized by small values of δ_a/δ, in this case $\delta_a/\delta < 0.05$. The Bradley model is assumed to apply when the elastic displacement δ_h associated with Hertz theory is much less than δ, where the criterion was selected as $\delta_h/\delta < 0.05$. Details of the DMT, M-D, and JKR adhesive contact models are discussed in the following sections.

DMT Model

The *Derjaguin approximation* (Derjaguin, 1934) relates the adhesion force $F_{s-s}(h)$ between two rigid spherical particles with radii r_1 and r_2 separated by a distance h to the interaction potential per unit area $w_{p-p}(h)$ between two planar surfaces with the same separation distance h, as given by (4.1.6). The approximation is based on discretizing the surface of each sphere into a series of concentric rings, each with width dr in the plane P normal to the line connecting the sphere centers, as shown in the close-up image in Figure 4.5. It is assumed that each of these rings, when projected onto the cross-sectional plane P, interacts with the opposing sphere in the same way as would a surface separated from an infinite plane by a distance Z, where Z is equal to the distance between the spheres at the ring centerline when evaluated along a line oriented normal to the plane P. If $F_{p-p}(Z)$ denotes the force per unit

Figure 4.5. Close-up figure showing the region between two spheres separated by a distance h. Indicated is a ring on the surface of Sphere 1 which when projected onto a plane P has width dr. The distance between the sphere surfaces at the centerline of the ring is denoted by $Z(r)$.

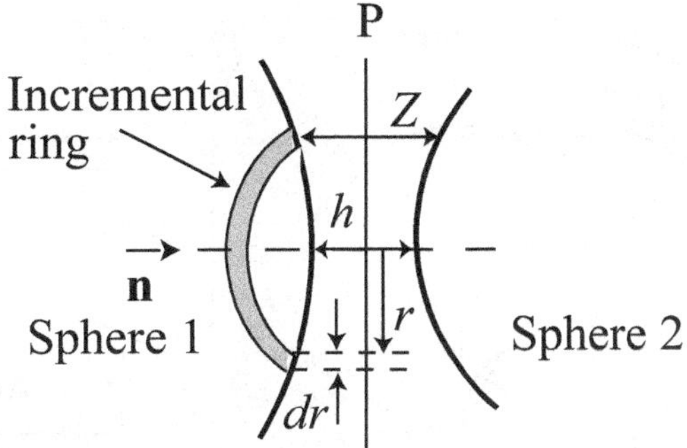

area exerted by a ring at separation distance $Z(r)$, then the total attractive force between the spheres can be written as

$$F_{s-s}(h) = \int_{Z=h}^{Z=\infty} 2\pi r F_{p-p}(Z) dr, \tag{4.2.4}$$

where r is the radial coordinate on the plane P. This approximation is valid provided that the distance h between the spheres is much less than either sphere radii. This limit will also allow us to approximate the separation distance $Z(r)$ between the spheres at radial distance r as

$$Z = h + \frac{r^2}{2}\left(\frac{1}{r_1} + \frac{1}{r_2}\right) = h + \frac{r^2}{2R}, \tag{4.2.5}$$

which is similar to the expression (3.1.4) but with the sign on the second term switched due to the fact that the particles do not yet overlap each other. Discretizing this expression with respect to r gives

$$dZ = (r/R)dr. \tag{4.2.6}$$

Substituting (4.2.6) into (4.2.4) gives

$$F_{s-s}(h) = 2\pi R \int_{Z=h}^{Z=\infty} F_{p-p}(Z) dZ. \tag{4.2.7}$$

Because $F_{p-p}(Z) = -dw_{p-p}/dZ$ and $w_{p-p}(\infty) = 0$, we can integrate (4.2.7) to obtain the final expression

$$F_{s-s}(h) = 2\pi R w_{p-p}(h). \tag{4.2.8}$$

If the separation distance is set equal to the equilibrium gap thickness δ for two spheres in contact, (4.1.12) and (4.2.8) can be used to write the sphere-sphere adhesion force as

$$F_{s-s}(\delta) = -4\pi R \gamma. \tag{4.2.9}$$

It is useful to define an effective radius c on the particle surface over which significant adhesion force acts. The force per unit area between two plane surfaces can be obtained from (4.1.6) as

$$p = -\frac{w_{p-p}}{dh} = -\frac{A_{12}}{6\pi h^3}. \tag{4.2.10}$$

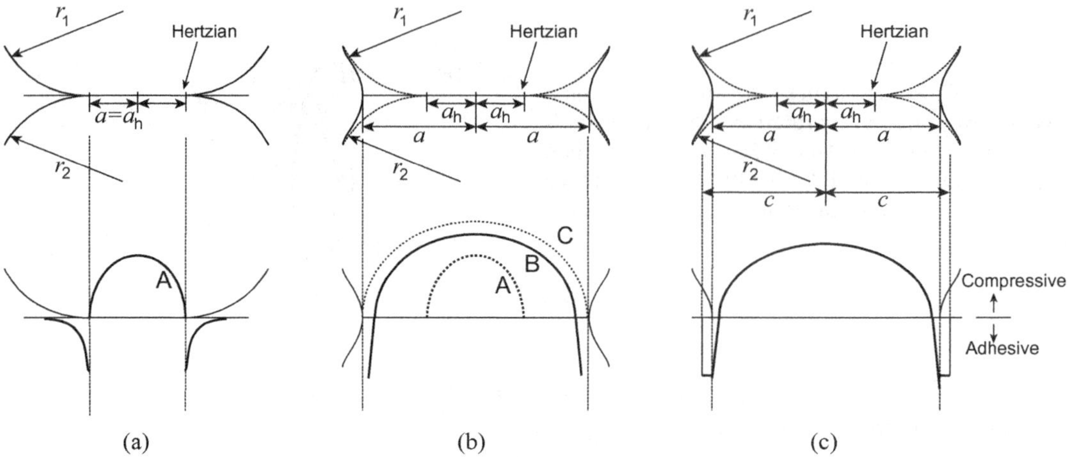

Figure 4.6. Schematic diagrams illustrating the stress distributions for various sphere-sphere adhesive contact models, including (a) DMT model; (b) JKR model; and (c) M-D model. The contact region radius is denoted by a and the radius that would be obtained with only the Hertz elastic forces is denoted by a_h. The radius of the region over which adhesive forces act is denoted by c.

Substituting (4.1.6) into (4.2.8) gives

$$F_{s-s} = -2\pi R \frac{A_{12}}{12\rho h^2}.$$ (4.2.11)

The effective area over which adhesion force acts is then obtained as

$$A_{eff} = \frac{F_{s-s}}{p} = \pi Rh.$$ (4.2.12)

Setting $A_{eff} = \pi c^2$ gives the effective adhesion force radius as

$$c = \sqrt{Rh}.$$ (4.2.13)

Recalling the solution (3.1.3) for contact region radius from the Hertz elastic theory as $a = \sqrt{R\delta_N}$ and taking the sphere separation distance h for contacting spheres to be on the order of magnitude of the equilibrium gap thickness δ gives

$$\frac{c}{a} = O\left(\frac{\delta}{\delta_N}\right)^{1/2} = O(\lambda_T^{-1/2}),$$ (4.2.14)

where we recall that the Tabor parameter $\lambda_T = O(\delta_N/\delta)$.

In Figure 4.6, different force profiles are sketched for the DMT, JKR, and M-D models of adhesive contact. The upper plots in this figure show the particle surface shape near the contact point (solid line) and the shape that the particle would have with only elastic forces (dashed line). The lower plots illustrate variation of the surface pressure as a function of radius in each model. The DMT model makes the assumption that the standard Hertz pressure distribution (3.2.1) applies within the contact region $r < a$ and that the adhesive force applies only outside of the contact region, within a region $a \geq r \geq c$. The net adhesive force can be obtained by evaluating an integral similar to (4.2.7) but with the lower limit of integration adjusted to equal the separation distance at the edge of the contact region, or $Z(a)$.

However, because (4.2.14) requires that $c/a \gg 1$ for cases where $\lambda_T \ll 1$, it follows that the contact region is very small compared to the region over which the adhesion force acts. Consequently, to first approximation the adhesive force can be evaluated for the entire sphere in (4.2.7). If we then add the Hertz elastic force to the adhesive force between two spheres, ignoring the elastic deformation in computation of the adhesion force, the total force acting on each particle of a colliding pair is given by

$$F_{ne} = \tfrac{4}{3} E a^3 / R - 4\pi R\gamma, \qquad (4.2.15)$$

where the contact radius $a = \sqrt{\delta_N R}$ in the Hertzian elastic rebound term. The pull-off force $F_{ne} = -F_C$ corresponds to that for which $a = 0$, which from (4.2.15) gives

$$F_C = 4\pi R\gamma. \qquad (4.2.16)$$

Equation (4.2.15) gives the relationship for the combined elastic and adhesive normal force in DMT theory. In comparison to numerical solutions based on integration of the Lennard-Jones potential, Muller et al. (1980) found that the DMT force expression compares well with the more exact results for $\lambda_T < 0.1$.

JKR Model

One of the earliest theoretical treatments of adhesive particle contact was proposed by Johnson, Kendall, and Roberts (1971) at the Cavendish Lab in Cambridge. The model is appropriate for relatively large, compliant particles for which the Tabor parameter λ_T is large compared to unity, implying that the length scale of elastic deformation is large compared to the length scale of the adhesive force. As a consequence, the JKR model assumes that the adhesive force acts only inside the contact region, such that $c \approx a$, as shown schematically in Figure 4.6b. The size of the contact region in the JKR model is larger than that in the nonadhesive Hertz model due to the additional compression caused by the adhesive force. The original derivation of JKR theory follows a type of thermodynamic approach. This approach involves estimation of the total energy U_T in the system as a function of contact radius a, and then determining the equilibrium conditions by setting $dU_T/da = 0$. The total energy of the system is made up of three terms – the stored elastic energy U_E, the mechanical energy U_M, and the surface energy U_S.

As shown in Figure 4.6b, under a normal loading $F_{ne} = F$, without the adhesive force there is a contact radius a_h predicted by the Hertz equation. The resulting expression for F from (3.2.8) and (3.2.9) is

$$F = \frac{4E a_h^3}{3R} = \frac{\kappa a_h^3}{R} = \kappa \sqrt{R}\,\delta_h^{3/2}, \qquad (4.2.17)$$

where $\delta_h = a_h^2/R$ and the coefficient $\kappa \; (\equiv 4E/3)$ is related to the stiffness K given in (3.2.9) by $\kappa = K/\sqrt{R}$. This condition corresponds to point A in the plot of normal load versus normal overlap in Figure 4.7.

If adhesive force is now introduced, the contact radius in the equilibrium condition will increase to a value a, which is larger than a_h, as shown in Figure 4.6b. The combined contact stresses at this new position arise from both adhesive and elastic forces, corresponding to point B in Figure 4.7. Finally, we can associate an effective

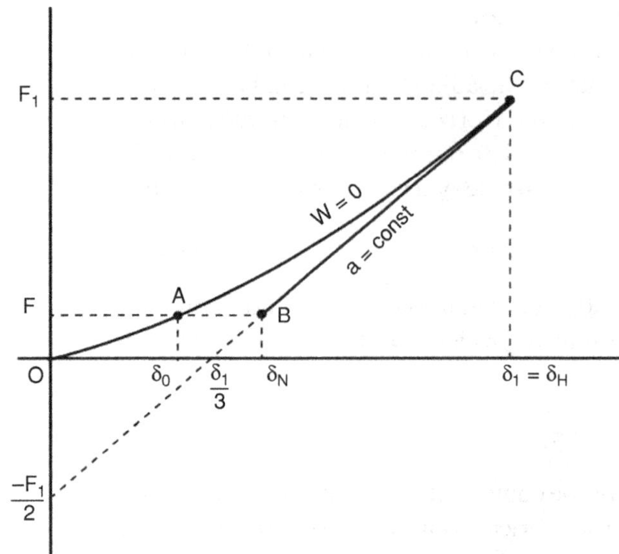

Figure 4.7. The loading path of JKR model. Hertz loading from O to C with $W = 0$, and then from C to B by keeping constant contact area a.

Hertz loading F_1 with the contact radius a (associated with point C in Figure 4.7), given by

$$F_1 = \frac{\kappa a^3}{R} = \kappa \sqrt{R} \delta_1^{3/2} \qquad (4.2.18)$$

where $\delta_1 = a^2/R$. The coordinates of points A, B, C in Figure 4.7 are (F, a_h, δ_h), (F, a, δ_N), and (F_1, a, δ_1), respectively. The stress profiles at points A, B, and C are plotted in Figure 4.6b, indicated by the different letters.

According to Figure 4.7, the JKR loading path can be decomposed into two steps: (1) a Hertz-type loading from point O to C with no adhesion ($\gamma = 0$), which requires energy U_{O-C}, and (2) unloading from point C to point B by keeping constant contact area a, releasing energy U_{C-B}. The energy required for the first step involving loading from O–C is given by

$$U_{O-C} = \int_0^{\delta_1} F d\delta_N = \frac{2}{5} \kappa R^{1/2} \delta_1^{5/2} = \frac{2}{5} \kappa \frac{a^5}{R^2}. \qquad (4.2.19)$$

The unloading process of B-C is similar to the problem of a flat punch that can be solved by Boussinesq theory (Maugis, 1999), giving

$$d\delta_N = \frac{2dF}{3a\kappa}. \qquad (4.2.20)$$

Integrating (4.2.20) and using (4.2.17) and (4.2.18) gives the overlap δ_N at Point B as

$$\delta_N = \delta_1 + \frac{2F - 2F_1}{3a\kappa} = \frac{a^2}{3R} - \frac{2a_h^3 a^{-1}}{3R}. \qquad (4.2.21)$$

Substituting (4.2.20) into (4.2.19), U_{C-B} can be calculated as

$$U_{C-B} = \int_{\delta_1}^{\delta} F d\delta_N = \int_{F_1}^{F} \frac{2F}{3a\kappa} dF = \frac{F^2 - F_1^2}{3a\kappa} = \frac{1}{3} \kappa \frac{a_h^6 a^{-1} - a^5}{R^2}. \qquad (4.2.22)$$

Using $U_E = U_{O-C} + U_{C-B}$, we obtain

$$U_E = \frac{1}{15}\kappa\frac{a^5}{R^2} + \frac{1}{3}\kappa\frac{a_h^6 a^{-1}}{R^2}. \tag{4.2.23}$$

The mechanical potential energy at point B under the normal force F is given by $U_M = -F\delta_N$. Applying (4.2.17) and (4.2.21) gives

$$U_M = -\frac{1}{3}\kappa\frac{a_h^3 a^2 + 2a_h^6 a^{-1}}{R^2}. \tag{4.2.24}$$

The surface energy U_S is given by

$$U_S = -2\pi\gamma a^2. \tag{4.2.25}$$

Adding the three different components of the energy, the total energy U_T is obtained as

$$U_T = \frac{1}{15}\kappa\frac{a^5}{R^2} - \frac{1}{3}\kappa\frac{a_h^6 a^{-1}}{R^2} - \frac{1}{3}\kappa\frac{a_h^3 a^2}{R^2} - 2\pi\gamma a^2. \tag{4.2.26}$$

Applying the equilibrium condition $dU_T/da = 0$ gives

$$\frac{dU_T}{da} = \frac{\kappa}{3a^2 R^2}\left[a^6 - 2\left(a_h^3 + 6\pi R\gamma\frac{R}{\kappa}\right)a^3 + a_h^6\right] = 0 \tag{4.2.27}$$

Using $F = a_h^3\kappa/R$ and $F_1 = a^3\kappa/R$, (4.2.27) can be written as

$$F_1^2 - 2(F + 6\pi R\gamma)F_1 + F^2 = 0. \tag{4.2.28}$$

Solving Equation (4.2.28) for F_1 and taking the positive sign gives

$$F_1 = F + 6\pi R\gamma + \sqrt{12\pi R\gamma F + (6\pi R\gamma)^2}. \tag{4.2.29}$$

Setting $F_{ne} = F$ and $F_1 = a^3\kappa/R$ gives an equation for the contact radius in the presence of adhesion as

$$a^3 = \frac{3R}{4E}[F_{ne} + 6\pi R\gamma + \sqrt{12\pi R\gamma F_{ne} + (6\pi R\gamma)^2}] \tag{4.2.30}$$

These results can be rearranged to obtain an expression for the normal overlap by dividing (4.2.28) by $6\pi R\gamma$ to write

$$\left(\frac{F_1}{6\pi R\gamma} - \frac{F}{6\pi R\gamma}\right)^2 = \frac{F_1}{3\pi R\gamma}. \tag{4.2.31}$$

Solving for $F - F_1$ from this equation and substituting into (4.2.21) gives

$$\delta_N = \delta_1 + \frac{2F - 2F_1}{3a\kappa} = \frac{a^2}{R} - \sqrt{\frac{16\pi a\gamma}{3\kappa}}. \tag{4.2.32}$$

The relation between the normal overlap and the contact radius can thus be written as

$$\delta_N = \frac{a^2}{R} - \sqrt{\frac{4\pi a\gamma}{E}}. \tag{4.2.33}$$

With zero applied load, the two colliding particles will approach an equilibrium state in which the elastic repulsion is balanced by the adhesive attraction of the particles. In this zero-load equilibrium state, the radius $a(t)$ of the contact region can be obtained by setting $F_{ne} = 0$ from (4.2.30) to obtain

$$a_0 = \left(\frac{9\pi \gamma R^2}{E} \right)^{1/3}. \tag{4.2.34}$$

For example, for two contacting particles with 2 μm radius ($R = 1$ μm), $E = 30$ GPa and $\gamma = 15$ mJ/m^2, the equilibrium contact radius a_0 is about 24 nm, accounting for only 2.4% of R.

Modified forms of these results, which are more convenient for DEM computations, were obtained by Chokshi et al. (1993) in which expressions for F_{ne} and δ_N are written in terms of the ratio a/a_0. The expression for F_{ne} is obtained by solving for F from (4.2.28) with $F_1 = a^3 \kappa / R$ and $F = F_{ne}$. The expression for δ_N is obtained simply by writing (4.2.33) in terms of a/a_0. The resulting system of equations is

$$\frac{F_{ne}}{F_C} = 4 \left(\frac{a}{a_0} \right)^3 - 4 \left(\frac{a}{a_0} \right)^{3/2} \tag{4.2.35}$$

and

$$\frac{\delta_N}{\delta_C} = 6^{1/3} \left[2 \left(\frac{a}{a_0} \right)^2 - \frac{4}{3} \left(\frac{a}{a_0} \right)^{1/2} \right]. \tag{4.2.36}$$

In these equations, F_C and δ_C are the critical force and overlap at the pull-off point assuming constant pulling force (or constant load). As the two particles move away, the contact will be maintained even for negative values of δ_N via *necking* of the particle material, until the critical pull-off point is reached, at which $F_{ne} = -F_C$ and $\delta_N = -\delta_C$. As the particles are pulled further apart, the contact will suddenly break. F_C and δ_C are given as

$$F_C = 3\pi R \gamma, \quad \delta_C = \frac{a_0^2}{2(6)^{1/3} R}. \tag{4.2.37}$$

The critical pull-off force F_C predicted by the JKR model is $3\pi R\gamma$, which is slightly lower than the value $4\pi R\gamma$ predicted by the DMT model. In the example of the 2 μm radius particles mentioned earlier, δ_C is only 0.16 nm, which is close to the minimum intermolecular center-to-center distance. Thus, although necking of the particle material occurs, it is in practice quite small for relatively stiff particles.

M-D Model

We have previously noted that the JKR model is suitable for "soft" materials with large surface energy and large particle size, corresponding to large values of the Tabor parameter λ_T. The DMT model is suitable for "hard" materials with a low surface energy and much smaller radius, corresponding to small values of the Tabor parameter. The adhesion map proposed by Johnson and Greenwood (1997) indicates roughly that the DMT model applies for $\lambda_T < 0.1$ while the JKR model applies for $\lambda_T > 3$.

An approximate theory valid for intermediate values of λ_T was developed by Maugis (1992), which includes the DMT and JKR theories as limiting cases. Due to the complexity of the Lennard-Jones potential for theoretical simulation, Maugis (1992) chose to use the much simpler *Dugdale approximation* of van der Waals

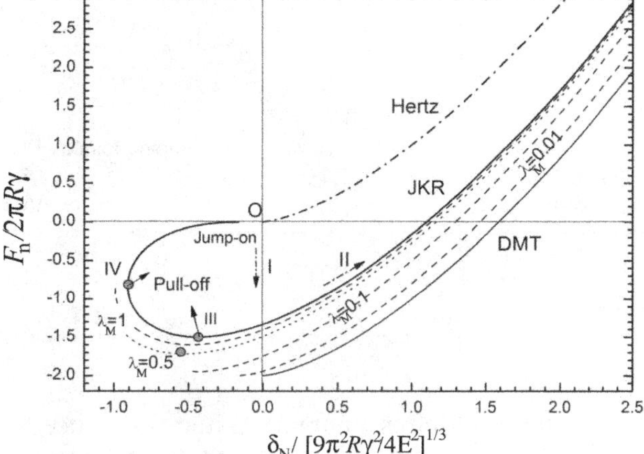

Figure 4.8. Relation between normal force and normal overlap for various adhesive contact models including JKR, DMT, and M-D models.

interactions. The resulting Maugis-Dugdale (M-D) model assumes a stress profile (shown in Figure 4.6c) that admits both repulsive and adhesive stresses in the contact region (like JKR), but assumes that the adhesive stress has a constant value σ_0 within the region outside of the contact region with radius $a \le r \le c$. Applying a step distribution of the adhesive stress within this outer region, the dimensionless normal force, normal overlap, and contact region radius in the M-D model are obtained as

$$\bar{F} = \bar{F}_h + \bar{F}_{ad} = \bar{a}^3 - \lambda_M\bar{a}^2[\sqrt{m^2 - 1} + m^2\cos^{-1}(1/m)] \tag{4.2.38a}$$

$$\bar{\delta}_N = \bar{\delta}_h + \bar{\delta}_{ad} = \bar{a}^2 - \frac{4}{3}\bar{a}\lambda_M\sqrt{m^2 - 1} \tag{4.2.38b}$$

$$\frac{\lambda_M\bar{a}^2}{2}[\sqrt{m^2 - 1} + (m^2 - 2)\cos^{-1}(1/m)]$$
$$+ \frac{4\lambda_M^2\bar{a}}{3}[\sqrt{m^2 - 1}\cos^{-1}(1/m) - m + 1] = 1 \tag{4.2.38c}$$

The normalization is done as $\bar{F} = F/2\pi R\gamma$, $\bar{a} = a[2E/3\pi R^2\gamma]^{1/3}$, and $\bar{\delta}_N = \delta_N[4E^2/9\pi^2R\gamma^2]^{1/3}$, and the subscripts h and ad represent contributions from the Hertzian elastic stress and the adhesive stress, respectively. The variable $m \equiv c/a$, where c is the radius over which adhesive force acts, and λ_M is the Maugis parameter, defined at the beginning of Section 4.2.1.

For given values of λ_M and $\bar{a}$, we can solve for m from (4.2.38c), and then substitute it into (4.2.38a) and (4.2.38b) to obtain $\bar{F}$ and $\bar{\delta}$. A plot of $\bar{F}$ versus $\bar{\delta}$ is given in Figure 4.8. When the pulling force is held constant, particle detachment occurs at $F_C = 3\pi R\gamma$ with $\bar{\delta}_N = 0.43$, corresponding to Point III in Figure 4.8. When the displacement is held constant, particle detachment occurs at $\bar{\delta}_N = 0.91$ with $F_{ne} = -1.66\pi R\gamma$, corresponding to Point IV in Figure 4.8. The force-displacement relationship for $\lambda_M \ge 0.5$ predicted by the M-D model agrees well with the prediction of the JKR model, while prediction of the M-D model for $\lambda_M \le 0.01$ is close to the prediction of the DMT model. The M-D model applies for all values of λ_M, but it is not commonly used in DEM computations because the solution of (4.2.38c)

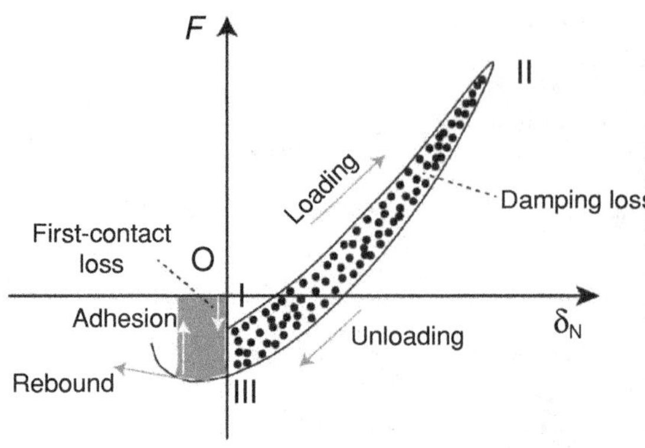

Figure 4.9. Loading curve illustrating sources of energy dissipation during the normal contact of adhesive spheres.

for *m* requires a numerical iteration procedure. However, an approximate fit to the numerical solutions of the M-D model has been proposed by Carpick et al. (1999), which gives analytical expressions for the force and contact area accurate to within about 1% of the M-D predictions for arbitrary values of λ_M.

For common materials, such as quartz in air, the work of adhesion due to van der Waals interaction is about 0.05J/m^2, the modulus parameter E is about 1.0 Gpa and the minimum separation δ is 0.165 nm, so the Tabor parameter λ_T is about 8.2 for $R = 1.0$ μm and 14.1 for $R = 5.0$ μm. This result suggests that the JKR model is most applicable for particles in the micrometer size range. Since $\lambda_M \propto R^{1/3}$, the Tabor (or Maugis) parameter decreases very slowly with particle radius and can still be fairly large even for particles in the nanometer size range. Application of the continuum JKR model (or any of these continuum models) becomes questionable as the particle size approaches 1 nm due to quantum effects for nanoparticles.

4.2.2 Normal Dissipation Force and Its Validation

The various normal contact models discussed in the previous section are developed for quasi-static deformations. Because particle collision processes are highly dynamic, the energy dissipation that arises due to dynamic effects must be incorporated into the normal elastic-adhesive model. Figure 4.9 illustrates the main energy dissipation processes that occur during the collision of adhesive particles. Both the viscoelastic damping force and the adhesive force contribute to energy loss. The energy loss from adhesion comes in part from the jump-on–pull-off behavior of the adhesive contact, which occurs at points with $\delta_N = 0$ and $\delta_N = -\delta_C$, respectively. At the jump-on point (Point I in Figure 4.9), the contact region area suddenly goes from zero to a finite value and the contact force suddenly goes from zero to a negative (adhesive) value. This sudden change leads to a *first-contact energy loss*, as indicated in Figure 4.9. The necking behavior of the material when the particles are pulled away from each other allows the adhesive force to act even when the overlap $\delta_N < 0$, leading to an energy loss indicated by the shaded region in Figure 4.9. The second part of the energy loss arises from the dynamic effect due to the viscoelasticity of materials, as discussed in Chapter 3.

Liu et al. (2011) examined the energy losses due to both adhesion and viscoelastic effects for an elastic material using the JKR approximation for normal pressure together with a two-component linear dashpot approximation, such as is given by

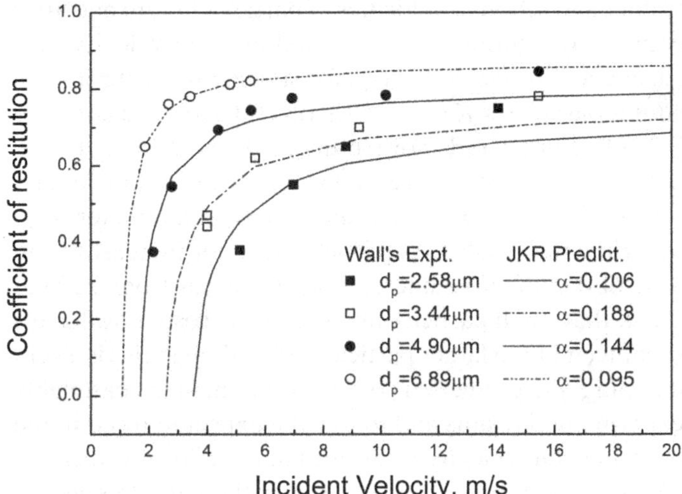

Figure 4.10. Coefficient of restitution versus normal incident velocity: comparison between the experiment (symbol) of Wall et al. (1990) for ammonium fluorescent microsphere impact with a flat surface and the prediction by the JKR model with linear damping given by Equation (4.2.40). [Reprinted with permission from Li et al. (2011).]

(3.3.1), for the attractive and repulsive components of the damping forces. This simple JKR-based damping model has been validated using classic particle/surface impact experiments such as those of Dahneke (1973) and Wall et al. (1990), in which both the coefficient of restitution and the critical sticking velocity of bouncing particles are well predicted. On the basis of the normal pressure distribution over the contact region predicted by the JKR model, Thornton and Ning (1998) used a cut-off corresponding to a limiting contact pressure for plastic deformation in order to derive expressions for the effect of plastic deformation on coefficient of restitution, critical sticking velocity, and so on for adhesive particle collision.

A particularly simple description of the normal collision of two particles with equal radius r_1 was given by Marshall (2009), which uses the JKR model result (4.2.35) for F_{ne} and the Tsuji et al. (1992) model given by (3.3.1) and (3.3.12) for the damping force F_{nd}. The governing equation for the normal overlap becomes

$$\frac{d^2\delta_N}{dt^2} + \alpha_{TTI}\sqrt{\frac{K}{M}}\delta_N^{1/4}\frac{d\delta_N}{dt} + \frac{4}{M}\left[\left(\frac{a}{a_0}\right)^3 - \left(\frac{a}{a_0}\right)^{3/2}\right] = 0 \qquad (4.2.39)$$

Dimensionless time, normal overlap, and contact region radius are defined by $\hat{t} = t(r_1^{1/2}K/M)^{1/2}$, $\hat{\delta} = \delta/r_1$ and $\hat{a} = a/r_1$, so that (4.2.39) and (4.2.36) reduce to

$$\frac{d^2\hat{\delta}}{d\hat{t}^2} + 2(2^{1/4})\alpha_{TTI}\hat{a}^{1/2}\frac{d\hat{\delta}}{d\hat{t}} + 4(2^{1/2})\hat{a}^3[1 - (\hat{a}_0/\hat{a})^{3/2}] = 0, \qquad (4.2.40a)$$

$$\hat{\delta} = 2\hat{a}^2\left[1 - \tfrac{2}{3}(\hat{a}_0/\hat{a})^{3/2}\right]. \qquad (4.2.40b)$$

The predictions of (4.2.40) are compared in Figure 4.10 with the experimental data reported by Wall et al. (1990) for ammonium fluorescein microspheres with diameters of 2–7 μm transported in an air jet at velocities ranging from 1 to 100 m/s and impacting on a flat surface (Li et al., 2011). The laser-Doppler technique was used for measuring both incident and rebound velocities of particles before and after impact. The relationship between restitution coefficient and incident velocity (i.e., the e-v_i curve) was obtained, and a critical velocity (v_c) for sticking was determined. In the experiments, the modulus parameter K ($= 4/3E$) is 1.80 GPa and the adhesive surface energy γ ($\approx \sqrt{\gamma_1\gamma_2}$) is 0.17 J/m^2. The Tabor parameter λ_T is estimated to be greater than 20 for all particle sizes used, so we are well within the range of validity of the JKR model. For each particle size examined, the damping parameter α_{TTI} is adjusted to best fit the experimental results. As particle diameter increases from 2.58 μm to 6.89 μm, the value of damping parameter α_{TTI} decreases from 0.206 to 0.095, corresponding to the fact that small particles in dispersed aerosol flow have a relatively lower restitution coefficient than larger particles. Even though this is a very simple theory for normal damping, the predictions of (4.2.40) remain in reasonably good agreement with the experimental results and follow the expected trend in the e-v_i curve. Particle impact with incident velocity v_i lower than the critical velocity v_c leads to a final capture of the particle (i.e., $e = 0$). The larger the particle size, the lower the value of v_c. For v_i values that are larger than, but still close to, v_c, the e-v_i curve exhibits a steep increase, but thereafter its slope rapidly decreases as v_i further increases until finally the restitution coefficient e asymptotes to a constant value.

4.2.3. Effect of Adhesion on Sliding and Twisting Resistance

Both sliding and twisting motions are relatively rare for small adhesive particles – rolling is generally the preferred deformation mode for agglomerates of adhesive particles (Dominik and Tielens, 1995, 1996; Marshall, 2007). It is therefore desirable to introduce relatively simple expressions for sliding and twisting resistance, similar to those used in the DEM model for nonadhesive particles introduced in Chapter 3. The standard sliding model for the case without adhesion is the spring-dashpot model proposed by Cundall and Strack (1979), for which the sliding force F_s is given by (3.5.5) when $|F_s| < F_{crit}$ and by the Amonton friction expression $F_s = -F_{crit}$ when $|F_s| \geq F_{crit}$. A detailed study of the effect of van der Waals adhesion on tangential sliding was given by Savkoor and Briggs (1977), and a relatively simple model was proposed by Thornton (1991) and Thornton and Yin (1991) which appears to agree reasonably well with experimental data. In this model, the only influence of van der Waals adhesion on sliding force is to modify the critical force F_{crit} at which sliding occurs, which in the presence of adhesion is given by

$$F_{crit} = \mu_F |F_{ne} + 2F_C|, \tag{4.2.41}$$

where F_C is the critical normal force given by (4.2.37) and μ_F is a friction coefficient. When particles are being pulled apart, the normal force approaches $-F_C$ at the point of separation, at which point the critical sliding force in (4.2.41) approaches $\mu_F F_C$.

The same model with twisting resistance as described in Chapter 3 can be used in the presence of adhesion, with the critical force F_{crit} given by the expression (4.2.41) used to obtain $M_{t,crit} = \frac{3\pi}{16} a F_{crit}$. For twisting moments with magnitude greater than $M_{t,crit}$, the torsional resistance is given by

$$M_t = -\frac{3\pi}{16} a\mu_f |F_n + 2F_C| \Omega_T/|\Omega_T|. \tag{4.2.42}$$

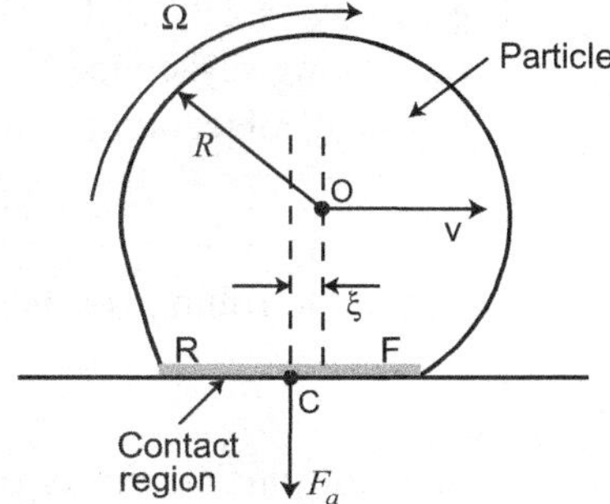

Figure 4.11. Schematic diagram illustrating lagging of the contact region behind a rolling particle due to adhesion force, such that the particle centroid O moves ahead of the centroid C of the contact region by an amount equal to the rolling displacement ξ.

4.2.4. Effect of Adhesion on Rolling Resistance

In Chapter 3, it was mentioned that the rolling resistance in low-velocity collisions of nonadhesive particles (e.g., granular materials) results mainly from the solid-phase viscoelastic deformation energy loss and the viscous fluid resistance. However, these kinds of classical sources of rolling friction such as microslip at the interface, inelastic or viscoelastic deformation of the involved materials, and large surface irregularities turn out to be less important for the micron or submicron particle sizes that are of primary concern in the fields of aerosol and colloidal science. The primary physical mechanism underlying rolling resistance for micron-sized and smaller particles arises from the asymmetry of the stress distribution resulting from the adhesion force.

A simple, but highly effective, model for adhesive rolling resistance was developed by Dominik and Tielens (1995). Although this model can be applied to the problem of any two particles colliding, it is most easily explained for the problem of a particle in contact with a plane surface. We recall with reference to Figure 4.6b that the stress distribution in the JKR model for normal collision of a particle with a flat surface is symmetric about the contact point, which is defined as the point on the surface directly below the particle centroid. As the particle starts to roll on the surface (illustrated in Figure 4.11), the part of the particle surface on the front (F) side of the particle moves downward and that on the rear (R) side of the particle moves upward. Due to the presence of adhesive force, the material points on the front side of the particle are continually "jumping on" to a state of contact with the planar surface, whereas material points on the rear side are continually "pulling off" of the surface. However, we know from Figure 4.8 that jump-on and pull-off do not occur at the same value of the normal overlap. Specifically, jump-on occurs approximately at $\delta_N = 0$, whereas pull-off occurs at $\delta_N = -\delta_C$. The fact that pull-off is delayed relative to jump-on due to the phenomenon of material necking gives rise to an asymmetry in the contact region during particle rolling. Specifically, the necking phenomenon allows the particle surface to remain in contact with the plate longer in the rear side of the particle than on the front side, causing the horizontal position of the center of the contact region to lag behind the horizontal position of the particle centroid by a distance ξ, called the *rolling displacement*, as illustrated in Figure 4.11.

The result (4.2.35) gives the total force in the JKR model as the sum of two terms, the first of which is the Hertz stress and the second of which is associated with the adhesive force, given by $F_a = 4F_C(a/a_0)^{3/2}$, where F_C is the critical pull-off force given by (4.2.37). The Dominik-Tielens model of adhesive rolling resistance gives the rolling resistance moment M_r as the product of the adhesive force F_a and the rolling displacement ξ, which serves as the lever arm, giving

$$M_r = -k_R \xi, \qquad (4.2.43)$$

where the rolling coefficient k_R is equal to F_a, or

$$k_R = 4F_C (a/a_0)^{3/2}. \qquad (4.2.44)$$

When a particle starts rolling from a state of rest, the contact region will initially remain fixed as the particle centroid moves by a distance ξ. When ξ increases to a critical value ξ_{crit}, the contact region will start moving at the same velocity as the particle centroid, but with the contact region centroid remaining behind the particle controid by a horizontal displacement equal to ξ_{crit}. Consequently, once $\xi \geq \xi_{crit}$, the rolling moment is given by

$$M_r = -k_R \xi_{crit}. \qquad (4.2.45)$$

The critical rolling displacement resistance ξ_{crit} can also be expressed in terms of a critical rolling angle $\theta_{crit} = \xi_{crit}/R$, where, for the example of a particle on a plane surface, R is equal to the particle radius.

The rolling displacement can be written in terms of a rolling velocity $\mathbf{v}_L$, defined in (3.6.13), as

$$\xi = \left(\int_{t_0}^{t} \mathbf{v}_L(\tau) d\tau \right) \cdot \mathbf{t}_R, \qquad (4.2.46)$$

where $\mathbf{t}_R = \mathbf{v}_L/|\mathbf{v}_L|$ is the direction of rolling. Experimental investigations of the effect of adhesion on particle rolling were presented by Peri and Cetinkaya (2005) and Ding et al. (2008). These demonstrate good agreement with the theoretical expressions and can be used to estimate typical values for the maximum rolling displacement ξ_{crit} for different materials. For instance, for polymer microspheres with diameter of nominally 8 μm, Ding et al. (2008) reported values of ξ_{crit} ranging between 70 to 245 nm, with corresponding values for the critical rolling angle ranging between 19 and 64×10^{-3} rad.

4.3. Electrical Double-Layer Force

The electrostatic force between two charged particles in a vacuum or in air obeys the inverse-square decay behavior expressed by Coulomb's law. Since this force decays slowly with distance and can act over distances of many times the particle diameter, it is considered a long-range interaction rather than an adhesive force. On the other hand, a charged particle immersed in an electrolyte (i.e., a solution of positive and negative ions) attracts a cloud of ions of opposite sign (counter-ions) around the particle, which act to screen the particle's charge. The resulting electrostatic potential decays rapidly to nearly zero within a few nanometers of the particle surface. If the size of the particle is on the micron scale, the presence of a particle charge together with electrolytic screening results in a short-range force with length scale much smaller than the particle diameter, which can easily be

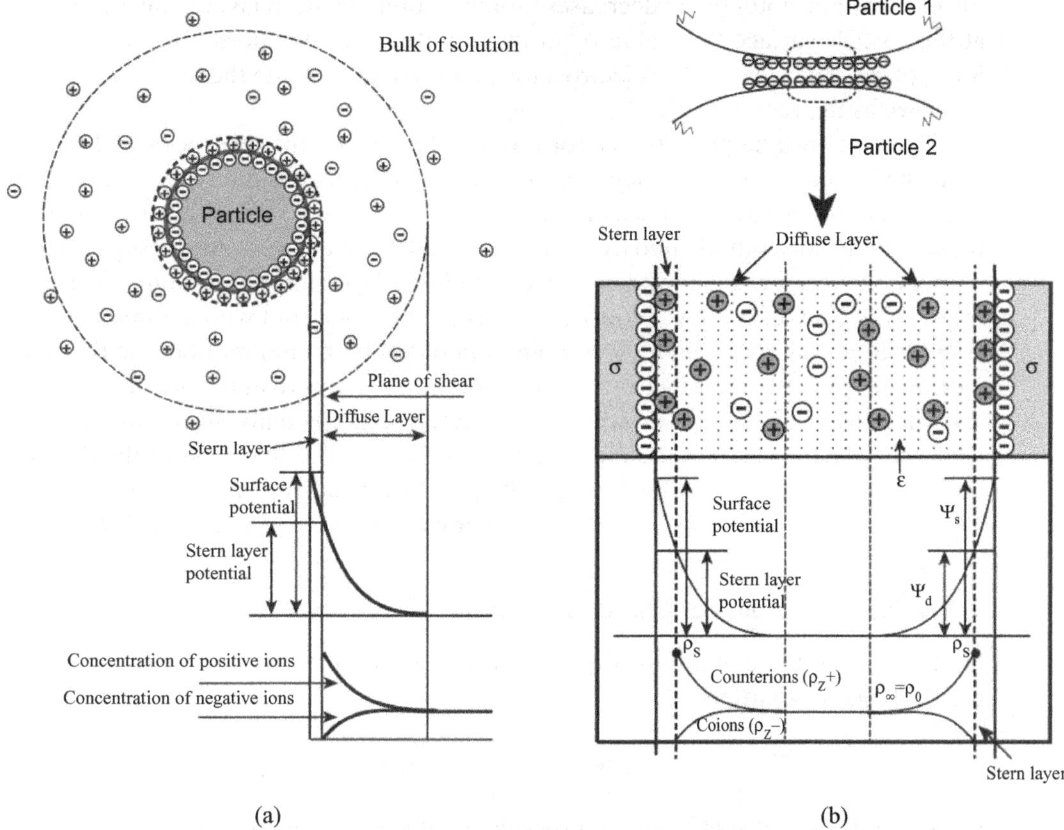

Figure 4.12. Electrical double layer theory: (a) isolated particle in solution; (b) two approaching particles in solution.

incorporated with the framework of adhesive contact models such as JKR or DMT. This current section, therefore, deals only with electrostatic force on a charged particle in an electrolytic solution, the so-called electrical double-layer force, and defers discussion of more general electrostatic forces until Chapter 8. Additional discussions of electrical double-layer forces can be found in books on colloid science and intermolecular forces, including those by Russel et al. (1989), Probstein (2003), Masliyah and Bhattacharjee (2006), and Israelachvili (1991).

4.3.1. Stern and Diffuse Layers

The free ions in the solution are either attracted to or repelled from the surface of a charged particle depending on the sign of the surface charge. For instance, if the particle surface is negatively charged, then positive ions within the solution will drift toward the particle, as shown in Figure 4.12. The motion of these positive ions toward the surface is limited by the finite ion size, so that only a fixed number of ions can fit in the layer immediately next to the surface. This layer of essentially motionless ions just outside of a charged surface is called the *Stern layer*. The width of the Stern layer is approximately equal to one ionic radius. Because the ions within the Stern layer are all of one sign, opposite to the sign of the particle surface charge, the electrostatic potential ψ decreases rapidly within the Stern layer from a value ψ_s at the particle surface to a value ψ_d at the outer edge of the Stern layer, called the Stern plane. The value of the electrostatic potential just outside the Stern layer (ψ_d) is known as the *zeta potential*.

The electrostatic potential is not reduced to zero by the fixed ions within the Stern layer, and as a consequence other counter-ions within the solution are also attracted to the particle surface. However, within this outer region the attraction of these counter-ions is mitigated by the thermal motion of the ions. As a result, a layer of mobile ions is formed in what is called the *diffuse layer*, located just outside of the Stern layer. The diffuse layer consists of ions of both signs, but with a dominance of counter-ions, that move under a combination of the bulk fluid motion, the thermal motion of the ions, and the attractive force driving the ions to and from the surface. The electrostatic potential decays approximately exponentially across the diffuse layer, approaching zero at the outer edge of this layer. The thickness of the diffuse layer is of the order of the Debye length λ_D, which is derived in the next section. The diffuse layer and the Stern layer together constitute the *electric double layer*.

4.3.2. Ionic Shielding of Charged Particles

Under equilibrium conditions, the ion concentration of the *i*th ion species follows the Boltzmann equation

$$n_i = n_{i\infty} \exp\left(-\frac{z_i e\psi}{k_B T}\right), \tag{4.3.1}$$

where z_i is the valence of the *i*th ion species, e is the elementary charge, ψ is the local value of the electrostatic potential, k_B is Boltzmann constant, and T is the absolute temperature. As an example, for the case of $CaCl_2$ salt, z_1 for the calcium ion is $+2$ and z_2 for the chloride ion is -1. The free charge density in the solution ρ_e can be expressed in terms of the ionic concentrations as

$$\rho_e = e \sum_i n_i z_i. \tag{4.3.2}$$

Equations (4.3.1) and (4.3.2) dictate how the electrostatic potential affects the free charge density ρ_e. However, the free charge density in turn changes the potential via Possion's equation

$$\varepsilon_f \nabla^2 \psi = -\rho_e, \tag{4.3.3}$$

where ε_f is the fluid electrical permittivity. Combining the three equations thus presented, we obtain the well-known Poisson–Boltzmann equation

$$\varepsilon_f \nabla^2 \psi = -\sum_i z_i e n_{i\infty} \exp\left(-\frac{z_i e\psi}{k_B T}\right). \tag{4.3.4}$$

When the surface potential is small, that is, ψ_s is below about 25 mV, the $z_i e\psi/k_B T$ term in (4.3.4) is much smaller than unity. The exponential term on the right-hand side of (4.3.4) can be linearized, which is known as the *Debye–Hückel approximation*, to obtain

$$\lambda_D^2 \nabla^2 \psi - \psi = 0. \tag{4.3.5}$$

The parameter λ_D has units of length and is known as the Debye length scale, defined by

$$\lambda_D = \left(\sum_i n_{i\infty} z_i^2 e^2 / \varepsilon_f k_B T \right)^{-1/2}. \tag{4.3.6}$$

The value of the Debye length scale is solely dependent on the properties of the solution, and it is independent of the surface charge or potential of the particle. For instance, for a symmetric (z:z) electrolyte solution (e.g., NaCl) with molar concentration M (mol/L), we obtain $n_+ = n_- = n_\infty = 1000 A_v M = 6.022 \times 10^{26} M$. Knowing e to be equal to 1.602×10^{-19} C and for water at $T = 298$ K and $\varepsilon_f = 6.95 \times 10^{-10}$ $\mathrm{C^2\,N^{-1}\,m^{-2}}$ (corresponding to a relative permittivity of $\varepsilon_r = 78.5$), the Debye length scale is obtained as

$$\lambda_D = \frac{0.304}{z\sqrt{M}} \times 10^{-9} \text{ m}. \tag{4.3.7}$$

Similarly, for (1:z) or (z:1) electrolytes with M (e.g. $CaCl_2$ or Na_2CO_3), we obtain

$$\lambda_D = \frac{0.304}{\sqrt{1+z}\sqrt{M}} \times 10^{-9} \text{ m}. \tag{4.3.8}$$

Consequently, the value of λ_D for a 0.01 M NaCl solution is 3.04 nm, and that of a 10^{-4} M NaCl solution increases to 30.4 nm. On the other hand, the value of λ_D for pure water at a pH of 7 is approximately 960 nm, which is sufficiently large that it might start to approach the size of the particles.

4.3.3. DLVO Theory

DLVO theory, which was developed independently during the Second World War by Derjaguin and Landau (1941) and Verwey and Overbeek (1948), examines the adhesive forces that arise when two charged particles collide with each other in an electrolytic solution. These forces result from a combination of van der Waals attractive force and an electrostatic repulsive force resulting from the overlap of the ionic diffuse layers of each particle. Because colloidal solutions usually deal with very small particles, with particle sizes ranging from approximately 0.1 to 1.0 μm, the treatment of adhesive and elastic forces in the standard DLVO theory follows the DMT model for adhesive contact mechanics. However, it is not difficult to extend the theory to apply to the JKR model for larger particles as needed.

We recall that the DMT model for adhesive contact is based on the Derjaguin approximation (4.2.8), which relates the adhesive force between two spheres with effective radius R and separation distance h to the interaction potential per unit area $w_{p-p}(h)$ between two planar surfaces separated by the same distance h, resulting in the expression $F_{s-s}(h) = 2\pi R w_{p-p}(h)$. This approximation is valid for arbitrary adhesive forces provided that the separation distance h satisfies $h \ll R$. In order to obtain $w_{p-p}(h)$, we consider the problem of an electrolyte in a gap of width h between two infinite planar surfaces, where the two surfaces are located at $x = 0$ and $x = h$. If we further consider the special case of an electrolyte consisting of two types of ions having equal and opposite valences (i.e., a z-z electrolyte), such as NaCl or KCl, the Poisson–Boltzmann equation (4.3.4) reduces to

$$\frac{d^2\psi}{dx^2} = \frac{2ez}{\varepsilon_f}n_\infty \sinh\left(\frac{ze\psi}{k_B T}\right), \tag{4.3.9}$$

where z and n_∞ denote the valance and the ambient ion concentration of the cation, respectively. When the dimensionless potential $\Psi \equiv ez\psi/k_B T \ll 1$, the $\sinh(\cdot)$ term in (4.3.9) can be linearized to obtain the one-dimensional Debye–Hückel approximation

$$\frac{d^2\psi}{dx^2} = \frac{2e^2 z^2 n_\infty}{\varepsilon_f k_B T}\psi = \kappa^2\psi, \tag{4.3.10}$$

where the inverse Debye length scale $\kappa \equiv 1/\lambda_D$ is a constant.

In solving these equations, we first consider the case of the diffuse layer along a single surface in an unbounded space (i.e., letting $h \to \infty$). In this case, the electrostatic potential within the diffuse layer is solved using the boundary conditions

$$\psi(0) = \psi_s, \quad \psi(\infty) = 0. \tag{4.3.11}$$

An exact solution to (4.3.9) is given by

$$\Psi(x) = 2\ln\left(\frac{1 + \exp(-\kappa x)\tanh(\Psi_s/4)}{1 - \exp(-\kappa x)\tanh(\Psi_s/4)}\right), \tag{4.3.12}$$

where Ψ_s is the dimensionless surface potential. For small surface potentials ($\Psi_s \ll 1$), (4.3.12) reduces to the solution of the Debye–Hückel approximation (4.3.10), given by

$$\Psi(x) \sim \Psi_s \exp(-\kappa x). \tag{4.3.13}$$

For large positive surface potentials ($\Psi_s \gg 1$), $\tanh(\Psi_s/4) \cong 1$ and (4.3.12) can be approximated by

$$\Psi(x) \sim 2\ln\left(\frac{1 + \exp(-\kappa x)}{1 - \exp(-\kappa x)}\right), \tag{4.3.14}$$

from which it follows that the potential is independent of the surface potential at a sufficient distance away from the surface. For this problem, a near-surface correction must be introduced in order to satisfy the boundary condition at $x = 0$. For large values of x, such that $\kappa x \gg 1$, the general solution (4.3.12) can be approximated using the binomial theorem and the Taylor series approximation $\ln(1 + z) \sim z$ for small z as

$$\Psi \sim 4\exp(-\kappa x)\tanh(\Psi_s/4). \tag{4.3.15}$$

This result demonstrates that the electrostatic potential always decays exponentially with x sufficiently far from the surface.

We now turn to the problem of determining the force per unit area that is exerted on two infinite parallel planar surfaces separated by a distance h whose diffuse layers overlap each other, as shown in Figure 4.13. The momentum equation

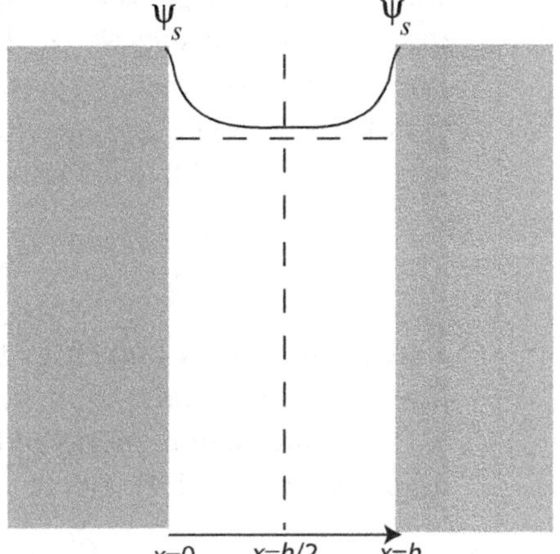

Figure 4.13. Sketch of the gap between two parallel charged surfaces in an electrolyte, showing the midplane and the variation of the electric potential ψ.

for this one-dimensional problem in the absence of fluid motion or gravity reduces to

$$\frac{dp}{dx} = \rho_e E_x, \tag{4.3.16}$$

where ρ_e is the free charge density defined by (4.3.2) and $E_x = -d\psi/dx$ is the x-component of the electric field vector. The product on the right-hand side of (4.3.16) is simply the electric field body force per unit mass acting on the fluid. Substituting (4.3.1) and (4.3.2) for ρ_e into (4.3.16) and writing $E_x = -d\psi/dx$ gives

$$\frac{dp}{dx} + \sum_i e z_i n_{i\infty} \exp\left(-\frac{z_i e \psi}{k_B T}\right) \frac{d\psi}{dx} = 0. \tag{4.3.17}$$

Integrating this equation over x and using a point in the bulk solution at which $\psi \to 0$ as a boundary condition yields a solution for pressure of the form

$$p(x) = k_B T \sum_i n_{i\infty} \left[\exp\left(-\frac{z_i e \psi}{k_B T}\right) - 1 \right]. \tag{4.3.18}$$

The result (4.3.18), when combined with the expression (4.3.1) for ion concentration, leads to the *contact-value theorem* (Israelachvili, 1991, 225), which gives the pressure within the gap as proportional to the ionic concentration evaluated at the midplane $x = h/2$, or

$$p = k_B T \left(\sum_i n_{mi}(h) - \sum_i n_{mi}(\infty) \right), \tag{4.3.19}$$

where the first term in parentheses in (4.3.19) is the sum of the midplane ionic concentration when the plates are a distance h apart and the second term is the sum of the ionic concentrations in the bulk electrolyte. The ionic concentration is related to the electric potential ψ by the Boltzmann equation (4.3.1), so for a (z:z) electrolyte we can write

$$p = k_B T n_\infty \left[\left(\exp\left(-\frac{e z \psi_m}{k_B T}\right) - 1 \right) + \left(\exp\left(\frac{e z \psi_m}{k_B T}\right) - 1 \right) \right] \simeq \frac{z^2 e^2 \psi_m^2 n_\infty}{k_B T}, \tag{4.3.20}$$

where the last expression assumes that the dimensionless electrostatic potential at the midplane satisfies $\Psi_m \ll 1$. The first term in brackets in (4.3.20) represents the cations and the second term is the anions, and the symbols z and n_∞ in this equation denote the valence and ionic concentration of the cations. Substituting the far-field solution (4.3.15) for $\psi(x)$ into (4.3.20) gives

$$p = 64 n_\infty k_B T e^{-\kappa h} \tanh^2 \left(\frac{ez\psi_s}{4k_B T} \right). \tag{4.3.21}$$

This expression is known as the *weak overlap approximation* because it uses the far-field approximation for the electrostatic potential.

The interaction energy of the surfaces $w_{p-p}(h)$ can be obtained by integrating p as indicated in (4.1.4), giving

$$w_{p-p}(h) = \int_h^\infty p(h)dh = (64 n_\infty k_B T / \kappa) e^{-\kappa h} \tanh^2 \left(\frac{ez\psi_s}{4k_B T} \right). \tag{4.3.22}$$

This result can be used to obtain the electrical repulsion force during collision of two charged spherical particles with effective radius R and separated by a distance h in an electrolyte simply by applying the Derjaguin approximation (4.2.8) to write

$$F_{s-s}(h) = 2\pi R w_{p-p}(h) = (128\pi R n_\infty k_B T / \kappa) e^{-\kappa h} \tanh^2 \left(\frac{ez\psi_s}{4k_B T} \right). \tag{4.3.23}$$

For small values of the dimensionless surface potential Ψ_s (corresponding to approximately $\psi_s < 25$ mV), a simpler form of (4.3.23) can be derived by linearizing the $\tanh(\cdot)$ term and using (4.3.6) to write $\kappa^2 = 2n_\infty z^2 e^2 / \varepsilon_f k_B T$, giving

$$w_{p-p}(h) \cong 2\varepsilon_f \psi_s^2 \kappa \exp(-\kappa h). \tag{4.3.24}$$

The corresponding sphere-sphere interaction force for this case is given by

$$F_{s-s}(h) \cong 4\pi \varepsilon_f \psi_s^2 \kappa R \exp(-\kappa h). \tag{4.3.25}$$

The electric double-layer interaction is quite different from the van der Waals interaction discussed in the previous section. Whereas the interaction potential between two plane surfaces for van der Waals force decays with distance in proportion to h^{-2}, the double-layer interaction potential decays exponentially on a length scale equal to the Debye length λ_D. Figure 4.14 plots the interaction energy per unit area for two parallel plane surfaces due to the electrical double layer alone, the van der Waals force plus the short-range repulsive force alone, and the combination of all three forces in a (1:1) electrolytic solution for different values of the surface electrical potential ψ_s. Because $\lambda_D \propto M^{-1/2}$, the DLVO interaction energy depends mainly on the electrolyte concentration M and the surface potential ψ_s.

The upper solid curve in Figure 4.14 represents a well-stabilized colloidal system, typical of a high surface potential in a dilute electrolyte. The long Debye length causes a strong long-range repulsive force that may extend to several nanometers, as far as 10 nm. The large energy barrier caused by the electrical repulsive force prevents the approach of two colloidal particles from short-range contact. If this barrier is larger than the thermal energy $k_B T$ of the colloidal particles, the system should be always stable in the absence of external forces (e.g, fluid shear forces), and there

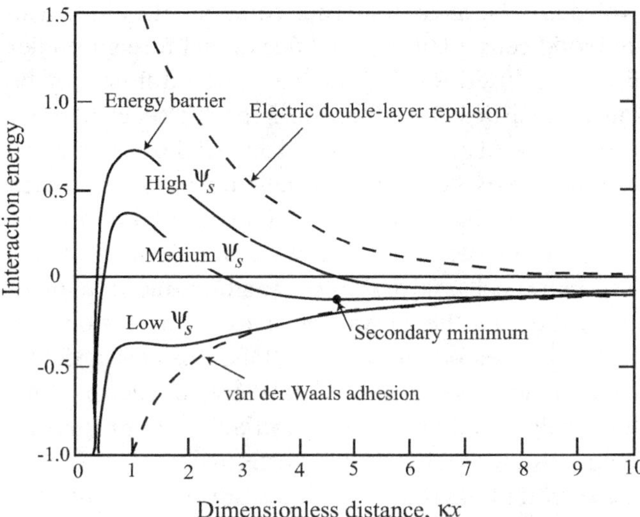

Figure 4.14. Schematic showing interaction energy per unit area between two parallel plates as a function of separation distance for the electrical double-layer alone (upper dashed line), for van der Waals attraction force alone (lower dashed line), and for the combined DLVO theory with different values of the surface potential.

will be little particle agglomeration. In such cases, computational methods such as Brownian or Langevin dynamics might be appropriate for modeling the system.

The lower solid curve in Figure 4.14 represents cases with low surface potential values in which the energy barrier falls below zero. In these cases, the electrical double-layer repulsive barrier is so small that the suspension will be unstable. In this regime, the colloidal particles will coagulate rapidly, as they will be attracted to the deep attractive "primary" energy minimum occurring at separation $h = \delta$ (not shown) where the attractive van der Waals force and the short-range repulsive force are in balance. DEM is suitable for prediction of the coagulation rate of the particles and for examining the microstructures of particle agglomerates or aggregates, provided that the condition $\lambda_D \ll R$ is met. In the presence of a fluid flow, the agglomerates will be broken up by fluid shear at the same time as they agglomerate following collisions.

There is also a transition region between the extremes of a stable and an unstable suspension in which a second energy minimum occurs, as illustrated by the middle solid curve in Figure 4.14. This secondary energy minimum is usually located several nanometers away from the peak of the energy barrier, and it is typically observed in cases where both the surface potential and the electrolyte concentration are relatively large. If the energy barrier is too high for a given colloidal particle to overcome, it may either be attracted to the weaker secondary minimum or remain totally dispersed in the solution. This condition is known as *kinetically stable*.

4.4. Protein Binding

A particularly interesting area of application of the DEM approach is in the area of transport of biological flows containing cells, such as blood. In addition to plasma and red blood cells (RBCs), which are the most numerous components, blood also contains various types of white blood cells, platelets, and occasional foreign bodies such as cancerous cells and liposomes used for drug delivery. Blood flow may be examined at a mesoscale level using a particle flow framework in which each of these discrete components is modeled as a type of particle (Longest and Kleinstreuer, 2003; Chesnutt and Marshall, 2009). Adhesion of these different cells and particles to each other and to the endothelial cells lining the blood vessel walls is of critical importance for many biological processes. For instance, in blood flows cell adhesion governs processes such as clotting of red blood cells, thrombosis formation due to platelet activation, white cell adhesion to the endothelial wall prior to extravasation into surrounding tissues, and formation of red blood cell rouleaux. Diseases such as sickle-cell disease or high shear flow due to the presence of artificial elements such as stints or artificial heart valves can adversely modify cell vessel wall adhesion properties, leading to serious health problems. Adhesion of bacterial cells is necessary for the growth of bacterial diseases in surroundings where surfaces are exposed to fluid flow. This is of particular importance in biofilms, which contribute to infections related to prosthetic heart valves, artificial joints, and catheters (Smith, 2005). The redistribution of cancer cells from one location to another, known as *metastasis*, is correlated with a reduction in the ability of tumor cells to adhere to each other (Cavallaro and Christofori, 2001).

Since blood particles (as we will loosely refer to the discrete entities suspended in blood) typically have radii in the range of 1 to 5 μm and are fairly compliant, the basic framework for the adhesive contact described by the JKR model is applicable to these problems. Although both van der Waals and electrical double-layer forces are present for these biological problems, the dominant adhesion force typically arises from phenomena such as entanglement in polymeric networks, osmotic forcing and protein binding. One prevalent mechanism for cell adhesion is the so-called ligand-receptor binding process, in which ligand proteins attached to the membrane of one cell bond to receptor proteins attached to the membrane of a second cell. The bonding occurs by intermolecular forces, such as ionic or hydrogen bonds and van der Waals forces, rather than by covalent bonds. The characteristic length scale over which protein binding occurs is on the order of 10 nm, which is similar to the length scale for van der Waals adhesion. Unlike many other adhesion forces, the protein connections involved in ligand-receptor binding are often specific to particular cell types. Opposing the attractive force from the ligand-receptor bonds is a nonspecific repulsive force due both to the fact that cell surfaces are negatively charged (and therefore repel each other) and to the steric repulsion caused by the presence of a layer of negatively charged glycoproteins coating the cell membranes (called the glycocalyx). An illustration of these components of cell adhesion is shown in Figure 4.15.

An early model of cell adhesion by ligand-receptor binding was the kinetic model proposed by Bell (Bell, 1978; Bell et al., 1984), which assumes that the force exerted by each protein bond can be modeled as a spring with spring constant σ and equilibrium length x_o. The energy released by compressing or stretching a bond to a length x_b is

$$E_s = -\frac{\sigma}{2}(x_b - x_0)^2. \tag{4.4.1}$$

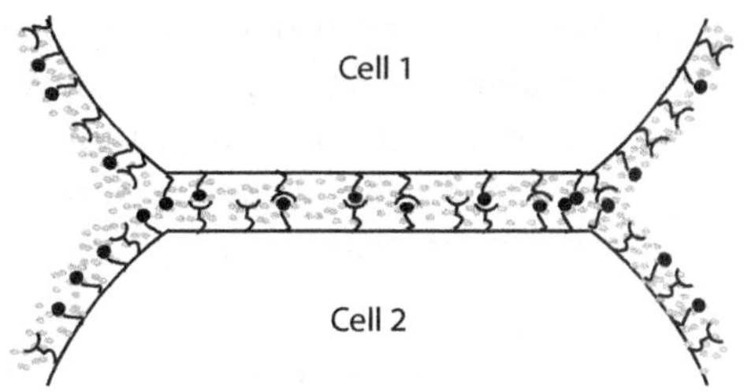

Figure 4.15. Schematic diagram showing ligand and receptor proteins (in black), and long-chain molecules that make up the glycocalyx layer (in grey).

The protein bonds are assumed to break at some maximum bond length, such that the adhesive force vanishes when the bond length exceeds the contact region gap size δ. The total energy associated with receptor-ligand bonds per unit area of the membrane is the product $N_b E_s$ of the single-bond energy E_s and the number density $N_b(t)$ of bonds joining the cell membranes. The formation and elimination of bonds is modeled by a kinetic equation for the bond number density $N_b(t)$ given by

$$\frac{dN_b}{dt} = k_f(N_{\ell 0} - N_b)(N_{r0} - N_b) - k_r N_b, \tag{4.4.2}$$

where N_{r0} and $N_{\ell 0}$ are the initial receptor and ligand densities on the membrane, respectively, and k_f and k_r are the forward and reverse reaction rate coefficients. The reaction rate coefficients vary with the length of the bond in accordance with (Dembo et al., 1988)

$$k_f = k_{f0}\exp\left[-\frac{\sigma_{ts}(x_b - x_0)^2}{2k_B T}\right], \quad k_r = k_{r0}\exp\left[\frac{(\sigma - \sigma_{ts})(x_b - x_0)^2}{2k_B T}\right], \tag{4.4.3}$$

where k_{f0} and k_{r0} are the initial forward and reverse equilibrium reaction rates, σ_{ts} is the "transition state spring constant," k_B is the Boltzmann constant, and T is the absolute temperature.

Because the gap thickness in the JKR model is assumed to remain approximately constant during the binding process, the rate coefficients in (4.4.3) are constants and the rate equation (4.4.2) has the form of a Ricotti equation for the number density $N_b(t)$. If an initial condition is assumed to be a state where the two cells have just collided, such that there are no bonds at the initial time t_0, an analytical solution for $N_b(t)$ is obtained as (Chesnutt and Marshall, 2009)

$$N_b(t) = \frac{2A\tanh\left[\frac{1}{2}(B^2 - 4AC)^{1/2}(t - t_0)\right]}{-B\tanh\left[\frac{1}{2}(B^2 - 4AC)^{1/2}(t - t_0)\right] + (B^2 - 4AC)^{1/2}}, \tag{4.4.4}$$

where $A \equiv k_f N_{\ell 0} N_{r0}$, $B \equiv -[k_f(N_{\ell 0} + N_{r0}) + k_r]$, and $C \equiv k_f$. The coefficients are restricted by the condition $B^2 - 4AC > 0$, which is satisfied for all cases provided that $N_{\ell 0} = N_{r0}$. The value of $N_b(t)$ asymptotes at long time to an equilibrium value $N_b(\infty)$ given by

$$N_b(\infty) = 2A/\left[-B + (B^2 - 4AC)^{1/2}\right]. \tag{4.4.5}$$

We make the usual JKR approximation that the adhesive force acts only over a circular contact region of radius $a(t)$, over which the surface separation thickness is taken as constant. Integration of the receptor-ligand binding energy per unit area, $N_b E_s$, over the contact region yields the net adhesion binding energy for a particle-particle collision event as

$$U_S(t) = 2\pi E_s \int_0^{a(t)} N_b(t, r) r \, dr = -\pi\sigma(x_b - x_0)^2 \int_0^{a} N_b(t, r) r \, dr. \tag{4.4.6}$$

The bond number density depends on radius r through the impact time $t_0(r)$, which is defined as the time at which the particle surfaces at radius r within the contact region first approach to within the equilibrium separation distance δ. The time scale for bond formation τ_B can be approximated from (4.4.4) as $\tau_B = 2/(B^2 - 4AC)^{1/2}$. When τ_B is much smaller than the contact time scale τ_C, the additional approximation can be made that the surface impact time $t_0(r)$ is nearly uniform over the contact region, such that (4.4.6) reduces to

$$U_S = -\frac{\pi\sigma}{2}(x_b - x_0)^2 N_b(t) a^2(t). \tag{4.4.7}$$

Recalling that 2γ is the work per unit area required to separate two surfaces, where γ is the surface energy density, it follows that $U_S = \pi a^2(2\gamma)$. Consequently, (4.4.6) is equivalent to assuming a time-dependent effective surface energy density $\gamma(t)$ as

$$\gamma(t) = -\frac{1}{2}\sigma(x_b - x_0)^2 \left[\int_0^1 N_b(t - t_0(s)) s \, ds \right], \tag{4.4.8}$$

which for $\tau_B \ll \tau_C$ reduces to the equilibrium value

$$\gamma = -\tfrac{1}{4}\sigma(x_b - x_0)^2 N_b(\infty). \tag{4.4.9}$$

Using the modified surface energy density expression (4.4.8) or (4.4.9), the expressions for contact forces of two cells colliding under ligand-receptor binding are the same as the expressions for normal force and sliding, twisting, and rolling resistance discussed in Section 4.2. Ligand-receptor binding can therefore be treated mathematically as a van der Waals adhesive contact process with time-dependent surface energy density, under the context of the JKR adhesive contact model.

There have been a variety of extensions to the above ligand-receptor binding model. A number of investigators have noted that the receptor proteins can move along the interface during the bonding process. A model for this motion was developed by Tozeren et al. (1989), who add a diffusive term to the right-hand side of (4.4.2). A model of this type was used by Agresar (1996) for two-dimensional simulations of the collision of two cells, whereas other simulation approaches neglect protein mobility (N'Dri et al., 2003). Experimental work on measurement of receptor-ligand binding kinetics was reported by Chesla et al. (1998) and Tachev et al. (2000). Competition between multiple ligand-receptor species was discussed by Zhu and Williams (2000) and Coombs et al. (2004), and a more comprehensive review of receptor-ligand binding was given by Zhu (2000).

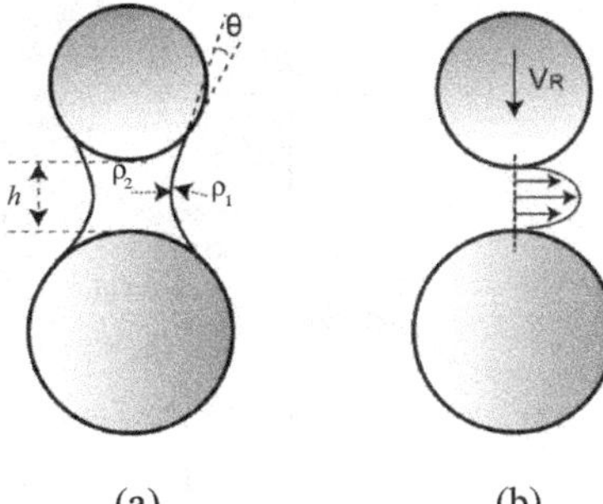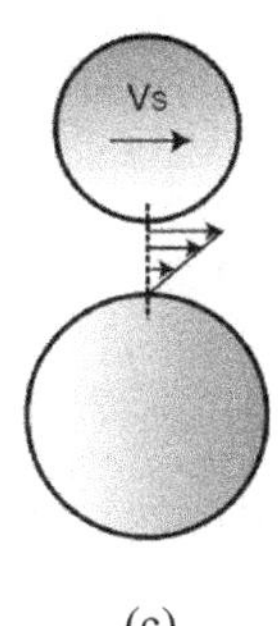

Figure 4.16. Liquid bridge forces for two elastic spheres: (a) capillary cohesion; (b) viscous effect due to squeezing; (c) viscous effect due to shearing.

4.5. Liquid Bridging Adhesion

In a humid environment, each particle of an aerosol or granular flow is surrounded by a thin liquid film due to condensation of water from the atmosphere. The addition of even a small amount of liquid in a dry particle system can lead to strong cohesive force between the particles, significantly changing the particle system behavior (Halsey and Levine, 1998; Fiscina et al., 2010). When two particles with thin liquid films collide, the liquid films join to form a "liquid bridge" stretching between the particles. This liquid bridge introduces a capillary force F_{cap} that pulls the two particles toward each other, leading to adhesion of the particles, as shown Figure 4.16a. In addition, the liquid film introduces an enhanced frictional force F_{visc} between the particles in both the normal and tangential directions due to the higher viscosity of the liquid filling the contact region compared to the surrounding gas, as illustrated in Figure 4.16b and c. The total normal liquid-bridge force is given by the sum of the capillary force and the viscous force. Computation of particle collision with liquid-bridging adhesion requires models for both the capillary and viscous forces on the particles, as well as a criterion for rupture of the liquid bridge at a critical separation distance. Much of the earlier literature examined the capillary force and bridge rupture condition for two static spheres, good summaries of which were given by Mehrotra and Sastry (1980) and Lian et al. (1993). The significance of the viscous force arising from relative motion of the two particles on the normal force and rupture condition was pointed out by Ennis et al. (1990).

4.5.1. Capillary Force

The problem of the capillary force F_{cap} acting between two equally sized spherical particles with radius r_p is examined. It is assumed that an axisymmetric liquid bridge connects the two spheres, with separation distance $h(t)$ between the spheres. The problem is described in terms of a cylindrical polar coordinate system (r, z). The radius of curvature of the liquid-gas interface profile in the r-z plane is denoted by ρ_1, which is assumed to be constant along the liquid-gas interface, and the radial position of the liquid bridge at the "neck" (or mid-plane between the spheres) is ρ_2. The solid-liquid contact angle is θ and the so-called half-filling angle is ϕ, as shown in Figure 4.17a.

In a static condition, an equal and opposite force acts on each particle which arises from two distinct effects: (1) the vertical component of the surface tension force

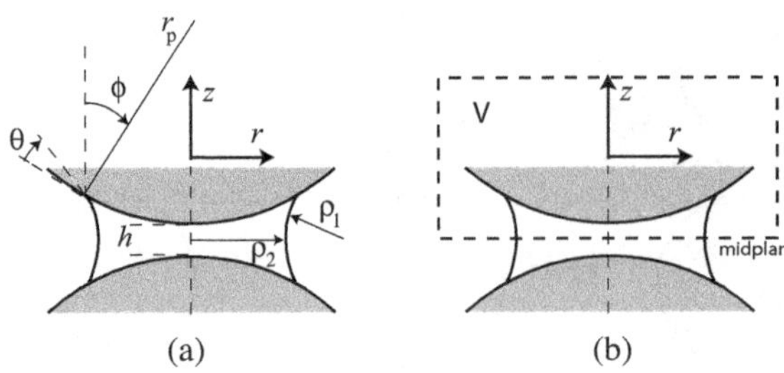

Figure 4.17. Close-up sketches showing (a) geometrical parameters for liquid bridge between two spherical particles of the same radius and (b) control volume V used for the "gorge approximation."

on each particle and (2) the difference in hydrostatic pressure across the liquid-gas interface due to the interface curvature. We refer to the sum of these two effects as the capillary force, which is given by

$$F_{cap} = -2\pi r_p \sigma \sin\phi \sin(\phi + \theta) - \pi r_p^2 \Delta p_I \sin^2\phi, \qquad (4.5.1)$$

where $\Delta p_I = p_{liq} - p_{gas}$ is the pressure jump over the gas-liquid interface and σ is the liquid-gas surface tension. We follow the convention that a positive force is repulsive as has been used in the rest of the book. The Young-Laplace formula gives

$$\Delta p_I = 2\sigma\bar{\kappa}, \qquad (4.5.2)$$

where $\bar{\kappa} \equiv 0.5(\rho_1^{-1} - \rho_2^{-1})$ is the mean curvature of the surface. The negative sign before the ρ_2^{-1} term in this equation results from the fact that the center of curvature for the tangent circles corresponding to ρ_1 and ρ_2 are on opposite sides of the interface. An approximate solution to (4.5.1) can be obtained using the so-called "gorge approximation" (Hotta et al., 1974; Lian et al., 1993). This approximation uses the control volume V shown in Figure 4.17b, which passes through the midplane between the spheres and encloses one of the spheres. Using this control volume, the force on the sphere can be written in terms of the pressure force and surface tension force exerted on the midplane as

$$F_{cap} = -2\pi\sigma\rho_2 - \pi\rho_2^2 \Delta p_I. \qquad (4.5.3)$$

Substituting (4.5.2) into (4.5.3) gives

$$F_{cap} = -\pi\sigma\rho_2 \left[1 + \frac{\rho_2}{\rho_1}\right]. \qquad (4.5.4)$$

The radii of curvature at the midplane can be written in terms of the angles ϕ and θ as (Pitois et al., 2000)

$$\rho_1 = \frac{h/2 + r_p(1 - \cos\phi)}{\cos(\phi + \theta)}, \qquad \rho_2 = r_p \sin\phi - [1 - \sin(\phi + \theta)]\rho_1. \qquad (4.5.5)$$

The static contact angle θ is a material property of the liquid-gas-solid system. The half-fill angle ϕ depends on the liquid bridge volume. For the general case of liquid bridges of arbitrary volume, it is necessary to use an iterative method to solve for ϕ for the given liquid volume, which can be used in (4.5.5) to obtain ρ_1 and ρ_2, and then substituted into (4.5.4) to obtain the capillary force. Mathematical details are provided in the work by Lian et al. (1993) and Hotta et al. (1974).

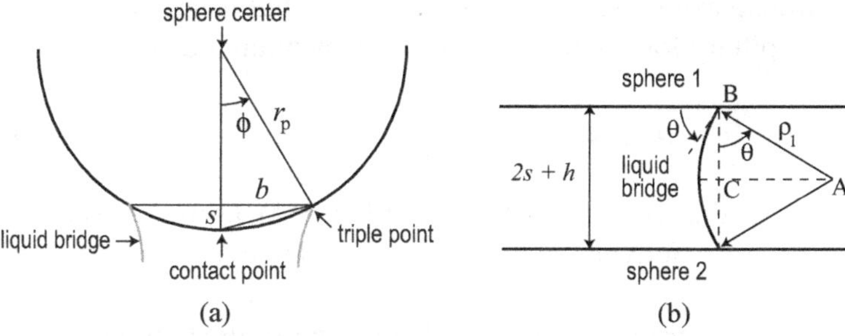

Figure 4.18. Sketches illustrating geometrical arguments for liquid bridge force derivation with small fill angle. (a) Relationship between length scales s and b. (b) Close-up of region between spheres, with sphere surfaces approximated as parallel lines.

A particularly useful limiting case is that where the volume of the liquid bridge is very small, so that the radial location $r = b$ of the liquid-solid-gas triple point on the particle satisfies $b \ll r_p$. This case corresponds to the small fill-angle approximation $\phi \ll 1$, and consequently $\rho_2 \ll r_p$. With reference to Figure 4.18a, the triple-point radius b and height s relative to the contact point can be written in general as $b = r_p \sin \phi$ and $s = r_p(1 - \cos \phi)$. For $\phi \ll 1$, a Taylor series expansion can be used to approximate these quantities to leading order in ϕ as

$$b = r_p \phi, \quad s = r_p \phi^2/2. \tag{4.5.6}$$

Eliminating ϕ from these two equations gives the useful formula

$$\pi b^2 \cong 2\pi r_p s. \tag{4.5.7}$$

A second useful formula can be obtained using the geometrical argument illustrated in Figure 4.18b. The top and bottom horizontal lines in this figure each represent a horizontal line passing through the triple point on each particle. The distance between these parallel lines is equal to $2s + h$. For small ϕ, the horizontal line is nearly tangent with the sphere surface at the triple point, so the angle between the horizontal line and the liquid-gas interface is approximately equal to the contact angle θ. Under this approximation, the angle at vertex B of the right triangle ABC shown in Figure 4.18b must also be equal to θ, with the consequence that the distance between the horizontal lines is given by

$$2s + h \cong 2\rho_1 \cos \theta. \tag{4.5.8}$$

For sufficiently small separation distance $h = O(s)$, it follows from (4.5.7) that

$$\rho_1/\rho_2 = O(s/b) = O(\phi). \tag{4.5.9}$$

Consequently, for small liquid volumes with $\phi \ll 1$, the force arising from the pressure reduction $\Delta p_l \cong \sigma/\rho_1$ dominates the surface tension force, as well as the pressure reduction force associated with the azimuthal curvature $1/\rho_2$, and the capillary force expression (4.5.4) reduces to

$$F_{cap} = -\pi \sigma \rho_2^2/\rho_1. \tag{4.5.10}$$

From (4.5.5) with $\rho_1 \ll \rho_2$, we can write $\rho_2 \cong r_p \sin\phi = b$, so that using (4.5.7) and (4.5.8) an expression for capillary force with small fill angle is obtained as

$$F_{cap} = -2\pi r_p \sigma \cos\theta \left(1 + \frac{h}{2s}\right)^{-1}. \qquad (4.5.11)$$

This is the same expression as that derived by Maugis (1999, 315); however, Maugis used a derivation based on a thermodynamic approach related to crack propagation theory (see also Maugis, 1987).

It is convenient to write the parameter s in (4.5.11) in terms of the liquid volume V_L per contact. If $Z(r)$ denotes the height between the sphere surfaces at a given radial position r, given by (4.2.5) for $h \ll r_p$, then V_L can be written as

$$V_L = 2\pi \int_0^b Z(r)r\,dr = 2\pi \int_0^b (h + r^2/r_p)r\,dr = \pi b^2 h + \frac{\pi b^4}{2r_p}, \qquad (4.5.12)$$

where we have used (4.5.7) to write $s = b^2/2r_p$. Equation (4.5.12) is a quadratic equation for b^2, which yields the solution

$$b^2 = hr_p \left[-1 + \left(1 + \frac{2V_L}{\pi h^2 r_p}\right)^{1/2}\right]. \qquad (4.5.13)$$

Substituting $s = b^2/2r_p$ into (4.5.11) with b^2 given by (4.5.13), an expression for capillary force between two equal spheres is obtained as

$$F_{cap} = -2\pi r_p \sigma G_f \cos\theta, \qquad (4.5.14)$$

where the coefficient G_f is defined by

$$G_f \equiv 1 - \left[1 + 2V_L/\pi r_p h^2\right]^{-1/2}. \qquad (4.5.15)$$

The term in brackets in (4.5.15) approaches zero as $h \to 0$, implying that $G_f \to 1$ for cases where the particles are touching. While we have derived the expression (4.5.14) for interaction of two equal spheres, it can readily be generalized to the case of interaction of two arbitrarily sized spheres simply by replacing $2s$ in (4.5.8) with the sum $s_1 + s_2$ of the triple-point heights from the two spheres. The resulting general expression has the same form as (4.5.14) and (4.5.15) with r_p replaced by $2R$, where R is the effective radius defined by (3.1.1). The liquid bridge volume can be normalized as $V^* = V_L/R_{12}^3$, where the harmonic mean diameter R_{12} is equal to twice the effective radius ($R_{12} = 2R$). Figure 4.19 compares the prediction from (4.5.14) with experimental data from Pitois et al. (2000) and Willett et al. (2000). Both theory and experiment show a monotonic decay in the capillary force as the separation distance $h(t)$ is increased. The expression (4.5.14) is found to be reasonably accurate both for collisions of particles with equal radii and for collisions of particles with unequal size over the full range of separation distances, up to the rupture distance.

Adding the capillary force (4.5.14), the Hertz contact force (3.2.8), and the dissipation force (3.3.1), the total normal contact force is given by

$$F_n = -4\pi R\sigma G_f \cos\theta + \tfrac{4}{3} E\sqrt{R}\delta_N^{3/2} - \eta_N \mathbf{v}_R \cdot \mathbf{n}. \qquad (4.5.16)$$

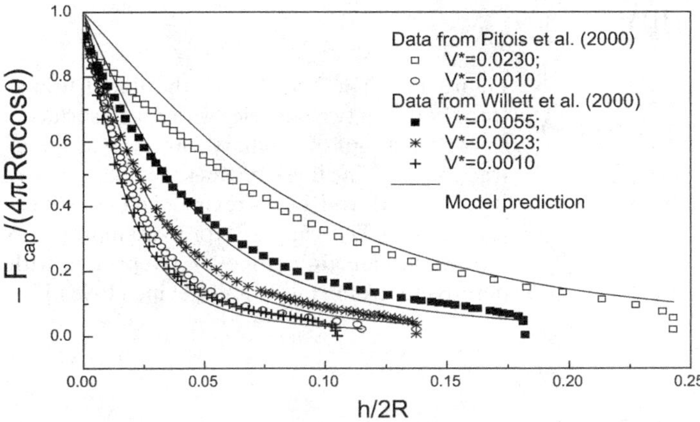

Figure 4.19. Comparison of predictions from Equation (4.5.16) with experimental data from Pitois et al. (2000) and Willett et al. (2000) for normalized liquid bridging force as a function of separation distance between two spheres. [Reprinted with permission from Li et al., 2011.]

The prediction (4.5.16) for normal contact force in the presence of a liquid bridge is plotted in Figure 4.19. The parameters used in the simulation are consistent with a case examined experimentally by Willett et al. (2000). The particle has a diameter of 2.4 mm, which is sufficiently large that van der Waals force is negligible. The liquid surface tension $\sigma = 0.206$ N/m, and the nondimensional liquid volume $V^* = 0.001$. Considering that the elastic modulus of most solids is of magnitude 10^9 Pa, the difference of length scale between deformation and adhesion is several orders of magnitude.

In order to compare liquid bridge adhesion with van der Waals adhesion, we also predict the JKR model results using the force expression (4.2.35) with the surface energy $\gamma = 0.206$ J/m^2 and no liquid bridge adhesion (Li et al., 2011). Results of two computations are shown in Figure 4.20, one with $E = 10^5$ Pa and one with $E = 10^6$ Pa. It is found that for $\delta_N > 0$, the force predictions with the combined theory given by (4.5.16) are very close to the JKR results from (4.2.35) with no liquid bridge. This agreement arises because the Hertz elastic rebound force dominates either the van der Waals or liquid-bridge adhesion forces in this regime. On the other hand, for $\delta_N < 0$, indicating that the spherical shapes of the particles do not overlap, the JKR

Figure 4.20. Prediction for the normal contact force in the presence of a liquid bridge using (4.5.16), with $d = 2.4$ mm and $\sigma = 0.206$ N/m. [Reprinted with permission from Li et al., 2011.]

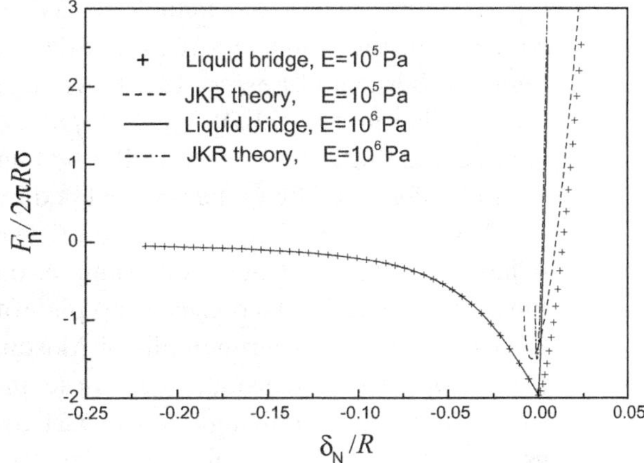

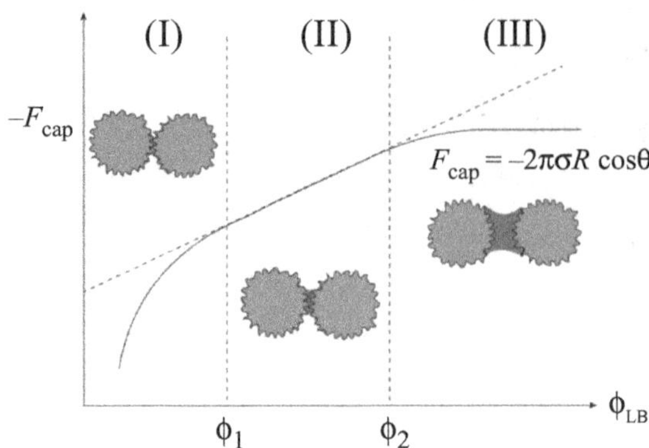

Figure 4.21. The behavior of the liquid bridge force between two particles with rough surfaces, shown on a plot of capillary force versus liquid fraction ϕ_{LB}. The three regimes include: I, asperity regime; II, roughness regime; and III, spherical regime. The capillary force asymptotes to a constant value in regime III. [Reprinted with permission from Halsey and Levine (1998).]

predictions and the predictions from the combined theory given by (4.5.16) deviate strongly. Specifically, JKR theory predicts that the particles will reach a critical state and snap apart from each other at a relatively small separation distance on the order of $h/R = -\delta_N/R \cong 0.01 - 0.02$. The case with liquid bridging, on the other hand, exhibits significant force out to $h/R = -\delta_N/R \cong 0.2$.

4.5.2. Effect of Roughness on Capillary Cohesion

The capillary force acting between two particles with rough boundaries is more complex than for smooth particles for cases where the liquid film thickness is less than the size of the surface asperities. The particle surface roughness can be characterized by a mean radial height ℓ_r and lateral separation distance ℓ_d of the surface asperities. We denote by ϕ_{LB} the ratio of the volume of liquid to the particle volume. Three kinds of liquid-bridge regimes are identified as a function of ϕ_{LB}: (1) the *asperity regime* ($\phi_{LB} < \phi_1$), where the cohesive force depends primarily on surface roughness in the neighborhood of the contact point; (2) the *roughness regime* ($\phi_1 \leq \phi_{LB} < \phi_2$), where the Laplace pressure in the liquid bridge between surface asperities is set by the asperity height; and (3) the *spherical regime* ($\phi_2 < \phi_{LB}$), where the liquid film thickness is larger than the size of the surface asperities and surface roughness no longer significantly affects the cohesive force. As illustrated in Figure 4.21, the cohesive force between the particles depends approximately on the cube root of the liquid volume fraction ϕ_{LB} in the asperity regime, it increases approximately linearly with ϕ_{LB} in the roughness regime, and it approaches a constant in the spherical regime (Halsey and Levine, 1998). Scaling estimates by Halsey and Levine (1998) indicate that in general, $\phi_1 \approx \ell_r\ell_d^2/V_p$ and $\phi_2 \approx \ell_r^2 R/V_p \approx (\ell_r/R)^2$, where V_p is the volume of a single particle. A similar scaling was observed in an experimental study of effect of a liquid film on a spouted bed by Zhu et al. (2013).

  The different regimes of rough particle interaction result in distinct macroscopic behaviors of the particles. For example, the effect of different liquid fractions on the critical angle θ_C of a conical sand pile was examined theoretically by Halsey and Levine (1998) and experimentally by Mason et al. (1999). Within the asperity regime ($\phi_{LB} < \phi_1$), the critical angle does not seem to be significantly affected by the liquid surface tension σ, implying that the particle behavior is much the same as for a dry granular system. In the spherical regime ($\phi_{LB} \geq \phi_2$), the cohesive forces between

Figure 4.22. Sketch illustrating the cylindrical region of radius b (shaded) used to approximate the liquid bridge for calculation of the viscous force.

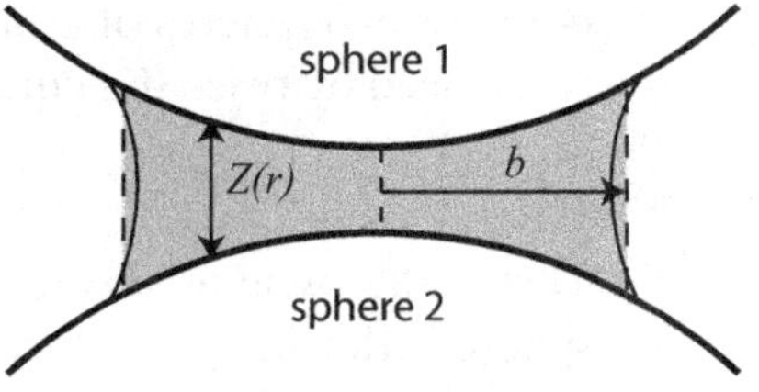

two particles are essentially the same as those between two smooth particles. In the roughness regime ($\phi_1 < \phi_{LB} < \phi_2$), the cohesive force due to the capillarity is approximately linear in the volume of the liquid bridge, so that using (4.5.14) we can write F_{cap} approximately as

$$F_{cap} = -4\pi R\sigma G_f \cos\theta \frac{\phi_{LB}}{\phi_2}. \tag{4.5.17}$$

4.5.3. Viscous Force

In addition to the capillary force, there is an additional viscous force as two particles move toward or away from each other due to the greater viscosity of the liquid compared to the surrounding gas. Following the approach of Matthewson (1988), the viscous force is evaluated using an approximation of the liquid bridge as a cylindrical region with radius b shown in Figure 4.22. If $Z(r)$ is the distance between the sphere surfaces at radius r, the pressure within the liquid bridge is given by the lubrication equation (2.4.1),

$$\dot{h} = \frac{1}{12\mu_L r}\frac{\partial}{\partial r}\left[rZ^3\frac{\partial p}{\partial r}\right], \tag{4.5.18}$$

where $\dot{h}$ is negative for approaching spheres. Integrating twice over r, the pressure within the liquid bridge is obtained as

$$p(r,t) - p_0 = 6\mu_L \dot{h}\int_b^r \frac{\xi}{Z^3(\xi)}d\xi. \tag{4.5.19}$$

For small separation distances $Z = h + r^2/r_p$, so performing the integration (4.5.19) gives

$$p(r,t) - p_0 = 3\mu_L r_p \dot{h}\int_{Z(b)}^{Z(r)} \frac{dZ}{Z^3} = \frac{3}{2}\mu_L r_p \dot{h}\left[\frac{1}{Z^2(b)} - \frac{1}{Z^2(r)}\right]. \tag{4.5.20}$$

Integrating the pressure over the region with area πb^2 wetted by the liquid bridge gives

$$F_{visc} = 2\pi\int_0^b [p(r) - p_0]rdr = -\frac{3\pi}{2h}\mu_L r_p^2 \dot{h}[1 - h/Z(b)]^2 \tag{4.5.21}$$

From (4.5.13) and (4.5.15), it follows that $1 - h/Z(b) = G_f$, so that (4.5.21) becomes

$$F_{visc} = -\frac{3}{2}\pi\mu_L r_p^2 \frac{\dot{h}}{h}G_f^2. \tag{4.5.22}$$

The result (4.5.22) is the same as the lubrication force expression (2.4.14) for unbounded spaces with the exception of the factor G_f. A similar derivation can be done for two spheres of arbitrary size, for which case the sphere radius r_p in (4.5.22) is replaced by twice the equivalent radius R and (4.5.22) becomes

$$F_{visc} = -6\pi\mu_L R^2 G_f^2 \dot{h}/h. \qquad (4.5.23)$$

This is the same as the formula derived by Matthewson (1988) for collision of a sphere with a flat plate, for which problem R is equal to the sphere radius.

The ratio of the viscous force to the capillary force is given by

$$\frac{F_{visc}}{F_{cap}} = \frac{3}{2}\frac{R}{h}\frac{\mu_L \dot{h}}{\sigma}\frac{G_f}{\cos\theta}. \qquad (4.5.24)$$

The ratio $\mu_L \dot{h}/\sigma$ is simply the capillary number Ca for the particle collision. The viscous force becomes particularly important for high approach velocities (large Ca) and small separation distance (small h/R).

4.5.4. Rupture Distance

The liquid bridge adhesion forces apply only up to the critical separation distance h_{rupt} at which the bridge ruptures. Lian et al. (1993) numerically computed the rupture distance in a static case by minimization of the free energy subject to the constraint of constant liquid volume. This paper notes that the computational results are well approximated by a simple approximate expression of the form

$$h_{rupt}/r_p = \left(1 + \frac{\theta}{2}\right)(V_L/r_p^3)^{1/3}. \qquad (4.5.25)$$

In an experimental study, Pitois et al. (2001) modified this expression to include effects of particle motion on rupture of the liquid bridge. Their experimental data are reported as a function of the capillary number $\mathrm{Ca} \equiv \mu_L |\mathbf{v}_R \cdot \mathbf{n}|/\sigma$, where $\mathbf{v}_R \cdot \mathbf{n}$ is the normal component of the particle relative velocity and μ_L is the liquid viscosity. Pitois et al. find that over the interval $0.001 \leq \mathrm{Ca} \leq 0.1$ the relative change in rupture distance in the dynamic case compared with the static value varies in proportion to $\mathrm{Ca}^{1/2}$, which yields a modified rupture criterion for dynamic contact as

$$h_{rupt}/r_p = \left(1 + \frac{\theta}{2}\right)(1 + \mathrm{Ca}^{1/2})(V_L/r_p^3)^{1/3}. \qquad (4.5.26)$$

4.5.5. Capillary Torque on a Rolling Particle

The presence of a liquid bridge might also introduce modifications to the resistance forces and torques that occur during collisions with oblique particle motion, resulting in motions such as sliding, rolling, and twisting. Although most of the literature on liquid bridge adhesion has been limited to normal particle interactions, there have been two recent papers that experimentally examined the problem of a particle rolling on a flat surface in the presence of a liquid film. The first of these papers, written by Bico et al. (2009), considers the rolling motion of a particle down an inclined plane that is covered with a thin film of transparent oil measuring 50–500 μm in thickness. Experiments are conducted with different sphere materials and sizes, lubricating oils with different viscosity and surface tension, and a range of slope angles for the inclined plane. Some "overhang" experiments are also conducted, where the sphere is underneath the plane and the gravitational force is countered by the capillary force of the liquid bridge. Although the presence of the liquid film both modifies the sliding friction coefficient of the particle on the planar surface and introduces an enhanced viscous force within the fluid layer, the experimental results also indicate a significant influence of surface tension on the particle rolling velocity. In fact, the experiments indicate that the particle rolling speed varies approximately as $v \propto \sigma^{-0.35}$ as the film surface tension is changed. The authors explain the effect of surface tension in these experiments by noting that in the presence of rolling, the meniscus surrounding the particle contact point becomes asymmetric, and they suggest that this asymmetry may give rise to a tangential capillary force and torque that opposes the rolling motion. Visualization of the wake structure in these experiments exhibits three different regimes as a function of the capillary number Ca. For slowly rolling particles with Ca < 1, the meniscus has an approximately circular shape. For more rapidly rolling particles with Ca > 1, the meniscus breaks up to form two ridges that trail behind the particle, leaving a tire-print type track in the liquid film. If the particle is overhanging underneath the inclined plane, the observed track has only one ridge.

Schade and Marshall (2011) reported a series of experiments conducted at low capillary numbers, where a particle mounted on a wire is allowed to roll on a translated glass plate at a fixed speed in the presence of a liquid film. The liquid properties and rolling velocity are varied. Because the centroid position of the particle was fixed in these experiments, it was possible to perform detailed imaging of the liquid film using both the laser-induced fluorescence (LIF) and particle-image velocimetry (PIV) methods in order to examine the meniscus asymmetry in detail. Although the meniscus profile in these experiments was approximately circular, distinct differences were observed in both the contact angles and radial locations of the contact points between the leading and trailing sides of the particle. A sketch illustrating the observed differences is given in Figure 1.12b. On the leading side of the particle (the side toward which the particle is rolling), the contact angle is increased relative to that in the static state and the contact point moves inward toward the contact point. On the trailing side of the particle, the contact angle decreases slightly relative to the ambient and the contact point moves outward away from the contact point.

The net effect of the contact point motion is that the meniscus slightly trails behind the particle, so that the horizontal position of the center of the meniscus is slightly behind the horizontal position of the particle centroid. As a consequence of this shift, the line of force of the attractive capillary pressure force, which arises from the meniscus curvature, no longer passes through the particle center, but rather passes slightly behind the particle center by a shift distance ξ. This backward shift gives rise to a torque that impedes the particle rolling motion, with magnitude equal approximately to the magnitude of the capillary force F_{cap}, given by (4.5.14), times the meniscus shift distance ξ. This mechanism for rolling resistance resulting from a backward shift in the region over which the adhesion force acts is analogous to that discussed with respect to the rolling resistance torque due to van der Waals adhesion in Section 4.2.4, with the difference that in the current case it is not the contact region itself that shifts, but the liquid meniscus.

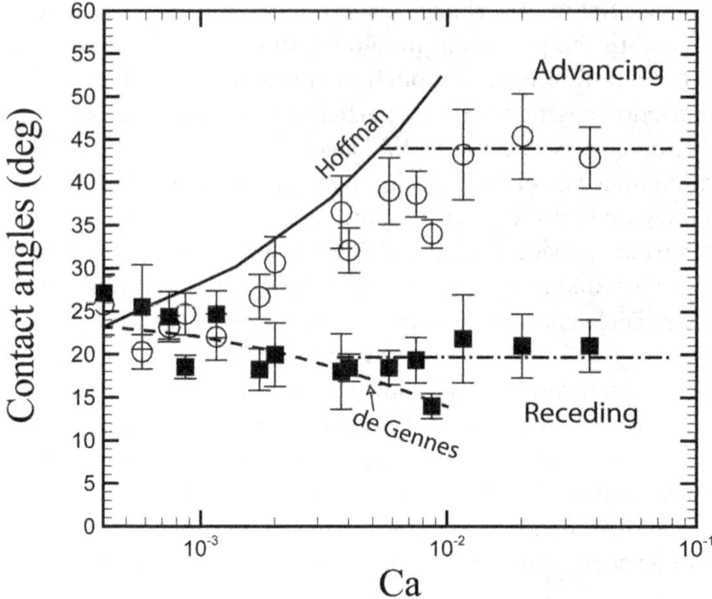

Figure 4.23. Plot showing the experimental data of Schade and Marshall (2011) for advancing contact angles (open symbols) and receding contact angles (filled symbols) as a function of capillary number, compared to the correlations of Hoffman (1975) for advancing contact angle (solid line) and of de Gennes (1986) for receding contact angle (dashed line). Data is obtained for different Reynolds and capillary numbers. Asymptotic values of the advancing and receding contact angles are indicated by dashed-dotted lines.

A second mechanism causing an additional torque on a rolling particle in the presence of a liquid film arises from the difference in contact angle between the particle leading and trailing sides. Measurements of these contact angles as a function of capillary number are presented in Figure 4.23. The contact angle at the leading edge of the particle compares well with the correlation of Hoffman (1975) for advancing contact angle on a flat surface for low capillary numbers ($Ca < 0.01$), but then it asymptotes to an approximately constant value at higher capillary numbers. Similarly, the contact angle at the trailing side compares well with the theoretical expression of de Gennes (1986) for capillary number below about 0.01, but asymptotes to a constant value at higher capillary numbers. It therefore appears that the asymmetry in the meniscus shape can be related to the classical contact angle hysteresis phenomenon, in which the contact angle of a meniscus on a moving plate can vary between an advancing and a receding value, both of which are dependent on the rolling speed.

4.6. Sintering Force

Sintering is a thermal process for bonding two or more particles into a single coherent solid structure via mass transport between the particles, some of which occurs at the atomic scale. Sintering processes play an important role in the consolidation of metal and ceramic powders and polymeric particles. These types of problems have taken on an increasingly important role in recent years with the development and widespread

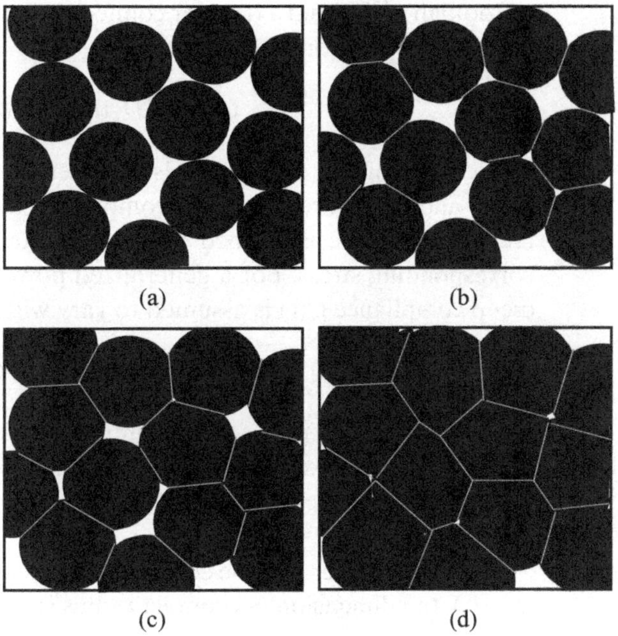

Figure 4.24. Sketch illustrating the four stages of sintering: (a) adhesion stage; (b) initial sintering stage; (c) intermediate sintering stage; (d) final sintering stage. Grain boundaries are indicated using a gray line.

application of 3D printing technologies. In the growing field of nanotechnology, coalescence of nanoparticles via sintering governs growth of the particle size and properties in certain flame synthesis and chemical-vapor decomposition synthesis techniques for nanoparticle production.

The sintering of a particle packing proceeds through a series of stages, as shown in Figure 4.24. The first image in this figure represents a collection of loose particles in the *adhesion stage*, in which particles first attach to each other (e.g., via one of the adhesion forces discussed previously in this chapter). Next, there is an *initial sintering stage* in which particle bonding occurs over the small contact regions, but the particles are still very distinct. As the bonds continue to grow, this progresses into an *intermediate sintering stage* in which the volume of bonds grows and begins to fill up the pore space. In the *final sintering stage* nearly all of the pores between the particles are filled with material (German, 1996). In the current section, we are concerned with initial-stage sintering, in which the material changes continuously from a particulate state to a monolithic state but still may be reasonably described as a discrete packing of particles. Three distinct types of sintering mechanisms have been identified: (1) elastic adhesive contact (i.e., JKR contact); (2) "zipping" contact growth driven by adhesive intersurface forces and accommodated by viscoelastic deformation; and (3) "stretching" contact growth driven by surface tension and accommodated by viscous flow (i.e., the classical viscous sintering).

4.6.1. Sintering Regime Map

A regime map showing regions where each sintering mechanism is dominant was developed by Lin et al. (2001) for polymeric particles. The material viscoelastic response is assumed to have a Maxwell form, which as shown in Figure 3.2 consists of a Hookian spring and a damper connected in series. The effective creep compliance $C(t)$ for the collision is defined as

$$C(t) = [D_1(t) + D_2(t)](1 - v^2), \tag{4.6.1}$$

where the Poisson modulus v is assumed to be the same for both materials and $D_1(t)$ and $D_2(t)$ are the creep compliance for the two spheres. It is recalled that creep compliance is defined as the instantaneous uniaxial strain divided by the corresponding stress. For a generalized power-law Maxwell material, the effective creep compliance $C(t)$ is assumed to vary with time as

$$C(t) = C_0 + C_1 t^m \tag{4.6.2}$$

where $C_0 = (1 - v_i^2)/E_i$ is the instantaneous compliance for sphere i, C_1 is a material constant, and the exponent m is a material constant in the interval $0 < m \leq 1$. For two identical particles, $C_0 = 1/2E$, where E is the effect elastic modulus defined in (3.1.1).

If the time after particle contact is sufficiently short that the material is unrelaxed ($t < t_0$), the dimensionless contact radius is given by the JKR model (4.2.34) as

$$\frac{a_0}{R} = \left(\frac{9\pi \gamma}{RE}\right)^{1/3} = \left(\frac{18\pi C_0 \gamma}{R}\right)^{1/3} \tag{4.6.3}$$

where R is the effective particle radius given in (3.1.1). For intermediate times in the interval $t_0 < t < t_{vis}$, the dimensionless contact radius can be estimated for viscoelastic adhesive contact (Lin et al., 2001) as

$$\frac{a}{R} = (63\pi^3)^{1/7} \left(\frac{\delta_C}{R}\right)^{2/7} \left(\frac{C_1 \gamma}{R} t\right)^{1/7} \tag{4.6.4}$$

where δ_C is the critical normal overlap from JKR theory, given by (4.2.37). Finally, if the time is large enough ($t > t_{vis}$), the contact radius can be obtained using the approach taken by Frenkel (1945), who equated the rate of surface tension work to the viscous flow energy dissipation rate to obtain

$$\frac{a}{R} = \left(\frac{2\sigma t}{\mu R}\right)^{1/2}. \tag{4.6.5}$$

Replacing the viscosity in this equation by the effective particle creep viscosity, $\mu_{c,N} = 1/(4C_1)$ and the surface tension σ by the surface energy density γ, we can write

$$\frac{a}{R} = \left(\frac{8C_1 \gamma}{R} t\right)^{1/2}. \tag{4.6.6}$$

For instance, for fly-ash particles in a coal combustion boiler, typical values for these parameters are $C_0 = 1 \times 10^{-10}$ Pa^{-1}, $C_1 = 1 \times 10^{-12}$ Pa^{-1} s^{-1}, $m = 1$, $\gamma = 0.5$ J/m^2, and $\delta_c/R = 0.01$. The change in dimensionless contact radius, a/R, is sketched as a heavy solid line in Figure 4.25. If $t < t_0$ (point A), the dimensionless contact radius is a constant which depends on the dimensionless parameter $C_0 \gamma / R$.

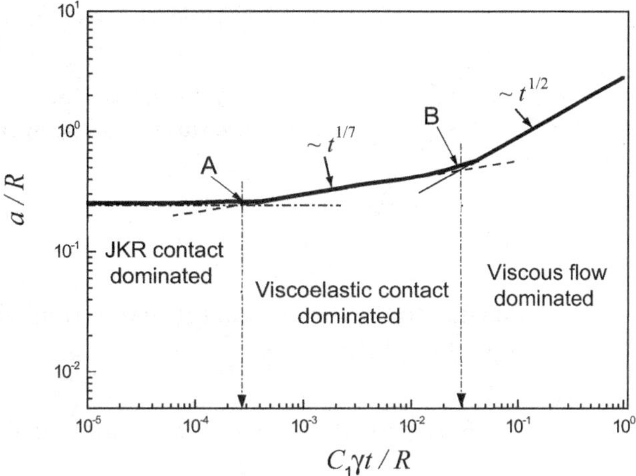

Figure 4.25. A regime map for three different sintering mechanisms, similar to that developed by Lin et al. (2001). Variation of the contact region radius for a typical sintering case is indicated by a solid line, which is nearly independent of time in the JKR-contact regime, increases in proportion to $t^{1/7}$ in the viscoelastic regime, and increases in proportion to $t^{1/2}$ in the viscous flow regime.

For intermediate times ($t_0 < t < t_{vis}$), the power law behavior for a/R exhibits only very slow time dependence, to the 1/7 power. For $t > t_{vis}$ (point B), a/R depends on the square root of the normalized time $C_1\sigma t/R$, indicative of viscous sintering.

4.6.2. Approximate Sintering Models

Several attempts have been made to apply the DEM approach to solve for sintering of a realistic random packing of particles, including Jagota and Scherer (1993, 1995), Luding et al. (2005), and Luding and Suiker (2008). In the model of Jagota and Scherer, the viscous sintering regime is assumed ($t > t_{vis}$ in Figure 4.25) and the normal contact force is modeled as

$$F_n = 3\pi \sigma a(t) f(a/R) - 6\pi \mu_{c,N} a(t) \frac{R}{|X_{ij}|} \mathbf{v}_R \cdot \mathbf{n}, \qquad (4.6.7)$$

where $\mu_{c,N}$ is the particle creep viscosity, σ is the surface tension, $a(t)$ is the contact radius, $|X_{ij}|$ is the distance between the centers of particles i and j, and $f(a/R)$ is a dimensionless function of contact radius which has order unity. The first and second terms on the right-hand side of (4.6.7) represent the sintering rate driven by surface tension and by viscous dissipation, respectively. Similarly, the model used by Jagota and Scherer for shear force transmission through the contact is related to both the difference in tangential velocity and the spin of the two particles through the tangential contact viscosity.

During the sintering process, the contact forces are often driven by the coupled grain-boundary and surface diffusion, instead of the viscous flow. A detailed numerical simulation comparing the grain boundary and surface diffusion mechanisms during sintering of two particles was reported by Bouvard and McMeeking (1996). The results of this study were used in the context of DEM by Parhami and McMeeking (1998) and by Martin et al. (2006) to perform simulations of diffusion-controlled particle sintering, including problems with both free and pressure-assisted sintering in three dimensions. In this work, the normal contact forces between two particles with equivalent radius R are expressed as a sum of both sintering tensile

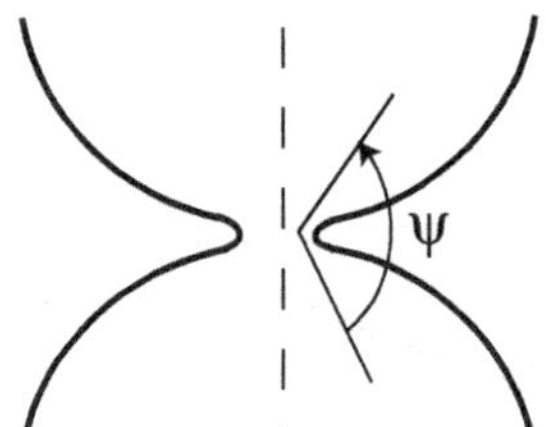

Figure 4.26. Sketch showing the definition of the dihedral angle between the particle surfaces at the neck.

force due to the surface energy and normal viscous force due to the grain boundary and surface diffusion as

$$F_n = \pi\gamma[8R(1 - \cos(\psi/2)) + a(t)\sin(\psi/2)] - \frac{k_B T}{8\Omega D_b \delta_b}\pi a^4(t)\frac{d\delta_N}{dt}, \qquad (4.6.8)$$

where $a(t)$ is the sintering contact radius, ψ is the dihedral angle between the particle surfaces at the neck (as shown in Figure 4.26), γ is the surface energy density, Ω is the atomic volume, and $D_b = D_{0b}\exp(-Q_b/RT)$ is the diffusion coefficient for vacancy transport in the grain boundary with thickness δ_b and activation energy Q_b. The sintering contact radius grows according to Coble's geometric model as (Coble, 1958)

$$\frac{da(t)}{dt} = \frac{R}{a(t)}\frac{d\delta_N}{dt}, \qquad (4.6.9)$$

with the initial value of $a(t)$ given by $a^2 = R\delta_N$ for elastic deformation or by $a^2 = BR\delta_N$ for plastic deformation, where B is a constant varying between 1 for linear hardening and 2.9 for perfectly plastic deformation. The normal contact force is set to zero once the contact radius has grown to the equilibrium value $a_{eq} = R\sin(\psi/2)$.

DEM models based on the normal contact force expression in the presence of sintering for the viscous flow regime, such as (4.6.7), and in the diffusion-dominated regime, such as (4.6.8), are applicable to problems involving quasi-static packing and consolidation of the sintering particles for cases without appreciable inertia. In a general flow field, the particles collide with a nonzero impact velocity and the elastic repulsive force between the particles must also be considered. For very small particles, such that DMT theory applies, the Hertz elastic force can be added to these two equations for a more complete description. In the general case of finite Tabor number, the particle deformation under collision will influence the adhesive force and so the two forces must be combined, such as was done in the JKR and M-D models for nonsintering adhesion.

4.6.3. Hysteretic Sintering Contact Model

An alternative modeling approach for sintering phenomena was developed by Luding and coworkers (2005, 2008) using a hysteric contact model, which can be regarded as an extension of the Walton and Braun (1986) model for elastic-plastic contact

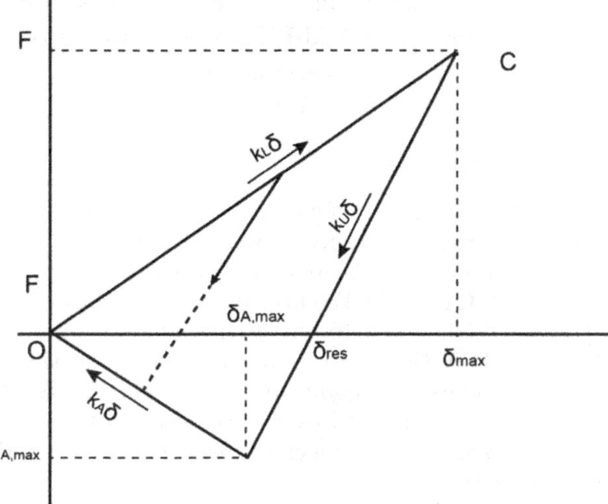

Figure 4.27. Load-displacement plot for the hysteretic sintering contact model of Luding et al. (2005).

force. The hysteretic sintering contact model is expressed using a force-displacement relation given by

$$F_n = \begin{cases} k_L\delta_N & \text{Loading} \\ k_U(\delta_N - \delta_{res}) & \text{Un/reloading} \\ -k_A\delta_N & \text{Unloading} \end{cases} \qquad (4.6.10)$$

where k_L and k_U are the spring constants during loading and unloading processes, respectively, such that $k_L \leq k_U$. As seen in Figure 4.27, a contact begins at $\delta_N = 0$, such that during initial compressive loading, the contact repulsive force increases with the overlap as $F_n = k_L\delta_N$ until a maximum overlap δ_{max} is attained. Beyond this time the particles start to move apart, so that δ_N decreases in time. The repulsive force between the particles decreases during this unloading process in proportion to $F_n = k_U(\delta - \delta_{res})$. For subsequent loading with the same contact, the spring constant remains equal to k_U. The residual overlap $\delta_{res} = (1 - k_L/k_U)\delta_{max}$ is attained when the repulsive force between the particles vanishes. As the particles continue to be pulled apart, there is an attractive force between them due to the material bonding at the particle contact. This attractive force increases with decrease in δ_N until a maximum attractive force is attained, given by $F_{A,max} = -k_A\delta_{A,max}$, where $\delta_{A,max} = k_U\delta_{res}/(k_U + k_A)$ and k_A represents the effects of tensile softening. The magnitude of the attractive contact force decreases as δ_N is decreased further according to $F_n = -k_A\delta_N$ until the load vanishes at $\delta_N = 0$. In the sintering process, the values of k_L, k_U and k_A are strongly temperature-dependent, and some related empirical expressions that can be used to estimate these parameters were given by Luding et al. (2005).

More extensive work on validation of DEM force models for sintering contact between two spherical particles is needed. An experimental set-up used to measure the critical sticking velocity of an incident particle in a vacuum column in which particles impact on a flat surface was developed by van Beek (2001). Abd-Elhady et al. (2005) extended this set-up for studying critical sticking velocity of an incident particle with a sintered particulate fouling layer, which is exposed to a high gas-side temperature ambience. Models for both normal and shear contact were proposed and tested by Abd-Elhady for sintered particles. Nevertheless, DEM modeling of sintered particles is still a relatively recent development.

REFERENCES

Abd-Elhady MS. Gas-side particulate fouling in biomass gasifiers. Ph.D. thesis, Eindhoven University of Technology, The Netherlands (2005).

Agresar G. A Computational Environment for the Study of Circulating Cell Mechanics and Adhesion. Ph.D. Dissertation, University of Michigan, Ann Arbor, 1996.

Bell GI. Models for the specific adhesion of cells to cells. *Science* **200**, 618–627 (1978).

Bell GI, Dembo M, Bongrand P. Cell adhesion, competition between nonspecific repulsion and specific bonding. *Biophysics Journal* **45**, 1051–1064 (1984).

Bico J, Ashmore-Chakrabarty J, McKinley GH, Stone HA. Rolling stones: The motion of a sphere down an inclined plane coated with a thin liquid film. *Physics of Fluids* **21**, 082103 (2009).

Bouvard D, McMeeking RM. The deformation of interparticle necks by diffusion controlled creep. *Journal of the American Ceramic Society* **79**, 666–672 (1996).

Carpick RW, Ogletree DF, Salmeron M. A general equation for fitting contact area and friction vs load measurements. *Journal of Colloid and Interface Science* **211**, 395–400 (1999).

Cavallaro U, Christofori G. Cell adhesion in tumor invasion and metastasis: Loss of the glue is not enough. *Biochimica et Biophysica Acta* **1552**, 39–45 (2001).

Chesla SE, Selvaraj P, Zhu C. (1998) Measuring two-dimensional receptor-ligand binding kinetics by micropipette. *Biophysics Journal* **75**, 1553–1572 (1998).

Chesnutt JKW, Marshall JS. Blood cell transport and aggregation using discrete ellipsoidal particles. *Computers & Fluids* **38**, 1782–1794 (2009).

Chokshi A, Tielens AGGM, Hollenbach D. Dust coagulation. *The Astrophysical Journal* **407**, 806–819 (1993).

Coble RL. Initial sintering of alumina and hematite. *Journal of the American Ceramic Society* **41**, 55–62 (1958).

Coombs D, Dembo M, Wofsy C, Goldstein B. Equilibrium thermodynamics of cell-cell adhesion mediated by multiple ligand-receptor pairs. *Biophysics Journal* **86**, 1408–1423 (2004).

Cundall PA, Strack ODL. A discrete numerical model for granular assembles. *Geotechnique* **29**(1), 47–65 (1979).

Dahneke B. Measurements of the bouncing of small latex spheres. *Journal of Colloid and Interface Science* **45**, 584–590 (1973).

de Gennes PG. Deposition of Langmuir-Blodgett layers. *Colloid & Polymer Science* **264**, 463–465 (1986).

Dembo M, Torney DC, Saxman K, Hammer D. The reaction-limited kinetics of membrane-to-surface adhesion and detachment. *Proceedings of the Royal Society of London B* **234**, 55–83 (1988).

Derjaguin BV. Analysis of friction and adhesion. IV. The theory of the adhesion of small particles. *Kolloid Z.* (in German) **69**(2), 155–164 (1934).

Derjaguin BV, Landau L. Theory of the stability of strongly charged lyophobic sols and of the adhesion of strongly charged particles in solutions of electrolytes. *Acta Physico Chemica URSS* **14**, 633–662 (1941).

Derjaguin BV, Muller VM, Toporov YP. Effect of contact deformations on the adhesion of particles. *Journal of Colloid and Interface Science* **53**(2), 314–326 (1975).

Ding W, Zhang H, Cetinkaya C. Rolling resistance moment-based adhesion characterization of microspheres. *Journal of Adhesion* **84**(12), 996–1006 (2008).

Dominik C, Tielens AGGM. Resistance to sliding on atomic scales in the adhesive contact of two elastic spheres. *Philosophical Magazine A* **73**(5), 1279–1302 (1996).

Dominik C, Tielens AGGM. Resistance to rolling in the adhesive contact of two elastic spheres. *Philosophical Magazine A* **72**(3), 783–803 (1995).

Ennis BJ, Li J, Tardos GI, Pfeffer R. The influence of viscosity on the strength of an axially strained pendular liquid bridge. *Chemical Engineering Science* **45**(10), 3071–3088 (1990).

Fiscina JE, Lumay G, Ludewig F, Vandewalle N. Compaction dynamics of wet granular assemblies. *Physical Review Letters* **105**, 048001 (2010).

Frenkel J. Viscous flow of crystalline bodies under the action of surface tension. *Journal of Physics* **9**, 385–391 (1945).

German RM. *Sintering Theory and Practice*. John Wiley & Sons: Wiley-Interscience, (1996).

Halsey TC, Levine AJ. How sandcastles fall. *Physical Review Letters* **80**, 3141–3144 (1998).

Hoffman RL. A study of the advancing interface. I. Interface shape in liquid-gas systems. *Journal of Colloid and Interface Science* **50**(2), 228–241 (1975).

Hotta K, Takeda K, Iinoya K. The capillary binding force of a liquid bridge. *Powder Technology* **10**, 231–242 (1974).

Israelachvili J. *Intermolecular and Surface Forces*, Academic Press, 2nd edition, London (1991).

Jagota A, Scherer GW. Viscosities and sintering rate of a two-dimensional granular composite. *Journal of the American Ceramic Society* **76**, 3123–3135 (1993).

Jagota A, Scherer GW. Viscosities and sintering rates of composite packings of spheres. *Journal of the American Ceramic Society* **78**, 521–528 (1995).

Johnson KL, Greenwood JA. An adhesion map for the contact of elastic spheres. *Journal of Colloid and Interface Science* **192**, 326–333 (1997).

Johnson KL, Kendall K, Roberts AD. Surface energy and the contact of elastic solids. *Proceedings of the Royal Society of London A* **324**, 301–313 (1971).

Li S-Q, Marshall JS, Liu G, Yao Q. Adhesive particulate flow: The discrete element method and its application in energy and environmental engineering. *Progress in Energy and Combustion Science* **37**(6), 633–668 (2011).

Lian G, Thornton C, Adams MJ. A theoretical study of the liquid bridge forces between two rigid spherical bodies. *Journal of Colloid and Interface Science* **161**, 138–147 (1993).

Lin YY, Hui CY, Jagota A. The role of viscoelastic adhesive contact in the sintering of polymeric particles. *Journal of Colloid and Interface Science* **237**, 267–282 (2001).

Liu GQ, Li SQ, Yao Q. A JKR-based dynamic model for the impact of micro-particle with a flat surface. *Powder Technology* **207**, 215–223 (2011).

Longest PW, Kleinstreuer C. Comparison of blood particle deposition models for non-parallel flow domains. *Journal of Biomechanics* **36**, 421–430 (2003).

Luding S, Suiker A. Self-healing of damaged particulate materials through sintering. *Philosophical Magazine* **88**, 3445–3457 (2008).

Luding S, Manetsberger K, Muellers J, A discrete model for long time sintering. *Journal of the Mechanics and Physics of Solids* **53**, 455–491 (2005).

Marshall JS. Particle aggregation and capture by walls in a particulate aerosol channel flow. *Journal of Aerosol Science* **38**, 333–351 (2007).

Marshall JS. Discrete-element modeling of particulate aerosol flows. *Journal of Computational Physics* **228**, 1541–1561 (2009).

Martin CL, Schneider LCR, Olmos L, Bouvard D. Discrete element modeling of metallic powder sintering. *Scripta Materialia* **55**, 425–428 (2006).

Masliyah JH, Bhattacharjee S. *Electrokinetic and Colloid Transport Phenomena*. John Wiley and Sons, New Jersey (2006).

Mason T G, Levine A J, Ertas D, Halsey TC. Critical angle of wet sandpiles. *Physical Review E* **60**(5), 5044–5047 (1999).

Matthewson MJ. Adhesion of spheres by thin liquid films. *Philosophical Magazine A* **57**(2), 207–216 (1988).

Maugis D. Adherence of elastomers: Fracture mechanics aspects. *Journal of Adhesion Science and Technology* **1**(2), 105–134 (1987).

Maugis D. Adhesion of spheres: The JKR-DMT transition using a Dugdale model. *Journal of Colloid and Interface Science* **150**(1), 243–269 (1992).

Maugis D. *Contact, Adhesion and Rupture of Elastic Solids*, Springer-Verlag, Berlin (1999).

Mehrotra VP, Sastry KVS. Pendular bond strength between unequal-sized spherical particles. *Powder Technology* **25**, 203–214 (1980).

Muller VM, Yushenko VS, Derjaguin BV. On the influence of molecular forces on the deformation of an elastic sphere and its sticking to a rigid plane. *Journal of Colloid and Interface Science* **77**(1), 91–101 (1980).

N'Dri NA, Shyy W, Tran-Son-Tay R. Computational modeling of cell adhesion and movement using a continuum-kinetics approach. *Biophysics Journal* **85**, 2273–2286 (2003).

Parhami F, McMeeking RM. A network model for initial stage sintering. *Mechanics of Materials* **27**, 111–124 (1998).

Peri MDM, Cetinkaya C. Rolling resistance moment of microspheres on surfaces. *Philosophical Magazine* **85**(13), 1347–1357 (2005).

Pitois O, Moucheront P, Chateau X. Liquid bridge between two moving spheres: An experimental study of viscosity effects. *Journal of Colloid and Interface Science* **231**, 26–31 (2000).

Pitois O, Moucheront P, Chateau X. Rupture energy of a pendular bridge. *European Physical Journal B* **23**, 79–86 (2001).

Probstein RF. *Physicochemical Hydrodynamics, An Introduction*, 2nd ed., Wiley Interscience, New York (2003).

Russel WB, Saville DA, Schowalter W R. *Colloidal Dispersions*. Cambridge University Press, Cambridge, UK (1989).

Savkoor AR, Briggs GAD. The effect of tangential force on the contact of elastic solids in adhesion. *Proceedings of the Royal Society of London A* **356**, 103–114 (1977).

Schade P, Marshall JS. Capillary effects on a particle rolling on a plane surface in the presence of a thin liquid film. *Experiments in Fluids* **51**(6), 1645–1655 (2011).

Smith AW. Biofilms and antibiotic therapy: Is there a role for combating bacterial resistance by the use of novel drug delivery systems? *Advanced Drug Delivery Reviews* **57**, 1539–1550 (2005).

Tabor D. Surface forces and surface interactions. *Journal of Colloid and Interface Science* **58**(1), 2–13 (1977).

Tachev KD, Angarska JK, Danov KD, Kralchevsky PA. Erythrocyte attachment to substrates: determination of membrane tension and adhesion energy. *Colloids and Surfaces B: Biointerfaces* **19**, 61–80 (2000).

Thornton C. Interparticle sliding in the presence of adhesion. *Journal of Physics D-Applied Physics* **24**, 1942–1946 (1991).

Thornton C, Ning Z. A theoretical model for the stick/bounce behaviour of adhesive, elastic-plastic spheres. *Powder Technology* **99**, 154–162 (1998).

Thornton C, Yin KK. Impact of elastic spheres with and without adhesion. *Powder Technology* **65**, 153–166 (1991).

Tozeren A, Sung KLP, Chien S. Theoretical and experimental studies on cross-bridge migration during cell disaggregation. *Biophysics Journal* **55**, 479–487 (1989).

Tsuji Y, Tanaka T, Ishida T. Lagrangian numerical simulation of plug flow of cohesionless particles in a horizontal pipe. *Powder Technology* **71**, 239–250 (1992).

van Beek MC. Gas-side fouling in heat-recovery boilers. Ph.D. Thesis, Eindhoven University of Technology, The Netherlands (2001).

Verwey EJW, Overbeek JTG. *Theory of the Stability of Lyophobic Colloids*. Elsevier, Amsterdam (1948).

Wall S, John W, Wang H, Goren SL. Measurements of kinetic energy loss for particles impacting surfaces. *Aerosol Science and Technology* **12**, 926–946 (1990).

Walton OR, Braun RL. Viscosity, granular temperature, and stress calculations for shearing assemblies of inelastic, frictional disks. *Journal of Rheology* **30**, 949–980 (1986).

Willett CD, Adams MJ, Johnson SA, Seville JPK. Capillary bridges between two spherical bodies. *Langmuir* **16**(24), 9396–9405 (2000).

Zhu C. Kinetics and mechanics of cell adhesion. *Journal of Biomechanics* **33**, 23–33 (2000).

Zhu C, Williams TE. Modeling concurrent binding of multiple molecular species in cell adhesion. *Biophysics Journal* **79**, 1850–1857 (2000).

Zhu RR, Li SQ, Yao Q. Effects of cohesion on the flow patterns of granular materials in spouted beds. *Physical Review E* **87**, 022206 (2013).

5 Fluid Forces on Particles

It is necessary in any simulation of a particulate flow to utilize a model for the forces exerted by the fluid on the particles. In development of such a model, it is usually assumed that the particles are immersed in a fluid flow that has characteristic length scales that are much greater than the particle diameter. If this is the case, then the fluid flow can be approximated from the perspective of a particle as a uniform flow plus a linear velocity gradient, which greatly simplifies the task of model development for the fluid forces on the particle. Even with this simplification, there is still a vast literature devoted to fluid forces on bodies of various shapes and sizes in different Reynolds number and Mach number regimes. The current chapter focuses on forces on spherical particles in an incompressible flow with particle Reynolds numbers less than about 10, and in some cases with particle Reynolds numbers less than unity. The problem of fluid flows with ellipsoidal particles is taken up in Chapter 7.

The first five sections of this chapter discuss various forces acting on a particle immersed in a fluid shear flow in isolation, without the presence of walls or other particles. The first two of these sections discuss the drag and lift forces that act on particles immersed in a steady velocity field, whereas the third section concerns forces that act in unsteady flows, including added mass force, inertial pressure-gradient force, and the so-called history force. The fourth section introduces the force due to Brownian motion. A scaling analysis examining the order of magnitude of these different forces relative to the dominant drag force is given in the fifth section. The remaining three sections of the chapter examine the effect of container walls and of other particles on the particle forces. The effect of a nearby plane wall, discussed in the sixth section, is a particularly important problem for adhesive particles, which tend to stick to walls immersed in the flow field. Surrounding particles also have an important impact on particle fluid forces as the particle concentration becomes large. The seventh section discusses the influence of surrounding particles on the drag force with a specified mean fluid flow. The eighth section describes a method called Stokesian dynamics, which is used to provide a rapid computation of the hydrodynamic interaction between particles in a flow, accounting for the modification of the fluid velocity field by the presence of surrounding particles. The final section discusses different types of particle interactions with acoustic fields.

5.1. Drag Force and Viscous Torque

In most particulate flows, there are two primary forces that are most important for controlling the particle dynamics, and these two forces are in approximate balance with each other as the particles move about. Additionally, there will be a number of secondary forces that are less important, although perhaps not completely negligible. Identification and accurate modeling of the primary forces is critical for solution of the particle transport problem, although secondary forces are often approximated with a less rigorous model or even neglected. For problems involving neutrally buoyant particles with diameter greater than about 1 μm, the primary forces are usually drag and particle inertia. This statement remains valid even for cases where the particle Reynolds number is small, for which case both the particle inertia and drag force will be small, but other forces will usually be even smaller. For nanometer-size particles, Brownian motion can become sufficiently large so as to exceed the particle inertia, in which case the particle drag is approximately balanced by the Brownian force. This is the assumption typically used in the Brownian dynamics method discussed in Section 2.3.2. For particles that are not neutrally buoyant, the reduced gravity force (which combines gravity and buoyancy) can also become important in certain particle size ranges.

The classical solution for drag force on a sphere was obtained by Stokes, given by (1.2.1). This solution assumes that the particle Reynolds number is small compared to unity, that the fluid flow relative to the particle is uniform, that the no-slip condition is satisfied on the surface of the particle, and that the space is unbounded with no solid surfaces present except for the particle itself. All of these approximations break down at various instances in particulate flow problems. A modified form of the Stokes expression for drag force $\mathbf{F}_d$ on a spherical particle can be written as

$$\mathbf{F}_d = -3\pi d\mu(\mathbf{v} - \mathbf{u})f, \tag{5.1.1}$$

where f is called the *friction factor*, $\mathbf{v}$ is the particle velocity, and $\mathbf{u}$ is the velocity that the fluid would have had at the particle centroid had the particle not been present. The difference $\mathbf{v} - \mathbf{u}$ is known as the *slip velocity*. The friction factor is used to account for breakdown of some of the approximations noted earlier, so that the original Stokes solution is recovered when $f = 1$. The friction factor can be decomposed as the product of various correction factors, given by

$$f = C_I C_C C_F, \tag{5.1.2}$$

where C_I accounts for effect of fluid inertia, C_F accounts for effects of crowding by surrounding particles, and C_C accounts for the effect of nonzero slip on the particle surface. The inertia and slip correction factors are discussed in the following, and the effect of surrounding particles is discussed in Section 5.7.

5.1.1. Effect of Flow Nonuniformity

Perhaps the first objection that one might make to application of the Stokes drag expression for a particle in a general fluid flow is that the incident flow field relative to the particle is not actually uniform, but rather changes with some characteristic length scale L. How important is nonuniformity of the incident flow to the particle drag force? This question can be addressed with reference to a more general solution due to Faxén (1922), who found that the force on a sphere of diameter d immersed in a flow field $\mathbf{u}(\mathbf{x})$ with small particle Reynolds number and small value of the parameter d/L is given by

$$\mathbf{F}_d = -3\pi d\mu(\mathbf{v} - \mathbf{u}) + \frac{\mu\pi}{8}d^3\nabla^2\mathbf{u}, \tag{5.1.3}$$

where both $\mathbf{u}$ and $\nabla^2\mathbf{u}$ are evaluated at the particle centroid. The second term in (5.1.3) is similar to a buoyancy (or pressure gradient) force. To see this, we recall that for a flow with small flow Reynolds number the fluid inertia term is negligible, so the Navier-Stokes equation reduces to the Stokes equation

$$0 = -\nabla p + \mu\nabla^2\mathbf{u}, \tag{5.1.4}$$

where p is fluid pressure and $\mathbf{u}$ is the fluid velocity field. Solving for $\nabla^2\mathbf{u} = (1/\mu)\nabla p$ from (5.1.4) and substituting into (5.1.3) gives

$$\mathbf{F}_d = -3\pi d\mu(\mathbf{v} - \mathbf{u}) + \frac{\pi}{8}d^3\nabla p. \tag{5.1.5}$$

The last term in (5.1.5) is proportional to the sphere volume times the ambient fluid pressure gradient to which the sphere is subjected, which is equal to the buoyancy force. We note that this analysis requires that the flow Reynolds number Re_f is small, which is substantially more restrictive than the requirement that the particle Reynolds number be small. A scaling analysis indicates that the ratio of the second term on the right-hand side of (5.1.3) to the first (Stokes) term is of order $(d/L)^2$. Applicability of the discrete-element method usually requires that d/L is small, so the second term in (5.1.3) is negligible. An exception might be flow near walls, which is discussed further in Section 5.6.

Nonuniformity of the flow field also introduces the possibility that nonzero rotation rate of the material elements in the flow surrounding the particle might induce a torque on the particle. This issue was also examined by Faxén (1922), who concluded that a viscous torque $\mathbf{M}_F$ acts on the particle that is proportional to the difference between the local fluid rotation rate $\omega/2$, where $\omega = \nabla \times \mathbf{u}$ is the fluid vorticity evaluated at the particle centroid, and the particle rotation rate $\boldsymbol{\Omega}$. The viscous torque is on the sphere is given by the formula

$$\mathbf{M}_F = \pi\mu d^3 \left(\frac{1}{2}\omega - \boldsymbol{\Omega}\right). \tag{5.1.6}$$

5.1.2. Effect of Fluid Inertia

For small particle Reynolds numbers ($\mathrm{Re}_p = \rho_f d\,|\mathbf{v} - \mathbf{u}|\,/\mu \ll 1$), one-third of the total drag force acting on a sphere is due to asymmetry in the pressure distribution and two-thirds is due to viscous shear stress on the sphere surface. As shown in Figure 5.1, the particle wake becomes increasingly pronounced as the particle Reynolds number approaches and exceeds unity. At a Reynolds number of about 24, a standing vortex ring is observed to form behind the sphere (Figure 5.1c). This vortex ring moves backward gradually as the Reynolds number is increased

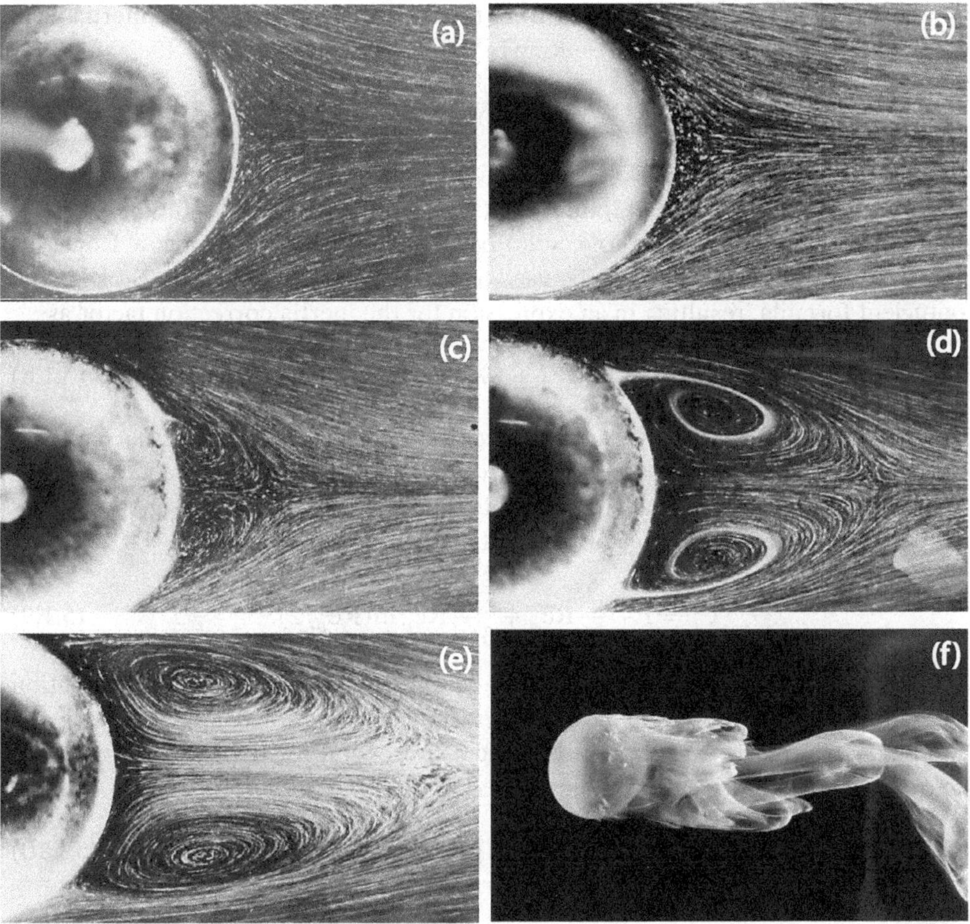

Figure 5.1. Flow visualization showing the wake behind a sphere at Reynolds numbers (a) 9.15; (b) 17.9; (c) 37.7; (d) 73.6; (e) 118; and (f) 550. Images (a)–(e) are the wakes of solid spheres visualized by suspended aluminum powder. Image (f) is the wake of an oil drop of carbon tetrachloride in water colored by fluorescein. [Reprinted with permission from Taneda (1956).]

(Figure 5.1d, e). At a Reynolds number of about 130, oscillations are first observed in the wake at the rear of the vortex ring, which at higher Reynolds numbers leads to periodic shedding of horseshoe-type vortex structures on alternating sides of the sphere (Figure 5.1f).

The wake development leads to increase in the rear-aft pressure asymmetry, causing the pressure drag on the particle to increase relative to the viscous drag. This increase in pressure drag leads to the so-called inertial correction of the Stokes drag expression, valid for particle Reynolds numbers that are not sufficiently small that the fluid inertia is negligible but also not sufficiently large that the pressure force completely dominates the drag force. The original Stokes expression (5.1.4) is obtained by neglecting the fluid inertia term in the Navier-Stokes equations, and this equation is valid in the region near the sphere for particle Reynolds numbers up to about unity. Oseen (1910) developed a zeroth-order correction of the Stokes equation valid in the region far away from the sphere by linearizing the inertia term in the Navier-Stokes equation, resulting in the Oseen equation

$$(\mathbf{U} \cdot \nabla)\mathbf{u} = -\nabla p + \mu \nabla^2 \mathbf{u}, \tag{5.1.7}$$

where $\mathbf{U}$ is the uniform fluid velocity at infinity and the sphere is assumed to be stationary. Matching the far-field solution of (5.1.7) to the near-field solution of (5.1.4) gives a uniformly valid asymptotic solution of the velocity field for low Reynolds number flow. This procedure can be used to obtain a correction for the Stokes drag coefficient formula, resulting in an expression for the inertia correction factor as

$$C_I = 1 + (3/16)\mathrm{Re}_p. \tag{5.1.8}$$

Proudman and Pearson (1957) used a singular perturbation scheme that matches solutions near and far from the sphere to obtain a first-order correction for the sphere flow field valid at finite Reynolds numbers. This perturbation scheme results in an expression for the inertia correction factor as

$$C_I = 1 + \frac{3}{16}\mathrm{Re}_p + \frac{9}{160}\mathrm{Re}_p^2 \ln(\mathrm{Re}_p/2). \tag{5.1.9}$$

Increase in particle Reynolds number up to about 100 is accompanied by formation of a standing vortex ring within the wake, resulting in formation of an elongated wake "bubble" (Figure 5.1c). A popular empirical expression for the inertial correction factor is that of Schiller and Naumann (1933),

$$C_I = 1 + 0.15\mathrm{Re}_p^{0.687}. \tag{5.1.10}$$

This expression remains valid to within 5% of the experimental data for particle Reynolds numbers up to about 800. There are many other empirical expressions for drag coefficient that are valid for even higher Reynolds numbers, but for adhesive particle flows the particle Reynolds numbers are typically small and so the expressions given here are usually adequate.

A plot comparing the predictions of expressions (5.1.8), (5.1.9), and (5.1.10) for the inertial correction factor is given in Figure 5.2a. The empirical Schiller-Naumann expression is a close fit to the experimental data. This correlation indicates that the inertial correction factor increases the drag by a factor of about 0.2 for $\mathrm{Re}_p = 1$, 0.5 for $\mathrm{Re}_p = 5$, and 1.7 for $\mathrm{Re}_p = 10$. The Oseen equation (5.1.8) follows reasonably close to the Schiller-Naumann expression for $\mathrm{Re}_p < 2$, but for larger Reynolds number it gives excessively high estimates for the sphere drag. The Proudman-Pearson expression (5.1.9) is closer to the Schiller-Naumann expression for $\mathrm{Re}_p < 2$ than is the Oseen expression, but at higher Reynolds numbers it also gives excessively high estimates for sphere drag. This deviation is not surprising because the Proudman-Pearson expression is derived from an asymptotic theory assuming a small particle Reynolds number. The drag coefficient $C_d = F_d / \frac{\pi}{8}\rho |\mathbf{v} - \mathbf{u}|^2 d^2$ is plotted as a function of particle Reynolds number in Figure 5.2b, comparing experimental data with the Stokes, Oseen, and Schiller-Naumann expressions for a wide range of Reynolds numbers. For DEM calculations, we recommend using the original Stokes expression (with $C_I = 1$) for particle Reynolds numbers up to about

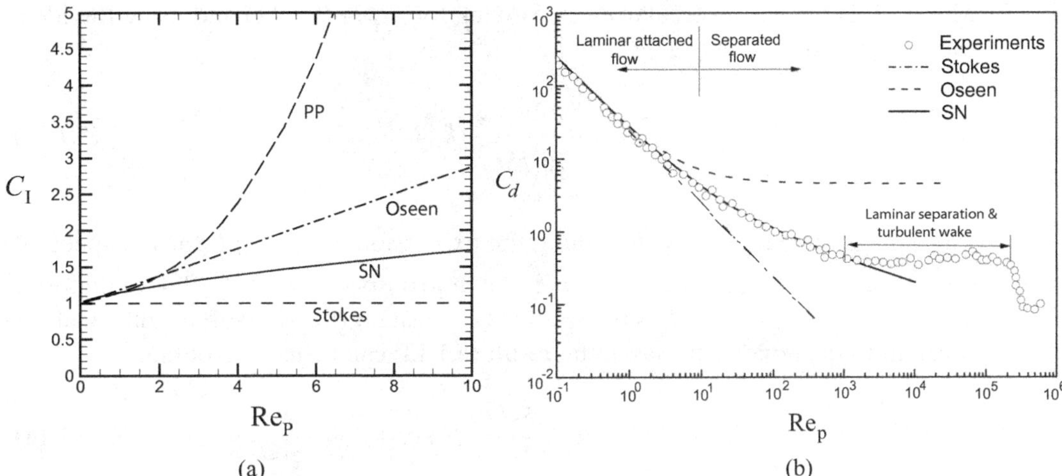

Figure 5.2. (a) Comparison of different expressions for the inertia correction factor C_I as a function of particle Reynolds number Re_p, including the Stokes expression ($C_I = 1$) (dashed line), the Oseen expression (5.1.8) (dashed-dotted line), the Proudman-Pearson expression (5.1.9) (long dashed line), and the Schiller-Naumann expression (5.1.10) (solid line). (b) Plot of drag coefficient for flow past a sphere including Stokes, Oseen, and Schiller-Naumann expressions and experimental data reported by White (2006).

$\mathrm{Re}_p = 1$, since this linear expression is most accommodating of an implicit numerical solution of the particle momentum equation (see further comments on numerical instability of the particle momentum equation in Chapter 10). For higher values of the particle Reynolds number, the Schiller-Naumann expression (5.1.10) is the best choice.

5.1.3. Effect of Surface Slip

Because particles of interest in adhesive particle flows are often quite small, it can sometimes happen in gas-particle flows that the particle diameter approaches the order of magnitude of the mean free path λ of the gas molecules. Mean free path is the average distance traveled by a molecule of the gas before collision with another gas molecule. Typical values of λ are about 68 nm for gases such as O_2 or N_2 at standard temperature and pressure. Kinetic theory can be used to relate the mean free path to the temperature and pressure of an ideal gas as

$$\lambda = \frac{k_B T}{\sqrt{2}\pi \sigma_M^2 p}, \tag{5.1.11}$$

where the Boltzmann constant $k_B = 1.380648813 \times 10^{-23}$ J/K is equal to the ratio of the universal gas constant to Avogadro's number, T and p are the absolute temperature and pressure, and σ_M is the molecular diameter. Kinetic theory also provides a relationship between molecular diameter and the fluid viscosity μ as

$$\mu = \frac{5}{16\sigma_M^2}\left(\frac{M_g k_B T}{\pi}\right)^{1/2}, \tag{5.1.12}$$

where M_g is the gas molecular mass. Solving for σ_M^2 in (5.1.11) and using the ideal gas law, the expression (5.1.12) can be written as

$$\mu = \frac{5\sqrt{2\pi}\,\rho_f\lambda}{16\sqrt{M_g}}(k_B T)^{1/2}, \tag{5.1.13}$$

where ρ_f is the gas density. Recalling the expression $c = \sqrt{\kappa R_g T}$ for the speed of sound in an ideal gas, where $\kappa = c_p/c_v$ is the ratio of specific heats ($\cong 1.4$ for diatomic molecules) and $R_g = k_B/M_g$ is the specific gas constant (the ratio of the universal gas constant to the molecular mass), the result (5.1.13) can be used to obtain

$$\mu = \frac{5\sqrt{2\pi}}{16\sqrt{\kappa}}\rho_f\lambda c. \tag{5.1.14}$$

Hence, the gas viscosity is directly proportional to the mean free path of the gas molecules. Equation (5.1.14) provides a useful formula for estimating mean free path based on macroscopic quantities such as gas viscosity, density, and speed of sound.

The no-slip condition on the particle surface begins to break down when the ratio of mean free path to particle diameter exceeds about 0.01. As discussed in Section 1.2, the ratio of mean free path to particle diameter is called the Knudsen number, given by

$$\mathrm{Kn} \equiv \frac{\lambda}{d}. \tag{5.1.15}$$

From (5.1.14), $\lambda \sim \mu/\rho_f c$, so (5.1.15) becomes

$$\mathrm{Kn} \sim \frac{\mu}{\rho_f c d} = \frac{\mathrm{Ma}}{\mathrm{Re}_p}, \tag{5.1.16}$$

where the Mach number $\mathrm{Ma} = U/c$ is the ratio of the characteristic fluid velocity scale U and the speed of sound in the fluid. From (5.1.16), it follows that particles with very small Reynolds numbers may exhibit significant effects due to flow rarefaction even in flows with small Mach numbers.

A solution for the effect of surface slip on the drag on a sphere at low Reynolds number was obtained by Basset (1888) using the common assumption that the slip velocity $\mathbf{u}_{\mathrm{slip}}$ is proportional to the wall shear stress τ_{wall}, or

$$G_S\mathbf{u}_{\mathrm{slip}} = \tau_{\mathrm{wall}}, \tag{5.1.17}$$

where G_S is the coefficient of sliding friction. Basset considered the case where the Knudsen number is sufficiently small that the continuum equation can still be used (approximately $\mathrm{Kn} < 0.1$). He solved the Stokes equation (5.1.4) with the boundary condition (5.1.17) to obtain an expression for the slip correction factor as

$$C_C = \frac{G_S d + 4\mu}{G_S d + 6\mu}. \tag{5.1.18}$$

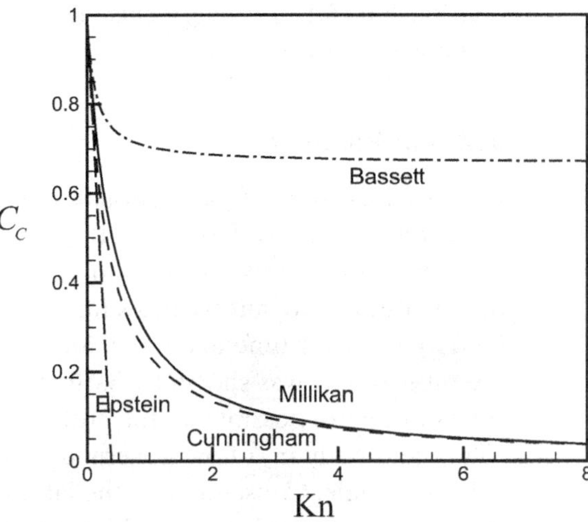

Figure 5.3. Plot comparing different expressions for the slip correction factor as a function of Knudsen number, including that of Basset (5.1.18) (dashed-dotted line), Epstein (5.1.19) (long dashed line), Cunningham (5.1.21) (short dashed line), and Millikan (1923) (5.1.22) (solid line).

For no slip $G_S \to \infty$ and $C_C \to 1$, whereas for perfect slip $G_S = 0$ and $C_C = 2/3$. Epstein (1924) used kinetic theory to derive an expression for slip correction factor that is valid over a broader range of Knudsen numbers, given by

$$C_C = 1 - \frac{2\mu}{G_S d}. \tag{5.1.19}$$

Schmitt (1959) argued that the ratio μ/G_S should be proportional to the mean free path, and proposed

$$\mu/G_S = A\lambda, \tag{5.1.20}$$

where $A \cong 1.3$. Using (5.1.20), the various expressions for C_C can be written as functions of the Knudsen number. A similar expression was derived earlier by Cunningham (1910), also using kinetic theory, as

$$C_C = \frac{1}{1 + 3.26 \text{Kn}}. \tag{5.1.21}$$

In recognition of this work, the slip correct factor is often called the Cunningham correction factor.

An empirical expression for the slip correction factor was obtained by Millikan (1923), based on his famous oil-drop experiments, as

$$C_C = \frac{1}{1 + \text{Kn}[A_1 + A_2 \exp(-A_3/\text{Kn})]}. \tag{5.1.22}$$

For air at standard temperature and pressure, $A_1 = 2.49$, $A_2 = 0.84$, and $A_3 = 1.74$. Using (5.1.22), $C_C \cong 0.99$ for 15 μm diameter particles in air at standard temperature and pressure, and $C_C \cong 0.87$ for 1 μm diameter particles. A plot comparing the different expressions for slip correction factor is shown in Figure 5.3. The Basset expression (5.1.18) gives values for C_C that are much greater than Millikan's empirical fit, where the latter is in good agreement with the experimental data. Epstein's expression is accurate for very small Knudsen numbers but quickly deviates from the experimental data as Kn increases above about 0.02. The simple Cunningham relationship (5.1.21) is very close to the Millikan correlation throughout the range of Kn values considered.

5.2. Lift Force

Particles placed in a unidirectional flow with a velocity gradient normal to the flow direction are observed to drift in the direction of the velocity gradient. A well-known demonstration of this drift was noted by Segré and Silberberg (1962), who observed that neutrally buoyant particles immersed in a fluid flowing through a circular tube tend to drift over time to a radial location that is midway between the tube wall and the tube axis. It was shown by Saffman (1956) and Bretherton (1962) that a lateral (lift) force is not possible for the problem of unidirectional flow past an isolated, rigid spheroidal body in a flow governed by the Stokes equations (5.1.4), whatever the velocity profile. Consequently, the lateral motion observed by Segré and Silberberg must involve the effect of fluid inertia, even though the inertia associated with the small particle sizes involved may be quite small compared with the other terms effecting particle motion.

The lateral force that acts on the particle is observed to depend on both the particle rotation rate and the velocity gradient of the flow in the vicinity of the particle. The problem is therefore usually divided into two separate problems. The first problem concerns a particle moving at a velocity $\mathbf{v}$ in a uniform shear flow with velocity $\mathbf{u} = u(y)\mathbf{e}_x$, where $\mathbf{v}$ and $\mathbf{u}$ are in the same direction. The second problem is that of a particle traveling at a velocity $\mathbf{v}$ and spinning with rotation rate $\mathbf{\Omega}$ through a fluid with uniform velocity $\mathbf{u} = U\mathbf{e}_x \mathbf{u}$. Both of these problems result in lateral forces on the particle, often called the *Saffman force* and the *Magnus force*, respectively, where these forces can be added together to obtain the total lateral force.

5.2.1. Saffman Lift Force

For the case of a particle traveling in the same direction as a plane shear flow with ambient velocity gradient $G = du/dy$, Saffman obtained an analytic solution for the magnitude F_ℓ of the lift force of the form

$$F_\ell = 6.46\mu(d/2)^2 v_s(G/\nu)^{1/2}, \tag{5.2.1}$$

where $v_s = |\mathbf{v} - \mathbf{u}|$ is the magnitude of the particle slip velocity and ν is the fluid kinematic viscosity. This asymptotic solution is valid under the assumptions that the particle Reynolds number $\mathrm{Re}_p = \rho_f v_s d/\mu$ is much less than the square root of the "shear Reynolds number" $\mathrm{Re}_G \equiv \rho_f d^2 G/\mu$, and that both Re_p and Re_G are much less than unity. It is noted that this leading-order solution is independent of the particle rotation rate. The lift force is oriented in a direction $\mathbf{t}_\ell$ that is parallel to the fluid velocity gradient, pointing from the low-velocity to the high-velocity side of the sphere.

The Saffman expression was extended by McLaughlin (1991) for cases where $\mathrm{Re}_p = O(\sqrt{\mathrm{Re}_G})$, but where both Re_p and Re_G are still assumed to be small, to obtain an analytical expression for lift force magnitude as

$$F_\ell = \frac{9J}{\pi}\mu(d/2)^2 v_s(G/\nu)^{1/2}, \tag{5.2.2}$$

where J is a function of the dimensionless parameter $\kappa \equiv \mathrm{Re}_G^{1/2}/\mathrm{Re}_p$. McLaughlin provided numerical computations for J as well as asymptotic solutions for large and small values of κ, given by

$$J = \begin{cases} 2.255 - 0.6463/\kappa^2 + O(1/\kappa^4) & \text{for } \kappa \gg 1 \\ -32\pi^2\kappa^5 \ln(1/\kappa^2) + O(\kappa^5) & \text{for } \kappa \ll 1 \end{cases}. \tag{5.2.3}$$

As $\kappa \to \infty$, $J \to 2.255$ and the Saffman result (5.2.1) is recovered. A curve fit to the numerical values of $J(\kappa)$ for $0.1 \leq \kappa \leq 20$ was reported by Mei (1992) as

$$J = 0.6765\{1 + \tanh[2.59\log_{10}\kappa + 0.191)]\}\{0.667 + \tanh[6(\kappa - 0.32)]\}. \tag{5.2.4}$$

Numerical simulations of lift force on a spherical particle in a linear shear flow were reported by Dandy and Dwyer (1990) for a fixed particle over Reynolds number range $0.1 \leq \mathrm{Re}_p \leq 100$. Mei (1992) developed a correlation for lift force based on the results of Saffman (1965) and the numerical results of Dandy and Dwyer (1990) which gives the ratio of the lift force at finite Reynolds number to the Saffman lift force (5.2.1) as

$$\frac{F_\ell}{F_{\ell,SAFF}} = \begin{cases} (1 - 0.3314\beta^{1/2})\exp(-\mathrm{Re}_p/10) + 0.3314\beta^{1/2} & \text{for } \mathrm{Re}_p < 40 \\ 0.0524(\beta\mathrm{Re}_p)^{1/2} & \text{for } \mathrm{Re}_p > 40 \end{cases} \tag{5.2.5}$$

where $\beta \equiv Gd/v_s = \mathrm{Re}_G/\mathrm{Re}_p$. Kurose and Komori (1999) and Bagchi and Balachandar (2002a) simulated flow past both fixed and rotating spheres in a linear shear flow for Reynolds numbers ranging from order unity to several hundred. Kurose and Komori observed that the sum of the lift force for a fixed particle in a linear shear flow and that for a rotating particle in a uniform flow is nearly equal to the lift force for a rotating particle in a linear shear flow provided that the particle Reynolds number is not too large compared to unity. This conclusion is also supported by the asymptotic study of Saffman (1965) at small particle Reynolds numbers.

A comparison of the numerical results of Dandy and Dwyer (1990), Kurose and Komori (1999), and Bagchi and Balachandar (2002a) for lift coefficient $C_L = 8F_\ell/\rho_f v_s^2 \pi d^2$ with $\beta = 0.4$ and both fixed spheres and freely rotating spheres is given in Figure 5.4, in comparison to the Saffman and McLaughlin analytic formulas and to the correlation of Mei (1992). For small particle Reynolds numbers, of order unity or less, the particle rotation rate appears to have little influence on the lift force and the data are well predicted by both the McLaughlin and Saffman solutions. The lift coefficient decreases rapidly for both fixed and rotating sphere conditions as the Reynolds number increases above order unity. The lift coefficient data at large Reynolds numbers generally fall between the predictions of the Saffman and McLaughlin formulas.

All of the developments for lift force discussed earlier are concerned with the problem of a particle in a linear shear flow. Shear flows can be decomposed into a local rotation plus a straining flow. It is of interest to examine the lift force for other flow fields and to see to what extent the lift force in such flows differs from that for a linear shear flow. The problem of a particle immersed in a rigid-body rotation, such as would occur within the core of a vortex or in a centrifuge, has been a focus of particular interest. This problem was examined using an asymptotic analysis by

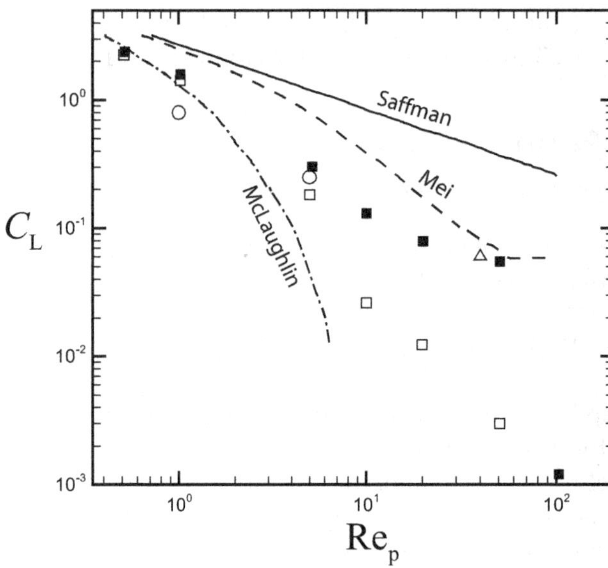

Figure 5.4. Plot of lift force coefficient versus particle Reynolds number comparing predictions of Saffman (solid line), McLaughlin (dashed-dotted line), and Mei (dashed line) for $\beta = 0.4$. The theoretical results are compared to numerical predictions with the same value of β from Bagchi and Balachandar (2002a) for spheres rotating with no net torque (filled squares) and nonrotating spheres (open squares), from Dandy and Dwyer (1990) for nonrotating spheres (delta), and from Kurose and Komori (1999) for nonrotating spheres (circles).

Herron et al. (1975) for small particle Reynolds numbers. Experimental results for the lift force on contaminated microbubbles, which behave similarly to solid spheres, immersed in vortices were reported by Sridhar and Katz (1995), who observed lift force between one and two orders of magnitude higher than predicted by expressions for linear shear flow. Bagchi and Balachandar (2002b) compared lift force for both fixed and freely rotating particles immersed in a vortex flow with that for particles immersed in a linear shear flow, both flows having the same vorticity, with Reynolds numbers in the range $10 \le \text{Re}_p \le 100$ based on numerical simulations using a high-resolution pseudospectral method. They also observed significantly higher lift force for the vortex flow at moderate values of the Reynolds number.

5.2.2. Magnus Lift Force

Problems involving colliding particles often exhibit large particle rotation rates due to the sliding resistance torque exerted on the particles during collisions. A spinning particle moving through a fluid is observed to experience a lateral force. This phenomenon, called the *Magnus effect*, is commonly observed in the motion of a curveball in baseball or in the various spin shots used in tennis. For flow with small particle Reynolds numbers, Rubinow and Keller (1961) showed that a sphere spinning with rotation rate $\mathbf{\Omega}$ and translating at a velocity $\mathbf{v}$ through an otherwise still fluid experiences a lift force $\mathbf{F}_m$ given by

$$\mathbf{F}_m = \pi \rho_f (d/2)^3 \mathbf{\Omega} \times \mathbf{v}. \tag{5.2.6}$$

Experimental results for lift force on a spinning sphere traveling in an otherwise stationary fluid were reported by Oesterlé and Dinh (1998), Tsuji et al. (1985), and Tri et al. (1990), and numerical results were given by Bagchi and Balachandar (2002a). These studies have found that although (5.2.6) gives the correct trends for the Magnus force even for Reynolds numbers as high as 100, experimental and numerical data are better fit if the expression in (5.2.6) is multiplied by a factor $C_S \le 1$. Values of

C_S proposed by different investigators include $C_S = 0.25$ (Tanaka et al., 1990), $C_S = 0.4$ (Tri et al., 1990), and $C_S = 0.55$ (Bagchi and Balachandar, 2002a).

The problem of a rotating particle immersed in a linear shear flow was considered analytically by Saffman (1965) and via numerical computation by Kurose and Komori (1999) and Bagchi and Balachandar (2002a). Saffman employed an asymptotic analysis to show that the total lift is given by

$$F_{\ell,TOT} = 6.46\mu(d/2)^2 v_s (G/v)^{1/2} + \pi\rho_f (d/2)^3 v_s \left(\Omega - \frac{11}{8\pi}G\right), \qquad (5.2.7)$$

where Ω is the magnitude of the particle rotation rate. The first term in (5.2.7) is the same as given in (5.2.1) for a fixed particle in a linear shear flow, and the part proportional to $\boldsymbol{\Omega}$ in the second term is the same as the Rubinow-Keller expression (5.2.6) for the Magnus force. The factor $11G/8\pi \cong 0.4377G$ in the last term in (5.2.7) is close to the value $G/2$ for the rotation rate of fluid elements in a flow with shear rate G.

It has been noted by a number of investigators that a spherical particle in a linear shear flow develops a spin that asymptotes at a torque-free rotation rate Ω_{st} which is different than the rotation rate $\Omega_f = G/2$ of fluid elements in the shear flow. Lin et al. (1970) employed an asymptotic method for small Reynolds numbers to show that

$$\frac{\Omega_{st}}{\Omega_f} = 1 - 0.0385\mathrm{Re}_G^{3/2}. \qquad (5.2.8)$$

Based on numerical simulations at higher Reynolds numbers, Bagchi and Balachandar (2002a) developed a correlation for this ratio as

$$\frac{\Omega_{st}}{\Omega_f} = \begin{cases} 1 - 0.03464\mathrm{Re}_p^{0.95} & \text{for } 0.5 < \mathrm{Re}_p \leq 5 \\ 1 - 0.0755\mathrm{Re}_p^{0.455} & \text{for } 5 \leq \mathrm{Re}_p \leq 200 \end{cases}. \qquad (5.2.9)$$

In the vanishing Reynolds number limit, $\Omega_{st} \to \Omega_f$, in agreement with the viscous torque expression (5.1.6).

5.3. Forces in Unsteady Flows

Several additional types of forces act on a particle if the fluid velocity at the particle centroid location varies in time. This can occur either due to an unsteady velocity field or due to movement of the particle in a nonhomogeneous velocity field.

5.3.1. Pressure-Gradient (Buoyancy) Force

Pressure-gradient force exists whenever a body is placed in an ambient pressure gradient. The general expression for the force F_p due to a pressure field p acting on the surface S of a body with outward unit normal $\mathbf{n}$ is

$$\mathbf{F}_p = -\int_S p\mathbf{n}\,da. \qquad (5.3.1)$$

Using the divergence theorem and assuming a constant pressure gradient ∇p, this expression reduces to

$$\mathbf{F}_p = -\int_V \nabla p \, dv = -V \nabla p. \tag{5.3.2}$$

Pressure-gradient force can arise from any type of pressure gradient. For instance, in a gravitational field the hydrostatic pressure variation is given by $\nabla p = -\rho_f g \mathbf{e}_z$, where g is the gravitational acceleration. Substituting this expression into (5.3.2) gives

$$\mathbf{F}_p = \rho_f g V \mathbf{e}_z, \tag{5.3.3}$$

which is the buoyancy force predicted by Archimedes' law. Combining (5.3.3) with the gravitational force acting on the particle yields the *reduced gravity force*

$$\mathbf{F}_g = (\rho_f - \rho_p) g V \mathbf{e}_z = -\rho_p g_R V \mathbf{e}_z, \tag{5.3.4}$$

where $g_R = (1 - \chi)g$ is called the reduced gravitational acceleration and $\chi = \rho_f / \rho_p$.

In many fluid flows of interest, the flow Reynolds number is large and, outside of boundary layers, the viscous term is small compared with the inertial term. In such cases, the pressure gradient associated with fluid acceleration is $\nabla p = -\rho_f D\mathbf{u}/Dt$, where D/Dt denotes the material derivative following a fluid material element. Substituting this expression into (5.3.2) gives the pressure-gradient force as

$$\mathbf{F}_p = \frac{\pi}{6} \rho_f d^3 \frac{D\mathbf{u}}{Dt}. \tag{5.3.5}$$

This expression is often known as the inertial pressure-gradient force to distinguish it from the buoyancy force due to the hydrostatic pressure gradient. An interesting and important example of the inertial pressure-gradient force occurs when a particle is exposed to a vortex flow field, as was discussed in Section 1.2.

5.3.2. Added Mass Force

The added mass force is a term used to describe the additional force that is required to accelerate a body immersed in a dense fluid compared to that required when the body is in a vacuum. This additional force makes the body respond as if it has a greater mass than it actually does, hence the term "added mass." The added mass arises from the fact that an object accelerating relative to the surrounding fluid must also increase the kinetic energy associated with motion of the surrounding fluid. This increased kinetic energy is derived from work that is performed by the "added mass force" as the body is accelerated.

The classical expression for added mass force $\mathbf{F}_a$ of a body traveling with speed $\mathbf{v}$ in an inviscid fluid with uniform velocity field $\mathbf{u}(t)$ is (Batchelor, 1967)

$$\mathbf{F}_a = -c_M \frac{\pi}{6} \rho_f d^3 \left(\frac{d\mathbf{v}}{dt} - \frac{d\mathbf{u}}{dt} \right), \tag{5.3.6}$$

where for a sphere the added mass coefficient is $c_M = 1/2$. The notation d/dt is the time derivative following the particle. In the case that the ambient fluid flow is not uniform, Auton et al. (1988) showed that the correct expression for force on a particle in an inviscid flow has the form

$$\mathbf{F} = \rho_f V \left[\frac{D\mathbf{u}}{Dt} - c_M \left(\frac{d\mathbf{v}}{dt} - \frac{D\mathbf{u}}{Dt} \right) + c_L (\mathbf{u} - \mathbf{v}) \times \omega \right], \tag{5.3.7}$$

where V is the particle volume, c_L is a lift coefficient, ω is the ambient fluid vorticity, and D/Dt denotes the material derivative following a fluid element. For a spherical particle, $c_M = c_L = 1/2$. We recognize several different forces within this expression. For instance, the first term within the brackets is simply the inertial pressure-gradient force given by (5.3.5). The third term is an inviscid lift force that acts on a particle in the presence of nonzero fluid vorticity. The ratio of the magnitude of this inviscid lift term and the magnitude of the Saffman lift (5.2.1) for a linear shear flow gives

$$\frac{F_{\ell,INV}}{F_{\ell,SAFF}} \cong 0.16 \mathrm{Re}_G^{1/2}. \tag{5.3.8}$$

Hence, for small values of Re_G the Saffman lift is substantially larger than the inviscid lift, and the latter may be safely neglected, at least for a solid particle. On the other hand, numerical simulations by Legendre and Magnaudet (1998) for a clean bubble at a moderate particle Reynolds number ($\mathrm{Re}_p = 500$) yielded a value of lift coefficient of $c_L = 0.49$, which is quite close to the inviscid result.

The second term in (5.3.7) provides an expression for added mass force that is valid for general ambient flow fields as

$$\mathbf{F}_a = -c_M \frac{\pi}{6} \rho_f d^3 \left(\frac{d\mathbf{v}}{dt} - \frac{D\mathbf{u}}{Dt} \right). \tag{5.3.9}$$

The only difference between the expression (5.3.6) for uniform flow and the expression (5.3.9) for general flows is the type of derivative acting on the fluid velocity $\mathbf{u}$. The difference between these two derivatives is given by

$$\frac{D\mathbf{u}}{Dt} - \frac{d\mathbf{u}}{dt} = (\mathbf{u} - \mathbf{v}) \cdot \nabla \mathbf{u}. \tag{5.3.10}$$

When this is substituted into (5.3.9), the difference between the expression (5.3.9) and the standard expression (5.3.6) is of the same order of magnitude as the inviscid lift force. While the expression (5.3.9) for added mass force was derived for inviscid flow, it was demonstrated in the numerical simulations by Rivero et al. (1991) that the expression remains accurate with $c_M = 1/2$ even in viscous flows over a wide range of Reynolds number and fluid accelerations (see also discussion in Magnaudet, 1997).

5.3.3. History Force

The Basset history force is associated with the response of the particle viscous shear force to acceleration of the flow relative to the particle. The basic effect can be interpreted in part by consideration of the classic Stokes problem of a flat plate lying along the x-axis in an unbounded fluid domain. At time $t = t'$ the plate is accelerated in the x-direction with a small velocity increment dU. The fluid above the plate is driven into motion in order to satisfy the no-slip condition. The Navier-Stokes equation for the fluid velocity $u(y, t)$ reduces to

$$\frac{\partial u}{\partial t} = \nu \frac{\partial^2 u}{\partial y^2}. \tag{5.3.11}$$

This problem admits a similarity solution of the form

$$u(y, t) = dU \operatorname{erf}\left[\frac{y}{2\sqrt{\nu(t - t')}}\right], \tag{5.3.12}$$

where $\operatorname{erf}(\cdot)$ is the error function. The shear stress at the plate surface ($y = 0$) varies as a function of time as

$$\tau_w = \mu \frac{\partial u}{\partial y}\bigg|_{y=0} = \left(\frac{\rho_f \mu}{\pi}\right)^{1/2} \frac{dU}{\sqrt{t - t'}}. \tag{5.3.13}$$

Because (5.3.11) is a linear equation, we can decompose a gradual velocity variation $U(t)$ into a series of small discrete jumps and write the total velocity as the sum of that due to each of these jumps. The wall shear stress at time t due to this series of discrete jumps can then be obtained from (5.3.13) as

$$\tau_w = \mu \frac{\partial u}{\partial y}\bigg|_{y=0} = \left(\frac{\rho_f \mu}{\pi}\right)^{1/2} \int_0^t \frac{dU/dt'}{\sqrt{t - t'}} dt'. \tag{5.3.14}$$

This example illustrates some of the physics associated with how acceleration of the particle relative to the velocity $\mathbf{u}$ of the surrounding fluid can lead to a time-varying viscous force that augments the steady-state force on the particle. For a spherical particle at low particle Reynolds number, an expression for this additional "history" force was obtained by Boussinesq (1885) and Basset (1888) as

$$\mathbf{F}_h = \mu d \int_{-\infty}^t K(t - t') \left(\frac{d\mathbf{u}}{dt'} - \frac{d\mathbf{v}}{dt'}\right) dt', \tag{5.3.15}$$

where the kernel function $K(t - t')$ is given by

$$K(t - t') = \frac{3}{2} \left[\frac{\pi \rho_f d^2}{\mu(t - t')}\right]^{1/2}. \tag{5.3.16}$$

This force is often referred to as the Boussinesq-Basset history force, or simply the Basset force. Druzhinin and Ostrovsky (1994) used this expression to study the effect of the Basset force for particle transport in both a vortex and a cellular flow field at low Stokes numbers. For the problems of drift of a heavy particle out of a vortex or drift of a light particle toward a vortex, Druzhinin and Ostrovsky found that the Basset force acts in the direction opposite to that of the particle drift, significantly decreasing the rate of drift compared with simulations with no Basset force. A later experimental and numerical study of this problem was reported by Candelier et al. (2004), who found that the difference in drift rate with and without the Basset force varies with the square root of the Stokes number. Although the steady-state viscous drag is normally the dominant force on the particle, particularly at low Reynolds numbers, Thomas (1992) observed in an investigation of particle motion through a shock wave that within the shock, the Basset force can be of a similar order of magnitude to the viscous drag.

Numerical studies conducted by Mei et al. (1991) and Rivero et al. (1991) for a flow with small-amplitude oscillations concluded that while the kernel (5.3.16) yields reasonable results at short times, it predicts values for the Basset force at long times that are too large. They trace this problem to the fact that in the far-field (Oseen) region of the sphere wake, advection becomes a more effective method for vorticity transport than diffusion due to the small velocity gradients. The Basset-Boussinesq expression given by (5.3.15) and (5.3.16) was derived using the Stokes equation (5.1.4), which assumes purely diffusion-controlled vorticity transport. Alternative expressions for history force were proposed by Mei and Adrian (1992) and Lovalenti and Brady (1993), giving an integration kernel that decays as t^{-2} at long time.

5.4. Brownian Motion

Observe what happens when sunbeams are admitted into a building and shed light on its shadowy places. You will see a multitude of tiny particles mingling in a multitude of ways... their dancing is an actual indication of underlying movements of matter that are hidden from our sight.... It originates with the atoms which move of themselves. Then those small compound bodies that are least removed from the impetus of the atoms are set in motion by the impact of their invisible blows and in turn cannon against slightly larger bodies. So the movement mounts up from the atoms and gradually emerges to the level of our senses, so that those bodies are in motion that we see in sunbeams, moved by blows that remain invisible (Lucretius, "On the Nature of Things," c. 60 BC).

It has been known since ancient times that very small bodies execute strange, random motions when suspended in a gas. The verse by the Roman poet and philosopher Lucretius presented here is remarkable not only for its understanding of the particulate nature of the gas but also for its interpretation of this random motion as being due to the collision of dust particles with the elementary particles that make up the fluid. A similar random motion was observed in the 1827 by the botanist Robert Brown (Brown, 1828), who attributed random motions of pollen particles suspended in water to the collisions of the particles with the water molecules. These random motions have played an important role not only in the understanding of the molecular nature of matter, but also as an important example in the development of the field of stochastic processes, with key early contributions on development of a mathematical theory of Brownian motion of particles by Einstein (1906) and Smoluchowski (1906).

The random force acting on a small particle immersed in a fluid that gives rise to Brownian motion is known as the Brownian force. In order to examine Brownian force, we consider a simplified model in which the particle momentum equation is reduced to three terms – the particle inertia, the Stokes drag, and the Brownian force. The resulting equation, given by

$$m\frac{d^2\mathbf{x}}{dt^2} + 3\pi d\mu \frac{d\mathbf{x}}{dt} = \mathbf{F}_b(t), \tag{5.4.1}$$

is a form of the *Langevin equation* discussed in Chapter 2. In this equation, $\mathbf{x}(t)$ is the position vector of the particle centroid and $\mathbf{F}_b(t)$ denotes the Brownian force.

For very small particles, it might be necessary to modify the drag term in (5.4.1) to include the slip correction factor C_C discussed in Section 5.1, but we will omit the effect of slip for the present discussion.

The Brownian force can be characterized on the basis of two assumptions. The first assumption is that the Brownian force is an uncorrelated random function that can be represented by a Gaussian probability distribution with mean and covariance given by

$$\langle \mathbf{F}_b(t) \rangle = 0 \tag{5.4.2}$$

$$\langle \mathbf{F}_b(t)\mathbf{F}_b^T(t+\tau) \rangle = \mathbf{B}\delta(\tau). \tag{5.4.3}$$

The notation $\langle \cdot \rangle$ denotes the ensemble average, defined by

$$\langle \mathbf{F}_b(t) \rangle = \frac{1}{N} \sum_{i=1}^{N} \mathbf{F}_b^{(i)}(t). \tag{5.4.4}$$

Here $\mathbf{F}_b^{(i)}(t)$ is the ith realization of a stochastic process having the same initial conditions. The notation $\delta(t)$ denotes the Dirac delta, defined such that $\delta(t) = 0$ for $t \neq 0$ and $\int_{-\infty}^{\infty} \delta(t)dt = 1$, and $\mathbf{B}$ is a constant, second-order tensor. The second assumption is that when the particle is in an equilibrium state (such that there is no mean motion), the particle kinetic energy is partitioned as

$$\frac{1}{2}m \langle \mathbf{v}\mathbf{v}^T \rangle = \frac{1}{2}k_B T \mathbf{I}, \tag{5.4.5}$$

where k_B is the Boltzmann constant, T is absolute temperature, m is the particle mass, and $\mathbf{I}$ is the identity tensor.

The Langevin equation (5.4.1) is a first-order linear ordinary differential equation for the particle velocity $d\mathbf{x}/dt$, which admits the solution

$$\frac{d\mathbf{x}}{dt} = \frac{1}{m} \int_{-\infty}^{t} \mathbf{F}_b(t') \exp[-3\pi d\mu(t-t')/m]dt'. \tag{5.4.6}$$

From (5.4.6) and (5.4.3), the velocity autocorrelation is obtained as

$$\mathbf{R}(\tau) \equiv \left\langle \frac{d\mathbf{x}}{dt}(t)\frac{d\mathbf{x}^T}{dt}(t+\tau) \right\rangle = \frac{\mathbf{B}}{6\pi \mu d m} \exp\left(-\frac{3\pi \mu d}{m}\tau\right). \tag{5.4.7}$$

Because $\mathbf{R}(0) = \langle \mathbf{v}\mathbf{v} \rangle$, (5.4.7) with $\tau = 0$ and (5.4.5) yield an expression for the constant tensor $\mathbf{B}$ as

$$\mathbf{B} = 6\pi \mu d k_B T \mathbf{I}. \tag{5.4.8}$$

The ensemble average of the square of the particle displacement $\langle \mathbf{x}\mathbf{x}^T \rangle$ is called the mean-square displacement (MSD). The autocorrelation function (5.4.7) can be used to show that the MSD over long time ($t \gg d^2/\nu$) approaches

$$\langle \mathbf{x}\mathbf{x}^T \rangle = \frac{2k_B T}{3\pi \mu d}t\mathbf{I}, \tag{5.4.9}$$

from which it follows that the averaged position of a particle increases in proportion to the square root of time.

Einstein (1906) and Smoluchowski (1906) showed that over long times, a set of particles undergoing Brownian motion diffuse outward from regions of high particle concentration in a manner as if the concentration $\phi(\mathbf{x}, t)$ obeyed a diffusion equation of the form

$$\frac{\partial \phi}{\partial t} = D_b \nabla^2 \phi. \tag{5.4.10}$$

The diffusion coefficient D_b is equal to the coefficient on the right-hand side of (5.4.9), or

$$D_b = \frac{k_B T}{3\pi \mu d}, \tag{5.4.11}$$

which is known as the Stokes-Einstein relation. In terms of the diffusion coefficient, the MSD in (5.4.9) can be written as

$$\langle \mathbf{x}\mathbf{x}^T \rangle = 2 D_b t \mathbf{I}. \tag{5.4.12}$$

Equation (5.4.12) is valid for long times, satisfying $t \gg \tau_p$ where $\tau_p = m/3\pi d\mu$ is the particle response time. In the opposite extreme of very short times, such that $t \ll \tau_p$, the particle dynamics are dominated by the particle inertia, and the corresponding particle motion changes from a diffusive nature with MSD proportional to t to a ballistic nature with MSD proportional to t^2. The short- and long-time behavior of particles under Brownian motion was examined in a recent experimental study by Li et al. (2010), which used a dual-beam optical tweezer to measure the MSD of a 3 μm silica bead as a function of time. The MSDs over short time intervals are found to be significantly different from the predictions from Einstein's theory of Brownian motion in a diffusive regime, but agree well with the predictions of the ballistic Brownian motion theory that the MSD is proportional to t^2.

For purposes of numerical solution, it is convenient to write (5.4.1) in terms of the velocity differential $d\mathbf{v}$ as

$$m d\mathbf{v} + 3\pi d\mu \mathbf{v} dt = B^{1/2} d\mathbf{W}. \tag{5.4.13}$$

In this equation, the scalar B is defined such that $\mathbf{B} = B\mathbf{I}$, so from (5.4.8)

$$B = 6\pi \mu d k_B T. \tag{5.4.14}$$

The vector $d\mathbf{W}$ can be written in component form as $d\mathbf{W} = dW_i \mathbf{e}_i$, where the components dW_i are independent random variables with Gaussian probability distribution having zero mean and variance dt. Comparing (5.4.13) with (5.4.1), the Brownian force $\mathbf{F}_b$ is given by

$$\mathbf{F}_b = B^{1/2} \frac{d\mathbf{W}}{dt}. \tag{5.4.15}$$

5.5. Scaling Analysis

The relative importance of the different forces discussed in this chapter can be examined using a scaling analysis, which we present for the case of a particle of diameter d moving in a linear shear flow with shear rate G. The major parameters used in this scaling analysis are the dimensionless particle diameter $\varepsilon_d \equiv d/L$, the flow Reynolds

Table 5.1. *Order of magnitude estimates for the different fluid forces on the particles, compared with the particle inertia or the particle drag, which are assumed to be of the same order of magnitude*

Force	Order of magnitude relative to particle inertia/drag
Drag force	1
Saffman lift force	$\mathrm{Re}_G^{1/2}$
Magnus lift force	Re_G
Added mass force	χ
Pressure-gradient force	χ
Basset history force	$(\mathrm{Re}_f \varepsilon_d \mathrm{St})^{1/2} = \mathrm{Re}_p^{1/2}$
Gravitational force	$(\varepsilon_d \mathrm{Fr}^2)^{-1}$
Brownian force	$(1/\varepsilon_d \mathrm{St}\,\mathrm{Pe}_f)^{1/2} = \mathrm{Pe}_p^{-1/2}$

number $\mathrm{Re}_f = \rho_f U L / \mu$, the shear Reynolds number $\mathrm{Re}_G = \rho_f G d^2 / \mu$, the density ratio $\chi = \rho_f / \rho_p$, the Stokes number $\mathrm{St} \equiv \rho_p d^2 U / 18 \mu L$, the particle Froude number $\mathrm{Fr} = U / \sqrt{g_R d}$, and the flow Peclet number associated with Brownian motion $\mathrm{Pe}_f = L U / D_b$. Here L and U are length and velocity scales associated with the flow field and D_b is the Brownian diffusion coefficient defined in (5.4.11). As is typical of a particle in the micrometer size range, it is assumed that fluid drag is the dominant force balanced by the particle inertia. For particle flows at low Stokes numbers, balancing the order of magnitude of particle inertia and drag results in the scaling $|\mathbf{v} - \mathbf{u}| = O(\mathrm{St} U)$ for the particle slip velocity. The scaling estimate also neglects particle collisions, instead concentrating on the forces acting on an isolated particle immersed in the fluid, and it assumes that the particle Reynolds number is small.

A listing of the order of magnitude of the different forces discussed in the chapter is given in Table 5.1. This table lists the ratio of the order of magnitude of each force divided by the order of magnitude of the particle inertia. The entry next to drag is unity, as drag is of the same order of magnitude as inertia. The ratio of the Saffman lift force (5.2.1) to the Stokes drag force gives

$$\frac{F_\ell}{F_d} = \frac{O\left[\mu d^2 v_S (G/\nu)^{1/2}\right]}{O(\mu d v_S)} = O\left[d(G/\nu)^{1/2}\right] = O\left(\mathrm{Re}_G^{1/2}\right). \tag{5.5.1}$$

Taking the ratio of the Magnus force expression (5.2.6) to the Stokes drag force and assuming that the particle rotation rate Ω is of the same order of magnitude as the shear rate G gives

$$\frac{F_m}{F_d} = \frac{O[\rho_f d^3 v_S \Omega]}{O(\mu d v_S)} = O[\rho_f d^2 G / \mu] = O(\mathrm{Re}_G) \tag{5.5.2}$$

The ratio of the reduced gravity force (5.3.4) to the Stokes drag, using the observation that $v_S = O(\mathrm{St} U)$, is obtained as

$$\frac{F_g}{F_d} = \frac{O(\rho_p g_R d^3)}{O(\mu d v_S)} = O\left(\frac{\varepsilon_d \mathrm{Re}_f}{\chi \mathrm{St}\,\mathrm{Fr}^2}\right). \tag{5.5.3}$$

Using the relationship (1.2.14) between the Stokes number and the flow Reynolds number, this expression becomes

$$\frac{F_g}{F_d} = O\left(\frac{1}{\varepsilon_d \mathrm{Fr}^2}\right). \tag{5.5.4}$$

The inertial pressure-gradient force and the added mass force have the same order of magnitude. In this case it is more convenient to divide the force by the order of magnitude of the particle inertia $F_i = O(md\mathbf{v}/dt) = O(\rho_p d^3 U/\tau_p)$, where τ_p is a particle time scale, to get

$$\frac{F_p}{F_i} = O\left(\frac{F_a}{F_i}\right) = \frac{O(\rho_f d^3 U/\tau_p)}{O(\rho_p d^3 U/\tau_p)} = O(\chi). \tag{5.5.5}$$

The ratio of the history force (5.3.15) to the drag force gives

$$\frac{F_h}{F_d} = \frac{O(\mu d^2 v_S/\sqrt{\nu\tau})}{O(\mu d v_S)}. \tag{5.5.6}$$

The time scale most suited for the viscous stresses on the particle surface is given by the time associated with the Stokes drag, or $\tau_S = d/v_S$. Substituting this time scale into (5.5.6) and using $v_S = O(\mathrm{St}U)$ for the slip velocity gives

$$\frac{F_h}{F_d} = O\left(\frac{dv_S}{\nu}\right)^{1/2} = O(\varepsilon_d \mathrm{Re}_f \mathrm{St})^{1/2}. \tag{5.5.7}$$

We recall from (1.2.13) that the combination $\varepsilon_d \mathrm{Re}_f \mathrm{St}$ in (5.5.7) has the same order of magnitude as the particle Reynolds number Re_p.

From (5.4.15), the Brownian force has order of magnitude $(B/\tau)^{1/2}$, where τ is an appropriate particle time scale. If we again use the Stokes drag time scale $\tau_S = d/v_S$, the ratio of the Brownian force to the drag force is given by

$$\frac{F_b}{F_d} = \frac{O(\mu d k_B T/\tau)}{O(\mu d v_S)} = O\left(\frac{k_B T}{\mu d^2 v_S}\right)^{1/2}. \tag{5.5.8}$$

Recalling the estimate $v_S = O(\mathrm{St}U)$, (5.5.8) can be written as

$$\frac{F_b}{F_d} = O\left(\frac{1}{\varepsilon_d \mathrm{St}\,\mathrm{Pe}_f}\right)^{1/2}. \tag{5.5.9}$$

A particle Peclet number can alternatively be defined using the particle diameter d and slip velocity v_S as length and velocity scales to write

$$\mathrm{Pe}_p \equiv v_S d/D_b = O(\varepsilon_d \mathrm{St}\,\mathrm{Pe}_f), \tag{5.5.10}$$

in terms of which (5.5.9) becomes

$$\frac{F_b}{F_d} = O\left(\mathrm{Pe}_p^{-1/2}\right). \tag{5.5.11}$$

Brownian force starts to become important when $\mathrm{Pe}_p = O(1)$. When $\mathrm{Pe}_p \ll 1$, the Brownian force will become one of the dominant forces, in which case the particle motion is governed primarily by a balance between the Stokes drag and the random Brownian force.

As a first example, we consider a dust particle with diameter $d = 10\,\mu\text{m}$ in an air flow with characteristic length and velocity scales $L = 1\,\text{cm}$ and $U = 1\,\text{m/s}$. Assuming that the dust is made of quartz, the fluid and particle densities are $\rho_f = 1.2$ kg/m^3 and $\rho_p = 2650$ kg/m^3. The shear rate is estimated as $G \cong U/L = 100\,\text{s}^{-1}$. The Brownian diffusion coefficient at room temperature ($T = 300$ K) is given by (5.4.11) as $D_b = 2.44 \times 10^{-12}$ m^2/s. For this example, the dimensionless parameters listed at the start of this section are given by

$$\chi = 0.000453, \quad \text{Re}_f = 667, \qquad \text{Re}_G = 6.7 \times 10^{-4}, \quad \text{St} = 0.082,$$

$$\varepsilon_d = 0.001, \qquad \text{Pe}_f = 4.1 \times 10^9, \qquad \text{Fr} = 101.$$

Using the order of magnitude estimates stated earlier, the typical particle Reynolds number and Peclet number can be estimated as

$$\text{Re}_p \cong 0.055, \quad \text{and} \quad \text{Pe}_p = 3.4 \times 10^5.$$

From the estimates in Table 5.1, for this example the force ratios are given approximately by

$$\frac{F_\ell}{F_d} \approx 0.026, \quad \frac{F_m}{F_d} \approx 0.0007, \quad \frac{F_a}{F_d} \approx \frac{F_a}{F_i} \approx 0.0004,$$

$$\frac{F_h}{F_d} \approx 0.23, \quad \frac{F_g}{F_d} \approx 0.1, \quad \frac{F_b}{F_d} \approx 0.0017.$$

We conclude from these estimates that Magnus lift, added mass force, pressure-gradient force, and Brownian force are all negligible. The Saffman lift force is also small and can probably be neglected. The history and gravity forces are significant enough that they will likely influence the analysis; however, both remain minor forces compared to drag and inertia.

In a second example, we consider a neutrally buoyant bacterial cell (with no flagella) having diameter $d = 1\,\mu\text{m}$ in a microfluidic flow of water through a microchannel with $L = 1\,\text{mm}$ and $U = 10\,\text{cm/s}$. The shear rate is estimated as $G \cong U/L = 100\,\text{s}^{-1}$, and the Brownian diffusion coefficient is $D_b = 4.4 \times 10^{-13}$ m^2/s. For this example, the dimensionless parameters listed at the start of this section are given by

$$\chi = 1, \quad \text{Re}_f = 100, \quad \text{Re}_G = 1.0 \times 10^{-7}, \quad \text{St} = 5.6 \times 10^{-6},$$

$$\varepsilon_d = 0.001, \quad \text{Pe}_f = 2.3 \times 10^8, \quad \text{Fr} = 31.9.$$

The typical particle Reynolds number and particle Peclet number are given by

$$\text{Re}_p \cong 5.6 \times 10^{-7} \quad \text{and} \quad \text{Pe}_p = 1.27.$$

The estimates in Table 5.1 give the force ratios for this example as

$$\frac{F_\ell}{F_d} \approx 3.2 \times 10^{-4}, \quad \frac{F_m}{F_d} \approx 10^{-7}, \quad \frac{F_a}{F_d} \approx \frac{F_a}{F_i} \approx 1,$$

$$\frac{F_h}{F_d} \approx 7.5 \times 10^{-4}, \quad \frac{F_g}{F_d} \approx 0, \quad \frac{F_b}{F_d} \approx 0.89.$$

In this example, the reduced gravity force is identically zero, and the Saffman and Magnus lift forces and the history force are all negligible. However, the added mass and pressure-gradient forces and the Brownian force are all of about the same order of magnitude as the particle inertia and drag force.

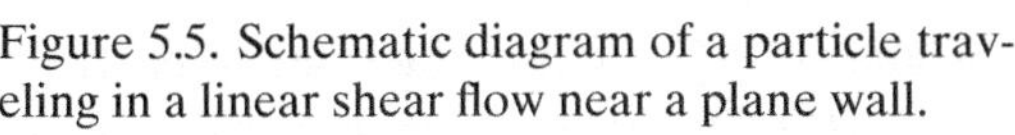
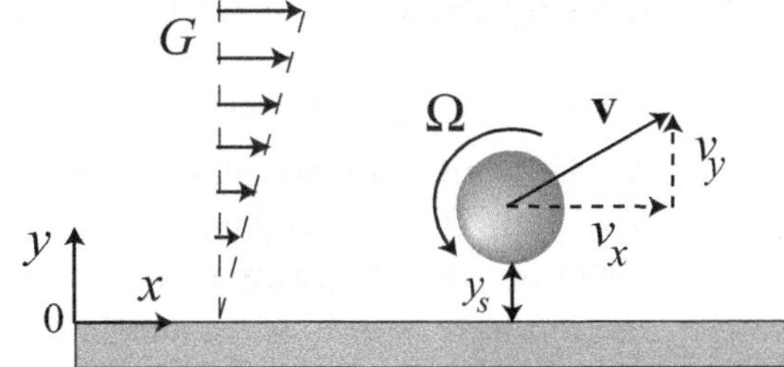

Figure 5.5. Schematic diagram of a particle traveling in a linear shear flow near a plane wall.

5.6. Near-Wall Effects

The presence of a wall in the flow field can significantly modify the hydrodynamic forces acting on particles close to the wall. We consider a particle of diameter d with centroid at a height $d/2 + y_s$ above a wall, where the offset distance y_s is on the order of the particle diameter or smaller. Assuming that the particle diameter is much less than the characteristic fluid length scale, the influence of the wall on the particle forces can be examined by considering the simplified problem of a particle above a flat, infinite wall subject to a linear shear flow in the x-direction with shear rate G. The particle centroid has translational velocity $\mathbf{v} = v_x\mathbf{e}_x + v_y\mathbf{e}_y$ and the particle rotates with rotation rate $\mathbf{\Omega} = \Omega\mathbf{e}_z$, where Ω is positive for a particle rotating in a counter-clockwise direction. A schematic diagram of this flow field is given in Figure 5.5.

The limiting case where the particle touches the plate ($y_s \rightarrow 0$) is of particular interest for adhesive particles, which are commonly drawn to surfaces in the flow field. Included in this limit is the important problem of a particle rolling along a surface, for which $\Omega = -2v_x/d$ and $v_y = 0$. As pointed out by Goldman et al. (1967a) and O'Neill and Stewartson (1967), the force and torque acting on a particle exhibit a logarithmic singularity as the separation distance y_s approaches zero for any problem with nonzero particle translation velocity or rotation rate. Hence, a perfectly smooth spherical particle in contact with a perfectly smooth, inclined wall cannot roll down the wall under any finite body force, according to the restrictions imposed by the Navier-Stokes equations and the no-slip boundary condition. This statement is, of course, at odds with the everyday observation that apparently smooth particles do in fact roll down apparently smooth walls. Because this force singularity is proportional to $\log(y_s)$, even very small values of the separation distance y_s are sufficient to allow particle motion with reasonable force magnitudes. It was demonstrated by Smart et al. (1993) that this singularity can be effectively resolved by accounting for the effects of surface roughness, such that the separation distance y_s is not allowed to decrease below the approximate height of the particle surface roughness elements.

5.6.1. Drag Force

The drag force on a spherical particle at low particle Reynolds number is assumed to vary linearly with the shear rate G, the particle translation velocities in the x- and y-directions, and the particle rotation rate, so that

$$F_{D,x} = \tfrac{3}{2}\pi\mu d^2 G C_G + 3\pi\mu d v_x C_U + \tfrac{3}{2}\pi\mu d^2 \Omega C_R \qquad (5.6.1)$$

$$F_{D,y} = 3\pi\mu d v_y C_V. \qquad (5.6.2)$$

Similarly, the torque on the particle can be expressed as

$$M_F = -\pi \mu d^3 G D_G - \pi \mu d^3 \Omega D_R - 2\pi \mu d^2 v_x D_U. \tag{5.6.3}$$

We define the dimensionless separation distance $\zeta \equiv 2y_s/d$ and the relative particle velocity in the x-direction, $v_S = v_x - G(1 + \zeta)d/2$. A variety of studies have been performed in order to obtain expressions for the coefficients $C_G, C_U, C_R, C_V, D_R, D_G$, and D_U in (5.6.1), (5.6.2), and (5.6.3) as functions of the dimensionless separation distance ζ. In the limit of a particle far away from the wall ($\zeta \to \infty$), the Stokes drag result is obtained for $C_G = -C_U = -C_V = D_R = -2D_G = 1$ and $C_R = D_U = 0$. This provides a limit that expressions for these coefficients with finite ζ must satisfy.

Most analyses examining particle forces and torques near a wall adopt one of two approaches. The first approach, valid for large separation distances ($\zeta \gg 1$), is to use some variation of the method of images to account for the reflection of the sphere over the wall surface, where the sphere is modeled by a superposition of a stokeslet, a doublet, and a uniform flow. For instance, Blake (1971) derived the image system of a stokeslet over a no-slip wall, and Happel and Brenner (1983, 288–297) utilized this technique to examine particle interaction with a plane wall as well as with other particles based on the earlier work of Brenner. The second approach, valid for small separation distances ($\zeta \ll 1$), involves the use of the lubrication approximation for the gap flow between the particles and the wall, which is matched to an outer solution for the flow in the region away from the wall. This approach was used by Goldman et al. (1967a) and O'Neill and Stewartson (1967), as well as numerous later investigators, to derive the forces on a particle that is nearly in contact with a wall.

Goldman et al. (1967a) examined a number of scenarios using lubrication theory that give limiting expressions for these coefficients valid for $\zeta \ll 1$. Starting with the problem of a nonrotating sphere ($\Omega = 0$) translated over a wall with no shear flow ($G = 0$), they derived limiting expressions for C_U and D_U as

$$C_U = \tfrac{8}{15} \ln(\zeta) - 0.9588, \tag{5.6.4a}$$

$$D_U = -\tfrac{1}{10} \ln(\zeta) - 0.1895. \tag{5.6.4b}$$

Goldman et al. (1967a) then considered the problem of a rotating particle with no translation ($v_x = 0$) near a wall, finding that for $\zeta \ll 1$ the coefficients can be written as

$$C_R = -\tfrac{2}{15} \ln(\zeta) - 0.2526, \tag{5.6.4c}$$

$$D_R = -\tfrac{2}{5} \ln(\zeta) + 0.3817. \tag{5.6.4d}$$

In a follow-up paper examining a stationary, nonrotating particle in a shear flow, Goldman et al. (1967b) used lubrication theory to show that the coefficients C_G and D_G approach the constant values

$$C_G = 1.7005 + O(\zeta) \tag{5.6.4e}$$

$$D_G = 0.9440 + O(\zeta) \tag{5.6.4f}$$

as the particle approaches the wall.

The lubrication theory results in (5.6.4a–f) are complemented by results using the method of reflections, which are valid for $\zeta \gg 1$, obtained by a variety of different investigators. For instance, Goldman et al. (1967b) showed that for large separation distances, C_G approaches

$$C_G = 1 + \frac{9}{16}\left(\frac{1}{1+\zeta}\right). \tag{5.6.5}$$

This result in fact provides a reasonable approximation for arbitrary values of ζ, with a maximum error of about 10% as $\zeta \to 0$. Empirical fits for C_G and C_U that are valid for arbitrary ζ were reported by Zeng et al. (2009) as

$$C_G = 1 + 0.138\exp(-\zeta) + \frac{9}{16}\left(\frac{1}{1+\zeta}\right), \tag{5.6.6}$$

$$C_U = -1.028 + \frac{0.07}{1+\zeta^2} + \frac{8}{15}\ln\left(\frac{\zeta}{1+0.948\zeta}\right). \tag{5.6.7}$$

The problem of a spherical particle moving toward or away from a wall in an otherwise quiescent fluid was examined by Brenner (1961), who derived an expression for C_V valid for arbitrary ζ using the method of reflections as

$$C_V = -\frac{4}{3}\sinh(\alpha)\sum_{n=1}^{\infty}\frac{n(n+1)}{(2n-1)(2n+3)}\left[\frac{2\sinh[(2n+1)\alpha] + (2n+1)\sinh(2\alpha)}{4\sinh^2[(n+1/2)\alpha] - (2n+1)^2\sinh^2\alpha} - 1\right], \tag{5.6.8}$$

where $\alpha \equiv \cosh^{-1}(1+\zeta)$. In the limit of small ζ, Cox and Brenner (1967) and Cooley and O'Neill (1968) derived the lubrication-theory result

$$C_V = -\frac{1}{\zeta} + \frac{1}{5}\ln(\zeta) - 0.97128. \tag{5.6.9}$$

Wakiya (1960) gave a solution valid for large ζ as

$$C_V = -\frac{1}{1 - (9/8)(1+\zeta)^{-1} + (1/2)(1+\zeta)^{-3}}. \tag{5.6.10}$$

Ziskind et al. (1998) showed that a close approximation to the results of the exact solution (5.6.8) is obtained by using (5.6.9) for $\zeta < 1$ and (5.6.10) for $\zeta > 1$.

Modification of the results presented here at finite particle Reynolds numbers has been examined using numerical simulation in a number of studies. The problem of a stationary particle attached to a wall ($\zeta = 0$) is of particular importance in adhesive particle flows in which particles adhere to a wall. This problem was examined by Sweeney and Finlay (2007) for a particle immersed in a Blasius boundary layer with $\mathrm{Re}_x = Ux/\nu = 1000$ and several different values of particle Reynolds number Re_p. An empirical solution to the drag force obtained in this study was reported, giving

$$C_G = 1.7005\left[1 - \frac{0.2817}{\mathrm{Re}_p^{0.0826}}\sin^{-1}(0.238\mathrm{Re}_p)\right]^{-1}, \tag{5.6.11}$$

where for this study G in (5.6.1) is replaced by the ambient velocity within the Blasius boundary layer at the particle centroid divided by the particle radius $d/2$. This result reduces to (5.6.4e) as $\mathrm{Re}_p \to 0$. Zeng et al. (2005, 2009) performed an extensive numerical study of finite Reynolds number effects for a particle in a linear shear flow. These studies examine both the problem of a stationary particle at different separation distances above the wall as well as a translating particle in a quiescent flow. The latter problem was also examined experimentally by Takemura and Magnaudet (2003) using clean and contaminated bubbles rising parallel to a plane wall and theoretically by Vasseur and Cox (1977) for small, but finite, particle Reynolds numbers. If the low-Reynolds number results in (5.6.6) and (5.6.7) are denoted by $C_{G,0}$ and $C_{U,0}$, respectively, Zeng et al. (2009) reported a correlation based on their numerical results valid for finite Reynolds numbers in the range $2 \le \mathrm{Re}_p \le 250$ as

$$C_G = C_{G,0}\left(1 + \alpha_1 \mathrm{Re}_p^{\beta_1}\right), \quad C_U = C_{U,0}\left(1 + \alpha_2 \mathrm{Re}_p^{\beta_2}\right), \tag{5.6.12}$$

where

$$\alpha_1 = 0.15 - 0.046(1 - 0.04\zeta^2)\exp(-0.35\zeta)$$
$$\beta_1 = 0.687 - 0.066(1 - 0.19\zeta^2)\exp(-0.535\zeta^{0.9})$$
$$\alpha_2 = 0.15[1 - \exp(-\sqrt{\zeta/2})]$$
$$\beta_2 = 0.687 + 0.313\exp(-\sqrt{2\zeta}).$$

[These expressions have been modified to account for different definitions of the parameter ζ between the current chapter and the paper of Zeng et al. (2009).] Equation (5.6.12) reduces to the Schiller and Naumann (1933) expression (5.1.10) for drag in an unbounded fluid as $\zeta \to \infty$.

The importance of particle rotation on the drag force at finite Reynolds numbers was examined using numerical simulations by Zeng et al. (2009), comparing drag on a stationary particle in a shear flow with that for a freely rotating particle. A similar numerical study was performed by Stewart et al. (2010) for a spherical particle rolling on a plane wall with different rotation rates, the results of which are shown in Figure 5.6. In both cases, it is found that particle rotation has negligible influence on the drag force on the particle.

5.6.2. Lift Force

The lift force is an inherently nonlinear function of shear rate, rotation rate, and particle translation velocity, as was also observed in Section 5.2 for a particle in an unbounded flow. Cherukat and McLaughlin (1994) solved both the problem of lift on a nonrotating sphere and a freely rotating sphere at low particle Reynolds numbers. Validity of their solution requires that the distance $\ell = d(1 + \zeta)/2$ between the particle centroid and the wall is less than both the characteristic length scale $\ell_S = \nu/v_S$ associated with the sphere motion relative to the flow and the length scale $\ell_G = (\nu/G)^{1/2}$ associated with the shear flow. Defining $\kappa \equiv 1/(1 + \zeta)$, the condition $\ell \ll \min(\ell_S, \ell_G)$ is equivalent to the requirement that $\mathrm{Re}_p \ll \kappa$ and $\mathrm{Re}_G^{1/2} \ll \kappa$. Cherukat and McLaughlin (1994) found that the lift force for this problem can be written as

$$F_\ell = \rho I v_S^2 d^2/4, \tag{5.6.13}$$

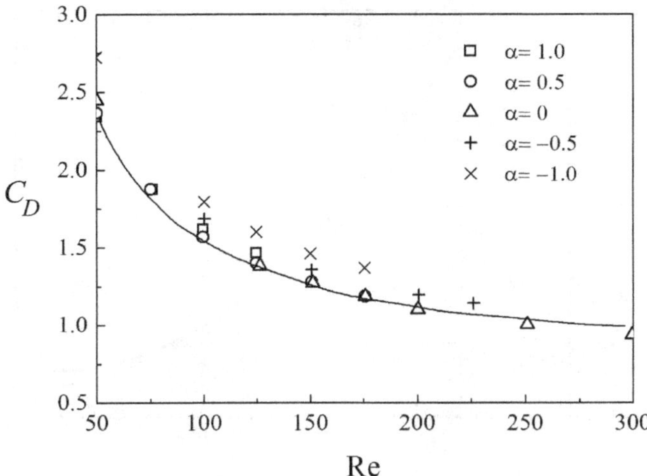

Figure 5.6. Numerical results of Stewart et al. (2010) (symbols) for drag on a particle rolling on a plane wall for different rotation rates (corresponding to different values of the parameter $\alpha = \Omega d/2v_x$), in comparison to the correlation of Zeng et al. (2009) (solid line). [Reprinted with permission from Stewart et al. (2010).]

where $v_S = v_x - (1 + \zeta)dG/2$ is the particle velocity relative to the shear flow velocity at the particle centroid. Defining $\Lambda_S \equiv Gd/2v_s$, polynomial approximations for the coefficient I in (5.6.13) are reported by Cherukat and McLaughlin as

$$I = [1.7716 + 0.2160\kappa - 0.7292\kappa^2 + 0.4854\kappa^3]$$
$$- \left[\frac{3.2397}{\kappa} + 1.1450 + 2.0840\kappa - 0.9059\kappa^2\right] \Lambda_G \qquad (5.6.14a)$$
$$+ [2.0069 + 1.0575\kappa - 2.4007\kappa^2 + 1.3174\kappa^3]\Lambda_G^2$$

for a nonrotating particle and as

$$I = [1.7631 + 0.3561\kappa - 1.1837\kappa^2 + 0.845163\kappa^3]$$
$$- \left[\frac{3.2139}{\kappa} + 2.6760 + 0.8248\kappa - 0.4616\kappa^2\right] \Lambda_G \qquad (5.6.14b)$$
$$+ [1.8081 + 0.879585\kappa - 1.9009\kappa^2 + 0.98149\kappa^3]\Lambda_G^2$$

for a freely rotating particle. The Cherukat-McLaughlin lift solutions for fixed and freely rotating spheres are also found to be quite close to each other, suggesting that sphere rotation rate has only a small effect on the lift force for a particle traveling in a shear flow near a wall. For instance, in Figure 5.7 the coefficient I is plotted as a function of Λ_G for two values of ζ using (5.6.14a) and (5.6.14b). In both cases, the solutions of I are quite close to each other.

The limit of a stationary particle resting on the plane wall was examined by Leighton and Acrivos (1985), for which the lift is given by $F_\ell = 9.22\rho(d/2)^4G^2$. The Cherukat and McLaughlin solution in this limit gives $F_\ell = I\rho(d/2)^4G^2$, where for $\kappa = 1$ and $\Lambda_G = -1$, (5.6.14a) gives $I = 9.28$. This result differs from the Leighton and Acrivos solution by 0.65%. The problem of a sphere traveling in contact with a plane wall is examined both theoretically and experimentally for cases with low

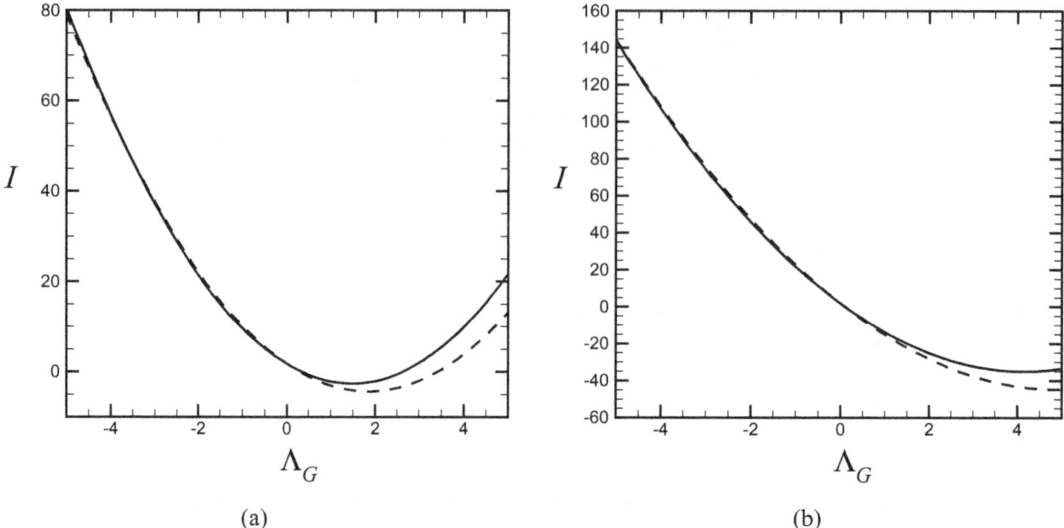

Figure 5.7. Plots of the Cherukat-McLaughlin coefficient I in the lift force for a particle in a shear flow near a wall as a function of Λ_G for cases with dimensionless separation distance (a) $\delta = 0.1$ and (b) $\delta = 4$ with both a nonrotating sphere (solid line) and a freely rotating sphere (dashed line).

particle Reynolds number by Krishnan and Leighton (1995) and King and Leighton (1997), including effects of surface roughness.

A number of investigators have examined the effect of finite Reynolds numbers on the lift force, including experiments by Takemura and Magnaudet (2003) for rising bubbles near a wall, calculations for a particle in a Blasius boundary layer by Sweeney and Finlay (2007), and numerical computations and experiments of a sphere rolling on a plane by Stewart et al. (2010), among others. Zeng et al. (2009) summarized these various correlations and evaluates these expressions based on their numerical solutions. However, as the correlations are numerous, fairly complex, and tend to produce widely scattered results depending on the specific conditions used, we defer to the original literature for further information on finite Reynolds number effects.

5.7. Effect of Surrounding Particles

Each particle in a particulate flow is surrounded by other particles. Unless the particle concentration is dilute, the presence of surrounding particles will influence the drag force for any given particle. A number of investigators have examined this "particle crowding" effect in different regimes. Some investigators have taken an empirical approach and have developed correlations for drag correction based on pressure drop data for flow through a fixed packed bed (Ergun, 1952) or a fluidized bed (Wen and Yu, 1966), or by measurement of particle terminal velocity during sedimentation (Richardson and Zaki, 1954). Other investigators develop the drag correlations by employing microscale simulations such as direct numerical simulation (DNS) or Lattice-Boltzmann method (LBM) simulations (Hill et al., 2001a,b; van der Hoef et al., 2005).

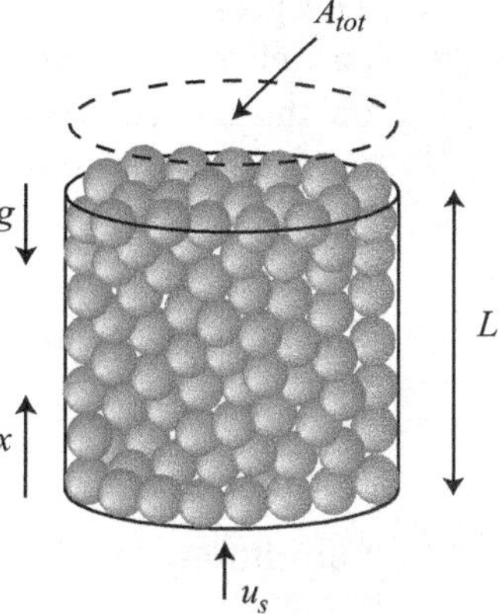

Figure 5.8. Schematic illustrating flow through a fluidized bed in a cylinder.

For definiteness, let us consider the problem of flow through a bed of particles contained in a cylinder with uniform cross-sectional area A_{tot}, through which a fluid flows with flow rate Q in the upward (positive x) direction, as shown in Figure 5.8. The fluid superficial velocity u_s in the x-direction is defined by the ratio of the flow rate to the total cross-sectional area of the flow, or $u_s = Q/A_{tot}$. If $\varepsilon \equiv 1 - \phi$ is the void fraction, the part of the cylinder cross-sectional area occupied by the fluid is εA_{tot}. The average velocity of the fluid phase within the flow is therefore given by $u_f = Q/\varepsilon A_{tot}$, so we can write

$$u_f = u_s/\varepsilon. \tag{5.7.1}$$

The total fluid force plus the gravitational force acting on an individual particle within the bed is given by

$$\mathbf{F} = \mathbf{F}_d - \rho_P V_p g \mathbf{e}_x - V_p \nabla p, \tag{5.7.2}$$

where $\mathbf{F}_d$ is the drag force on an individual particle and $V_p = \pi d^3/6$ is the particle volume. The pressure gradient ∇p in (5.7.2) can arise from a number of mechanisms, including a hydrostatic contribution due to gravity, a contribution due to acceleration $a_0 = Du_s/Dt$ of the fluid superficial velocity, and an additional contribution due to the flow of fluid through the bed. The first and last of these mechanisms lead to negative values of pressure gradient for upward flow in the bed. Separating out the first two of these mechanisms, we can decompose the pressure gradient as

$$\nabla p = -\rho_f(g - a_0)\mathbf{e}_x + \nabla p_d, \tag{5.7.3}$$

where ∇p_d is the part of the pressure gradient due to the drag force between the particles and the fluid. Substituting (5.7.3) into (5.7.2) gives

$$\mathbf{F} = (\mathbf{F}_d - V_p \nabla p_d) + (\rho_f - \rho_P)V_p g \mathbf{e}_x - \rho_f \mathbf{a}_0 V_p. \tag{5.7.4}$$

The first term on the right-hand side of (5.7.4) represents the total force on the particle that arises from interaction with the fluid, and this combination is denoted by

$$\hat{\mathbf{F}}_d \equiv \mathbf{F}_d - V_p \nabla p_d. \tag{5.7.5}$$

The pressure gradient ∇p_d can be determined using the control volume V for a particle bed of length L shown in Figure 5.8. Equating the pressure force $-\Delta p_d A_{tot}$ on the control volume to the volumetric force $n_p \hat{F}_d$, where n_p is the number of particles in the control volume, gives

$$\Delta p_d = -n_p \hat{F}_d / A_{tot}. \tag{5.7.6}$$

The number of particles per unit volume within the particle bed is given by ϕ/V_p, so that

$$n_p = A_{tot} L \phi / V_p. \tag{5.7.7}$$

Substituting (5.7.7) into (5.7.6) gives

$$\nabla p_d = \Delta p_d / L = -\phi \hat{F}_d / V_p. \tag{5.7.8}$$

Substituting (5.7.8) into (5.7.5) gives a relationship between F_d and $\hat{F}_d$ as

$$\hat{F}_d = F_d / \varepsilon. \tag{5.7.9}$$

The literature on fluid flow through particulate beds contains a number of apparent inconsistencies in the various correlations used for the friction factor f defined in (5.1.1). These differences can generally be traced back to whether the average fluid velocity u_f or the superficial fluid velocity u_s is used, as related by (5.7.1), or whether the form $\mathbf{F}_d$ or $\hat{\mathbf{F}}_d$ of the drag force is reported. The force on the particles depends in general on the difference between the average fluid velocity u_f and the particle velocity, where we denote the average fluid velocity in vector form simply as $\mathbf{u}$. We note that although $\mathbf{u}$ is defined as the fluid velocity at a particle centroid in the absence of the particle, it does include the effects of surrounding particles on the fluid velocity field.

In continuum models for particulate flows, the fluid-solid interaction force per unit volume, $\mathbf{f}_{gs}$, is commonly assumed to be proportional to the difference between the volume-averaged particle velocity $\mathbf{v}$ and the average fluid velocity $\mathbf{u}$, so we can write

$$\mathbf{f}_{gs} = -\beta(\mathbf{v} - \mathbf{u}). \tag{5.7.10}$$

The coefficient β in this equation is called the interphase momentum transfer coefficient. Discrete models, such as DEM, require an expression for drag force $\mathbf{F}_d$ on individual particles as a function of $\mathbf{v} - \mathbf{u}$. Dividing $\mathbf{f}_{gs}$ by the number of particles per unit volume, ϕ/V_p, gives

$$\mathbf{F}_d = -\frac{\beta V_p}{\phi}(\mathbf{v} - \mathbf{u}) = -\frac{\beta V_p}{1 - \varepsilon}(\mathbf{v} - \mathbf{u}). \tag{5.7.11}$$

Comparing (5.7.11) with (5.1.1), the friction factor f for the drag force can be written as

$$f = \frac{\beta d^2}{18\mu(1-\varepsilon)}. \tag{5.7.12}$$

We write an expression for the entire friction factor f, rather than just the crowding correction factor C_F, in this section because the inertial term is generally included in these correlations. Also, the conditions considered in this section generally have very little slip, so we can take $C_C \cong 1$.

5.7.1. Flow through Packed Beds

A well-known correlation for flow through a packed bed of particles was proposed by Ergun (1952), which reduces to the classic Kozeny-Carman relationship (Kozeny, 1927; Carman, 1937) for small particle Reynolds numbers and to the Burke-Plummer expression (Burke and Plummer, 1928) for high Reynolds numbers. The Ergun correlation gives the pressure gradient ∇p_d due to drag force in the particle bed as

$$\nabla p_d = \frac{\Delta p_d}{L} = -150\frac{(1-\varepsilon)^2}{\varepsilon^3}\frac{\mu u_s}{d^2} - 1.75\frac{1-\varepsilon}{\varepsilon^3}\frac{\rho_f u_s^2}{d}. \qquad (5.7.13)$$

From (5.7.8), (5.7.9), and (5.7.11), we have

$$\mathbf{F}_d = -\frac{\varepsilon V_p}{1-\varepsilon}\nabla p_d = \frac{\beta V_p}{1-\varepsilon}u_f. \qquad (5.7.14)$$

Substituting (5.7.13) into (5.7.14) and solving for β gives

$$\beta = -\frac{\varepsilon}{u_f}\nabla p_d = -\frac{\varepsilon^2}{u_s}\nabla p_d = 150\frac{(1-\varepsilon)^2}{\varepsilon}\frac{\mu}{d^2} + 1.75(1-\varepsilon)\frac{\rho_f|\mathbf{u}-\mathbf{v}|}{d}, \qquad (5.7.15)$$

where we have used (5.7.1) to write the superficial velocity as $u_s = \varepsilon u_f$. Substituting into (5.7.12) gives the friction factor as

$$f = \frac{\beta d^2}{18\mu(1-\varepsilon)} = 8.33\frac{(1-\varepsilon)}{\varepsilon} + 0.097\mathrm{Re}_p, \qquad (5.7.16)$$

where the particle Reynolds number is defined as $\mathrm{Re}_p = \rho_f d\,|\mathbf{u}-\mathbf{v}|\,/\mu$. The crowding correction factor given in (5.7.16) includes finite Reynolds number effects, so an additional inertial correction factor should not be used. The Ergun correlation fits the packed bed experimental data well over a large range of Reynolds numbers, but it is based on data obtained for a relatively small range of void fraction values, ranging between approximately 0.44 and 0.53.

5.7.2. Flow through Fluidized Beds

Much greater variation in void fraction values can be achieved using fluidized particle beds. For an equilibrium fluidized bed, both the particle velocity $\mathbf{v}$ and the total force $\mathbf{F}$ vanish and the pressure gradient is given by

$$\frac{dp}{dz} = -[(1-\varepsilon)\rho_p + \varepsilon\rho_f]g, \qquad (5.7.17)$$

where the term in brackets is the averaged density of the mixture. Substituting (5.7.17) into (5.7.2) and setting the total force equal to zero gives

$$\mathbf{F}_d = (\rho_P - \rho_f)V_p g\varepsilon\mathbf{e}_x. \qquad (5.7.18)$$

Measurement of void fraction of the fluidized bed for different fluid flow rates Q can thus be used with (5.7.18) to obtain data for particle drag force as a function of average fluid velocity.

The term multiplying ε on the right-hand side of (5.7.18) is the drag force F_{dt} acting on an isolated sphere of diameter d falling at its terminal velocity u_t in an otherwise stationary fluid. Consequently, we can write (5.7.18) as

$$F_d = \varepsilon F_{dt}. \tag{5.7.19}$$

The result (5.7.19) relates the magnitude of the drag force on a particle within the fluidized bed subject to an upward superficial velocity u_s to that acting on an isolated particle falling at a terminal velocity u_t in a stationary fluid. Using the Stokes formula (5.1.1), we can therefore write

$$F_d = 3\pi d\mu u_f f, \quad F_{dt} = 3\pi d\mu u_t, \tag{5.7.20}$$

where $u_f = u_s/\varepsilon$ is the mean fluid velocity in the particle bed. Substituting the two equations in (5.7.20) into (5.7.19) gives

$$f = \left(\frac{u_t}{u_f}\right)\varepsilon = \left(\frac{u_t}{u_s}\right)\varepsilon^2. \tag{5.7.21}$$

Richardson and Zaki (1954) showed experimentally that the ratio u_s/u_t can be well fit by an expression of the form

$$\frac{u_s}{u_t} = \varepsilon^n, \tag{5.7.22}$$

where the exponent n is a function of the particle Reynolds number Re_p evaluated using velocity u_t. For small values of Re_p this exponent approaches a value $n = 4.65$. Substituting (5.7.22) into (5.7.21) with $n = 4.65$ gives the friction factor as

$$f = \varepsilon^{-2.65}C_I, \tag{5.7.23}$$

where C_I is an appropriate inertial correction factor.

The data from experimental studies spanning a large range of particle Reynolds numbers were collected by Di Felice (1994), who noted that although (5.7.23) performs well for both large and small particle Reynolds numbers, it exhibits significant errors at intermediate Reynolds numbers. Di Felice fit available data over a particle Reynolds number range from 0.01 to 10^4 by an empirical expression of the form

$$f = \varepsilon^{1-\zeta}C_I, \quad \zeta = 3.7 - 0.65\exp\left(-\tfrac{1}{2}[1.5 - \ln(\mathrm{Re}_p)]^2\right). \tag{5.7.24}$$

This expression approaches the Wen-Yu expression for low particle Reynolds number, if the exponent 2.65 in (5.7.23) is rounded to 2.7.

While the Wen-Yu expression (5.7.23) has been validated over a wide range of values of the void fraction ε, the large value of the exponent makes the expression for C_F sensitive to the value of ε as void fraction decreases to significantly less than unity. The Ergun correlation gives accurate predictions for low void fractions, but it does not approach the correct limit $f \to 1$ as $\varepsilon \to 1$. Gidaspow (1994) therefore proposed combining these two correlations as

$$f = \xi f_{F,E} + (1-\xi)f_{F,WY}, \tag{5.7.25}$$

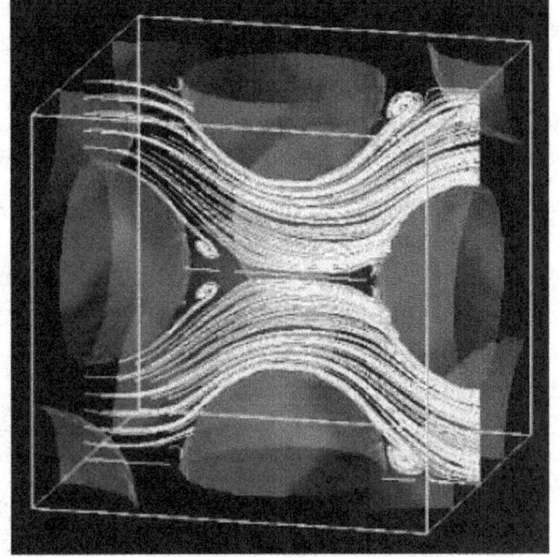

Figure 5.9. Fluid streamlines for flow around particles in a face-centered cubic array computed using the Lattice-Boltzmann method. [Reprinted with permission by Hill et al. (2001b).]

where ξ is a stitching function. Here $f_{F,E}$ denotes the Ergun correlation (5.7.16) and $f_{F,WY}$ denotes the Wen-Yu correlation (5.7.23). Gidaspow used a stitching function of the form

$$\xi = \begin{cases} 0 & \text{for } \varepsilon > 0.8 \\ 1 & \text{for } \varepsilon \leq 0.8 \end{cases}. \qquad (5.7.26)$$

This choice leads to a discontinuity in the particle drag expression at $\varepsilon = 0.8$, the magnitude of which becomes increasingly large as the particle Reynolds number increases. Later authors proposed continuous stitching functions that join these two expressions over an interval in ε. For instance, Huilin et al. (2003) used an inverse tangent function

$$\xi = \frac{\tan^{-1}[262.5(0.8 - \varepsilon)]}{\pi} + 0.5, \qquad (5.7.27)$$

and Dahl and Hrenya (2005) used a linear interpolation

$$\xi = \begin{cases} 0 & \varepsilon \geq 0.8 \\ 8 - 10\varepsilon, & 0.7 \leq \varepsilon < 0.8. \\ 1 & \varepsilon < 0.7 \end{cases} \qquad (5.7.28)$$

One problem with this combined approach is that it results in an abrupt change in the value of f in the region $0.7 \leq \varepsilon \leq 0.8$. The magnitude of this change becomes progressively larger as the particle Reynolds number increases.

5.7.3. Simulations

Simulations using the Lattice-Boltzmann method (LBM) for flow through random and ordered arrays of stationary spheres were reported by Hill et al. (2001a,b), from which correlations for friction factor are reported for small and moderate particle Reynolds numbers. Examples of streamlines of the flow around the particles in a face-centered cubic array are shown in Figure 5.9. Extensions to the Hill et al. drag correlations to apply to a broader range of values of Re_p and ε were proposed by

Benyahia et al. (2006) and Beetstra et al. (2007). A review of various expressions for particle friction factor was given by Deen et al. (2007).

The LBM simulation study by van der Hoef et al. (2005) for a wide range of void fractions with low particle Reynolds numbers, satisfying $\mathrm{Re}_p < 0.2$, was conducted for both arrays formed of uniform-size particles and arrays with particles of two different sizes. For beds with uniform particle size, van der Hoef et al. (2005) showed that their own data as well as the low Reynolds number simulation data of Ladd (1990) obtained with a particle multipole approach and that of Hill et al. (2001a) obtained with LBM can be accurately fit using a simple expression for friction factor of the form

$$f = 10\frac{1-\varepsilon}{\varepsilon} + \varepsilon^3(1 + 1.5\sqrt{1-\varepsilon}). \tag{5.7.29}$$

The first term in this expression is the same as that proposed by Carman (1937) based on experimental data for packed beds.

The interphase momentum transfer coefficient β can be expressed in terms of a dimensionless drag F^* as

$$\beta = 18\frac{\mu(1-\varepsilon)}{d^2}\varepsilon^2 F^*, \tag{5.7.30}$$

where F^* is related to the friction coefficient f by $f = F^*\varepsilon^2$. The expressions proposed by Benyahia et al. (2006), based on the simulation data of Hill et al. (2001a,b) as well as known limiting forms of the drag function, give F^* for a wide range of Reynolds number and void fraction values as follows:

Low Reynolds number region $(\mathrm{Re}_p < \mathrm{Re}_{TRANS})$

$$F^* = \begin{cases} F_0 + \frac{F_1\varepsilon^2}{4}\mathrm{Re}_p^2, & \text{when } \varepsilon < 0.99 \\[2mm] 1 + \frac{3\varepsilon}{16}\mathrm{Re}_p, & \text{when } \varepsilon \geq 0.99 \end{cases} \tag{5.7.31a}$$

High Reynolds number region $(\mathrm{Re}_p \geq \mathrm{Re}_{TRANS})$

$$F^* = F_2 + \frac{F_3\varepsilon}{2}\mathrm{Re}_p. \tag{5.7.31b}$$

The coefficients F_0, F_1, F_2, and F_3 in these expressions are functions of void fraction ε and volumetric concentration $\phi = 1 - \varepsilon$, and are given by

$$F_0 = \begin{cases} (1-w)\dfrac{1+3\sqrt{\phi/2}+2.109\phi\ln\phi+17.14\phi}{1+0.681\phi-8.48\phi^2+8.16\phi^3} + w\dfrac{10\phi}{\varepsilon^3}, & \text{when } 0.6 < \varepsilon < 0.99 \\[4mm] \dfrac{10\phi}{\varepsilon^3}, & \text{when } \varepsilon \leq 0.6 \end{cases} \tag{5.7.32a}$$

$$F_1 = \begin{cases} \dfrac{1}{40}\sqrt{\dfrac{2}{\phi}}, & \text{when } 0.9 \leq \varepsilon < 0.99 \\[3mm] 0.11 + 0.000051\exp(11.6\phi), & \text{when } \varepsilon < 0.9 \end{cases} \tag{5.7.32b}$$

$$
F_2 = \begin{cases} (1-w)\dfrac{1+3\sqrt{\phi/2}+2.109\phi\ln\phi+17.89\phi}{1+0.681\phi-11.03\phi^2+15.41\phi^3} + w\dfrac{10\phi}{\varepsilon^3}, & \text{when } \varepsilon > 0.6 \\[4mm] \dfrac{10\phi}{\varepsilon^3}, & \text{when } \varepsilon \le 0.6 \end{cases}
$$

$$(5.7.32c)$$

$$
F_3 = \begin{cases} 0.9351\phi + 0.03667, & \text{when } \varepsilon > 0.9047 \\[4mm] 0.0637 + 0.212\phi + \dfrac{0.0232}{\varepsilon^5}, & \text{when } \varepsilon \le 0.9047 \end{cases}
$$

$$(5.7.32d)$$

where

$$
w \equiv \exp[-10(0.4-\phi)/\phi] \tag{5.7.32e}
$$

The transition point between the low and high Reynolds number regions in the drag laws presented here can be determined by equating F^* at the transition point, giving

$$
\mathrm{Re}_{TRANS} = \frac{F_3 + \sqrt{F_3^2 - 4F_1(F_0 - F_2)}}{\varepsilon F_1}. \tag{5.7.33}
$$

We note that these expressions differ slightly from those given by Benyahia et al. (2006) due to a difference in the definition of the particle Reynolds number, where in the current book we use $\mathrm{Re}_p = \rho_f d\,|\mathbf{u}-\mathbf{v}|/\mu$.

Figure 5.10 shows a comparison of the friction factor computed using different correlations plotted as a function of void fraction, for values of particle Reynolds numbers of $\mathrm{Re}_p = 0.1, 1, 10,$ and 100. For example, a $100\ \mu\mathrm{m}$ diameter dust particle in air with a slip velocity of $|\mathbf{v}-\mathbf{u}| = 1$ m/s has a Reynolds number of $\mathrm{Re}_p = 6.7$, whereas a $10\ \mu\mathrm{m}$ diameter red blood cell in blood plasma with a slip velocity of 1 mm/s has a Reynolds number of $\mathrm{Re}_p = 0.003$. The void fraction is plotted from the value for a dense packing (approximately 0.4) to unity. The correlations plotted include (1) the Gidaspow combined Ergun-Wen-Yu correlation (5.7.25) with the Huilin-Gidaspow stitching function (5.7.27), (2) the Di Felice correlation (5.7.24), (3) the low Reynolds number van der Hoef et al. correlation (5.7.29), and (4) the Benyahia et al. modification of the Hill-Koch-Ladd correlation (5.7.31) and (5.7.32). The Schiller-Naumann inertial correction factor (5.1.10) is used together with the Wen-Yu and Di Felice correlations to obtain the combined friction factors. The Gidaspow correlation is found to smoothly bridge between the Ergun and Wen-Yu correlations for the three lower values of Reynolds number, but for $\mathrm{Re}_p = 100$ there is a large jump in friction factor value at the joining point. The Di Felice correlation agrees well with the Wen-Yu correlation for all Reynolds numbers examined. The van der Hoef and Hill-Koch-Ladd correlations agree closely with each other for the three lower Reynolds numbers, indicating that the particle Reynolds number does not have a major effect on the friction factor for $\mathrm{Re}_p \le 10$.

Because the crowding correction factor is based on the particle concentration, it is valid only in a time-averaged sense. An obvious shortcoming of (5.7.30) or (5.7.31) is that these equations take no account of the effect of a sharp gradient in the local concentration, such as might occur near the top part of a packed bed or along the sides of a particle agglomerate. Furthermore, the discrete element method

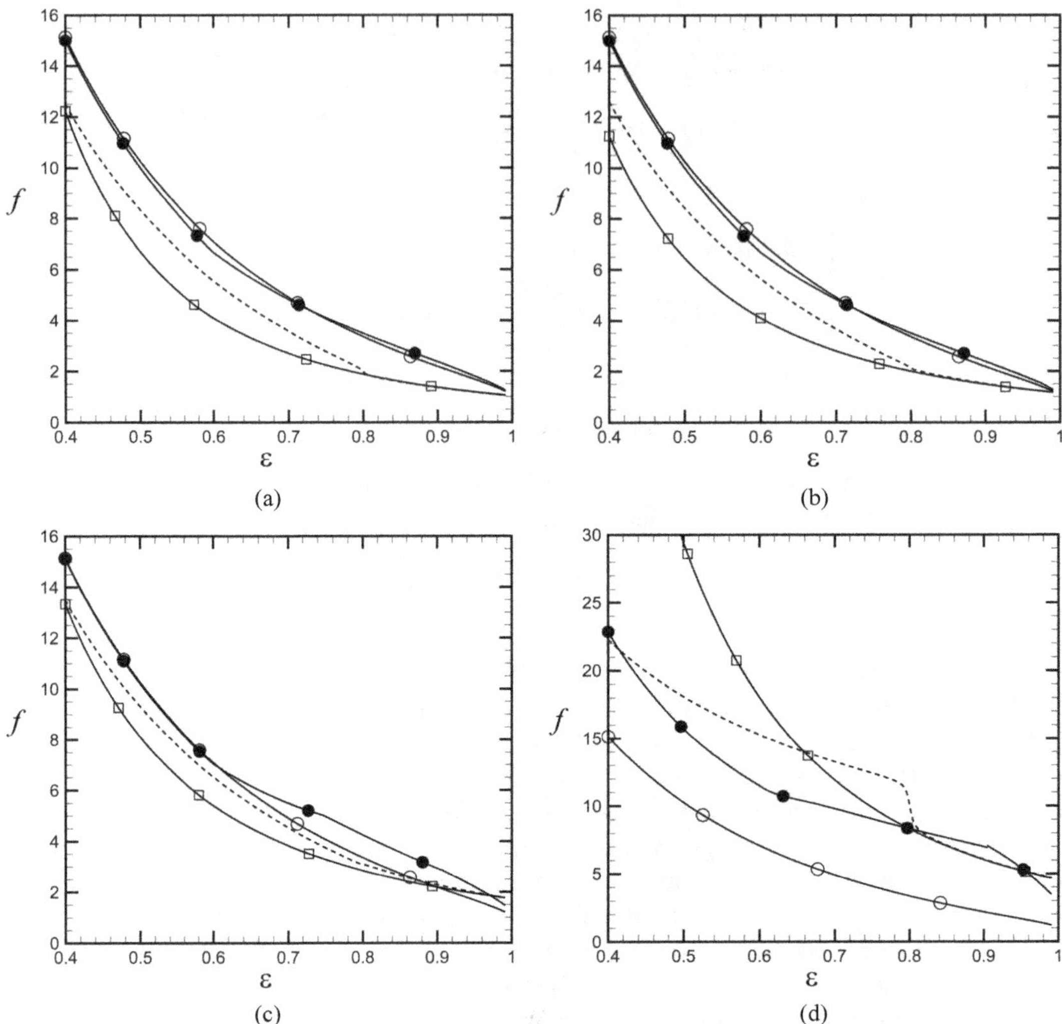

Figure 5.10. Comparison of friction factor predictions from different correlations as a function of void fraction ε for particle Reynolds numbers (a) 0.1; (b) 1; (c) 10; and (d) 100. Correlations used include the Gidaspow combined correlation (5.7.25) with the Huilin-Gidaspow stitching function (dashed line), Di Felice (open squares), van der Hoef et al. (2005) low Reynolds number (open circles), and Benyahia et al.'s (2006) modification of the Hill et al. (2001b) correlation (filled circles).

is based on evolution of the instantaneous particle velocity, so use of a time- or space-averaged correlation for the drag correction in the particle evolution equations will smear over important instantaneous effects. The problem of developing a simple analytical expression that yields a correction for the instantaneous particle drag force, accounting for the time-varying positions and velocities of surrounding particles, is an outstanding and important research challenge.

5.7.4. Effect of Particle Polydispersity

In practice, most particulate flows involve particles with a range of diameters, rather than a single diameter. Such systems are said to be polydisperse. In polydisperse systems, the number of particles n_i per unit volume of species i is given by $\phi_i/V_{p,i}$, where ϕ_i is the volume concentration of particle diameter d_i and where $V_{p,i} = \pi d_i^3/6$. An interphase momentum transfer coefficient β_i can be defined such that the drag force on a particle with size d_i is given by

$$\mathbf{F}_{d,i} = -\frac{\beta_i V_{p,i}}{\phi_i}(\mathbf{v} - \mathbf{u}), \tag{5.7.34}$$

where the void fraction satisfies $\varepsilon = 1 - \sum \phi_i$. An expression for the β_i coefficient for particle species i is obtained from the Ergun equation (5.7.15) as

$$\beta_i = \frac{\mu\phi_i}{\varepsilon d_i^2}[150(1 - \varepsilon) + 1.75\varepsilon \mathrm{Re}_{p,i}], \tag{5.7.35}$$

where $\mathrm{Re}_{p,i} \equiv \rho_f d_i|\mathbf{v}_i - \mathbf{u}|/\mu$. The Wen-Yu friction factor depends only on the void fraction, and so it is not changed for a polydisperse system.

For polydisperse systems, it is convenient to define a diameter ratio y_i by

$$y_i = \frac{d_i}{\langle d \rangle}. \tag{5.7.36}$$

Here, $\langle d \rangle$ denotes the Sauter diameter, which is defined in terms of the number N_i of particles with diameter d_i as

$$\langle d \rangle = \frac{\sum N_i d_i^3}{\sum N_i d_i^2}. \tag{5.7.37}$$

A computational study using LBM for bidisperse systems at low Reynolds numbers was reported by van der Hoef et al. (2005). Based on the assumption that the drag force $\mathbf{F}_{d,i}$ acting on particles of species i is proportional to the particle diameter d_i, these authors argued that the friction factor f_i for particles of species i can be written using a weighted average as

$$f_i = \left[\varepsilon y_i + (1 - \varepsilon)y_i^2 + 0.064\varepsilon y_i^3\right]f_{MD}, \tag{5.7.38}$$

where f_{MD} is the friction factor for a monodisperse system with the same void fraction, given by (5.7.29).

5.8. Stokesian Dynamics

As forces and torques are exerted on particles when particles move relative to a fluid, so too are forces and torques exerted from the particles back onto the fluid. In the case of dense particle flows, in which the momentum coupling parameter defined in (1.2.26) is not small, the effects of the particles on the fluid can lead to significant flow modification. In some cases this flow modification influences the entire flow field, whereas in other cases this modification may be confined to the region surrounding an agglomerate or a cluster of particles and not significantly influence the rest of the flow.

5.8.1. Example for Falling Cluster of Particles

A series of experiments and computations investigating the effect of two-way coupling on the motion of a cluster of particles was performed by Nitsche and

Batchelor (1997), who studied the problem of a "droplet" (or cluster) of particles falling in an otherwise stationary fluid. The particles were 0.9 mm diameter glass beads immersed in glycerine, which were initially grouped into clusters with diameter of approximately 0.7 cm and allowed to settle under gravitational force. The settling velocity was measured and compared against computations, as well as against the predicted settling velocity for isolated particles. The computations were performed by balancing the drag force on each particle with the gravitational force, so that each particle travels at its terminal velocity. By using the Stokes drag expression (5.1.1), with $f = 1$, the relative velocity between the fluid and a particle is obtained as

$$\mathbf{v} - \mathbf{u} = -\frac{mg_R}{3\pi\mu d}\mathbf{e}_z, \qquad (5.8.1)$$

where g_R is the reduced gravitational acceleration. However, instead of setting the fluid velocity $\mathbf{u} = 0$ at the particle centroids, Nitsche and Batchelor accounted for the induced velocity by the other particles on the fluid velocity. This was done by writing the fluid velocity at any point $\mathbf{x}$ in the flow field as a sum of the velocity induced by the different particles, such that the fluid velocity $\mathbf{u}_i$ evaluated at the centroid of particle i can be written as

$$\mathbf{u}_i = \sum_{j\neq i}\mathbf{W}(\mathbf{x}_i, \mathbf{x}_j) \cdot \mathbf{F}_{d,j}, \qquad (5.8.2)$$

where $\mathbf{F}_{d,j}$ is the Stokes drag force on particle j and $\mathbf{W}(\mathbf{x}_i, \mathbf{x}_j)$ is the Oseen tensor, defined by

$$\mathbf{W}(\mathbf{x}, \mathbf{x}_0) = \frac{1}{8\pi\mu}\left[\frac{\mathbf{I}}{r} + \frac{(\mathbf{x} - \mathbf{x}_0)(\mathbf{x} - \mathbf{x}_0)^T}{r^3}\right]. \qquad (5.8.3)$$

In (5.8.3), $\mathbf{I}$ is the identity tensor and $r \equiv |\mathbf{x} - \mathbf{x}_0|$ is the distance between points $\mathbf{x}$ and $\mathbf{x}_0$.

The experimental results of Nitsche and Batchelor show that the particle cluster falls as a nearly spherical "blob," which leaves behind a trail of particles (Figure 5.11a). The motion of particles backward into the tail region is due to the fact that the fluid streamlines penetrate into the particle blob, as shown in Figure 5.11b. These outer streamlines carry particles backward toward the rear of the blob, feeding the particles that are entrained into the tail region. Nitsche and Batchelor found that the blob of particles settles downward at a faster rate than would an individual particle. In fact, if $v_0 = mg_R/3\pi\mu d$ is the settling velocity under Stokes flow of an isolated particle, Nitsche and Batchelor showed that an approximate expression for the settling velocity v_b of a blob of N_b particles with particle diameter d and blob radius R_b is given by

$$\frac{v_b}{v_0} \cong 1 + \frac{3}{5}N_b\frac{d}{R_b}. \qquad (5.8.4)$$

This approximation agrees closely with the computational data. The increased settling velocity of the particle blob is due to the effect of surrounding particles in dragging the fluid downward, so as to produce an average downward fluid velocity $\mathbf{u}$ in (5.8.1).

Additional experimental studies of falling particle clouds by Noh and Fernando (1993) and Metzger et al. (2007) reported the presence of a transition point in the

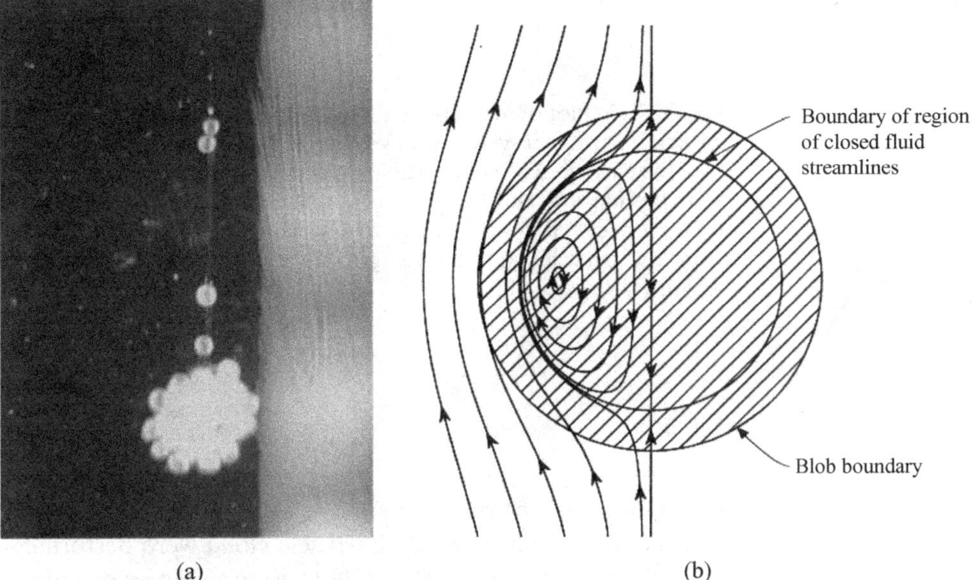

Figure 5.11. Results showing the sedimentation of a "blob" of particles under gravity: (a) photograph of particle blob with trailing particles; (b) diagram showing fluid streamlines relative to the falling particle blob. [Reprinted with permission from Nitsche and Batchelor (1997).]

cloud structure as it falls. For falling clouds under low flow Reynolds number conditions, Metzger et al. described a process in which the cloud transitions from an initially spherical shape into a torus, which eventually breaks up into two or more subclouds in a repeating cascade process (Figure 5.12). This process occurs over a

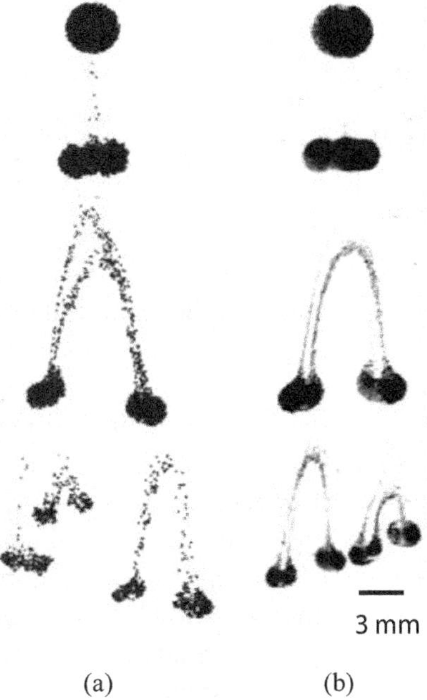

Figure 5.12. Results showing deformation and break-up of an initially spherical particle cloud: (a) simulation using the Stokesian dynamics method and (b) experiment. [Reprinted with permission from Metzger et al. (2007).]

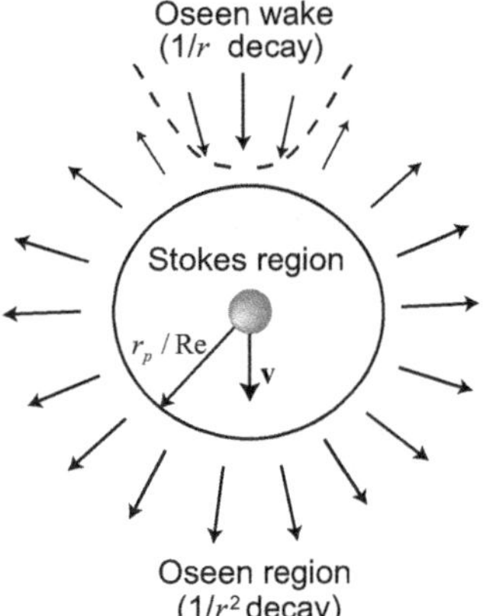

Figure 5.13. Plot of velocity field due to the Oseen solution about a spherical particle traveling with velocity **v**, drawn in a frame traveling with the particle. The velocity field reduces to the Stokes flow solution near the particle and has the form of a potential source flow with $O(1/r^2)$ decay and a thin wake region with $O(1/r)$ decay far away from the particle (at distances greater than the inertial screening length r_p/Re_p).

long fall length scale and requires that the system contain a large number of particles. Numerical simulations of the transitions in a falling particle cloud were performed by Subramanian and Koch (2008) and Pignatel et al. (2011) using an *Oseen dynamics* method, which is similar to the Stokesian dynamics approach employed by Nitsche and Batchelor (1997) but uses the full Oseen solution for hydrodynamic interaction of the particles (Proudman and Pearson, 1957).

The Oseen solution for the flow field generated by a spherical particle with radius r_p translating with a velocity $U\mathbf{e}_x$ relative to the surrounding fluid at low particle Reynolds number Re is given in spherical coordinates, with the polar axis ($\theta = 0$) coincident with the direction of the ambient flow, by

$$
\begin{aligned}
u_r &= \frac{Ur_p^2}{r^2}\left\{ -\frac{r_p}{2r}\cos\theta - \frac{3(1-\cos\theta)r}{4r_p}\exp\left[-\frac{\mathrm{Re}r(1+\cos\theta)}{2r_p} \right] \right. \\
&\quad \left. + \frac{3}{2\mathrm{Re}}\left(1 - \exp\left[-\frac{\mathrm{Re}r(1+\cos\theta)}{2r_p} \right] \right) \right\} \\
u_\theta &= Ur_p^2\left(\frac{r_p\sin\theta}{4r^3} - \frac{3\sin\theta}{4r_p r}\exp\left[-\frac{\mathrm{Re}r(1+\cos\theta)}{2r_p} \right] \right).
\end{aligned}
\tag{5.8.5}
$$

This solution approaches the Stokes solution for flow past a sphere for distances r much smaller than the *inertial screening length* r_p/Re. At very large distances from the particle, the velocity field approximates a point source with velocity magnitude decaying as $O(1/r^2)$, except within a thin wake region near $\theta = \pi$. Within this wake region the velocity is directed inward toward the sphere and the velocity magnitude decays with distance as $O(1/r)$. A sketch showing the different features of the Oseen flow is given in Figure 5.13. Unlike the Stokes solution, the Oseen flow solution is uniformly valid throughout the flow domain. Use of the oseenlet to model particle hydrodynamic interactions, instead of the sum of the stokeslet and the potential doublet, allows us to extend the Stokesian dynamics method to flows in which the flow Reynolds number is not small, provided that the particle Reynolds number remains small.

5.8.2. General Theory

In general, the velocity induced by a stress distribution $\mathbf{t}(\mathbf{x})$ defined on the surface S of the particle at a point $\mathbf{x}$ located outside of the particle is given for a Stokes flow by

$$u_i(\mathbf{x}) = -\frac{1}{8\pi\mu}\int_S G_{ij}(\mathbf{x}, \mathbf{x}')t_j(\mathbf{x}')\,da'. \tag{5.8.6}$$

Here we use tensor indices for the components of the second-order tensor $\mathbf{G}$ and the vectors $\mathbf{u}$ and $\mathbf{t}$. The integration in (5.8.6) is with respect to the primed position vector. The tensor $\mathbf{G}$ is related to the Oseen tensor $\mathbf{W}$ by $G_{ij} = 8\pi\mu W_{ij}$. The stress vector for a Newtonian fluid is given by

$$\mathbf{t} = -p\mathbf{n} + 2\mu\mathbf{D}\cdot\mathbf{n}, \tag{5.8.7}$$

where $\mathbf{n}$ is the outward unit normal of S, p is pressure, and $\mathbf{D}$ is the rate of deformation tensor, defined as the symmetric part of the velocity gradient tensor.

The basic idea of the Stokesian dynamics method is to expand the integral in (5.8.6) in terms of a multipole expansion about the centroid ξ of the particle, so as to write fluid velocity at a point $\mathbf{x}$ as (Pozrikidis, 1992, 45)

$$u_i(\mathbf{x}) = -\frac{1}{8\pi\mu}\left[G_{ij}(\mathbf{x}, \xi)\int_S t_j(\mathbf{x}')\,da' + \frac{\partial G_{ij}}{\partial \xi_k}(\mathbf{x}, \xi)\int_S (x'_k - \xi_k)t_j(\mathbf{x}')\,da' + \cdots\right].$$
$$\tag{5.8.8}$$

Usually in Stokesian dynamics this expansion is only used for points located sufficiently distant from the particle that the expansion can be truncated after two terms. Of course, in this case the multipole expansion only accounts for the far-field flow and as particles move toward each other we must also include lubrication forces, such as those discussed in Section 2.4. The first integral in the multipole expansion (the zeroth moment) is simply the net force exerted on the particle. This term is the same as the stokeslet term (or point force) that was used in the example calculation discussed in Section 5.8.1. The second integral in the multipole expansion (the first moment) is the flow due to a force doublet. If the force doublet is decomposed into symmetric and antisymmetric parts, the symmetric part can be written in terms of a quantity called a stresslet, and the antisymmetric part can be written in terms of a point torque, or a rotlet. The advantage of the multipole expansion is that the integrals in (5.8.8) depend only on the stress and location on the particle surface (relative to the centroid position) and not on the position of the point $\mathbf{x}$ at which we seek to know the velocity. For this reason, these integrals can be evaluated only once each time step and used to find the velocity anywhere in the flow field. A summary of the basic theory of Stokesian dynamics is given in the review article by Brady and Bossis (1988).

Because Stokesian dynamics calculations usually do not account for contact forces between the particles, it can sometimes happen that particles come too close together. In this case, because the stokeslet and higher order terms (such as stresslet and rotlet) are singular at the particle centroid, the computations can quickly break down. To address this problem investigators have developed regularized versions of the stokeslet that are not singular (e.g., Cortez, 2001) by distributing the force over a finite region. If used together with a DEM computation, however, these singularities would not be a problem since the contact forces used in the DEM approach keep the particles sufficiently far away from each other.

The Stokesian dynamics method requires solution of a matrix equation for the fluid velocity at each particle centroid. This can be seen by considering the expression (5.8.2) for the fluid velocity and noting that the Stokes equation for $\mathbf{F}_d$ is proportional to the fluid velocity $\mathbf{u}$ evaluated at the center of each particle. For a system of N particles, direct solution of this matrix problem requires $O(N^3)$ calculations per time step, and an efficient iterative matrix solver still requires $O(N^2)$ calculations. To address these computational requirements, an accelerated approach using Fourier transforms has been proposed by Sierou and Brady (2001) which reduces the computational cost per time step to $O(N \ln N)$.

A significant limitation of the Stokesian dynamics method is that the flow Reynolds number must be sufficiently small that the Stokes equation (5.1.4) can be used to solve for the flow field through the entire flow domain. This is obviously a much more restrictive requirement than simply requiring that the particle Reynolds number be small, and it limits the Stokesian dynamics method to problems such as microfluidics or similar flows in small-scale, confined regions. The force-coupling method proposed by Lomholt and Maxey (2003) seeks to ease this restriction by distributing the particle force to a grid and then solving the flow field on the grid using the full Navier-Stokes equation. This approach is more time consuming than Stokesian dynamics, but it considerably extends the types of problems for which such an approach can be used and still retains many of the advantages of the Stokesian dynamics method. Alternatively, the Oseen dynamics method used by Subramanian and Koch (2008) for a falling particle cloud is valid for arbitrary flow Reynolds numbers, assuming only small particle Reynolds numbers.

The presence of walls in the flow domain introduces another complicating feature of the Stokesian dynamics method. For simple flow geometries, such as a flat surface or a sphere, analytical expressions can be used for the image of a stokeslet over the wall (Blake, 1971; Bossis et al., 1991). For more complex geometries, a boundary element method can be used to compute the Stokes flow within the domain (Pozrikidis, 1992). Complications can arise when using a boundary element method if particles move close to the wall, for which case it might be difficult to resolve the particle image set when the particle is much smaller than the panel used to discretize the flow domain. In such cases the pseudoimage method, described in Section 8.4 with reference to particle-induced electric fields, might be of use.

5.9. Particle Interactions with Acoustic Fields

Acoustic fields are used in a number of applications to accelerate the rate of particle agglomeration. For instance, in coal power plants it has been found effective to expose the exhaust stream to high intensity acoustic excitation just downstream of the combustion chamber in order to increase formation of agglomerations of micron- and submicron-size particles upstream of the electrostatic precipitators (Gallego-Juárez et al., 1999; Liu et al., 2009). This acoustic pretreatment significantly enhances performance of the precipitators due to the fact that the net fluid drag on an agglomerate is substantially less than the drag on the sum of the individual particles in isolation, as discussed in Section 5.7. In a related application, de Sarabia et al. (2003) investigated the effect of ultrasonic forcing on agglomeration due to liquid bridge forces for particulates in diesel exhaust streams. Very intense acoustic excitation has also been used as a method to break up particle agglomerates. For instance, Ding and Pacek (2008) examined use of ultrasound to break up nanoparticle colloidal agglomerates with different ionic strengths. Acoustic excitation has been used by numerous investigators for cleaning of particles from a planar surface. For instance, Chen and Wu (2010) and Furhmann et al. (2013) showed that exposure of a flat plate to acoustic waves in air significantly enhances the effectiveness of different aerodynamic particle removal approaches. Ultrasonic excitation is a common method used in cleaning silicon wafers, ceramic membranes, and other sensitive surfaces in liquids (Brereton and Bruno, 1994; Kuehn et al., 1996; Lamminen et al., 2004). As noted by Kim et al. (2009) and Keswani et al. (2009), ultrasonic cleaning not only moves the attached particles directly, but it leads to formation of cavitation bubbles on the surface which dislodge nearby particles.

Acoustic waves are also used to manipulate particles on a surface. For instance, an ultrasound trap is a device that focuses particles onto the nodal points of an acoustic pressure field as part of a biological assay process (Kuznetsova et al., 2007; Sobanski et al., 2001). Acoustic radiation force produces a particle drift at a rate that varies with particle size. This observation has been used by a number of investigators for development of acoustic particle separation devices in microfluidic systems (Kapishnikov et al., 2006; Laurell et al., 2007; Nam et al., 2011).

Acoustic-particle interaction effects are usually categorized as first-order or second-order, with the former depending linearly on the imposed acoustic velocity field and the latter depending nonlinearly on the acoustic velocity field. The dominant first-order effects leading to particle agglomeration are *orthokinetic motion* and the *acoustic wake effect* (or hydrodynamic interaction) (Dong et al., 2006; Song et al., 1994; Sheng and Shen, 2006). Orthokinetic motion arises from the oscillatory motion of particles of different sizes induced by the acoustic wave motion. The acoustic wake effect arises from the hydrodynamic interaction of one particle with the wake of another particle.

The second-order acoustic-particle interactions give rise to a steady force on the particle called the acoustic radiation pressure, which can be used to trap particles into velocity nodes or antinodes of the acoustic wave field. The seminal theoretical work on acoustic radiation pressure was the paper by King (1934), who derived an expression for radiation pressure on a rigid particle suspended freely in an inviscid fluid with arbitrary particle diameter relative to the acoustic wavelength. Most subsequent work builds on King's theory by releasing various of his assumptions. The effect of sphere elasticity was examined by Hasegawa and Yosioka (1969), as well as in subsequent papers by the Hasegawa and colleagues. Effects of fluid viscosity are particularly important both because small suspended particles typically have small particles Reynolds numbers and because the presence of viscosity introduces new physical effects, such as acoustic streaming around the particle, which can have a substantial influence on the radiation force. Theoretical studies of radiation pressure on a particle in a viscous fluid in various limits are reported in a series of papers by Danilov (Danilov, 1985; Danilov and Mironov, 1984, 2000) and in a series of more comprehensive papers by Doinikov (Doinikov, 1994, 1997). The radiation pressure is often neglected for studies of particle agglomeration because in many cases the time scale required for particle drift under this force is long compared to the particle agglomeration time scale associated with the two first-order forces discussed in the previous paragraph (Song, 1990).

5.9.1. Orthokinetic Motion

In the presence of an acoustic wave, particles oscillate back and forth at the same frequency as the acoustic wave velocity. The particle phase is different than that of the acoustic wave velocity, however, due to effect of particle inertia. To see this, consider a simplified problem introduced by Tiwary and Reethof (1986) in which the fluid velocity oscillates as

$$u = U_0 \sin(\omega t), \tag{5.9.1}$$

where ω is the acoustic angular frequency. For sufficiently small particles in an aerosol, it is reasonable to neglect effects such as added mass and Basset force to write the particle moment equation simply in terms of the particle inertia and fluid drag, giving

$$m\frac{dv}{dt} = 3\pi\mu d(u - v), \tag{5.9.2}$$

where we have used scalars for the fluid velocity $u(t)$ and particle velocity $v(t)$, because the motion in this example is assumed to be one-dimensional. In this equation it is assumed that the particle Reynolds number is sufficiently small that the Stokes drag expression applies and that the particle diameter is much smaller than the acoustic wave length. The particle velocity oscillates with the same frequency as the acoustic wave, but with a different phase and amplitude, so we can write

$$v = \eta_p U_0 \sin(\omega t - \phi_p), \tag{5.9.3}$$

where the particle entrainment coefficient η_p is the ratio of the particle velocity amplitude to the fluid velocity amplitude and the phase factor ϕ_p is the phase delay between the particle and fluid oscillations. Substituting (5.9.3) into (5.9.2) and solving for η_p and ϕ_p gives

$$\eta_p = 1/\sqrt{1 + (\omega\tau_p)^2}, \quad \phi_p = \tan^{-1}(\omega\tau_p), \tag{5.9.4}$$

where $\tau_p = m/3\pi\mu d$ is the particle time scale. The product $\omega\tau_p$ can be regarded as an acoustic Stokes number, where we see that in the limit $\omega\tau_p \to 0$ the particle amplitude and phase approach those of the fluid.

Orthokinetic particle collisions occur because the amplitude and phase of the particle motion in response to acoustic excitation depend on the particle diameter, as $\omega\tau_p$ varies with diameter in proportion to d^2. In general, the amplitude of particle oscillation decreases as $\omega\tau_p$ increases. The relative velocity between two nearby particles (i.e., particles that are sufficiently close to each other to experience essentially the same fluid velocity) with diameters d_1 and d_2 is given by

$$U_{rel,12} = [\eta_1 \cos(\omega t - \phi_1) - \eta_2 \cos(\omega t - \phi_2)]U_0, \tag{5.9.5}$$

Figure 5.14. Schematic diagram showing the acoustic agglomeration volume surrounding a large particle oscillating under an acoustic wave in the presence of a smaller particle. [Reprinted with permission from Li et al. (2011).]

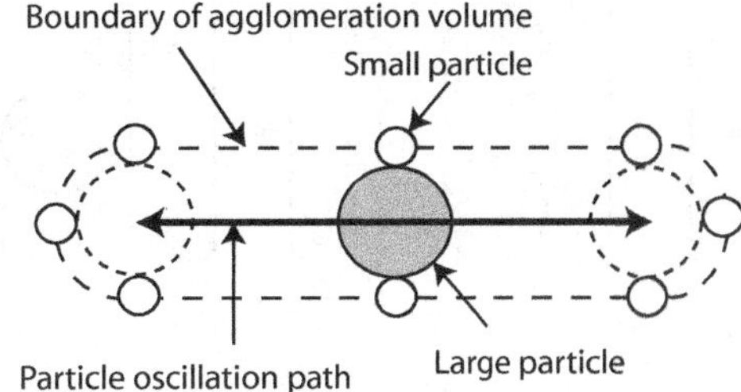

where η_i and ϕ_i are the entrainment coefficient and phase for particle i. In the frame of one particle, any second particle located within an *agglomeration volume* surrounding the first particle, as shown in Figure 5.14, would collide with the first particle, assuming that there is no deflection of the particle path due to hydrodynamic interaction. In a typical DEM simulation of acoustic particle interaction, the fluid time step is chosen to be sufficiently small to resolve the acoustic wave oscillation. In this case, the motion of individual particles is accurately resolved in time, such that orthokinetic collisions are obtained as a result of the simulation and no special modeling is required. However, this will often require that the time scale used for the fluid transport must be set extremely small.

Use of a time-averaged modeling approach allows much larger fluid time steps to be used which do not capture the acoustic wave motion; instead, the effect of acoustic waves is incorporated in the particle transport model. In a population balance approach, the effect of acoustic waves is usually accounted for by modifying the collision kernel so that particle collision occurs in a certain fraction f_C of cases where two particles enter into their respective agglomeration volumes. The fraction f_C is called the *collision efficiency*, and it can be written as a function of the relative acoustic Stokes number St_{12} as (Dong et al., 2006)

$$f_C = \left(\frac{\mathrm{St}_{12}}{\mathrm{St}_{12} + A}\right)^B, \tag{5.9.6}$$

where A and B are constants, given empirically by $A = 0.65$ and $B = 3.7$. For two particles with diameters d_1 and d_2, such that $d_1 > d_2$, the relative acoustic Stokes number is defined by

$$\mathrm{St}_{12} \equiv \frac{\rho_p \eta_{12} U_0 d_2^2}{18 \mu d_1}, \tag{5.9.7}$$

where

$$\eta_{12} \equiv \frac{\omega(\tau_1 - \tau_2)}{\sqrt{1 + (\omega\tau_1)^2}\sqrt{1 + (\omega\tau_2)^2}} \tag{5.9.8}$$

is the relative entrainment between the two particles.

5.9.2. Acoustic Wake Effect

Two idealized types of hydrodynamic interactions between particles under an imposed acoustic field are illustrated in Figure 5.15. In the first type, the line connecting the centroids of two interacting particles is perpendicular to the wave propagation

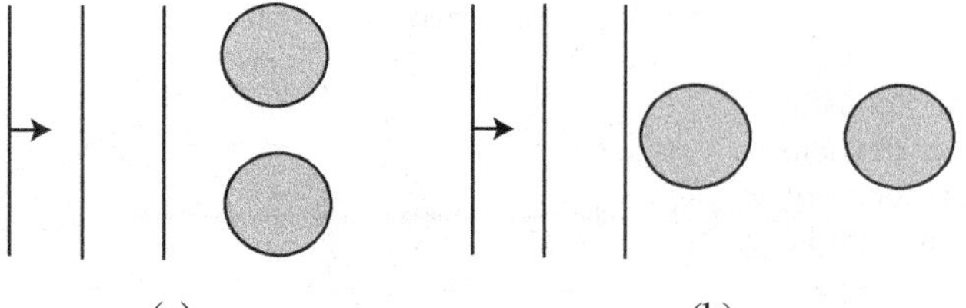

Figure 5.15. Configurations of particles and acoustic wave field for hydrodynamic interaction between particles with connecting line (a) orthogonal to the wave direction and (b) parallel to the wave direction.

direction (Figure 5.15a), whereas in the second type this line is parallel to the wave propagation direction (Figure 5.15b). For the first type of interaction, the velocity relative to the particles induced by the acoustic wave field is squeezed between the two particles, leading to a reduction in pressure in the region between the particles. This pressure reduction results in an agglomeration force that acts to draw the particles toward one another. However, Shaw (1978) showed that this type of hydrodynamic interaction generally has a small influence on the particles. The second type of hydrodynamic interaction occurs when one particle travels into the wake of an upstream particle. The viscous wake of the leading particle causes the downstream particle to be exposed to a decreased relative fluid velocity compared with that of the leading particle, allowing the downstream particle to travel slightly faster than the leading particle. In the second half of the acoustic wave cycle the roles of the two particles are reversed, but still with the effect that the particles are drawn toward each other.

The tendency of particles that lie in each other's wakes in the presence of an acoustic wave to be attracted to each other is called the *acoustic wake effect*. An interesting experiment that visualizes the acoustic wake effect was developed by Hoffman and Koopmann (1997) in which two particles suspended in a fluid settle side-by-side under gravity. At a given time, the particles are subjected to a horizontal acoustic wave. The particles continue to settle downward, but the acoustic wake effect makes them very gradually become attracted to each other as they oscillate back and forth in the horizontal direction, so that the particle path forms a tuning-fork pattern such as that shown in Figure 5.16.

In general, orthokinetic motion is the dominant cause of particle collisions at low acoustic frequencies and for particles with significantly different diameters. The acoustic wake effect dominates at high acoustic frequencies and for particles of similar size. A model that accounts for the combined action of orthokinetic motion and acoustic wake effect on the collision kernel used in the population balance model

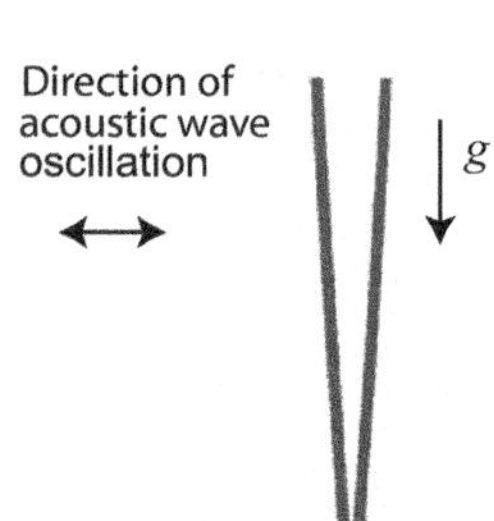

Figure 5.16. DEM simulation of the tuning-fork pattern observed for a pair of microspheres settling vertically under gravity and subject to a horizontal planar acoustic wave field.

was reported by Song et al. (1994). The acoustic wake effect can be implemented very simply in a DEM approach using an expression by Dianov et al. (1968) for the velocity reduction on downstream particles due to the acoustic wake of an upstream particle, which is derived based on the Oseen linearization of the convective term in the Navier-Stokes equation for low particle Reynolds numbers. The derivation makes the additional assumption that the acoustic frequency is sufficiently low that $|\frac{\partial u}{\partial t}| \ll |v\nabla^2 u|$, which is valid provided that the acoustic Reynolds number satisfies $\omega r_p^2/v \ll 1$. The Dianov et al. theory, modified by González-Gómez et al. (2000), gives an expression for the change in fluid velocity in spherical coordinates at a location (r, θ) due to a particle located at the origin subject to a relative "slip" velocity $w = u - v$ in the x-direction as

$$\Delta u_r = \frac{A_0}{r^2} - \frac{A_0}{r^2}\left[1 + \frac{r}{2v}(|w| + w\cos\theta)\right]\exp\left[-\frac{r}{2v}(|w| - w\cos\theta)\right], \quad (5.9.9)$$

$$\Delta u_\theta = -\frac{A_0 w}{2rv}\exp\left[-\frac{r}{2v}(|w| - w\cos\theta)\right]\sin\theta, \quad (5.9.10)$$

where

$$A_0 = \frac{3vr_p}{2}\left(1 + \frac{3r_p|w|}{8v}\right). \quad (5.9.11)$$

The technique is the same as the Oseen dynamics approach discussed in the previous section. By determining the relative velocity of all particles surrounding a given particle, the equations given here can be used to determine the effect of nearby particle wakes on the fluid velocity incident to the particle, which when summed yield the net fluid velocity relative to the give particle. Details on implementation of this approach in the context of a DEM simulation were given by González-Gómez et al. (2000). This modified DEM approach can be used to predict acoustic attraction and agglomeration of particles, and predictions obtained with this approach (e.g., Figure 5.16) compare well with experimental data by Hoffmann and Koopmann (1996) and González et al. (2002).

REFERENCES

Auton TR, Hunt JCR, Prud'homme M. The force on a body in inviscid unsteady non-uniform rotational flow. *Journal of Fluid Mechanics* **197**, 241–257 (1988).

Bagchi P, Balachandar S. Effect of free rotation on the motion of a solid sphere in linear shear flow at moderate Re. *Physics of Fluids* **14**(8), 2719–2737 (2002a).

Bagchi P, Balachandar S. Shear versus vortex-induced lift force on a rigid sphere at moderate Re. *Journal of Fluid Mechanics* **473**, 379–388 (2002b).

Basset AB. *A Treatise on Hydrodynamics*, Vol. 2, Deighton, Bell and Co., Cambridge (1888) [reprinted by Dover Publ., New York, 1961].

Batchelor GK. *An Introduction to Fluid Dynamics*, Cambridge University Press, Cambridge (1967).

Beetstra R, van der Hoef MA, Kuipers JAM. Drag force of intermediate Reynolds number flow past mono- and bidisperse arrays of spheres. *AIChE Journal* **53**, 489–501 (2007).

Benyahia S, Syamlal M, O'Brien TJ. Extension of Hill-Koch-Ladd drag correlation over all ranges of Reynolds number and solids volume fraction. *Powder Technology* **162**, 166–174 (2006).

Blake JR. A note on the image system for a stokeslet in a no-slip boundary. *Mathematical Proceedings of the Cambridge Philosophical Society* **70**, 303–310 (1971).

Bossis G, Meunier A, Sherwood JD. Stokesian dynamics simulations of particle trajectories near a plane. *Physics of Fluids A* **3**(8), 1853–1858 (1991).

Boussinesq JV. Sur la résistance qu'oppose un fluide indéfini au repos, sans pesanteur, au mouvement varié d'une sphère solide qu'il mouille sur toute sa surface, quand les vitesses restent bien continues et assez faibles pour que leurs carrés et produits soient négligeables. *Comptes Rendu de l'Academie des Sciences* **100**, 935–937 (1885).

Brady J, Bossis G. Stokesian dynamics. *Annual Review of Fluid Mechanics* **20**, 111–157 (1988).

Brenner H. The slow motion of a sphere through a viscous fluid towards a plane surface. *Chemical Engineering Science* **16**, 242–251 (1961).

Brereton GJ, Bruno BA. Particle removal by focused ultrasound. *Journal of Sound and Vibration* **173**(5), 683–698 (1994).

Bretherton FP. The motion of rigid particles in a shear flow at low Reynolds number. *Journal of Fluid Mechanics* **14**, 284–304 (1962).

Brown R. A brief account of microscopical observations made in the months of June, July and August, 1827, on the particles contained in the pollen of plants; and on the general existence of active molecules in organic and inorganic bodies. *Philosophical Magazine* **4**, 161–173 (1828).

Burke SP, Plummer WB. Gas flow through packed columns. *Ind. Eng. Chem.* **20**, 1196–1200 (1928).

Candelier F, Angilella JR, Souhar M. On the effect of the Boussinesq-Basset force on the radial migration of a Stokes particle in a vortex. *Physics of Fluids* **16**(5), 1765–1776 (2004).

Carman PC. Fluid flow through granular beds. *Transactions of the Institute of Chemical Engineers* **15**, 150–166 (1937).

Chen D, Wu J-W. Dislodgement and removal of dust-particles from a surface by a technique combining acoustic standing wave and airflow. *Journal of the Acoustical Society of America* **127**, 45–50 (2010).

Cherukat P, McLaughlin JB. The inertial lift on a rigid sphere in a linear shear flow field near a flat wall. *Journal of Fluid Mechanics* **263**, 1–18 (1994).

Cooley MDA, O'Neill ME. On the slow rotation of a sphere about a diameter parallel to a nearby plane wall. *Journal of the Institute of Mathematics and its Applications* **4**, 163–173 (1968).

Cortez R. The method of regularized stokeslets. *SIAM Journal of Scientific Computing* **23**(4), 1204–1225 (2001).

Cox RG, Brenner H. The slow motion of a sphere through a viscous fluid towards a plane surface. II Small gap widths, including inertial effects. *Chemical Engineering Science* **22**(12), 1753–1777 (1967).

Cunningham E. On the velocity of steady fall of spherical particles through fluid medium. *Proceedings of the Royal Society of London A* **83**, 357–365 (1910).

Dahl SR, Hrenya CM. Size segregation in gas-solid fluidized beds with continuous size distributions. *Chemical Engineering Science* **60**, 6658–6673 (2005).

Dandy DS, Dwyer HA. A sphere in shear flow at finite Reynolds number: effect of shear on particle lift, drag, and heat transfer. *Journal of Fluid Mechanics* **216**, 381–410 (1990).

Danilov SD. Average force acting on a small sphere in a travelling-wave field in a viscous fluid. *Soviet Physics Acoustics* **31**(1), 26–28 (1985).

Danilov SD, Mironov MA. Radiation pressure force acting on a small particle in a sound field. *Soviet Physics Acoustics* **30**(4), 280–283 (1984).

Danilov SD, Mironov MA. Mean force on a small sphere in a sound field in a viscous fluid. *Journal of the Acoustical Society of America* **107**(1), 143–153 (2000).

Deen NG, Van Sint Annaland M, Van der Hoef MA, Kuipers JAM. Review of discrete particle modeling of fluidized beds. *Chemical Engineering Science* **62**, 28–44 (2007).

Dianov DB, Podol'skii AA, Turubarov VI. Calculation of the hydrodynamic interactions of aerosol particles in a sound field under Oseen flow conditions. *Soviet Physics Acoustics* **13**(3), 314–319 (1968).

Di Felice R. The voidage function for fluid-particle interaction systems. *International Journal of Multiphase Flow* **20**, 153–159 (1994).

Ding P, Pacek AW. De-agglomeration of goethite nano-particles using ultrasonic comminution device. *Powder Technology* **187**(1), 1–10 (2008).

Doinikov AA. Acoustic radiation pressure on a rigid sphere in a viscous fluid. *Proceedings of the Royal Society of London A* **447**, 447–466 (1994).

Doinikov AA. Acoustic radiation force on a spherical particle in a viscous heat-conducting fluid. I. General formula. *Journal of the Acoustical Society of America* **101**(2), 713–721 (1997).

Dong S, Lipkens B, Cameron TM. The effects of orthokinetic collision, acoustic wake, and gravity on acoustic agglomeration of polydisperse aerosols. *Journal of Aerosol Science* **37**, 540–553 (2006).

Druzhinin OA, Ostrovsky LA. The influence of Basset force on particle dynamics in two-dimensional flows. *Physica D* **76**, 34–43 (1994).

Einstein A. Über die von der molekularkinetischen Theorie der Wärme geforderte Bewegung von in ruhenden Flüssigkeiten suspendierten Teilchen. *Annalen der Physik* **17**, 549–560 (1906).

Epstein PS. On the resistance experienced by spheres in their motion through gases. *Physical Review* **23**, 710–733 (1924).

Ergun S. Fluid flow through packed columns. *Chemical Engineering Progress* **48**(2), 89–94 (1952).

Faxén H. Der Widerstand gegen die Bewegung einer starren Kugel in einer zähen Flüssigkeit, die zwischen zwei parallelen ebenen Wänden eingeschlossen ist. *Annalen der Physik* **373**(10), 89–119 (1922).

Furhmann A, Marshall JS, Wu J-R. Effect of acoustic levitation force on aerodynamic particle removal from a surface. *Applied Acoustics* **74**, 535–543 (2013).

Gallego-Juárez JA, de Sarabia ERF, Rodríguez-Corral G, Hoffman TL, Gálvez-Moraleda JC, Rodríguez-Maroto JJ, Gómez-Moreno FJ, Bahillo-Ruiz A, Martín-Espigares M, Acha M. Application of acoustic agglomeration to reduce fine particle emissions from coal combustion plants. *Environmental Science and Technology* **33**, 3843–3849 (1999).

Gidaspow D. *Multiphase Flow and Fluidization*. Academic Press, San Diego (1994).

Goldman AJ, Cox RG, Brenner H. Slow viscous motion of a sphere parallel to a plane wall. I. Motion through a quiescent fluid. *Chemical Engineering Science* **22**(4), 637–651 (1967a).

Goldman AJ, Cox RG, Brenner H. Slow viscous motion of a sphere parallel to a plane wall. II. Couette flow. *Chemical Engineering Science* **22**(4), 653–660 (1967b).

González I, Hoffmann TL, Gallego JA. Visualization of hydrodynamic particle interactions: validation of a numerical model. *Acta Acustica* **88**, 19–26 (2002).

González-Gómez I, Hoffman TL, Gallego-Juárez JA. Theory and calculation of sound induced particle interactions of viscous origin. *Acta Acustica* **86**, 784–797 (2000).

Happel J, Brenner H. *Low Reynolds Number Hydrodynamics*. Martinus Nijhoff Publ., The Hague, Netherlands, (1983).

Hasegawa T, Yosioka K. Acoustic-radiation force on a solid elastic sphere. *Journal of the Acoustical Society of America* **46**, 1139–1143 (1969).

Herron IH, Davis SH, Bretherton FP. On the sedimentation of a sphere in a centrifuge. *Journal of Fluid Mechanics* **68**(2), 209–234 (1975).

Hill RJ, Koch DL, Ladd AJC. The first effects of fluid inertia on flows in ordered and random arrays of spheres. *Journal of Fluid Mechanics* **448**, 213–241 (2001a).

Hill RJ, Koch DL, Ladd AJC. Moderate-Reynolds-number flows in ordered and random arrays of spheres. *Journal of Fluid Mechanics* **448**, 243–278 (2001b).

Hoffman TL, Koopmann GH. Visualization of acoustic particle interaction and agglomeration: theory and experiments. *Journal of the Acoustical Society of America* **99**(4), 2130–2141 (1996).

Hoffman TL, Koopmann GH. Visualization of acoustic particle interaction and agglomeration: theory evaluation. *Journal of the Acoustical Society of America* **101**(6), 3421–3429 (1997).

Huilin L, Gidaspow D, Bouillard J, Wentie L. Hydrodynamic simulation of gas-solid flow in a riser using kinetic theory of granular flow. *Chemical Engineering Journal* **95**, 1–13 (2003).

Kapishnikov S, Kantsler V, Steinberg V. Continuous particle size separation and size sorting using ultrasound in a microchannel. *Journal of Statistical Mechanics: Theory and Experiment* P01012 (2006).

Keswani M, Raghavan S, Deymier P, Verhaverbeke S. Megasonic cleaning of wafers in electrolytic solutions: possible role of electro-acoustic and cavitation effects. *Microelectronic Engineering* **86**, 132–139 (2009).

Kim W, Kim T-H, Choi J, Kim H-Y. Mechanism of particle removal by megasonic waves. *Applied Physics Letters* **94**, 081908 (2009).

King LV. On the acoustic radiation pressure on spheres. *Proceedings of the Royal Society of London A* **147**(861), 212–240 (1934).

King MR, Leighton DT, Measurement of the inertial lift on a moving sphere on contact with a plane wall in a shear flow. *Physics of Fluids* **9**(5), 1248–1255 (1997).

Kozeny J. Ueber kapillare Leitung des Wassers im Boden. *Sitzber. Akad. Wiss. Wien* **136**, 271–306 (1927).

Krishnan GP, Leighton DT. Inertial lift on a moving sphere in contact with a plane wall in a shear flow. *Physics of Fluids* **7**(11), 2538–2545 (1995).

Kuehn TH, Kittelson TB, Wu Y, Gouk R. Particle removal from semiconductor wafers by megasonic cleaning. *Journal of Aerosol Science* **47**, S427–S428 (1996).

Kurose R, Komori S. Drag and lift forces on a rotating sphere in a linear shear flow. *Journal of Fluid Mechanics* **384**, 183–206 (1999).

Kuznetsova LA, Bazou D, Coakley WT. Stability of 2-D colloidal particle aggregates held against flow stress in an ultrasound trap. *Langmuir* **23**, 3009–3016 (2007).

Ladd AJC. Hydrodynamic transport coefficients for random dispersions of hard spheres. *Journal of Chemical Physics* **93**, 3484–3494 (1990).

Lamminen MO, Walker HW, Weavers LK. Mechanisms and factors influencing the ultrasonic cleaning of particle-fouled ceramic membranes. *Journal of Membrane Science* **237**, 213–223 (2004).

Laurell T, Petersson F, Nilsson A. Chip integrated strategies for acoustic separation and manipulation of cells and particles. *Chemical Society Reviews* **36**, 492–506 (2007).

Legendre D, Magnaudet J. The lift force on a spherical bubble in a viscous linear shear flow. *Journal of Fluid Mechanics* **368**, 81–126 (1998).

Leighton D, Acrivos A. The lift on a small sphere touching a plane in the presence of a simple shear flow. *Journal of Applied Mathematics and Physics (ZAMP)* **36**, 174–178 (1985).

Li S-Q, Marshall JS, Liu G, Yao Q. Adhesive particulate flow: The discrete element method and its application in energy and environmental engineering. *Progress in Energy and Combustion Science* **37**(6), 633–668 (2011).

Li T, Kheifets S, Medellin D, Raizen MG. Measurement of the instantaneous velocity of a Brownian particle. *Science* **328**, 1673–1675 (2010).

Lin JC, Perry JH, Schowalter WR. Simple shear flow round a rigid sphere: Inertial effects and suspension rheology. *Journal of Fluid Mechanics* **44**, 1–17 (1970).

Liu J, Zhang G, Zhou J, Wang J, Zhao W, Cen K. Experimental study of acoustic agglomeration of coal-fired fly ash particles at low frequencies. *Powder Technology* **193**(1), 20–25 (2009).

Lomholt S, Maxey MR. Force-coupling method for particulate two-phase flow: Stokes flow. *Journal of Computational Physics* **184**, 381–405 (2003).

Lovalenti PM, Brady JF. The hbydrodynamic force on a rigid particle undergoing arbitrary time-dependent motion at small Reynolds number. *Journal of Fluid Mechanics* **256**, 561–605 (1993).

Lucretius, "On The Nature of Things," translated by WE Leonard, reprinted by Dover, New York (2004).

Magnaudet JM. The forces acting on bubbles and rigid particles. *Proceedings of the Fluids Engineering Division Annual Meeting*, ASME (FEDSM97-3522), Vancouver, June 22–26 (1997).

McLaughlin JB. Inertial migration of a small sphere in linear shear flows. *Journal of Fluid Mechanics* **224**, 261–274 (1991).

Mei R. An approximate expression for the shear lift force on spherical particles at finite particle Reynolds number. *International Journal of Multiphase Flow* **18**(1), 145–147 (1992).

Mei R, Adrian RJ. Flow past a sphere with an oscillation in the free-stream velocity and unsteady drag at finite Reynolds number. *Journal of Fluid Mechanics* **237**, 323–341 (1992).

Mei R, Lawrence CJ, Adrian RJ. Unsteady drag on a sphere at finite Reynolds number with small-amplitude fluctuations in the free-stream velocity. *Journal of Fluid Mechanics* **233**, 613–631 (1991).

Metzger B, Nicolas M, Guazzelli E. Falling clouds of particles in viscous fluids. *Journal of Fluid Mechanics* **580**, 283–301 (2007).

Millikan RA. The general law of fall of small spherical body through a gas, and its bearing upon the nature of molecular reflection from surfaces. *Physical Review* **22**, 1–23 (1923).

Nam J, Lee Y, Shin S. Size-dependent microparticles separation through standing surface acoustic waves. *Microfluidics and Nanofluidics* **11**, 317–326 (2011).

Nitsche JM, Batchelor GK. Break-up of a falling drop containing dispersed particles. *Journal of Fluid Mechanics* **340**, 161–175 (1997).

Noh Y, Fernando HJS. The transition in the sedimentation pattern of a particle cloud. *Physics of Fluids A* **5**(12), 3049–3055 (1993).

Oesterlé B, Dinh TB. Experiments on the lift of a spinning sphere in a range of intermediate Reynolds numbers. *Experiments in Fluids* **25**, 16–22 (1998).

O'Neill ME, Stewartson K. On the slow motion of a sphere parallel to a nearby plane wall. *Journal of Fluid Mechanics* **27**(4), 705–724 (1967).

Oseen CW. Über die Stokes'sche Formel und über eine verwandte Aufgabe in der Hydrodynamik. *Arkiv för Matematik Astronomi och Fysik* **6**, 1–20 (1910).

Pignatel F, Nicolas M, Guazzelli E. A falling cloud of particles at a small but finite Reynolds number. *Journal of Fluid Mechanics* **671**, 34–51 (2011).

Pozrikidis C. *Boundary Integral and Singularity Methods for Linearized Viscous Flows.* Cambridge University Press, Cambridge UK (1992).

Proudman I, Pearson JRA. Expansions at small Reynolds numbers for the flow past a sphere and a circular cylinder. *Journal of Fluid Mechanics* **2**(3), 237–262 (1957).

Richardson JF, Zaki WN. Sedimentation and fluidization. *Part I. Transactions of the Institution of Chemical Engineers* **32**, 35–53 (1954).

Rivero M, Magnaudet J, Fabre J. Quelques résultats nouveaux concernant les forces exercées sur une inclusion sphérique par un écoulement accéléré. *C.R. Acad. Sci. Paris Série II* **312**, 1499–1506 (1991).

Rubinow SI, Keller JB. The transverse force on a spinning sphere moving in a viscous fluid. *Journal of Fluid Mechanics* **11**, 447–459 (1961).

Saffman PG. On the motion of small spheroidal particles in a viscous liquid. *Journal of Fluid Mechanics* **1**, 540–553 (1956).

Saffman PG. The lift on a small sphere in a slow shear flow. *Journal of Fluid Mechanics* **22**, 385–400 (1965).

Saffman PG. Corrigendum to "The lift force on a small sphere in a slow shear flow." *Journal of Fluid Mechanics* **31**, 624 (1968).

de Sarabia ERF, Elvira-Segura L, González-Gómez I, Rodríguez-Maroto JJ, Muñoz-Bueno R, Dorronsoro-Areal JL. Investigation of the influence of humidity on the ultrasonic agglomeration of submicron particles in diesel exhausts. *Ultrasonics* **41**, 277–281 (2003).

Schiller L, Naumann A. Über die gundlegenden Berechungen bei der Schwerkraftaufbereitung. *Zeitschrift des Vereines Deutscher Ingenieure* **77**, 318–320 (1933).

Schmitt K. Grundlegende Untersuchungen zum Thermalpräzipitator. *Staub* **19**, 416–421 (1959).

Segré G, Silberberg A. Behavior of macroscopic rigid spheres in Poiseuille flow. *Journal of Fluid Mechanics* **14**, 136–157 (1962).

Shaw DT. Acoustic agglomeration of aerosol. In *Recent Developments in Aerosol Science*, DT Shaw, editor, John Wiley & Sons, New York, pp. 279–319 (1978).

Sheng C, Shen X. Modelling of acoustic agglomeration processes using the direct simulation Monte Carlo method. *Journal of Aerosol Science* **37**, 16–36 (2006).

Sierou A, Brady JF. Accelerated stokesian dynamics simulations. *Journal of Fluid Mechanics* **448**, 115–146 (2001).

Smart JR, Beimfohr S, Leighton DT. Measurement of the translational and rotational velocities of a non-colloidal sphere rolling down a smooth inclined plane at low Reynolds number. *Physics of Fluids A* **5**(1), 13–24 (1993).

Smoluchowski M. Zur kinetischen Theorie der Brownschen Molekularbewegung und der Suspensionen. *Annalen der Physik* **21**, 756–780 (1906).

Sobanski MA, Tucker CR, Thomas NE, Coakley WT. Sub-micron particle manipulation in an ultrasonic standing wave: applications in detection of clinically important biomolecules. *Bioseparation* **9**, 351–357 (2001).

Song L. Modeling of acoustic agglomeration of fine aerosol particles. Ph.D. dissertation, Pennsylvania State University, College Station, Pennsylvania (1990).

Song L, Kooperman GH, Hoffman TL. An improved theoretical model of acoustic agglomeration. *Journal of Vibration and Acoustics* **116**, 208–214 (1994).

Sridhar G, Katz J. Drag and lift forces on microscopic bubbles entrained by a vortex. *Physics of Fluids* **7**(2), 389–399 (1995).

Stewart BE, Thompson MC, Leweke T, Hourigan K. Numerical and experimental studies of the rolling sphere wake. *Journal of Fluid Mechanics* **643**, 137–162 (2010).

Subramanian G, Koch DL. Evoluation of clusters of sedimenting low-Reynolds-number particles with Oseen interactions. *Journal of Fluid Mechanics* **603**, 63–100 (2008).

Sweeney LG, Finlay WH. Lift and drag forces on a sphere attached to a wall in a Blasius boundary layer. *Journal of Aerosol Science* **38**, 131–135 (2007).

Takemura F, Magnaudet J. The transverse force on clean and contaminated bubbles rising near a vertical wall at moderate Reynolds number. *Journal of Fluid Mechanics* **495**, 235–253 (2003).

Tanaka T, Yamagata K, Tsuji Y. Experiment on fluid forces on a rotating sphere and spheroid. *Proceedings of the 2nd KSME-JSME Fluids Engineering Conference*, Vol. 1, p. 366 (1990).

Taneda S. Experimental investigation of the wake behind a sphere at low Reynolds numbers. *Journal of the Physical Society of Japan* **11**(10), 1104–1118 (1956).

Thomas PJ. On the influence of the Basset history force on the motion of a particle through a fluid. *Physics of Fluids A* **4**(9), 2090–2093 (1992).

Tiwary R, Reethof G. Hydrodynamic interaction of spherical aerosol particles in a high intensity acoustic field. *Journal of Sound and Vibration* **108**(1), 33–49 (1986).

Tri BD, Oesterle B, Deneu F. Premiers resulats sur la portance d'une sphere en rotation aux nombres de Reynolds intermediaies. *C.R. Acad. Sci. Ser. H: Mec. Phys. Chim. Sci. Terre Univers* **311**, 27 (1990).

Tsuji Y, Morikawa Y, Mizuno O. Experimental measurement of the Magnus force on a rotating sphere at low Reynolds numbers. *Journal of Fluids Engineering* **107**, 484–488 (1985).

Van der Hoef MA, Beetstra R, Kuipers JAM. Lattice-Boltzmann simulations of low-Reynolds-number flow past mono- and bidisperse arrays of spheres: Results for the permeability and drag force. *Journal of Fluid Mechanics* **528**, 233–254 (2005).

Vasseur P, Cox RG. The lateral migration of spherical particles sedimenting in a stagnant bounded fluid. *Journal of Fluid Mechanics* **80**(3), 561–591 (1977).

Wakiya S. Research report 9. Faculty of Engineering, Niigata University, Japan (1960).

Wen CY, Yu YH. Mechanics of fluidization. *Chemical Engineering Progress Symposium Series* **62**(62), 100–111 (1966).

White FM. *Viscous Fluid Flow*. 3rd ed, McGraw-Hill, New York (2006).

Zeng L, Balachandar S, Fischer P. Wall-induced forces on a rigid sphere at finite Reynolds number. *Journal of Fluid Mechanics* **536**, 1–25 (2005).

Zeng L, Najjar F, Balachandar S, Fischer P. Forces on a finite-sized particle located close to a wall in a linear shear flow. *Physics of Fluids* **21**, 0.33302 (2009).

Ziskind G, Fichman M, Gutfinger C. Effects of shear on particle motion near a surface: application to resuspension. *Journal of Aerosol Science* **29**(3), 323–338 (1998)

6   Particle Dispersion in Turbulent Flows

Turbulent particle dispersion is a key aspect of a large number of important processes, including droplet and smog transport in the atmosphere, dust contamination in electronics manufacturing, particulate combustion processes, and automotive and truck exhaust. In some of these processes the particles have very little interaction with each other, for which case our interest primarily focuses on modeling the time variation of the bulk concentration field, measured on a scale much larger than the turbulent flow eddies. On the other hand, certain particle transport problems involve particles that do interact with each other. For instance, droplet collision and merger in a cloud is a basic process that governs the size of water droplets, and eventually determines onset of precipitation. Collision of particles in a dust cloud can cause electrons or ions to be stripped from one or the other particle, causing charging of the two colliding particles via the mechanism of tribocharging. For instance, the frequent large-scale dust storms on Mars are a critical component of the planetary weather system, causing the fine dust particles that make up the Martian soil to become highly charged. In both of these examples, the rate of particle tribocharging and the rate of droplet merger are dependent on the collision rate of particles in a turbulent flow. Such problems depend critically on the local particle concentration, and are sensitive to variation of the concentration field on length scales on the order of the turbulent eddies throughout the inertial range.

Development of agglomerates of adhesive particles in turbulent flows is sensitive to the turbulent flow in two different ways. The inertia imparted on the particles by the turbulent fluctuations leads to enhanced collision rate, and hence increases the rate of agglomerate formation. On the other hand, the shear stress fluctuations associated with the turbulent flow can tear apart the agglomerates if sufficiently strong. Turbulent flow with adhesive particles is therefore particularly sensitive to local fluctuations in particle concentration and to the distribution of energy with length scale within the turbulent flow.

6.1. Particle Motion in Turbulent Flows

Because it is not possible to directly simulate all scales of most turbulent flows, it is necessary to introduce some model for the fluctuating fluid velocity field of a turbulent flow. For many engineering applications, the turbulent flow is modeled using a Reynolds-averaged Navier-Stokes (RANS) model, which solves for the mean velocity field and certain measures that characterize the turbulent fluctuations. For instance, in the well-known k-ε model (Launder and Sharma, 1974), the spatial variations of the turbulent kinetic energy (which we will denote by q) and the dissipation rate ε are obtained from two additional partial differential equations. From these two parameters, estimates of the Lagrangian integral length scale ℓ_L and time scale τ_L can be obtained as

$$\ell_L = u_0^3/2\varepsilon, \ \tau_L = 4q/3C_0\varepsilon, \tag{6.1.1}$$

where u_0 is a characteristic velocity scale of the energy-containing motion that is related to the turbulent kinetic energy q by $u_0 = (2q/3)^{1/2}$ and C_0 is the Kolmogorov constant with value $C_0 \cong 6.5$. The turbulent eddies range in size from this integral scale, associated with the most energetic eddies, to the Kolmogorov length scale $\eta = (v^3/\varepsilon)^{1/4}$, where v is the fluid kinematic viscosity, which is the scale associated with energy dissipation. Eddy length scales ℓ significantly larger than the Kolmogorov scale are said to lie in the *inertial range*, for which the fluid viscosity is not an important parameter and the flow can be scaled as if it were inviscid. Throughout this inertial range, the eddy velocity scale u can be related to its length scale ℓ by $u \sim (\varepsilon\ell)^{1/3}$.

While RANS models are still prevalent for applications, large eddy simulation (LES) approaches are becoming more common, particularly for relatively simple engineering flow problems. In LES, the larger scales of the flow are directly computed, ranging from the length scale of the computational domain to a scale ℓ_{GRID} associated with the grid resolution, which ideally is selected such that $\ell_{GRID} < \ell_L$. Particle transport by the resolved eddies in the LES computation can be directly computed using the DEM method discussed in Chapter 2, with the LES velocity field used to obtain the fluid velocity $\mathbf{u}$ in the force expressions in Chapter 5. Nevertheless, there is a significant amount of particle transport at scales smaller than ℓ_{GRID} in most real turbulent flows, which cannot be simulated by the LES method. It is thus necessary for both RANS and LES models to introduce an additional model that describes necessary statistical aspects of the subgrid-scale turbulent fluctuations required to obtain a reasonable prediction for particle transport by the unresolved turbulence.

Particles are transported relative to the mean-flow streamlines in turbulent flows by two very different mechanisms. The first mechanism (usually called particle dispersion) occurs when material regions of the particulate fluid are exchanged between different turbulent eddies. This is the same process by which turbulence enhances diffusion of heat, mass, and momentum, and it is well known to cause an enhancement in the rate at which particles are spread outward and mixed by the fluid flow. The second mechanism (called particle clustering) is related to the drift of particles across streamlines of the instantaneous flow field in regions of high streamline curvature. A principal cause of this drift in turbulent flows is the action of turbulent eddies in excluding particles that are more dense than the surrounding fluid via centrifugal force, causing the particles to cluster in the interstitial region between the eddies (Bec et al., 2005, 2007; Falkovich and Pumir, 2004; Grits et al., 2006). The equations governing this mechanism are discussed in Section 1.2.2. The local particle concentration can fluctuate via this mechanism by large amounts, with reports of local concentration values as large as 25 times the mean concentration (Squires

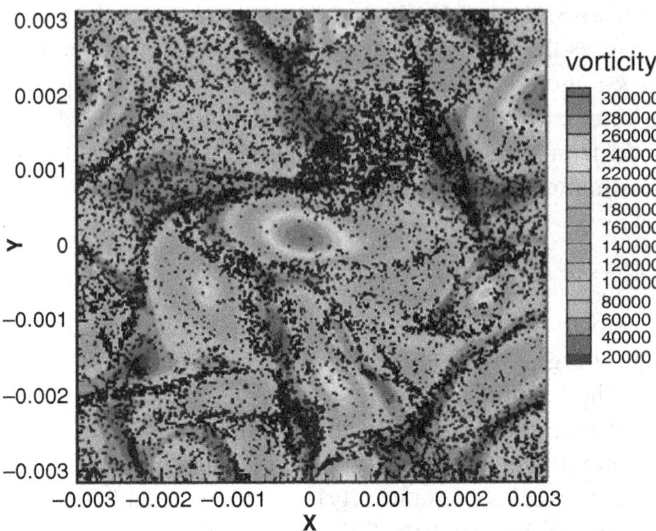

Figure 6.1. Small-scale clustering of particles in interstitial space between vortices in homogeneous turbulence. [Reprinted with permission from Garcia (2009).]

and Eaton, 1991). Figure 6.1 shows predictions of a DNS study of homogeneous turbulence by García (2009) in which heavy particles are thrown out of the turbulent eddies and collect in high-concentration regions between the eddies. This figure illustrates the antidiffusive characteristic of the particle clustering mechanism, which leads to formation of regions with highly heterogeneous concentration values.

The nature of particle transport by a given turbulent eddy varies significantly with the length scale ℓ and velocity scale u of the eddy. The effect of the eddy on nearby particles can be characterized by the *eddy Stokes number*, defined by

$$\mathrm{St}_\ell \equiv mu/3\pi\mu\,d\ell, \tag{6.1.2}$$

where d and m are the particle diameter and mass, respectively, and μ is the fluid viscosity. The eddy Stokes number, which is a function of eddy size, differs from the more general Stokes number St for a turbulent flow, which is usually based on the integral size and velocity scale. Using the inertial-range scaling $u \sim (\varepsilon\ell)^{1/3}$ and recalling that a basic characteristic of inertial range turbulence is that the dissipation rate is independent of scale (Tennekes and Lumley, 1972), the eddy Stokes number is found to vary with ℓ approximately as $\mathrm{St}_\ell \sim \ell^{-2/3}$. Consequently, larger eddies have smaller eddy Stokes numbers, and vice versa, for a given particle size.

Let us suppose that a critical value of eddy size ℓ_{crit} exists within the inertial range for which $\mathrm{St}_\ell(\ell_{crit}) = 1$. For eddy sizes $\ell \gg \ell_{crit}$, the eddy Stokes number is small and the particles are simply carried with the fluid flow. On the other hand, small-scale fluctuations, with $\ell \ll \ell_{crit}$, do not have much influence on the particle motion, but instead the force on the particles induced by these small velocity fluctuations is filtered out by the particle inertia (Ayyalasomayajula et al., 2008). To illustrate these limits, consider a simple velocity field that varies sinusoidally in time as

$$\mathbf{u} = u_x\mathbf{e}_x = U\sin(\omega t)\mathbf{e}_x, \tag{6.1.3}$$

where U is the fluid velocity scale and the angular frequency ω induced by an eddy is set equal to the ratio ℓ/U of eddy size to eddy velocity scale. If the particle motion in this example is controlled by the combination of the particle inertia and drag, then

$$m\frac{d\mathbf{v}}{dt} = -3\pi\,\mu\,d(\mathbf{v} - \mathbf{u}). \tag{6.1.4}$$

Nondimensionalizing velocity and time as $v_x' = v_x/U$ and $t' = t\omega$, (6.1.4) becomes

$$\frac{dv_x'}{dt'} + \frac{1}{\mathrm{St}_\omega}v_x' = \frac{1}{\mathrm{St}_\omega}\sin(t'), \tag{6.1.5}$$

where St_ω is the Stokes number computed using (6.1.2) with $\ell = U/\omega$. Integrating this equation in time and letting t' be sufficiently large that $\exp(-t'/\mathrm{St}_\omega) \cong 0$, the particle dimensionless velocity and acceleration are obtained as

$$v_x'(t') = \frac{\sin(t') - \mathrm{St}_\omega\cos(t')}{1 + \mathrm{St}_\omega^2}, \quad a_{p,x}'(t') = \frac{\cos(t') + \mathrm{St}_\omega\sin(t')}{1 + \mathrm{St}_\omega^2}. \tag{6.1.6}$$

For small Stokes numbers ($\mathrm{St}_\omega \ll 1$), the leading order solution is simply $v_x' \sim \sin(t') = u_x'$, indicating that the particle moves with the local velocity field. Taking the square of the velocity and acceleration in (6.1.6) and averaging over the oscillation period gives

$$\frac{\langle v_x'^2\rangle}{\langle u_x'^2\rangle} = \frac{\langle a_{p,x}'^2\rangle}{\langle a_{f,x}'^2\rangle} = \frac{1}{1 + \mathrm{St}_\omega^2}, \tag{6.1.7}$$

where the brackets denote the average value over the oscillation period and $u_x' = \sin(t')$ and $a_{f,x}' = \cos(t')$ are the dimensionless fluid velocity and acceleration. The result (6.1.7) shows that the kinetic energy of the particle per unit mass is damped by a factor $1/\mathrm{St}_\omega^2$ for large Stokes numbers relative to that of the fluid.

6.2. Particle Drift Measure

It has long been recognized that fluid acceleration plays a key role in determination of particle drift relative to the surrounding fluid flow. To understand why this is so, we return to the example of a particle governed by its own inertia and the Stokes drag force, for which the momentum equation is given by (6.1.4). A particle drift velocity $\mathbf{v}_d$ is defined by $\mathbf{v}_d \equiv \mathbf{v} - \mathbf{u}$, which from (6.1.4) is given by

$$\mathbf{v}_d = -\frac{m}{3\pi\,\mu\,d}\frac{d\mathbf{v}}{dt}. \tag{6.2.1}$$

Nondimensionalizing velocity by the root-mean square fluctuation velocity scale u_0 and time by the ratio ℓ_L/u_0, where ℓ_L is the Lagrangian integral length scale defined in (6.1.1), the drift velocity expression in (6.2.1) admits the dimensionless form

$$\mathbf{v}_d' = -\mathrm{St_L}\frac{d\mathbf{v}'}{dt'}, \tag{6.2.2}$$

where primes denote dimensionless variables and St_L is the Stokes number based on Lagrangian integral scaling. For small values of the Stokes number, it was discussed in Section 1.2.4 that the particle nearly follows the fluid streamlines, with a drift velocity $\mathbf{v}_d$ that scales as $O(\text{St}_{\text{L}}\, u_0)$. The particle acceleration term can be expanded using the definition of the drift velocity as

$$\frac{d\mathbf{v}'}{dt'} = \frac{D\mathbf{u}'}{Dt'} + \frac{d\mathbf{v}'_d}{dt'} + \mathbf{v}'_d \cdot \nabla'\mathbf{u}', \tag{6.2.3}$$

where D/Dt and d/dt denote material time derivative following a fluid element and following a particle, respectively. The first term on the right-hand side of (6.2.3) is $O(1)$, and the second and third terms are both $O(\text{St}_{\text{L}})$. Substituting (6.2.3) into (6.2.2) and retaining only leading-order terms in the Stokes number, an approximate equation for the particle drift velocity is obtained as

$$\mathbf{v}'_d = -\text{St}\,\mathbf{a}' + O(\text{St}^2), \tag{6.2.4}$$

where $\mathbf{a}' \equiv D\mathbf{u}'/Dt'$ is the dimensionless fluid acceleration. This result is the basis of the "fast-Eulerian method" proposed by Ferry and Balachandar (2001).

The result (6.2.4) implies that although a particle at low Stokes number may travel predominantly with the fluid flow, it also drifts relative to the fluid in a direction opposite to the direction of the fluid acceleration vector and at a rate proportional to the product of the Stokes number and the acceleration. An example illustrating this result was discussed in Section 1.2.2 with respect to the problem of particles thrown out of a vortex by the centrifugal force. Of course, the notion of centrifugal force is tied to solution of the governing equations in a rotating coordinate frame. In an inertial frame, as used in obtaining (6.2.4), the fluid rotating around the vortex has a centripetal acceleration $\mathbf{a}$ pointing inwards toward the vortex center. According to (6.2.4), the particle will drift in a direction opposite to the fluid acceleration, moving the particle radially outward from the vortex center.

Because of the important role that acceleration plays on particle dispersion, the statistics of the turbulence acceleration field is a matter of considerable interest. The velocity field in a turbulent flow is well known to exhibit a nearly Gaussian probability distribution; however, a wide range of experimental investigations (Voth et al., 1998; La Porta et al., 2001; Mordant et al., 2004; Ayyalasomayajula et al., 2006) and computational studies (Yeung et al., 2007; Biferale et al., 2004; Sawford et al., 2003) have demonstrated that the acceleration field exhibits a so-called super-statistical distribution characterized by fat tails (Reynolds, 2003b; Beck, 2008). Data for the acceleration probability density function (PDF) from La Porta et al. (2001) is shown in Figure 6.2, along with a Gaussian distribution having the same variance. For accelerations more than one order of magnitude smaller than the most probable acceleration, there is a strong deviation between these two distributions. A best-fit curve to the data is also plotted, which gives the probability distribution function $P(a)$ as

$$P(a) = C\exp[-a^2/(1 + |a\beta/\sigma|^\gamma \sigma^2)], \tag{6.2.5}$$

where for these data $\beta = 0.539$, $\gamma = 1.588$, and $\sigma = 0.508$. The constant C is selected such that the integral over the probability distribution equals unity, which for these data gives $C = 0.786$.

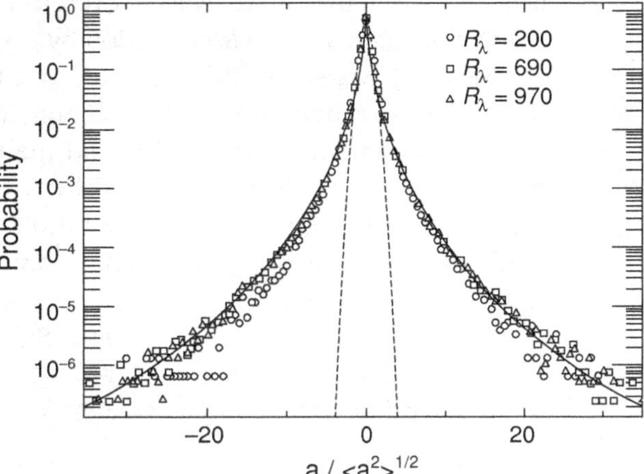

Figure 6.2. Plot showing the probability density function of the turbulence acceleration, comparing experimental measurements of La Porta et al. (2001) at different microscale Reynolds numbers to a best fit curve (solid line) and to a Gaussian distribution with the same variance (dashed line). [Reprinted with permission from La Porta et al. (2001).]

The relationship between the turbulence acceleration field and the turbulent eddy structures has been investigated by Lee and Lee (2005), Reynolds et al. (2005), and Mordant et al. (2004). These studies have demonstrated clearly that the magnitude of the turbulent acceleration field is dominated by the centripetal acceleration surrounding coherent vortex structures. An example showing this dominant feature of the acceleration field is given in Figure 6.3, from Lee and Lee (2005), in which contours of the enstrophy field ($E_s = \omega \cdot \omega/2$) are plotted for a turbulent flow over a cross-section of a coherent vortex structure. The pressure gradient vector is plotted

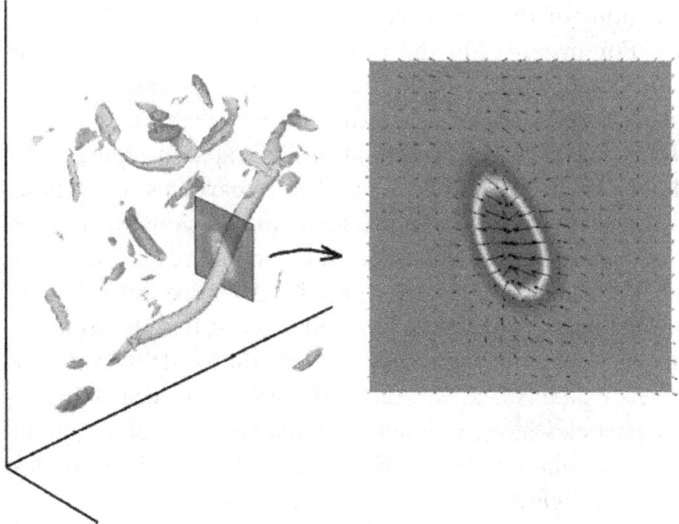

Figure 6.3. Plot from a direct numerical simulation of homogeneous turbulence, showing a vortex structure detected by iso-surface of enstrophy ($E = \omega \cdot \omega/2$) and enstrophy contour lines in its cross-section. The pressure gradient is indicated by arrows, which for inviscid flow is proportional to the acceleration vector. [Reprinted with permission from Lee and Lee (2005).]

using arrows in this figure, where it is recalled that for an inviscid flow the pressure gradient is related to the acceleration by $\rho\,\mathbf{a} = -\nabla p$.

Sala and Marshall (2013) pointed out that particle motion relative to a fluid streamline is related only to the component of the particle drift that is normal to the fluid streamline. Using (6.2.4) and the unit vector along the streamline, given by $\mathbf{s} \equiv \mathbf{u}/|\mathbf{u}|$, the rate of particle drift relative to the streamline is found to be proportional to $\mathrm{St}\,\mathbf{a} \times \mathbf{s}$. Taking the ratio of the magnitude of this vector with the fluid velocity magnitude gives a dimensionless particle drift measure φ_s as

$$\varphi_s(\mathbf{x}, t) = \mathrm{St_L}\,\frac{|\mathbf{a}' \times \mathbf{u}'|}{|\mathbf{u}'|^2}, \tag{6.2.6}$$

where the primes denote dimensionless variables.

6.3. Particle Collision Models

Many processes in which particles interact with each other start with particle collision. Whether we are concerned with droplet merger, particle agglomeration, particle sintering, or particle tribocharging, each of these processes is initiated by a particle collision event. Consequently, the ability to accurately predict particle collision rate in turbulent flows is of great importance.

6.3.1. Collision Mechanisms

As noted in Section 1.3 with reference to filtration applications, there are several different physical mechanisms that give rise to particle collisions. A summary and a classification of these different mechanisms were developed in the review article by Meyer and Deglon (2011). Two of the most important mechanisms have already been discussed in earlier sections of this book with respect to collisions between particles and larger bodies. For instance, in the *orthokinetic* (shear) mechanism illustrated in Figure 6.4a, the particles follow the fluid streamlines and collisions occur due to the finite particle size as particles are carried past each other by the local fluid shear. In the *accelerative-correlated* mechanism illustrated in Figure 6.4b, particles are mostly advected by the local fluid flow, but drift of particles across fluid streamlines in regions of strong streamline curvature leads to collisions with other particles. The *accelerative-independent* mechanism illustrated in Figure 6.4c occurs at large Stokes numbers, in which particles that receive impulse from distant turbulent eddies can travel substantial distances before collision, and as a result at the collision location the particle velocities are essentially uncorrelated with the local fluid velocity field. The *differential settling* mechanism illustrated in Figure 6.4d occurs due to the fact that larger, heavier particles have a higher terminal settling velocity than smaller, lighter particles. In a flow dominated by particle settling, the heavier particles can overtake the slower-moving lighter particles, leading to collisions. For very small particles, with diameters less than about 1 μm, the *perikinetic* (or Brownian motion) mechanism illustrated in Figure 6.4e can also be significant, in which particles move in a zig-zagging manner under the random forces caused by collisions with individual fluid molecules. Since the random paths of the particles are uncorrelated, they sometimes cross or come sufficiently close to each other that the particles

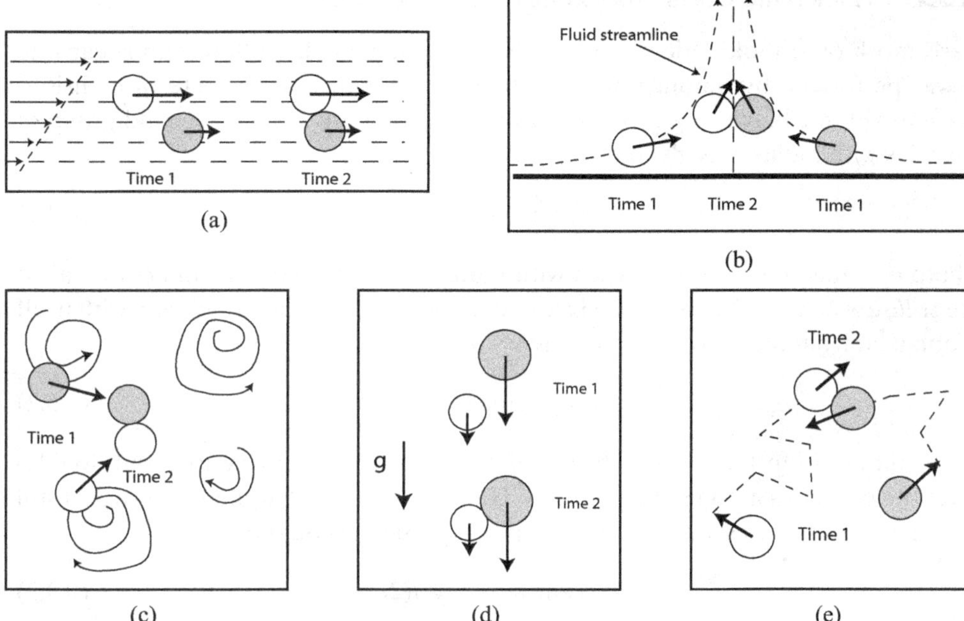

Figure 6.4. Illustrations of five different collision mechanisms: (a) orthokinetic; (b) accelerative-correlated; (c) accelerative-independent; (d) differential sedimentation; and (e) perikinetic.

collide. A summary of these different collision mechanisms and of the regime where each mechanism dominates is given in Table 6.1. Of particular interest in the current chapter are the orthokinetic, accelerative-correlated, and accelerative-independent mechanisms dealing with the effect of turbulence on particle collision rate.

Table 6.1. *Summary of collision mechanisms and regimes where they are valid or dominant*

Mechanism	Description	Regime where valid or dominant
Perikinetic	Particles collide under Brownian motion due to impulse from collision with fluid molecules	Small particle Peclet number ($\mathrm{Pe}_p < O(1)$)
Orthokinetic	Particles follow fluid streamlines at centroid and collide due to finite size	Small Stokes number ($\mathrm{St}_K \ll 1$)
Accelerative-correlated	Particles mostly follow fluid streamlines but drift at curves due to finite inertia, leading to collisions	Intermediate Stokes number
Accelerative-independent	Particles collide after traveling several eddy length scales, with velocity that has low correlation with local fluid velocity	Large Stokes number ($\mathrm{St}_K \gg 1$)
Differential sedimentation	Particles of different sizes and masses fall under gravity at different terminal velocities, such that faster particles run into slower particles	Small Froude number ($\mathrm{Fr} \leq O(1)$)

6.3.2. Orthokinetic Collisions (Small Stokes Numbers)

Early work on orthokinetic particle collision in laminar shear flow with shear rate $\dot{\gamma}$ was performed by Smoluchowski (1917), who found that the rate of collisions per unit volume between a group of particles with radius r_1 and a second group of particles with radius r_2 is given by

$$\dot{n}_{c12} = \tfrac{4}{3}\dot{\gamma}n_1 n_2(r_1 + r_2)^3 = \alpha_{12}n_1 n_2, \tag{6.3.1}$$

where n_i is the number of particles with radius r_i per unit volume and α_{12} is called the *collision kernel*. The rate of collisions of a set of particles with radius r_i with itself is obtained from the Smoluchowski theory as

$$\dot{n}_{cii} = \tfrac{2}{3}\dot{\gamma}n_i^2(2r_i)^3, \tag{6.3.2}$$

which differs from (6.3.1) by a factor of $1/2$, which is necessary to avoid double-counting collisions among like particles. Hu and Mei (1998) argued that (6.3.2) still contains self-collisions, and that the proper expression should be

$$\dot{n}_{cii} = \tfrac{2}{3}\dot{\gamma}n_i(n_i - 1/V)(2r_i)^3, \tag{6.3.3}$$

where V is the volume of the particulate flow domain. For large volumes and large numbers of particles, $n_i V \gg 1$ and (6.3.3) reduces to (6.3.2).

Saffman and Turner (1956) derived a well-known expression for orthokinetic particle collision rate in turbulent flows. The derivation made a number of significant simplifications, including: (1) the particles are neutrally buoyant spheres with diameter much smaller than the turbulence Kolmogorov length scale $\eta = (v^3/\varepsilon)^{1/4}$, (2) the particle response time is small compared with the Kolmogorov time scale, (3) the turbulence is isotropic and has a large Reynolds number, (4) there are no hydrodynamic interactions between the particles, (5) the concentration field is uniform, (6) Stokes drag is the only force acting on the particles, and (7) the gradient of the fluctuating fluid velocity is normally distributed. The second of these assumptions implies that the Stokes number St_K based on the Kolmogorov length and velocity scaling, defined by

$$\mathrm{St}_K \equiv mu_K/3\pi\mu d\eta, \tag{6.3.4}$$

must be small compared with unity. In this equation, the Kolmogorov velocity scale is defined by $u_K = (v\varepsilon)^{1/4}$ and ε is the turbulence dissipation rate per unit mass.

The argument put forth by Saffman and Turner examines transport over a spherical surface S with radius $\sigma \equiv r_1 + r_2$, centered at the centroid of particle 1, as shown in Figure 6.5. If w_r denotes the radial component of the velocity of particle 2 relative to particle 1 and $\mathbf{x}$ is a point on the surface S, then collision occurs for all $\mathbf{x}$ values for which $w_r(\mathbf{x}) < 0$. The collision kernel is given by

$$\alpha_{12} = -\left\langle \int_{S(w_r < 0)} w_r \, da \right\rangle, \tag{6.3.5}$$

where the brackets $\langle\cdot\rangle$ denote the expected value, or mean. Assuming that the particles travel at the same velocity as the surrounding fluid, so that w_r can be

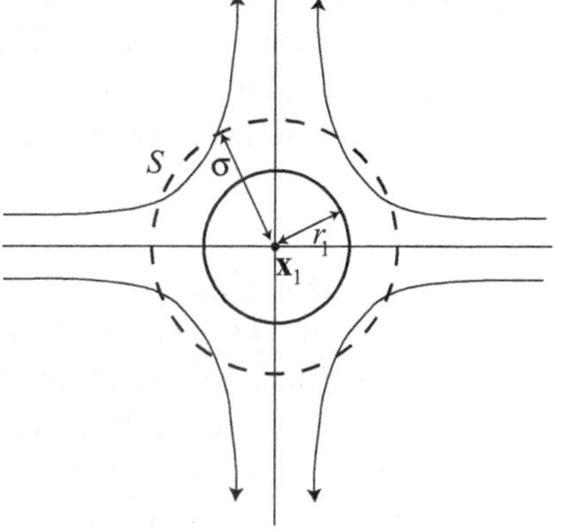

Figure 6.5. Schematic showing definition of the sphere surface S of radius $\sigma = r_1 + r_2$ surrounding particle 1, along with streamlines of a local straining flow relative to the particle.

interpreted as the fluid velocity relative to particle 1, the continuity equation for an incompressible flow requires that

$$\int_{S(w_r<0)} w_r \, da + \int_{S(w_r>0)} w_r \, da = 0. \tag{6.3.6}$$

From this requirement, it follows that

$$-\int_{S(w_r<0)} w_r \, da = \frac{1}{2} \int_S |w_r| \, da, \tag{6.3.7}$$

where the integration on the right-hand side is over the entire sphere. Substituting (6.3.7) into (6.3.5) gives

$$\alpha_{12} = \frac{1}{2} \int_S \langle |w_r| \rangle \, da = 2\pi (r_1 + r_2)^2 \langle |w_r| \rangle. \tag{6.3.8}$$

From the assumption that the turbulence is isotropic, so that $\langle w_x^2 \rangle = \langle w_y^2 \rangle = \langle w_z^2 \rangle$, it follows that

$$\langle |w_r| \rangle = \langle |\sin^2 \theta \cos^2 \varphi \, w_x^2 + \sin^2 \theta \sin^2 \varphi \, w_y^2 + \cos^2 \theta \, w_z^2|^{1/2} \rangle = \langle |w_x| \rangle. \tag{6.3.9}$$

Saffman and Turner assumed that the particle is much smaller than the Kolmogorov scale and use a Taylor series to write $w_x = \sigma (\partial u/\partial x)$, where u is the x-component of the turbulence fluctuation velocity. The usual scaling relationship for inertial-range isotropic turbulence gives the mean-square velocity gradient as $\langle (\partial u/\partial x)^2 \rangle = \varepsilon/15\nu$ (Tennekes and Lumley, 1972, 66–67). We recall that if X is a normally distributed random variable with zero mean, then $|X|$ has a half-normal distribution with mean value given by $\langle |X| \rangle = (2/\pi)^{1/2} \langle X^2 \rangle^{1/2}$. Applying this result to the velocity gradient, the expressions given earlier can be combined to write $\langle |w_r| \rangle = \sigma (2\varepsilon/15\pi \nu)^{1/2}$, so that (6.3.8) becomes

$$\alpha_{12} = (8\pi/15)^{1/2}(r_1 + r_2)^3 (\varepsilon/\nu)^{1/2}. \tag{6.3.10}$$

This expression is similar to an earlier expression by Camp and Stein (1943) based on the same assumptions, in which the collision kernel was obtained as

$$\alpha_{12} = \tfrac{4}{3}(r_1 + r_2)^3 (E_{tot}/\nu)^{1/2}. \tag{6.3.11}$$

Using the generalized version of this expression proposed by Pedocchi and Piedra-Cueva (2005) (see also Clark, 1985), we define E_{tot} as the total energy dissipation rate per unit mass, which can be associated with the mean-flow energy dissipation rate $E_m = 2\nu\, \bar{\mathbf{D}} : \bar{\mathbf{D}}$ and the turbulence energy dissipation rate ε by

$$E_{tot} = E_m + \varepsilon. \tag{6.3.12}$$

Here, $\bar{\mathbf{D}} = (1/2)[\nabla\bar{\mathbf{u}} + (\nabla\bar{\mathbf{u}})^T]$ is the rate of deformation tensor for the mean velocity field $\bar{\mathbf{u}}$ and $\bar{\mathbf{D}} : \bar{\mathbf{D}}$ denotes the scalar product of $\bar{\mathbf{D}}$ with itself. The traditional overbar is used here for the mean velocity field and related quantities in the turbulent flow for brevity. For turbulence with zero mean flow, (6.3.11) is similar to the Saffman-Turner expression, with a difference in the coefficient value of about 3%. For a laminar shear flow of the form $\bar{\mathbf{u}} = \dot{\gamma}\, y\, \mathbf{e}_x$, (6.3.11) reduces to the Smoluchowski expression (6.3.1).

6.3.3. Accelerative-Independent Collisions (Large Stokes Numbers)

In the opposite extreme of a flow with high values of St_K, an expression was proposed by Abrahamson (1975) in which particles move with random, independent motions as they are flung about by distant eddies of the turbulent flow. Because the particle Stokes number is large, the particles travel long distances with a velocity that is relatively uncorrelated to the local fluid velocity. Because of the intended application to large Stokes number particles with independent motion, the relative radial velocity in the Abrahamson expression is estimated not using a Taylor expansion from the local fluid velocity, as done by Saffman and Turner, but instead using the mean-square velocity for each particle. Again using the expression $\langle |X| \rangle = (2/\pi)^{1/2}\langle X^2 \rangle^{1/2}$ for a normally distributed random variable X, we can write $\langle |w_r| \rangle = (2/\pi)^{1/2}[\langle v_1^2 \rangle + \langle v_2^2 \rangle]^{1/2}$, where v_1 and v_2 are velocity components of particles 1 and 2, respectively, in one direction (say, the x-direction). The turbulence is assumed to be isotropic, so the mean-square velocity magnitude for particle 1 is given by $\langle v_{1m}^2 \rangle = 3\langle v_1^2 \rangle$, and similarly for the second particle. Substituting this approximation for the mean absolute value of the radial velocity into (6.3.8) gives the Abrahamson collision kernel as

$$\alpha_{12} = (8\pi/3)^{1/2}(r_1 + r_2)^2 \sqrt{\langle v_{1m}^2 \rangle + \langle v_{2m}^2 \rangle}. \tag{6.3.13}$$

6.3.4. Accelerative-Correlated Collisions (Intermediate Stokes Numbers)

Many researchers have proposed expressions for particle collision rate at intermediate Stokes numbers, which is dominated by the accelerative-correlated mechanism associated with gradual particle drift across the fluid streamlines. Early work on this problem for small particles was reported in the second part of the paper by Saffman and Turner (1956), followed by work from Williams and Crane (1983), Yuu (1984), Kruis and Kusters (1997), and others. The essential problem can be expressed by writing the collision kernel in the form

$$\alpha_{12} = (8\pi/3)^{1/2}(r_1 + r_2)^2 \sqrt{w_{shear}^2 + w_{accel}^2}, \tag{6.3.14}$$

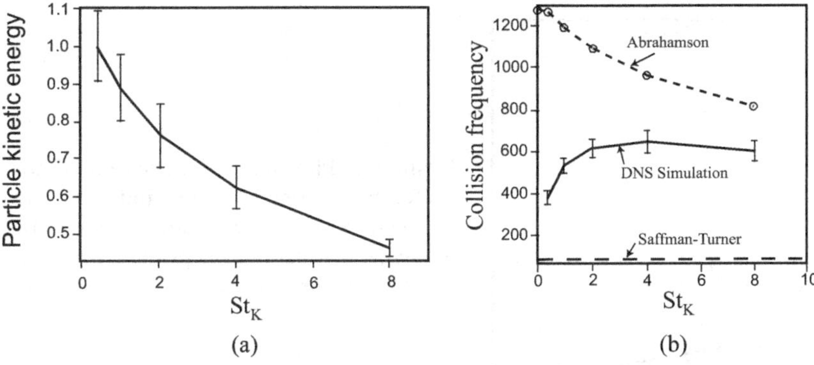

Figure 6.6. Plots showing DNS results for variation of (a) particle kinetic energy and (b) collision frequency as a function of Stokes number based on Kolmogorov scaling. The latter plot also compares DNS results to predictions of the Saffman-Turner theory (6.3.10) and of the Abrahamson kinetic theory (6.3.13). [Reprinted with permission from Sundaram and Collins (1997).]

where w_{shear} is the relative velocity between the particles due to the fluid shear flow and w_{accel} is the relative velocity due to particle dispersion. Setting $w_{accel} = 0$ and $w_{shear} = (r_1 + r_2)\sqrt{\varepsilon/5\nu}$ reduces to the Saffman-Turner result (6.3.10) for orthokinetic collisions at small Stokes numbers, and setting $w_{shear} = 0$ and $w_{accel} = \sqrt{\langle v_{1m}^2 \rangle + \langle v_{2m}^2 \rangle}$ reduces to the Abrahamson result (6.3.13) for accelerative-independent collisions at high Stokes numbers. A wide range of expressions have been derived giving estimates of w_{shear} and w_{accel} at intermediate values of the Stokes number, many of which were summarized by Meyer and Deglon (2011). These expressions, many of which are fairly complex, differ from each other primarily by the choice of particle-fluid interaction forces that are included in the model and the assumptions made about the particle size relative to turbulence length scales.

A key challenge in collision rate modeling at intermediate values of the Stokes number arises from the need to account for the occurrence of particle clustering in the region between turbulent eddies, as discussed in the first section of this chapter. This tendency of the particles to cluster is referred to as "preferential concentration" in the literature, as statistical models of particle motion in the turbulent flow must account for the fact that particles have a higher probability of being located within these interstitial regions than within the eddy structures. In order to provide data for turbulent particle collision rate and to examine the validity of various approximations, direct numerical simulations have been performed by Sundaram and Collins (1997), Wang et al. (1998), and Zhou et al. (1998) for homogeneous turbulence with a single particle size and by Zhou et al. (2001) for a mixture of particles of two different sizes. Simulation results from Sundaram and Collins (1997) are shown in Figure 6.6 for particle kinetic energy and collision rate as a function of Stokes number St_K. The particle kinetic energy decreases with increase in Stokes number, which is consistent with the filtering effect of the particle inertia as discussed in Section 6.1. The collision frequency obtained from direct numerical simulation is observed in Figure 6.6b to be higher than the value predicted by the Saffman-Turner expression (6.3.10) for orthokinetic collision and lower than the value predicted by the Abrahamson kinetic theory approximation (6.3.13) for accelerative-independent

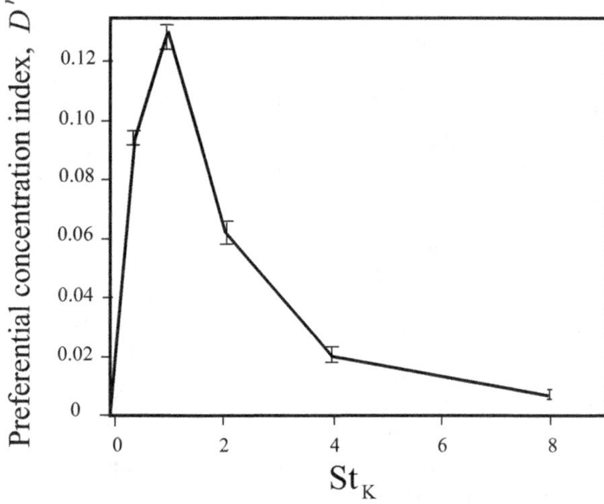

Figure 6.7. Plot showing variation of the preferential concentration index D^r as a function of Stokes number based on Kolmogorov scaling. [Reprinted with permission from Sundaram and Collins (1997).]

collisions. However, at low Stokes numbers the DNS results decrease substantially, moving toward the Saffman-Turner expression predictions, and at high Stokes numbers the DNS results appear to approach the Abrahamson expression predictions, although in both cases the Stokes number becomes neither high enough nor low enough for the results to converge to these limits.

Sundaram and Collins (1997) developed a measure of the extent of particle clustering by examining the statistics of the probability distribution $P(m)$ of the number of particles, denoted by m, located within a grid cell at a given time. The particle number distribution is obtained by examining all grid cells within the flow domain at a given time. For a randomly distributed system of noninteracting particles, the variable m exhibits a Poisson distribution $P^r(m) = M^m e^{-M}/m!$, where $M \equiv \langle m \rangle$ is the expected value of the random variable m, given by the ratio of the total number of particles to the total number of grid cells. A measure D^r of particle preferential concentration is defined based on the difference between the actual distribution of m, denoted by $P(m)$, and the Poisson distribution for randomly distributed particles as

$$D^r \equiv \sum_m [P(m) - P^r(m)]^2, \quad \mathrm{m} = 0, 1, \ldots \quad (6.3.15)$$

A plot showing DNS results for D^r as a function of Stokes number is given in Figure 6.7. The preferential concentration measure goes to zero for both very large and very small Stokes numbers, and attains a maximum value at $St_K = 1$. This observation is consistent with previous results of Wang and Maxey (1993) and Squires and Eaton (1991).

Jin et al. (2010) reported results of a comparison between a large eddy simulation (LES) of particle collision rate in homogeneous turbulence and a full direct numerical simulation of the same problem. Results are also compared to a filtered set of the direct numerical simulation data in order to examine separately whether differences are due to spatial resolution or other errors associated with LES. This study concludes that the LES and filtered DNS approaches provide reasonably accurate predictions of the particle collision rate when $St_K > 3$, but that these approaches exhibit significant error for small and intermediate values of St_K. The error in collision rate prediction was as much as 60% for $St_K = 0.5$. Of course, the accuracy of LES predictions in general will depend on the filter size (or, equivalently, the grid size) relative to the Kolmogorov length scale for the specific problem under examination. The reason for the difference between LES and DNS predictions at moderate and low values of Stokes number was attributed by Jin et al. (2010) to the development of preferential concentration (or clustering) of particles by the small-scale eddies.

6.4. Dynamic Models for Particle Dispersion

In the context of a simulation using the discrete-element method, diffusion, dispersion, and collision of particles are directly computed by solving the equations of motion of each particle given an appropriate fluid flow. The DEM approach may be used together with a direct numerical simulation of turbulence, in which case no additional assumptions are necessary, beyond those typically used to obtain the various forces on the particles in DEM (as discussed in Chapters 3–5). In particular, with DNS we do not need to introduce any model for the turbulent fluctuation field, at least for a computation employing the one-way interaction assumption, because it is directly computed. On the other hand, other turbulence simulation approaches, such as LES and RANS, do not simulate all of the scales of the turbulent motion, but instead introduce "closure models" for the effect of these unresolved scales on quantities such as stress and diffusion rate in the larger scales of the turbulent flow. Because these unresolved scales may also have a significant influence on the particle transport and on particle collisions, it is necessary to introduce some model of the unresolved scales of turbulent motion that can be added to the resolved motion for use in transporting the particulate phase. This model of the unresolved turbulent motion is not expected to be accurate in time and space (as this would be the same as a DNS), but rather it is expected to possess the necessary statistical properties so as to produce the same statistical measures of the particle dynamics as the actual turbulence. In this and the next section, we shall simply refer to this model of the unresolved turbulence scales as a *subgrid-scale fluctuation model.*

The choice of an appropriate subgrid-scale model is dependent on exactly which aspect of the particle transport the user desires to simulate. In particular, two very different categories of problems might be delineated – those in which the particle dynamics depend on interaction with other particles (type 1) and those in which it does not (type 2). In the latter (type 2) case, the focus in model development is to introduce subgrid-scale fluctuations that have the correct kinetic energy and time scale as seen by individual traveling particles. In the former (type 1) case, it is also essential that the fluctuations seen by two nearby particles are correlated in such a manner that both particles see similar fluctuations. This requirement of spatial correlation of fluctuations of nearby particles is of key importance, for instance, in prediction of particle collision and adhesion processes. Put another way, the subgrid-scale fluctuation field for particle transport problems of the first type must produce the phenomenon of particle preferential concentration (due to clustering of particles in the interstitial spaces between the eddies), whereas this is not essential for problems of the second type. As a consequence, whereas the turbulent fluctuations for problems of the second type must have the correct energy and time scales, fluctuations for problems of the first type must also have the correct spatial structure.

The current section deals with problems of the second type, in which particle interaction is not important, and specifically with a class of models that are called stochastic Lagrangian models (SLMs). These are very powerful and useful models for addressing problems such as particle diffusion, and other problems where the kinetic energy and time scales of individual particles are of central importance. The most common SLM approach uses a stochastic differential equation of the Langevin form to approximate the fluctuating fluid velocity field of a turbulent flow. The properties of the Langevin equation for modeling turbulent flow fluctuations were examined in detail by Thomson (1987), and the resulting SLM is sometimes referred to as the Thomson model. In this model, the governing equation for one component of the Lagrangian fluctuation velocity $\mathbf{u}^+(t)$ is approximated by solution of the stochastic differential equation

$$du^+ = -\frac{u^+}{\tau_L}\,dt + \left(\frac{2u'^2}{\tau_L}\right)^{1/2} d\xi, \tag{6.4.1}$$

where τ_L is the Lagrangian integral time scale given in (6.1.1) and the differential $d\xi$ is a Gaussian-distributed random variable with zero mean and variance equal to the time step dt. The solution of (6.4.1) for $u_i^+(t)$ has zero mean and variance $\langle u_i^+(t)^2\rangle = u_i'^2$, where u' is the root-mean-square of the Eulerian fluctuation velocity.

A component $u^+(t)$ of the Lagrangian velocity has zero mean and variance equal to $2q/3$, where q is the turbulent kinetic energy per unit mass. Two basic statistical measures of the time variation of the Lagrangian velocity field are the autocorrelation function $\rho(s)$ and the second-order structure function $D(s)$, defined by

$$\rho(s) \equiv \langle u^+(t+s)\,u^+(t)\rangle/\langle u^{+2}\rangle, \quad D(s) \equiv \langle |u^+(t+s) - u^+(t)|^2\rangle. \tag{6.4.2}$$

The structure function is the variance of the velocity change over a time interval of length s, and the autocorrelation function is related to the Lagrangian integral time scale as $\tau_L = \int_0^\infty \rho(s)\,ds$. These two measures are related to each other by

$$D(s) = 2u'^2[1 - \rho(s)]. \tag{6.4.3}$$

The Langevin equation (6.4.1) can be used to derive analytic solutions for these two measures as (Pope, 2000)

$$\rho(s) = \exp(-s/\tau_L), \quad D(s) = C_0 \varepsilon \tau_L \left[1 - \exp(-s/\tau_L)\right], \tag{6.4.4}$$

where for convenience we take $s \geq 0$. The result for $\rho(s)$ agrees well with results from direct numerical simulation of homogeneous turbulence by Yeung and Pope (1989). The solution for $D(s)$ follows from (6.4.3) after using the expression (6.1.1) for τ_L. For time increments $s \ll \tau_L$, (6.4.4) reduces to the asymptotic approximation

$$D(s) \sim C_0 s\varepsilon, \tag{6.4.5}$$

in agreement with Kolmogorov's similarity hypothesis (Kolmogorov, 1962; Biferale et al., 2008).

Pope (2002) noted three shortcomings of the Langevin model (6.4.1): (1) it contains a single time scale τ_L; (2) it has no Reynolds number dependence; and (3) the slope of $\rho(s)$ is discontinuous at $s = 0$. These shortcomings are particularly apparent when one compares the prediction (6.4.4) for the second-order structure function with data from direct numerical simulation. Whereas the ratio $D(s)/\varepsilon s$ obtained from the prediction (6.4.4) is independent of Reynolds number, values obtained from the DNS data of Yeung et al. (2006) for isotropic turbulence exhibit strong variation with the microscacle Reynolds number $\mathrm{Re}_\lambda = u'\lambda/\nu$, where $\lambda = 15\nu\,u'^2/\varepsilon$ is the Taylor microscale (Pope, 2011). These shortcomings in the Langevin stochastic Lagrangian model can be addressed by solving a stochastic differential equation for the fluctuating acceleration field $a^+(t)$, instead of velocity, and then solving for velocity using $du^+/dt = a^+$. This approach was first demonstrated by Sawford (1991), whose acceleration-based stochastic Lagrangian model can be written in a form suggested by Pope (1992) as

$$da^+ = -\left(1 + \frac{\eta_S}{\tau_S}\right)\frac{a^+}{\eta_S}dt - \frac{u^+}{\tau_S\eta_S}dt + \left[2a'^2\left(\frac{1}{\eta_S} + \frac{1}{\tau_S}\right)\right]^{\frac{1}{2}}d\xi, \qquad (6.4.6)$$

where $a'^2 = u'^2/\tau_c\eta_c$ is the square of the root-mean-square acceleration fluctuation and the time scales $\tau_S = 4\tau_L/\hat{C}_0$ and $\eta_S = \hat{C}_0\tau_\eta/2a_0$ are related to the integral time scale τ_L and Kolmogorov time scale $\tau_\eta = (\nu/\varepsilon)^{1/2}$, respectively. The coefficient a_0 is obtained by an empirical fit to DNS data as $a_0 = 0.13\mathrm{Re}_\lambda^{0.64}$. The coefficient $\hat{C}_0$ is also obtained by a fit to DNS data, giving $\hat{C}_0 = 7$. The use of two time scales in the Sawford model introduces a Reynolds number dependency associated with the ratio of these time scales, addressing the first two of the shortcomings listed by Pope (2002).

Another statistical measure of accuracy of the subgrid-scale fluctuation model is its ability to predict the correct accleration probability density function in turbulent flows, as shown based on experimental data in Figure 6.2. The Sawford model yields a probability density function with Gaussian form, which is inconsistent with the fat-tail form of the experimentally observed function. Modified acceleration-based stochastic Lagrangian models, developed by Reynolds (2003a) and by Lamorgese et al. (2007), are designed to yield the correct form of the acceleration probability density function. Reynolds proposed a stochastic differential equation for both the acceleration fluctuation a^+ and the local turbulence dissipation rate ε^+, which is defined such that $\langle\varepsilon^+\rangle = \varepsilon$ is the total dissipation rate per unit mass. He assumed that the logarithm of the normalized dissipation rate $\chi_\varepsilon \equiv \ln(\varepsilon^+/\varepsilon)$ can be modeled by an Uhlenbeck-Ornstein process of the form

$$d\chi_\varepsilon = -(\chi_\varepsilon - \langle\chi_\varepsilon\rangle)\frac{dt}{\tau_\chi} + \left(\frac{2\sigma_\chi^2}{\tau_\chi}\right)^{\frac{1}{2}}dW \qquad (6.4.7)$$

where dW is a Gaussian random variable with zero mean and variance dt. Here, χ_ε is a Gaussian-distributed random variable that is normalized such that $\langle\exp(\chi_\varepsilon)\rangle$ is unity, so its mean is given by $\langle\chi_\varepsilon\rangle = -\sigma_\chi^2/2$. The time scale τ_χ is given by $\tau_\chi = 2\sigma_u^2/C_0\langle\varepsilon\rangle$ and the variance is fit from DNS data in terms of the microscale Reynolds number Re_λ as $\sigma_\chi^2 = -0.354 + 0.289\ln(\mathrm{Re}_\lambda)$. Reynolds solves (6.4.7) together with a stochastic

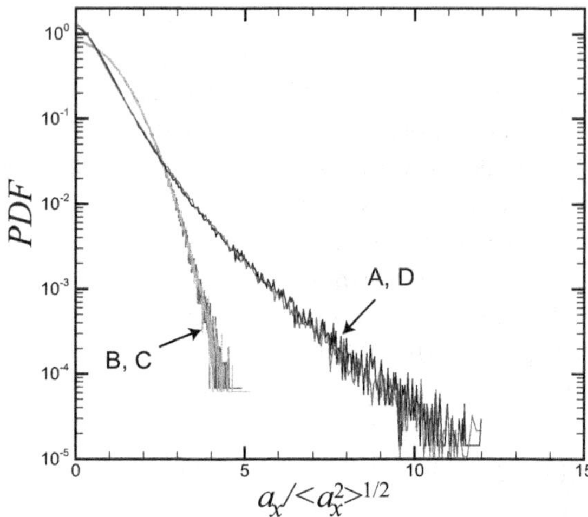

Figure 6.8. Probability density function for the x-component of acceleration comparing DNS data (A) with SLM predictions using the Langevin equation (B), the Sawford model (C), and the Reynolds model (D). [Reprinted with permission from Sala (2012).]

differential equation for the acceleration fluctuation of the form

$$da^+ = -\left(\frac{1}{\tau_S} + \frac{1}{\eta_S} - \frac{1}{\sigma_{a|\varepsilon}}\frac{d\sigma_{a|\varepsilon}}{dt}\right) a^+ \, dt - \frac{u^+}{\tau_S \eta_S} dt$$

$$+ \sqrt{2\sigma_u^2 \left(\frac{1}{\tau_S} + \frac{1}{\eta_S}\right)\frac{1}{\tau_S \eta_S}} \, d\xi, \tag{6.4.8}$$

where the time scales τ_S and η_S have the same meaning as defined earlier with reference to Sawford's model. The variance of velocity and acceleration are given by $\sigma_u^2 = 2q/3$ and $\sigma_a^2 = a_0^* \varepsilon \sqrt{\varepsilon/v}$, respectively. The constant $a_0^* = 3.3$ is obtained by fitting to the DNS data of Yeung and Pope (1989). The conditional acceleration variance $\sigma_{a/\varepsilon}$ is given by

$$\sigma_{a|\varepsilon} = a_0^* \, (\varepsilon/\langle\varepsilon\rangle)^{3/2} a_\eta^2, \tag{6.4.9}$$

where $a_\eta = (\langle\varepsilon\rangle^3/v)^{1/4}$ is the Kolmogorov acceleration scale. The differential $d\xi$ is also a Gaussian random variable with zero mean and variance dt, but it is independent of the variable dW. A plot is shown in Figure 6.8 comparing direct numerical simulation results for the acceleration probability density function with predictions from the Langevin equation (6.4.1), the Sawford model (6.4.6), and the Reynolds model (6.4.7), (6.4.8), and (6.4.9). The Reynolds model predictions are found to be in excellent agreement with the DNS data.

The use of stochastic Lagrangian models for solving for diffusion of heavy particles in a turbulent flow was examined by Sawford and Guest (1991), among many later investigators, by coupling the stochastic differential equation (6.4.1) for the fluid velocity fluctuations to the particle momentum equation. These authors particularly examined the ability of the SLM approach for dealing with the so-called inertia and crossing-trajectories effects discussed by Csanady (1963). The *inertia effect* is associated with the observation that particle inertia has the effect of filtering out fluctuations with high eddy Stokes number, as discussed in Section 6.1. As a consequence, the energy spectrum of the particle will differ from that of the surrounding

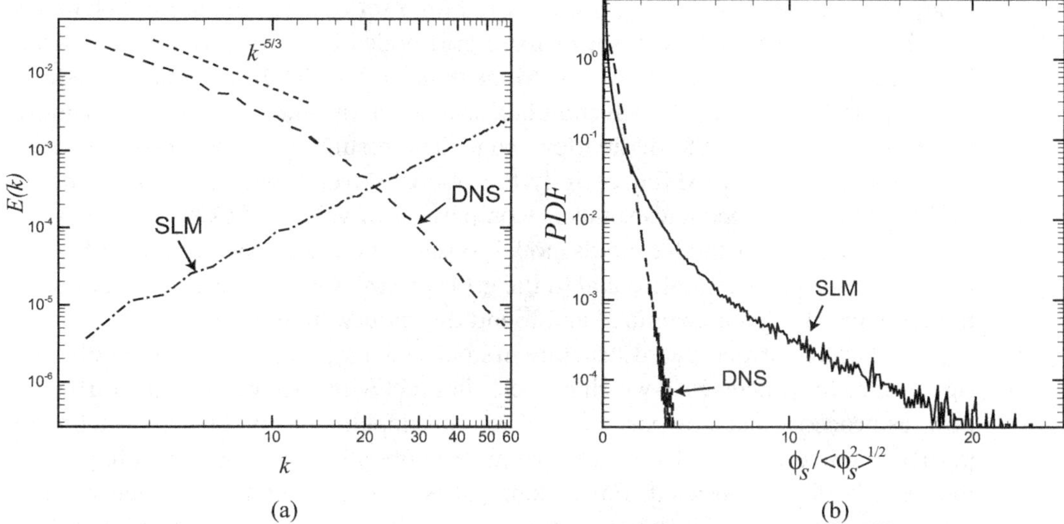

Figure 6.9. Comparison of (a) the energy spectrum and (b) the probability density function of the dispersion measure $\varphi_s/\langle\varphi_s^2\rangle^{1/2}$ between direct numerical simulation (DNS) for homogeneous turbulence and the Reynolds stochastic Lagrangian method (SLM). [Figure (b) reprinted with permission from Sala (2012).]

turbulent fluid. The *crossing-trajectories effect* is associated with the fact that particle drift across fluid streamlines, as discussed in Section 6.2, causes the time scale that a particle remains in a given turbulent eddy to be smaller than the time scale of a typical fluid element within the eddy. Whereas the fluid element time scale is associated with decay of the eddy or exchange of fluid between eddies, the particle time scale has the additional factor that particles drift relative to the surrounding fluid. This particle drift might occur either as a result of the particle's own momentum or from imposition of an external force on the particle, such as gravity.

6.5. Dynamic Models for Particle Clustering

The discussion of stochastic Lagrangian models in the previous section focused on the statistics of fluid element motion in time under this model, using measures such as the temporal second-order structure function defined in (6.4.2). Different levels of stochastic models were introduced, each of which is found to accurately predict progressively more detailed temporal stochastic measures in comparison to direct numerical simulation studies. On the other hand, there was no discussion of the spatial structure of the turbulence generated by these stochastic Lagrangian models. In general, SLM predictions do not possess the correct spatial structure because the random forcing terms used for the different fluid elements are not spatially correlated. As a consequence, the velocity field generated by stochastic Lagrangian models, when interpolated back to a grid, is found to consist of eddies with a size close to the grid scale.

For instance, a comparison between the energy spectrum of a direct numerical simulation for forced homogeneous turbulence and a solution of the Reynolds stochastic Lagrangian model is shown in Figure 6.9a. In this calculation, the stochastic

Lagrangian model (6.4.7)–(6.4.9) was solved on a set of Lagrangian fluid elements and then interpolated onto a set of fixed grid nodes using the conservative scattered data interpolation method of Monaghan (1985). The DNS results exhibit the expected $E(k) \propto k^{-5/3}$ wavenumber scaling in the inertial range, and more rapid energy decay in the dissipation range. The results of the Reynolds stochastic Lagrangian model exhibit entirely the opposite trend, with increasing energy as the grid cell size becomes smaller. Comparison of velocity fields shows that the velocity generated by the Reynolds model consists of random fluctuations with characteristic size approximately equal to the grid cell size. Other stochastic Lagrangian models have also been examined and found to produce similar results.

The lack of correct spatial structure in stochastic Lagrangian models has obvious significance for processes in which particles interact with each other in the turbulent flow, as processes such as particle collision are dependent on correct modeling of particle relative velocity. However, spatial structure also has a significant impact on the particle dispersion itself. For instance, it is argued in Section 6.2 that particle drift across fluid streamlines can be measured by the φ_s parameter defined in (6.2.6), which is associated with the cross product of the acceleration and velocity fields. While the Reynolds stochastic Lagrangian method has been shown to agree well with the probability density functions for both the fluid velocity and acceleration obtained from direct numerical simulations (as shown, for instance, in Figure 6.8), it does not follow that this method also agrees well with the statistics of the $\mathbf{a} \times \mathbf{u}$ field. A plot of the probability density function of the dispersion measure $\varphi_s / \langle \varphi_s^2 \rangle^{1/2}$ is given in Figure 6.9b, comparing results from a direct numerical simulation with those of the Reynolds stochastic Lagrangian method. From the definition of φ_s in (6.2.6), this ratio is independent of Stokes number and entirely depends upon turbulent fluid velocity field. The Reynolds model exhibits a superstatistical distribution with a fat tail for this dispersion measure similar to that for the acceleration field. By contrast, the DNS result in Figure 6.9b decays much more rapidly, and is closer in nature to a Gaussian distribution. This observation seems to suggest that the part of the acceleration field that gives rise to the fat-tailed acceleration distribution does not have an impact on particle dispersion, as characterized by drift of particles across fluid streamlines. The major reason for the difference in the distribution of φ_s in Figure 6.9b is the fact that in the direct numerical simulations, regions of high acceleration magnitude and regions of high values of the φ_s measure both tend to be located in regions immediately surrounding coherent turbulent vortices, as illustrated in Figure 6.3, whereas in a stochastic Lagrangian model these quantities are randomly distributed across the field with no coherent structure.

The difficulty with use of the stochastic Lagrangian method for modeling turbulent dispersion and clustering of interacting particles was noted by Ayyalasomayajula et al. (2008), who emphasized the importance of spatial structure in the turbulence model. Ayyalasomayajula et al. made a first step toward development of a structure-based approach by proposing a very simple two-dimensional vortex model for simulating particle dispersion. The model positioned the vortex structures on a two-dimensional Cartesian array, where each vortex structure was a line vortex of infinite length. The strength of the vortex structures was varied in a stochastic manner in accordance to the Lagrangian integral time scale τ_L, and the spacing between the vortices was set to be equal to the integral length scale ℓ_L. Even though this two-dimensional model is quite simple, Ayyalasomayajula et al. (2008) found that it exhibits several desirable features, such as prediction of the fat-tailed type of probability density function for acceleration observed experimentally by La Porta et al. (2001) and others, as shown in Figure 6.2.

This concept of developing a stochastic model based on vortex structures that generate the turbulent fluctuations was developed further in recent work by Sala (2012) and Sala and Marshall (2013) for homogeneous turbulence. These authors proposed a stochastic vortex structure (SVS) model, in which the turbulent fluctuating vorticity field is approximated by a set of three-dimensional vortex structures that have certain features set based on integral-scale length and time scales and other features selected as random variables with a prescribed probability distribution function. The total number of vortex structures in the computational domain, N, is set so that the average distance between the vortex centers is proportional to the Lagrangian integral length scale ℓ_L. The vortex structures have finite length, which is also proportional to ℓ_L, and a core radius that is proportional to the Kolmogorov length scale $\eta = (v^3/\varepsilon)^{1/4}$. Each vortex structure has a finite life span that is set equal to the integral time scale τ_L. The maximum vortex strength Γ_{max} is proportional to the ratio ℓ_L^2/τ_L. The constants of proportionality for these scaling parameters are specified so as to yield the prescribed kinetic energy; however, sensitivity tests indicate that the various flow measures are not sensitive to different combinations of these constants giving the same kinetic energy.

The flow field is initiated for this model by placing N vortex structures in the computational domain, where the vortex centroid position and orientation are selected randomly using a uniform probability distribution. For homogeneous turbulence the position of each vortex structure is fixed in time, because there is no mean flow. The initial vortex structures are each assigned an initial age τ_{0n}, chosen as a random variable for which the ratio τ_{0n}/τ_L has a uniform distribution between 0 and 1. The age $\tau_n(t)$ of the nth vortex structure increases with time as

$$\tau_n = \tau_{0n} + t - t_{0n}, \tag{6.5.1}$$

where t_{0n} is the time at which the vortex structure is introduced. When $\tau_n \geq \tau_L$, the nth vortex is removed from the flow field and a new vortex is introduced with $t_{0n} = t$ and $\tau_{0n} = 0$. This new vortex is again located with centroid position and vortex orientation selected as random variables with uniform probability distribution. The vortex strength $\Gamma_n(t)$ is specified as

$$\Gamma_n(t) = \Gamma_{\mathrm{max}} A_n \begin{cases} 5\tau_n/\tau_L & \text{for } 0 \leq \tau_n/\tau_L < 0.2 \\ 1 & \text{for } 0.2 \leq \tau_n/\tau_L \leq 0.8 \\ 1 - 5(\tau_n/\tau_L - 0.8) & \text{for } 0.8 < \tau_n/\tau_L \leq 1 \end{cases} \tag{6.5.2}$$

so that the strength ramps up gradually near the beginning of the vortex life, remains constant throughout the middle part of the vortex life, and ramps back down near the end of the vortex life. The coefficient A_n is a normally distributed random variable with zero mean and unit variance, which is set once for each vortex at the time that the vortex is initialized.

Computational results comparing scaling measures for the SVS velocity field with DNS are given in Figure 6.10. In Figure 6.10a, the energy spectrum predicted

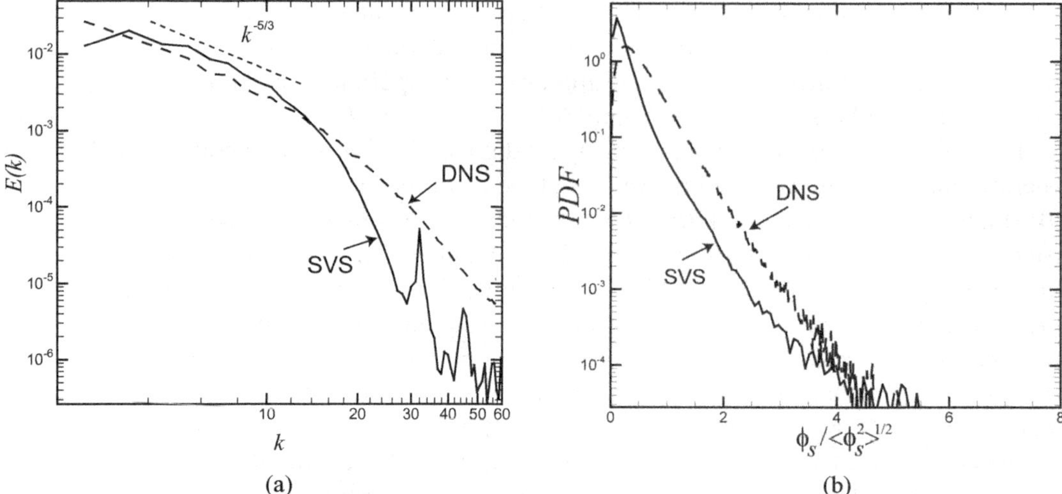

(a)　　　　　　　　(b)

Figure 6.10. (a) Comparison of the energy spectrum between direct numerical simulation (DNS) and the stochastic vortex structure (SVS) model. (b) The probability density function of the dispersion measure $\varphi_s/\langle\varphi_s^2\rangle^{1/2}$ between direct numerical simulation (DNS) for homogeneous turbulence and the stochastic vortex structure method. [Figure (b) reprinted with permission from Sala (2012).]

by SVS is observed to be very close to DNS for wavenumbers within the inertial range, for which the standard $E \sim k^{-5/3}$ scaling is observed. For high wavenumbers in the dissipation range, the SVS energy spectrum falls off more rapidly than DNS, which is expected, as the SVS model is based on integral scaling. The SVS model agrees well with DNS results for probability density function of the acceleration fluctuations, as shown by Sala (2012). The method also generates predictions for the dispersion measure $\varphi_s/\langle\varphi_s^2\rangle^{1/2}$ that agree reasonably well with DNS data, as shown in Figure 6.10b. Investigation of the flow field indicates that the regions with largest magnitude of this dispersion measure are located in a ring surrounding the large-scale vortex structures.

Figure 6.11 plots the number of particle collisions as a function of time for a DEM computation of particle transport in homogeneous turbulence with a Stokes number

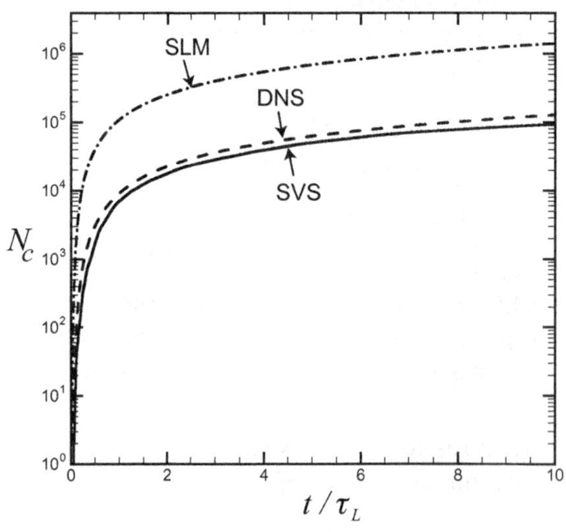

Figure 6.11. Comparison of the prediction for number of particle collisions as a function of dimensionless time from direct numerical simulation (DNS) for homogeneous turbulence in a $(2\pi)^3$ domain, the stochastic Lagrangian model (SLM) of Reynolds (2003), and the stochastic vortex structure (SVS) model, for a computation with Stokes number $St_L = 0.5$ based on integral-scale variables. [Reprinted from Sala and Marshall (2013).]

$St_L = 0.5$ based on the integral time and length scales. The plot compares results obtained using direct numerical simulation, the stochastic vortex structure model of Sala (2012) and Sala and Marshall (2013), and the stochastic Lagrangian model of Reynolds (2003a). The turbulent flow was allowed to attain a quasi-equilibrium state in a preliminary computation with no particles, and then the particles were added and the flow computation was restarted with the same initial conditions and values of computational and flow parameters for the three computations. The data in Figure 6.11 indicate that the Reynolds SLM approach predicts nearly an order of magnitude more particle collisions by the end of the computation than DNS, whereas the SVS model predictions are very close to the DNS results. This plot is presented in a semi-log form in order to more clearly denote the differences between the different models; however, the computed results for number of collisions increase nearly linearly in time. Although the models used by Ayyalasomayajula et al. (2008) and Sala and Marshall (2013) were applied to homogeneous turbulence and have not yet been tested over a large range of particle transport and flow conditions, the basic approach employed in the stochastic vortex structure model appears to have promise for future development in treating dispersion problems with interacting particles.

REFERENCES

Abrahamson J. Collision rates of small particles in a vigorously turbulent fluid. *Chemical Engineering Science* **30**, 1371–1379 (1975).

Ayyalasomayajula S, Gylfason A, Collins LR, Bodenschatz E, Warhaft Z. Lagrangian measurements of inertial particle accelerations in grid generated wind tunnel turbulence. *Physical Review Letters* **97**, 144507 (2006).

Ayyalasomayajula S, Warhaft Z, Collins LR. Modeling inertial particle acceleration statistics in isotropic turbulence. *Physics of Fluids* **20**, 095104 (2008).

Bec J, Celani A, Cencini M, Musacchio S. Clustering and collisions of heavy particles in random smooth flows. *Physics of Fluids* **17**, 073301 (2005).

Bec J, Biferale L, Cencini M, Lanotte A, Musacchio S, Toschi F. Heavy particle concentration in turbulence at dissipative and inertial scales. *Physical Review Letters* **98**, 084502 (2007).

Beck C. Superstatistics: Theoretical concepts and physical applications. In *Anomalous Transport: Foundations and Applications*, R. Klages, G. Radons, I. M. Sokolov, editors, Wiley-VCH Verlag GmbH & Co. KGaA, Weinheim, Germany, 2008.

Biferale L, Boffetta G, Celani A, Devenish BJ, Lanotte A, Toschi F. Multifractal statistics of Lagrangian velocity and acceleration in turbulence. *Physical Review Letters* **93**(6), 064502 (2004).

Biferale L, Bodenschatz E, Cencini M, Lanotte AS, Ouellette NT, Toschi F, Xu H. Lagrangian structure functions in turbulence: A quantitative comparison between experiment and direct numerical simulation. *Physics of Fluids* **20**, 065103 (2008).

Camp TR, Stein P. Velocity gradients and internal work in fluid motion. *Boston Society of Civil Engineers* **30**, 219–237 (1943).

Clark MM. Critique of Camp and Stein's RMS velocity gradient. *Journal of Environmental Engineering* **111**(3), 741–754 (1985).

Csanady GT. Turbulent diffusion of heavy particles in the atmosphere. *Journal of the Atmospheric Sciences* **20**, 201–208 (1963).

Falkovich G, Pumir A. Intermittent distribution of heavy particles in a turbulent flow. *Physics of Fluids* **16**(7), L47–L50 (2004).

Ferry J, Balachandar S. A fast Eulerian method for disperse two-phase flow. *International Journal of Multiphase Flow* **27**, 1199–1226 (2001).

García M. Développement et validation du formalisme Euler-Lagrange dans un solveur parallèle et non-structuré pour la simulation aux grandes échelles – TH/CFD/09/1. PhD thesis,

Université de Toulouse – Ecole doctorale : Mécanique, Energétique, Génie civil, Procédés (2009).

Grits B, Pinsky M, Khain A. Investigation of small-scale droplet concentration inhomogeneities in a turbulent flow. *Meteorol Atmos Phys* **92**, 191–204 (2006).

Hu KC, Mei R. Particle collision rate in fluid flows. *Physics of Fluids* **10**(4), 1028–1030 (1998).

Jin G, He G-W, Wang L-P. Large-eddy simulation of turbulent collision of heavy particles in isotropic turbulence. *Physics of Fluids* **22**, 055106 (2010).

Kolmogorov AN. A refinement of previous hypotheses concerning the local structure of turbulence in a viscous incompressible fluid at high Reynolds number. *Journal of Fluid Mechanics* **13**, 82–85 (1962).

Kruis FE, Kusters KA. The collision rate of particles in turbulent flow. *Chemical Engineering Communications* **158**, 210–230 (1997).

La Porta A, Voth GA, Crawford AM, Alexander J, Bodenschatz E. Fluid particle acceleration in fully developed turbulence. *Nature* **409**, 1017–1019 (2001).

Lamorgese AG, Pope SB, Yeung PK, Sawford BL. A conditionally cubic-Gaussian stochastic Lagrangian model for acceleration in isotropic turbulence. *Journal of Fluid Mechanics* **582**, 423–448 (2007).

Launder BE, Sharma BI. Application of the energy dissipation model of turbulence to the calculation of flow near a spinning disc. *Letters in Heat and Mass Transfer* **1**(2), 131–138 (1974).

Lee S, Lee C. Intermittency of acceleration in isotropic turbulence. *Physical Review E* **71**(5 Pt 2), 056310 (2005).

Meyer CJ, Deglon DA. Particle collision modeling: A review. *Minerals Engineering* **24**, 719–730 (2011).

Monaghan JJ. Extrapolating B splines for interpolation. *Journal of Computational Physics* **60**(2), 253–262 (1985).

Mordant N, Crawford AM, Bodenschatz E. Three-dimensional structure of the Lagrangian acceleration in turbulent flows. *Physical Review Letters* **93**(21), 214501 (2004).

Pedocchi F, Piedra-Cueva I. Camp and Stein's velocity gradient formalization. *Journal of Environmental Engineering* **131**(10), 1369–1376 (2005).

Pope SB. A stochastic-Lagrangian method for acceleration in turbulent flows. *Physics of Fluids* **14**(7), 2360–2375 (1992).

Pope SB. *Turbulent Flows*. Cambridge University Press, Cambridge (2000).

Pope SB. A stochastic Lagrangian model for acceleration in turbulent flows. *Physics of Fluids* **14**(7), 2360–2375 (2002).

Pope SB. Simple models of turbulent flows. *Physics of Fluids* **23**, 011301 (2011).

Reynolds AM. On the application of nonextensive statistics to Lagrangian turbulence. *Physics of Fluids* **15**(1), L1–L4 (2003a).

Reynolds AM. Superstatistical mechanics of tracer-particle motions in turbulence. *Physical Review Letters* **91**(8), 084503 (2003b).

Reynolds AM, Mordant N, Crawford AM, Bodenschatz E. On the distribution of Lagrangian accelerations in turbulent flows. *New Journal of Physics* **7**, 58 (2005).

Saffman PG, Turner JS. On the collision of drops in turbulent clouds. *Journal of Fluid Mechanics* **1**, 16–30 (1956).

Sala K. Analysis of stochastic methods for predicting particle dispersion in turbulent flows. MS Thesis, University of Vermont, Burlington, Vermont. (2012).

Sala K, Marshall JS. Stochastic vortex structure method for modeling particle clustering and collisions in homogeneous turbulence. *Physics of Fluids* **25**(10), 103301 (2013).

Sawford B. Reynolds number effects in Lagrangian stochastic models of turbulent dispersion. *Physics of Fluids* **3**(6), 1577–1586 (1991).

Sawford BL, Guest FM. Lagrangian statistical simulation of the turbulent motion of heavy particles. *Boundary-Layer Meteorology* **54**, 147–166 (1991).

Sawford BL, Yeung PK, Borgas MS, Vedula P, La Porta A, Crawford AM, Bodenschatz E. Conditional and unconditional acceleration statistics in turbulence. *Physics of Fluids* **15**(11), 3478–3489 (2003).

Smoluchowski M von. Versuch einer mathematischen Theorie der Koagulationkinetik kollider lösungen. *Z. Phys. Chem.* **92**, 129–168 (1917).

Squires KD, Eaton JK. Preferential concentration of particles by turbulence. *Physics of Fluids A* **3**, 1169–1178 (1991).

Sundaram S, Collins LR. Collision statistics in an isotropic particle-laden turbulent suspension. I. Direct numerical simulations. *Journal of Fluid Mechanics* **335**, 75–109 (1997).

Tennekes H, Lumley JL. *A First Course in Turbulence.* MIT Press, Cambridge, MA (1972).

Thomson DJ. Criteria for the selection of stochastic models of particle trajectories in turbulent flows. *Journal of Fluid Mechanics* **180**, 529–556 (1987).

Voth GA, Satyanarayan K, Bodenschatz E. Lagrangian acceleration measurements at large Reynolds numbers. *Physics of Fluids* **10**(9), 2268–2280 (1998).

Wang L-P, Maxey MR. Settling velocity and concentration distribution of heavy particles in homogeneous isotropic turbulence. *Journal of Fluid Mechanics* **256**, 27–68 (1993).

Wang L-P, Wexler AS, Zhou Y. On the collision rate of small particles in isotropic turbulence. I Zero-inertia case. *Physics of Fluids* **10**(1), 266–276 (1998).

Williams JJE, Crane RI. Particle collision rate in turbulent flow. *International Journal of Multiphase Flow* **9**(4), 421–435 (1983).

Yeung PK, Pope SB. Lagrangian statistics from direct numerical simulations of isotropic turbulence. *Journal of Fluid Mechanics* **207**, 531–586 (1989).

Yeung PK, Pope SB, Sawford BL. Reynolds number dependence of Lagrangian statistics in large numerical simulations of isotropic turbulence. *Journal of Turbulence* **7**, N58 (2006).

Yeung PK, Pope SB, Kurth EA, Lamorgese AG. Lagrangian conditional statistics, acceleration and local relative motion in numerically simulated isotropic turbulence. *Journal of Fluid Mechanics* **582**, 399–422 (2007).

Yuu S. Collision rate of small particles in a homogeneous and isotropic turbulence. *AICHE Journal* **30**(5), 802–807 (1984).

Zhou Y, Wexler AS, Wang L-P. On the collision rate of small particles in isotropic turbulence. II. Finite inertia case. *Physics of Fluids* **10**(5), 1206–1216 (1998).

Zhou Y, Wexler AS, Wang L-P. Modelling turbulent collision of bidisperse inertial particles. *Journal of Fluid Mechanics* **433**, 77–104 (2001).

7

7   Ellipsoidal Particles

Although the assumption that particles are spherical is frequently a useful approximation, there are many applications in which the deviation of the particulate matter from a spherical shape plays a major role in the system dynamics. For instance, red blood cells have the shape of biconcave disks with a flattened center. This shape allows blood to easily flow within blood vessels with concentrations of 40–45% by volume without jamming. Fuel particles are often of a nonspherical shape, particularly for biowaste combustion processes (Džiugys and Peters, 2001). Soil particles are often highly irregular in shape, particularly for soils formed of minerals with anisotropic crystal structure or for soils in dry regions where there is not much water erosion (Knuth et al., 2012). Key work on modeling of aerosol systems formed of nonspherical particles was reported by Gallily and Cohen (1979) and Fan and Ahmadi (2000). Such models can be used for aerosol dynamics problems such as inhalation of small asbestos fibers suspended in air, or modeling of snow fall or dispersion of mineral or soot aerosols. Liquid crystal phase transitions can be modeled as bifurcations in a mixture of rod- and platelike particles (Camp and Allen, 1996). Solis and Martin (2010) have recently reported that a solution of metallic platelets subjected to a biaxial oscillating magnetic field can exhibit an amazing variety of transitions as a function of the forcing frequency, including formation of vortex arrays that produce a flow field reminiscent of natural convection flows, only without the need for a temperature difference.

In cases such as those discussed here, the nonspherical particles can be reasonably approximated as ellipsoids. Indeed, in a large number of cases two of the ellipsoid axes will be approximately equal and the particles can be further approximated as spheroids. Hence, cylindrical particles of grass biofuel or short carbon nanotubes in suspension might be treated as high aspect ratio prolate spheroids (Yin et al., 2003), whereas red blood cells or platelets might be treated as oblate spheroids (Chesnutt and Marshall, 2009). Although inexact, this idealization allows us to introduce important features of nonspherical particles into the flow computation, such as the higher packing limits, shape effects on particle rolling, particle alignment both within the flow and within agglomerates (Chesnutt and Marshall, 2010), and effects on force chain transmission of stress through particle collisions (Campbell, 2011).

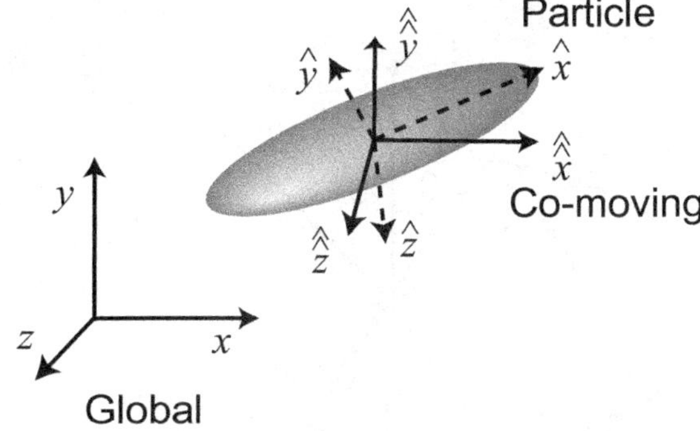

Figure 7.1. Coordinate frames for transport of an ellipsoidal particle: global, particle, and co-moving frames.

7.1. Particle Dynamics

In describing the motion of ellipsoidal particles, it is useful to employ different reference frames for handling different aspects of the dynamics of the system. Specifically, we introduce a global *inertial frame* $\mathbf{x} = [x \quad y \quad z]$, a *particle frame* $\hat{\mathbf{x}} = [\hat{x} \quad \hat{y} \quad \hat{z}]$, and a *co-moving frame* $\hat{\hat{\mathbf{x}}} = [\hat{\hat{x}} \quad \hat{\hat{y}} \quad \hat{\hat{z}}]$, all of which are assumed for convenience to be Cartesian frames. The origin of both the co-moving and particle frames is the particle centroid. The axes of the co-moving frame are parallel to the corresponding axes of the inertial frame, and hence do not move in time. The axes of the particle frame are fixed to the particle, and therefore rotate relative to the inertial frame as the particle rotates. These three reference frames are illustrated in Figure 7.1.

The equation of the surface of an ellipsoidal particle in the particle frame is

$$\frac{\hat{x}^2}{a_E^2} + \frac{\hat{y}^2}{b_E^2} + \frac{\hat{z}^2}{c_E^2} - 1 = 0, \tag{7.1.1}$$

where a_E, b_E, and c_E are semiaxis lengths in the $\hat{x}$, $\hat{y}$, and $\hat{z}$ directions, respectively. Since the co-moving and particle frames are both Cartesian and have the same origin, they can be mapped into each other using a coordinate rotation of the form

$$\hat{\mathbf{x}} = A\hat{\hat{\mathbf{x}}}, \tag{7.1.2}$$

where $\mathbf{A}$ is an orthonormal matrix (such that the transpose of $\mathbf{A}$ is equal to its inverse). The rotation of the particle frames in three dimensions can be decomposed into a sequence of three separate rotations, each about a single angle. These angles, denoted by φ, θ, and ψ, are called the Euler angles, and their definition is illustrated in Figure 7.2. The intersection of the co-moving $\hat{\hat{x}} - \hat{\hat{y}}$ plane and the rotating $\hat{x} - \hat{y}$ plane is called the *line of nodes*. The particle frame can then be rotated back onto the co-moving frame by a series of three steps. The first step is a rotation of the particle frame $(\hat{x}, \hat{y}, \hat{z})$ by an angle ψ about the $\hat{z}$ axis, so that at the end of the rotation the $\hat{x}$ axis lies on the line of nodes. The second step is a rotation of the particle frame by an angle θ about the new $\hat{x}$ axis, so that at the end of the rotation the $\hat{z}$ and $\hat{\hat{z}}$ axes are coincident. The third step involves a rotation of the particle frame by an angle φ about the new location of the $\hat{z}$ axis, at the end of which all the axes of the particle frame will be coincident with those of the co-moving frame.

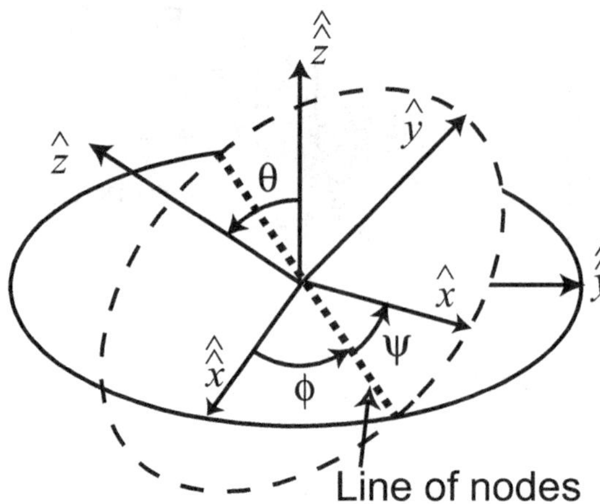

Figure 7.2. Illustration showing the co-moving and particle coordinate systems and the associated Euler angles φ, θ, and ψ. The intersection of the $\hat{\tilde{x}} - \hat{\tilde{y}}$ plane and the $\hat{x} - \hat{y}$ plane is the dotted line called the *line of nodes*.

The rotation matrix **A** can be expressed in terms of the Euler angles as

$$\mathbf{A} = \begin{bmatrix} \cos\psi\cos\varphi - \cos\theta\sin\theta\sin\psi & -\sin\psi\cos\theta - \cos\theta\sin\varphi\cos\psi & \sin\theta\sin\varphi \\ \cos\psi\sin\varphi + \cos\theta\cos\varphi\sin\psi & -\sin\psi\sin\varphi + \cos\theta\cos\varphi\cos\psi & -\sin\theta\cos\varphi \\ \sin\theta\sin\psi & \sin\theta\cos\psi & \cos\theta \end{bmatrix}.$$

$$(7.1.3)$$

However, it is often easier to express **A** in terms of a set of four *Euler parameters* (or quaternions) ε_1, ε_2, ε_3, and η, defined by

$$\varepsilon_1 = \cos\frac{\varphi - \psi}{2}\sin\frac{\theta}{2}, \quad \varepsilon_2 = \sin\frac{\varphi - \psi}{2}\sin\frac{\theta}{2},$$

$$\varepsilon_3 = \sin\frac{\varphi + \psi}{2}\cos\frac{\theta}{2}, \quad \eta = \cos\frac{\varphi + \psi}{2}\cos\frac{\theta}{2},$$

$$(7.1.4)$$

where by trigonometric identity

$$\varepsilon_1^2 + \varepsilon_2^2 + \varepsilon_3^2 + \eta^2 = 1. \tag{7.1.5}$$

The rotation matrix **A** can be written in terms of the Euler parameters as

$$A = \begin{bmatrix} 1 - 2\left(\varepsilon_2^2 + \varepsilon_3^2\right) & 2(\varepsilon_1\varepsilon_2 + \varepsilon_3\eta) & 2(\varepsilon_1\varepsilon_3 - \varepsilon_2\eta) \\ 2(\varepsilon_2\varepsilon_1 - \varepsilon_3\eta) & 1 - 2\left(\varepsilon_3^2 + \varepsilon_1^2\right) & 2(\varepsilon_2\varepsilon_3 + \varepsilon_1\eta) \\ 2(\varepsilon_3\varepsilon_1 + \varepsilon_2\eta) & 2(\varepsilon_3\varepsilon_2 - \varepsilon_1\eta) & 1 - 2\left(\varepsilon_1^2 + \varepsilon_2^2\right) \end{bmatrix}. \tag{7.1.6}$$

The rotation rate **Ω** of a particle can be written in terms of its components in the particle frame as

$$\mathbf{\Omega} = \Omega_{\hat{x}}\mathbf{e}_{\hat{x}} + \Omega_{\hat{y}}\mathbf{e}_{\hat{y}} + \Omega_{\hat{z}}\mathbf{e}_{\hat{z}}. \tag{7.1.7}$$

The rate of change of the Euler parameters can be expressed in terms of the components of **Ω** in (7.1.7) as (Hughes, 1986, 26)

$$\begin{bmatrix} d\varepsilon_1/dt \\ d\varepsilon_2/dt \\ d\varepsilon_3/dt \\ d\eta/dt \end{bmatrix} = \frac{1}{2}\begin{bmatrix} \eta\Omega_{\hat{x}} - \varepsilon_3\Omega_{\hat{y}} + \varepsilon_2\Omega_{\hat{z}} \\ \varepsilon_3\Omega_{\hat{x}} + \eta\Omega_{\hat{y}} - \varepsilon_1\Omega_{\hat{z}} \\ -\varepsilon_2\Omega_{\hat{x}} + \varepsilon_1\Omega_{\hat{y}} + \eta\Omega_{\hat{z}} \\ -\varepsilon_1\Omega_{\hat{x}} - \varepsilon_2\Omega_{\hat{y}} - \varepsilon_3\Omega_{\hat{z}} \end{bmatrix}. \tag{7.1.8}$$

The equations of motion for a particle consist of the linear momentum equation for the particle velocity $\mathbf{v}$ and the angular momentum equation for the rotation rate $\mathbf{\Omega}$. However, because it is much easier to write the equations for the torques acting on the particle in the particle coordinate frame, we utilize a component form of the angular momentum equation expressed in the rotating particle coordinate frame. The resulting equations of motion are given by

$$m\,\frac{d\mathbf{v}}{dt} = \mathbf{F}_F + \mathbf{F}_A, \tag{7.1.9}$$

$$I_{\hat{x}}\frac{d\Omega_{\hat{x}}}{dt} - \Omega_{\hat{y}}\Omega_{\hat{z}}\left(I_{\hat{y}} - I_{\hat{z}}\right) = M_{F,\hat{x}} + M_{A,\hat{x}}, \tag{7.1.10a}$$

$$I_{\hat{y}}\frac{d\Omega_{\hat{y}}}{dt} - \Omega_{\hat{z}}\Omega_{\hat{x}}\left(I_{\hat{z}} - I_{\hat{x}}\right) = M_{F,\hat{y}} + M_{A,\hat{y}}, \tag{7.1.10b}$$

$$I_{\hat{z}}\frac{d\Omega_{\hat{z}}}{dt} - \Omega_{\hat{x}}\Omega_{\hat{y}}\left(I_{\hat{x}} - I_{\hat{y}}\right) = M_{F,\hat{z}} + M_{A,\hat{z}}, \tag{7.1.10c}$$

where the mass moments of inertia of an ellipsoidal particle in the particle frame are $I_{\hat{x}} = \frac{1}{5}m(b_E^2 + c_E^2)$, $I_{\hat{y}} = \frac{1}{5}m(a_E^2 + c_E^2)$, and $I_{\hat{z}} = \frac{1}{5}m(a_E^2 + b_E^2)$. The moments in (7.1.10) are written in terms of the components in the particle coordinate frame.

7.2. Fluid Forces

In this section we examine the fluid forces and torques acting on an isolated ellipsoidal particle traveling relative to an external fluid flow at low particle Reynolds number, $\mathrm{Re}_P \equiv \rho_f|\mathbf{v} - \mathbf{u}|d/\mu < O(1)$, where $d = 2(a_E b_E c_E)^{1/3}$ is the effective particle diameter, ρ_f is the density of the fluid, $\mathbf{u}$ is the undisturbed fluid velocity at the particle centroid, and μ is the fluid viscosity. An expression for the hydrodynamic drag of an ellipsoidal particle was derived by Oberbeck (1876) for a uniform Stokes flow and by Brenner (1964) for an arbitrary external Stokes flow. Following Happel and Brenner (1963, 220–222) and retaining only the first-order term in the fluid velocity (neglecting derivatives of the fluid velocity), we can write

$$\mathbf{F}_d = \mu\hat{\mathbf{K}} \cdot (\mathbf{u} - \mathbf{v}), \tag{7.2.1}$$

where $\hat{\mathbf{K}}$ is the particle frame translation tensor. The particle-frame translation tensor for an ellipsoid is a diagonal matrix given by (see Gallily and Cohen, 1979)

$$\hat{\mathbf{K}} = 16\pi\, a_E b_E c_E \left(\frac{\mathbf{e}_{\hat{x}} \otimes \mathbf{e}_{\hat{x}}}{\chi_0 + a_E^2\alpha_0} + \frac{\mathbf{e}_{\hat{y}} \otimes \mathbf{e}_{\hat{y}}}{\chi_0 + b_E^2\beta_0} + \frac{\mathbf{e}_{\hat{z}} \otimes \mathbf{e}_{\hat{z}}}{\chi_0 + c_E^2\gamma_0}\right), \tag{7.2.2}$$

where $\mathbf{e}_{\hat{x}}$, $\mathbf{e}_{\hat{y}}$, and $\mathbf{e}_{\hat{z}}$ are unit vectors in the particle coordinate system and $\otimes$ denotes the tensor (or dyadic) product. The coefficients α_0, β_0, γ_0, and χ_0 are given by

$$\chi_0 = a_E b_E c_E \int_0^\infty \frac{d\lambda}{\Delta}, \qquad \alpha_0 = a_E b_E c_E \int_0^\infty \frac{d\lambda}{(a_E^2 + \lambda)\,\Delta},$$

$$\beta_0 = a_E b_E c_E \int_0^\infty \frac{d\lambda}{(b_E^2 + \lambda)\,\Delta}, \quad \gamma_0 = a_E b_E c_E \int_0^\infty \frac{d\lambda}{(c_E^2 + \lambda)\,\Delta}, \tag{7.2.3}$$

where

$$\Delta = \left[(a_E^2 + \lambda)(b_E^2 + \lambda)(c_E^2 + \lambda)\right]^{1/2}. \tag{7.2.4}$$

An expression for the torque acting on an ellipsoidal particle in the particle frame at small particle Reynolds numbers was derived by Jeffery (1922) and can be expressed in terms of the coefficients in (7.2.3) as

$$M_{F,\hat{x}} = \frac{16\pi\mu\, a_E b_E c_E}{3\left(b_E^2\beta_0 + c_E^2\gamma_0\right)}\left[\left(b_E^2 - c_E^2\right)d_{\hat{z}\hat{y}} + \left(b_E^2 + c_E^2\right)\left(w_{\hat{z}\hat{y}} - \Omega_{\hat{x}}\right)\right], \tag{7.2.5a}$$

$$M_{F,\hat{y}} = \frac{16\pi\mu\, a_E b_E c_E}{3\left(c_E^2\gamma_0 + a_E^2\alpha_0\right)}\left[\left(c_E^2 - a_E^2\right)d_{\hat{x}\hat{z}} + \left(c_E^2 + a_E^2\right)\left(w_{\hat{x}\hat{z}} - \Omega_{\hat{y}}\right)\right], \tag{7.2.5b}$$

$$M_{F,\hat{z}} = \frac{16\pi\mu\, a_E b_E c_E}{3\left(a_E^2\alpha_0 + b_E^2\beta_0\right)}\left[\left(a_E^2 - b_E^2\right)d_{\hat{y}\hat{x}} + \left(a_E^2 + b_E^2\right)\left(w_{\hat{y}\hat{x}} - \Omega_{\hat{z}}\right)\right], \tag{7.2.5c}$$

where

$$d_{\hat{i}\hat{j}} \equiv \frac{1}{2}\left(\frac{\partial u_{\hat{i}}}{\partial x_{\hat{j}}} + \frac{\partial u_{\hat{j}}}{\partial x_{\hat{i}}}\right), \qquad w_{\hat{i}\hat{j}} \equiv \frac{1}{2}\left(\frac{\partial u_{\hat{i}}}{\partial x_{\hat{j}}} - \frac{\partial u_{\hat{j}}}{\partial x_{\hat{i}}}\right)$$

are components of the rate of deformation tensor and the vorticity tensor, respectively, in the particle frame. The first term in the brackets in (7.2.5) is associated with rotation of the ellipsoidal particles by the straining of the external flow, and the second term is associated with particle rotation caused by the difference between the particle rotation rate and the rotation rate of the external fluid.

The coefficients in (7.2.3) can be further simplified for the case where the particles are approximated as being spheroidal in shape, for which two axes of the ellipsoid have the same length. For definiteness, we assume that $b_E = c_E$ and that the $\hat{x}$ coordinate axis is coincident with the particle symmetry axis. The particle aspect ratio is defined by $\beta \equiv a_E/b_E$, so that prolate and oblate spheroids correspond to cases with $\beta > 1$ and $\beta < 1$, respectively. We further define a dimensionless coefficient $\bar{\chi}_0 \equiv \chi_0/b_E^2$. Evaluation of the integrals in (7.2.3) for spheroidal particles gives

$$\alpha_0 = 2(1 - \beta_0), \quad \beta_0 = \gamma_0 = (-\beta^2 + \bar{\chi}_0/2)/(1 - \beta^2), \tag{7.2.6}$$

where the coefficient $\bar{\chi}_0$ is given by

$$\bar{\chi}_0 = -\frac{\beta}{(\beta^2 - 1)^{1/2}}\ln\left[\frac{\beta - (\beta^2 - 1)^{1/2}}{\beta + (\beta^2 - 1)^{1/2}}\right] \quad \text{for } \beta > 1 \tag{7.2.7a}$$

and

$$\bar{\chi}_0 = \frac{2\beta}{(1 - \beta^2)^{1/2}}\left[\frac{\pi}{2} - \tan^{-1}\left(\frac{\beta}{(1 - \beta^2)^{1/2}}\right)\right] \quad \text{for } \beta < 1. \tag{7.2.7b}$$

Expressions for a variety of other forces acting on ellipsoidal particles, as well as corrections to the drag force, have been derived in the literature. For instance, a slip correction factor for ellipsoidal particles at nonzero Knudsen numbers was derived by Dahneke (1973). A summary and evaluation of correlations for modification of the drag force to account for finite Reynolds numbers was given by Chhabra et al. (1999). An extension of Saffman's expression for lift force on a particle in a shear flow was developed by Harper and Chang (1968) for a three-dimensional body of arbitrary shape. Fan and Ahmadi (2000) gave expressions for the Brownian force on an ellipsoidal particle, and they also discussed related literature on slip correction factors for a rotating ellipsoidal particle.

7.3. Collision Detection and Contact Point Identification

One of the most challenging aspects of simulating motion of ellipsoidal particles is the problem of contact detection. A contact detection algorithm must perform three separate tasks:

(1) Determine when two particles collide with each other,
(2) Determine the contact points on each particle, along with the associated surface unit normal vectors at the contact points, and
(3) Determine the amount of overlap between the particles.

For spherical particles these tasks are trivial, but for ellipsoidal particles they are far from simple. Moreover, it is important that the contact detection algorithm performs these tasks in a manner that is accurate, stable, and efficient. Accuracy is vital to ensure that all contacts are identified as soon as they occur and no false contacts are reported. If the contact detection algorithm misses a collision when it first occurs and then identifies it at a later time, the amount of overlap can become very large and the resulting rebound velocities can exceed the impact velocity, leading to instability in the numerical calculation. Correct identification of the contact point and unit normal is important to ensure that momentum is conserved during the collision. Computational efficiency of the contact detection algorithm is essential because contact detection typically takes up the largest fraction of the overall computational time for calculations with nonspherical particles. DEM simulations with ellipsoidal particles can take an order of magnitude, or more, longer than simulations with the same number of spherical particles, and this difference is entirely controlled by the efficiency of the contact detection scheme.

Because contact detection is so time consuming, it is worthwhile to first reduce the number of particle pairs that must be inspected by using simple, inexpensive tests to see if it is possible for contact to occur. It is convenient to retain a short list for each particle of other particles that are sufficiently close to it that a collision may have occurred during the time step. This "collision list" might contain 20 or so particles that are close to a given particle, and collisions of the given particle are only considered with particles contained in this list. In a second test, the distance $\Delta_{ij} = |\mathbf{x}_i - \mathbf{x}_j|$ between a given particle i and each particle j in its collision list is determined. The pair interaction is sorted into one of the following categories:

$$\text{Case A: } \Delta_{ij} > \max(a_{Ei} + a_{Ej}, b_{Ei} + b_{Ej}), \tag{7.3.1}$$

$$\text{Case B: } \Delta_{ij} < \min(a_{Ei} + a_{Ej}, b_{Ei} + b_{Ej}), \tag{7.3.2}$$

$$\text{Case C: } \max(a_{Ei} + a_{Ej}, b_{Ei} + b_{Ej}) \geq \Delta_{ij} \geq \min(a_{Ei} + a_{Ej}, b_{Ei} + b_{Ej}). \tag{7.3.3}$$

In Case A, the collision cannot occur. In Case B, the collision definitely occurs and it remains to identify the collision point and amount of particle overlap. In Case C, it is possible that the collision may have occurred, and so we must investigate further using the full contact detection algorithm.

Numerous algorithms have been proposed in the literature for contact detection of ellipsoidal particles, as well as for the related two-dimensional problem of contact of elliptical particles. A summary follows of some of the major approaches used for

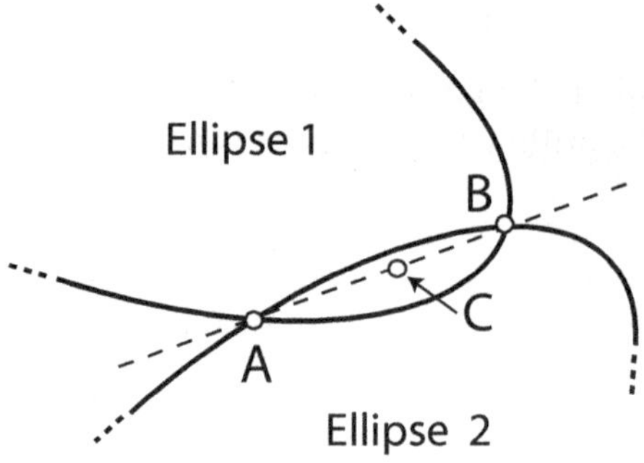

Figure 7.3. Illustration of the contact detection algorithm of Ting (1992) for two-dimensional elliptical particles. The intersection points of the ellipses are denoted A and B, and the midpoint of the line segment AB is the contact point C.

this purpose. The discussion is based on the review by Lin and Ng (1995), updated to include more recent results.

7.3.1. Two-Dimensional Algorithms

Many of the early contact detection algorithms were developed for two-dimensional problems involving colliding elliptical particles, but do not extend well to ellipsoidal particles in three-dimensions. For instance, Ting (1992) proposed a very simple contact detection algorithm illustrated in Figure 7.3, in which the equations for the surface of two ellipses are solved simultaneously to find the (x,y) positions of the two intersection points, labeled A and B. If there is no real-valued solution of this system of equations then the ellipses are not in contact. If a real-valued solution exists, then the midpoint C of the line segment AB is chosen as the contact point. This method requires finding the roots of a fourth-degree polynomial for each particle pair at each time step. It suffers from a high degree of ill-conditioning that can lead to problems with round-off errors if the amount of overlap of the ellipses is small. The biggest fault of this method, however, is the fact that it is quite complicated to extend to three dimensions.

In an alternative approach proposed for arbitrarily shaped two-dimensional particles by Hogue and Newland (1994), each particle is represented in polar form where the particle surface is discretized by a set of line segments and the particle interior is composed of the union of a set of triangles, as shown in Figure 7.4. In order to check for particle collision, each vertex point of a neighboring particle is examined to see if it lies within a triangle of the given particle. Determination of whether a point P is inside a triangle T is performed by drawing lines from the point to each vertex of the triangle, as illustrated for point P in Figure 7.4. Three subtriangles

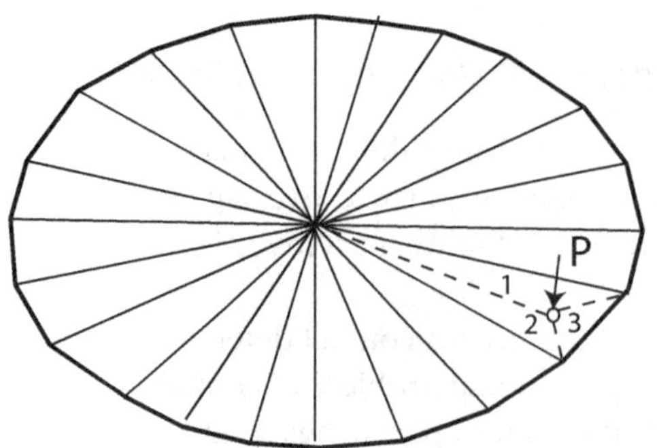

Figure 7.4. Triangularization of the particle using the polar representation proposed in the contact detection algorithm of Hogue and Newland (1994).

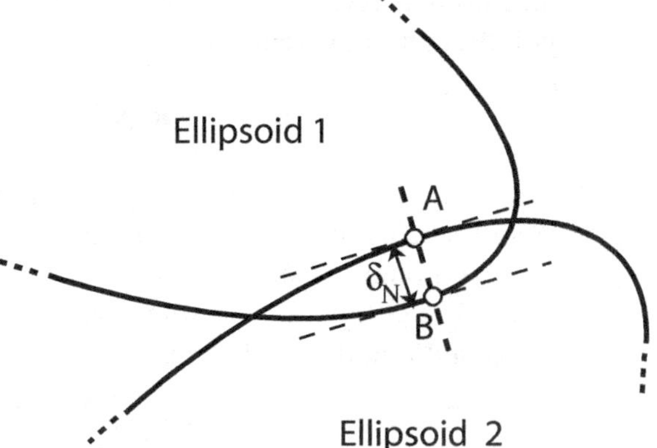

Figure 7.5. Illustration of contact point identification using the common normal method.

are formed by drawing lines from the point P to two vertices of the triangle T, as indicated by the dashed lines in Figure 7.4. If the sum of the area of these three subtriangles is greater than the total area of triangle T, then P is outside of triangle T. If the sum of the subtriangle areas is equal to the total area of triangle T, then P is inside T or lying on the surface of T. Although it is possible to extend this scheme to a tetrahedral discretization of three-dimensional particles, the method has a number of drawbacks that make it undesirable to do so for particles with smoothly curving surfaces, such as ellipsoidal particles. Specifically, because the maximum amount of particle overlap is usually very small, even slight displacement of the particle surface by the discretization method can have significant consequences for accurate detection of particle contact. Also, because each vertex of a neighboring particle must be checked to see if it lies in each triangle of the given particle, the method can be quite time consuming, even if the set of vertices and triangles that must be examined is further reduced. This problem becomes even more significant in three dimensions.

7.3.2. Algorithms Based on a Common Normal Vector

The unit normal vectors to the surfaces of colliding ellipsoidal particles at the contact point must be parallel, because the particle surfaces have the same tangent plane at the contact point. Among other things, this requirement is necessary to ensure that each particle receives an equal and opposite force from the collision, as required by Newton's second law. An illustration of the common unit normal algorithm for contact detection is shown in Figure 7.5. The algorithm seeks to find two points, A and B, located on ellipsoid 2 and ellipsoid 1, respectively, such that the unit normal vectors to the ellipsoid surface at the two points are oriented along the same line. The equations of the two ellipsoids are written in terms of the global coordinates (x, y, z) as

$$f_1(x, y, z) = 0, \quad f_2(x, y, z) = 0. \tag{7.3.4}$$

The position vectors of points A and B are denoted by $\mathbf{x}_2$ and $\mathbf{x}_1$, respectively, corresponding to the coordinate positions (x_2, y_2, z_2) and (x_1, y_1, z_1). The outward unit normal to a surface $f(x, y, z) = 0$ is given by $\mathbf{n} = \nabla f / |\nabla f|$. Because the outward unit normal vectors of the two ellipsoids must point in opposite directions at the points A and B, we can write

$$\mathbf{n}_1(\mathbf{x}_1) + \mathbf{n}_2(\mathbf{x}_2) = \frac{\nabla f_1(\mathbf{x}_1)}{\Delta_1} + \frac{\nabla f_2(\mathbf{x}_2)}{\Delta_2} = 0, \tag{7.3.5}$$

where $\Delta_1 \equiv |\nabla f_1|$ and $\Delta_2 \equiv |\nabla f_2|$. Not only must the unit normal vectors at points A and B be parallel to each other, but they must also be parallel to a line connecting the two points, as shown in Figure 7.5. The unit tangent vector along this line is given by $\mathbf{t} = (\mathbf{x}_2 - \mathbf{x}_1)/\Delta$, where $\Delta \equiv |\mathbf{x}_2 - \mathbf{x}_1|$ is the distance between the points. This condition then reduces to

$$\mathbf{t} - \mathbf{n}_2 = \frac{\mathbf{x}_2 - \mathbf{x}_1}{\Delta} - \frac{\nabla f_2(\mathbf{x}_2)}{\Delta_2} = 0. \tag{7.3.6}$$

The vector equations (7.3.5) and (7.3.6) provide a set of six nonlinear equations for the six coordinates of points A and B. After solving for these coordinates, contact is detected if point A is inside ellipsoid 1 and point B is inside ellipsoid 2, or in mathematical terms

$$f_1(\mathbf{x}_2) \le 0 \quad \text{and} \quad f_2(\mathbf{x}_1) \le 0. \tag{7.3.7}$$

Once the contact points are found and contact is decided to have occurred, the normal overlap δ_N is set equal to the contact point separation distance Δ .

An evaluation of the common unit normal contact detection algorithm was presented by Lin and Ng (1995). They found that although this algorithm can be used to accurately detect contact and identify contact points in both two and three dimensions, it is neither as accurate nor as efficient as algorithms based on geometric potentials, which are reviewed in the next section.

7.3.3. Algorithms Based on Geometric Level Surfaces

The geometric concept of *level surfaces* of an ellipsoidal shape defines a family of similar surfaces emanating outward and inward from the surface of an ellipsoid, as shown in Figure 7.6a for an ellipsoid with surface E. The contact point of one ellipsoid (E_1) in a pair is found by examining its intersection with the level surfaces of the other ellipsoid (E_2), and vice versa. The method for obtaining these contact points goes by many names in the literature, different versions of which can be found in papers by Ting (1992), Lin and Ng (1995), and Perram et al. (1984). The current discussion follows the presentation of Chesnutt and Marshall (2009), and is based on the geometric arguments of Alfano and Greer (2003), Chan (2001), and Schneider and Eberly (2003).

The surface of a general ellipsoid can be written as a quadratic equation of the form

$$P_k(\mathbf{x}) = \mathbf{x}^T \mathbf{A}_k \mathbf{x} + \mathbf{b}_k^T \mathbf{x} + c_k = 0, \tag{7.3.8}$$

where $\mathbf{x}$ is the position vector and k has the value 1 or 2 for ellipsoids E_1 and E_2, respectively. This equation can be expressed in a four-dimensional space using a generalized position vector $\mathbf{X} = [x \quad y \quad z \quad 1]$ in the canonical form

$$\mathbf{X}\mathbf{Q}\mathbf{X}^T = 0, \tag{7.3.9}$$

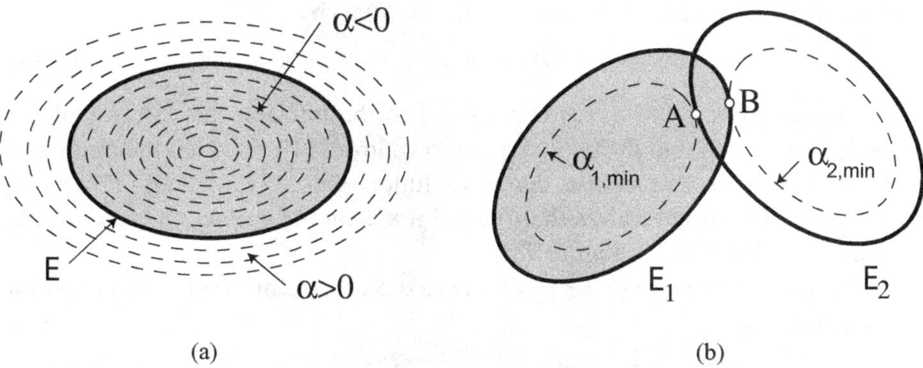

Figure 7.6. (a) Level surfaces of ellipsoid with surface E, where interior and exterior correspond to $\alpha < 0$ and $\alpha > 0$, respectively. (b) Contact points correspond to the point on each ellipsoid surface that yields the smallest value of the potential α of the opposing level-surface ellipsoid family.

where $\mathbf{Q}$ is the 4×4 symmetric characteristic matrix of the ellipsoid. In terms of the coefficients q_{ij} of $\mathbf{Q}$, the coefficients in (7.3.8) can be written as

$$\mathbf{A}_k = \begin{bmatrix} q_{11} & q_{12} & q_{13} \\ q_{12} & q_{22} & q_{23} \\ q_{13} & q_{23} & q_{33} \end{bmatrix}, \tag{7.3.10a}$$

$$\mathbf{b}_k = 2[q_{14} \quad q_{24} \quad q_{34}]^T, \tag{7.3.10b}$$

$$c_k = q_{44}. \tag{7.3.10c}$$

Walls are treated as degenerate ellipsoids, where (7.3.9) reduces to the equation of a plane by setting q_{11}, q_{22}, q_{33}, q_{12}, q_{13}, and q_{23} equal to zero. If any vector $\mathbf{X}$ exists such that (7.3.9) is satisfied for two different ellipsoids, with different characteristic matrices $\mathbf{Q}_1$ and $\mathbf{Q}_2$, then the corresponding point $\mathbf{x}$ lies on both ellipsoids.

In order to determine if two ellipsoids intersect, equation (7.3.9) for ellipsoid E_1 is multiplied by a scalar λ. Subtracting the same equation for another ellipsoid E_2 yields

$$\mathbf{X}(\lambda \mathbf{Q}_1 - \mathbf{Q}_2)\mathbf{X}^T = 0. \tag{7.3.11}$$

A nontrivial solution of (7.3.11) exists if and only if the matrix $(\lambda \mathbf{Q}_1 - \mathbf{Q}_2)$ is singular, from which it follows that λ must be an eigenvalue of the matrix $\mathbf{Q}_1^{-1}\mathbf{Q}_2$. The corresponding characteristic equation for λ requires finding the roots of the fourth degree polynomial

$$\det\left(\lambda \mathbf{I} - \mathbf{Q}_1^{-1}\mathbf{Q}_2\right) = 0. \tag{7.3.12}$$

This polynomial admits four roots. As demonstrated by Alfano and Greer (2003) and proved by Chan (2001), the two ellipsoids intersect at more than one point (without completely penetrating through each other) if and only if two of the roots of (7.3.12) are complex conjugates of each other.

Once contact detection has been verified, it remains to identify the contact points. For this purpose, the set of level surfaces of ellipsoid E_2 is parameterized in terms of a parameter α (called the geometric potential) by

$$P_2(\mathbf{x}) = \alpha, \tag{7.3.13}$$

where the regions $\alpha < 0$ and $\alpha > 0$ correspond to the interior and exterior of E_2, respectively. In order to find the contact point on ellipsoid E_1, we seek the minimum value of α for which a vector $\mathbf{x}$ exists that is a solution of both (7.3.13) and $P_1(\mathbf{x}) = 0$, where the latter condition ensures that the point $\mathbf{x}$ is on the surface of ellipsoid E_1. This condition is illustrated in Figure 7.6b.

In order to minimize α in (7.3.13) subject to the constraint $P_1(\mathbf{x}) = 0$, a function $F(\mathbf{x}, \tau)$ is defined as

$$F(\mathbf{x}, \tau) = P_2(\mathbf{x}) + \tau P_1(\mathbf{x}), \tag{7.3.14}$$

where τ is a Lagrange multiplier. Differentiating $F(\mathbf{x}, \tau)$ with respect to τ gives

$$\frac{\partial F}{\partial \tau} = P_1. \tag{7.3.15}$$

Setting this τ-derivative equal to zero recovers the constraint $P_1(\mathbf{x}) = 0$. Setting the gradient of F to zero gives

$$\nabla P_2 + \tau \nabla P_1 = 0 \tag{7.3.16}$$

for some τ, which implies that the gradients of P_1 and P_2 are parallel. Using (7.3.8) to obtain the gradients of P_1 and P_2, (7.3.16) becomes

$$2(\mathbf{A}_2 + \tau \mathbf{A}_1)\mathbf{x} + (\mathbf{b}_2 + \tau \mathbf{b}_1) = 0. \tag{7.3.17}$$

Solving (7.3.17) for $\mathbf{x}$ gives

$$\mathbf{x} = -\frac{1}{2}(\mathbf{A}_2 + \tau \mathbf{A}_1)^{-1}(\mathbf{b}_2 + \tau \mathbf{b}_1) = \frac{1}{\Phi(\tau)}\mathbf{y}(\tau), \tag{7.3.18}$$

where $\Phi(\tau) = \det(\mathbf{A}_2 + \tau \mathbf{A}_1)$ and the components of the vector $\mathbf{y}(\tau)$ are cubic polynomials in τ. Substituting (7.3.18) into the constraint $P_1(\mathbf{x}) = 0$ yields a sixth-degree polynomial in τ given by

$$\mathbf{y}(\tau)^T \mathbf{A}_1 \mathbf{y}(\tau) + \Phi(\tau) \mathbf{b}_1^T \mathbf{y}(\tau) + \Phi^2(\tau) c_1 = 0. \tag{7.3.19}$$

After calculating the real roots $\tau_n (n \leq 6)$ of this polynomial, the corresponding points $\mathbf{x}$ on the ellipsoid E_1 are calculated from (7.3.18) and the corresponding value of the potential α is obtained from (7.3.13). The contact point on ellipsoid E_1 corresponds to the root that yields the minimum value of the potential α, denoted as $\alpha_{2,\min}$. The procedure is then repeated with the ellipsoids switched to obtain the corresponding contact point on ellipsoid E_2.

It is noted that after the initial collision has occurred, two adhesive particles may remain in contact even if the ellipsoidal surfaces of the particles do not overlap. This occurs by the action of the adhesive force deforming the boundary of the particles as they are pulled apart, which is referred to as particle necking. For this reason, it is necessary to maintain a list of particle collision pairs at the previous time step and to repeat the contact point identification process described here for all particle pairs on this list, even if the ellipsoidal surfaces are no longer in contact. Only if the normal overlap δ_N has a negative value that is less than the critical overlap $-\delta_C$ identified in Section 4.2 are the particles determined to have broken apart and dropped from the collision list.

7.4. Contact Forces

Collision and adhesion forces and torques acting on two colliding particles consist of components that are both normal and tangential to the particle surface at the contact point. A unique unit normal vector $\mathbf{n}$ is defined for a contact between particles i and j as the average of the outward unit normal of particle i at its contact point and the inward unit normal of a colliding particle j at its contact point. Similar to the case of spherical particles, the normal force $-F_n\mathbf{n}$ for ellipsoidal particles results from a combination of elastic repulsion, dissipation, and adhesive force. Because the unit normal does not necessarily pass through the centroid of the ellipsoidal particles, the normal force produces a torque $-F_n\mathbf{r}_i \times \mathbf{n}$ on the particle, where $\mathbf{r}_i$ is the vector from the centroid of particle i to the contact point on particle i. The sliding motion of the particles gives rise to a sliding force $F_s\mathbf{t}_S$ and corresponding torque $F_s\mathbf{r}_i \times \mathbf{t}_S$, where $\mathbf{t}_S$ is a unit vector in the direction of relative motion of the particle surfaces at the contact point projected onto a plane orthogonal to $\mathbf{n}$. A torque also acts on the particles due to resistance from twisting motion, given by $M_t\mathbf{n}$. Additionally, as for spherical particles, rolling motion is resisted by a rolling torque $M_r\mathbf{t}_R \times \mathbf{n}$, where $\mathbf{t}_R$ is the direction of rolling velocity, to be defined later in this section. Combining these various forces and torques gives the net collision and adhesion force $\mathbf{F}_A$ and torque $\mathbf{M}_A$ acting on an ellipsoidal particle i as

$$\mathbf{F}_A = -F_n\mathbf{n} + F_s\mathbf{t}_S, \quad \mathbf{M}_A = -F_n\mathbf{r}_i \times \mathbf{n} + F_s\mathbf{r}_i \times \mathbf{t}_S + M_r\mathbf{t}_R \times \mathbf{n} + M_t\mathbf{n} \tag{7.4.1}$$

7.4.1. Geometry of Colliding Particles

Any point P_1 on the surface of an ellipsoid E_1 admits a curvature within any planar slice of the surface passing through P_1. Since there are infinitely many such slices, there are infinitely many curvatures κ of the ellipsoid surface that can be defined at P_1. The principal curvatures are given by $\kappa_1' = 1/R_1'$ and $\kappa_1'' = 1/R_1''$, where R_1' and R_1'' are the maximum and minimum values of the radius of curvature of any planar slices passing through P_1, respectively. If P_1 and P_2 are the contact points for two colliding ellipsoids E_1 and E_2, and we similarly define principal curvatures $\kappa_2' = 1/R_2'$ and $\kappa_2'' = 1/R_2''$ at point P_2 on ellipsoid E_2, then the principal relative radii of curvature R' and R'' are defined by

$$\frac{1}{R'} \equiv \frac{1}{R_1'} + \frac{1}{R_2'}, \quad \frac{1}{R''} \equiv \frac{1}{R_1''} + \frac{1}{R_2''}. \tag{7.4.2}$$

The effective radius R_e at the contact point is defined as the geometric mean of the principal relative radii of curvature, or

$$R_e = (R'R'')^{1/2}. \tag{7.4.3}$$

The effective radius R_e plays a similar role to the radius of the same name denoted by R for spherical particles, defined in (3.1.1).

The normal overlap δ_N is defined as the distance between the contact points of two colliding particles. The normal overlap is taken to have a positive sign for intersecting particles and a negative sign for nonintersecting (e.g., necking) particles.

The sum of the displacements normal to the contact surface for two colliding particles can be written as a function of local coordinates x and y as

$$u_{z1} + u_{z2} = \delta_N - Ax^2 - By^2, \tag{7.4.4}$$

where $x = y = 0$ at the contact point and the coordinate z measures distance normal to the tangent plane within the contact region. The coefficients A and B in (7.4.4) are related to the principal radii of curvature defined in (7.4.2) by (Johnson, 1985, 85)

$$A = \frac{1}{2R'}, \quad B = \frac{1}{2R''}. \tag{7.4.5}$$

7.4.2 Hertz Theory for Ellipsoidal Particles

The contact region that forms on collision of nonadhesive ellipsoidal bodies has an elliptical shape, with semimajor and semiminor axes a and b, respectively. The pressure distribution within the contact region can be written in terms of the local coordinates x and y in the direction tangent to the contact surface as

$$p = p_0[1 - (x/a)^2 - (y/b)^2]^{1/2}, \tag{7.4.6}$$

where the total elastic normal force F_{ne} on the particle is given by

$$F_{ne} = \tfrac{2}{3} p_0 \pi \, ab. \tag{7.4.7}$$

Solution for the elastic displacement field associated with the pressure distribution (7.4.6) using a normal force prescribed over an elliptical region on the surface of an elastic half-space yields expressions for the coefficients A and B and the normal overlap δ_N as (see Johnson, 1985, for details)

$$A = \frac{p_0}{E} \frac{b}{a^2 e_c^2} [K(e_c) - E(e_c)], \tag{7.4.8}$$

$$B = \frac{p_0}{E} \frac{b}{a^2 e_c^2} [(a^2/b^2)E(e_c) - K(e_c)], \tag{7.4.9}$$

$$\delta_N = \frac{p_0}{E} b\, K(e_c), \tag{7.4.10}$$

where $e_c \equiv (1 - b^2/a^2)^{1/2}$ is the ellipticity of the contact region and $K(e_c)$ and $E(e_c)$ are the complete elliptic integrals of the first and second kinds, respectively, defined by

$$K(e_c) \equiv \int_0^{\pi/2} \frac{d\theta}{\sqrt{1 - e_c^2 \sin^2 \theta}}, \quad E(e_c) \equiv \int_0^{\pi/2} \sqrt{1 - e_c^2 \sin^2 \theta}\, d\theta. \tag{7.4.11}$$

An equation for the aspect ratio of the elliptical contact region is obtained by taking the ratio of A and B using (7.4.8) and (7.4.9), and using (7.4.5), to write

$$\frac{B}{A} = \frac{R'}{R''} = \frac{(a/b)^2 E(e_c) - K(e_c)}{K(e_c) - E(e_c)}. \tag{7.4.12}$$

Because e_c is a function of a/b, the entire right-hand side of (7.4.12) is a function only of aspect ratio of the contact region. The ratio R'/R'' is known from the location of the contact point on the two ellipsoids, so (7.4.12) provides a nonlinear equation for the contact region aspect ratio a/b. In addition to aspect ratio, specification of the geometry of the contact region also requires that we specify a mean contact region radius, given by

$$a_e \equiv (ab)^{1/2}. \tag{7.4.13}$$

To obtain an expression for a_e, we take the cube of (7.4.13) to write

$$a_e^3 = (ab)^{3/2} = (b/a)^{3/2} a^3. \tag{7.4.14}$$

Using (7.4.8) and (7.4.9), we can write

$$(AB)^{1/2} = \frac{1}{2R_e} = \frac{p_0}{E}\frac{b}{a^2 e_c^2}C(e_c), \tag{7.4.15}$$

where $C(e_c) = \{[(a/b)^2 E(e_c) - K(e_c)]\,[K(e_c) - E(e_c)]\}^{1/2}$. Solving for p_0 from (7.4.7) and plugging into (7.4.15) gives

$$\frac{1}{2R_e} = \frac{3F_{ne}}{2\pi E}\frac{C(e_c)}{a^3 e_c^2}. \tag{7.4.16}$$

Solving for a^3 in (7.4.16) and substituting into (7.4.14), and then taking the cube root of the resulting equation, gives the mean contact region radius as

$$a_e = \left(\frac{3F_{ne}R_e}{4E}\right)^{1/3} F_1(e_c), \tag{7.4.17}$$

where $F_1(e_c) = (4C(e_c)/\pi e_c^2)^{1/3}(b/a)^{1/2}$ is a function only of the contact region ellipticity e_c. The result for spherical particles is recovered as $b/a \to 1$, which corresponds to $F_1(0) = 1$.

The equation for the normal overlap can be written in terms of the normal elastic force F_{ne} by solving for p_0 from (7.4.7) and substituting into (7.4.10) to obtain

$$\delta_N = \frac{3F_{ne}}{2\pi Ea}K(e_c). \tag{7.4.18}$$

Because $a = (ab)^{1/2}(a/b)^{1/2}$, the result (7.4.17) can be used in (7.4.18) to write

$$\delta_N = \left(\frac{9F_{ne}^2}{16E^2 R_e}\right)^{1/3} F_2(e_c), \tag{7.4.19}$$

where $F_2(e_c) = (2/\pi)(b/a)^{1/2}K(e_c)/F_1(e_c)$. The spherical particle result is recovered as $e_c \to 0$, for which $F_2(0) = 1$. Solving for F_{ne} from (7.4.19) gives

$$F_{ne} = K_H \delta_N^{3/2}, \tag{7.4.20}$$

where the stiffness coefficient K_H is

$$K_H = \frac{4E R_e^{1/2}}{3}[F_2(e_c)]^{-3/2}. \tag{7.4.21}$$

Equation (7.4.20) is the same result for elastic normal force as obtained in Section 3.2 for spherical particles, but the stiffness coefficient definition differs by the factor $F_3(e_c) \equiv [F_2(e_c)]^{-3/2}$, which accounts for the effect of contact region ellipticity.

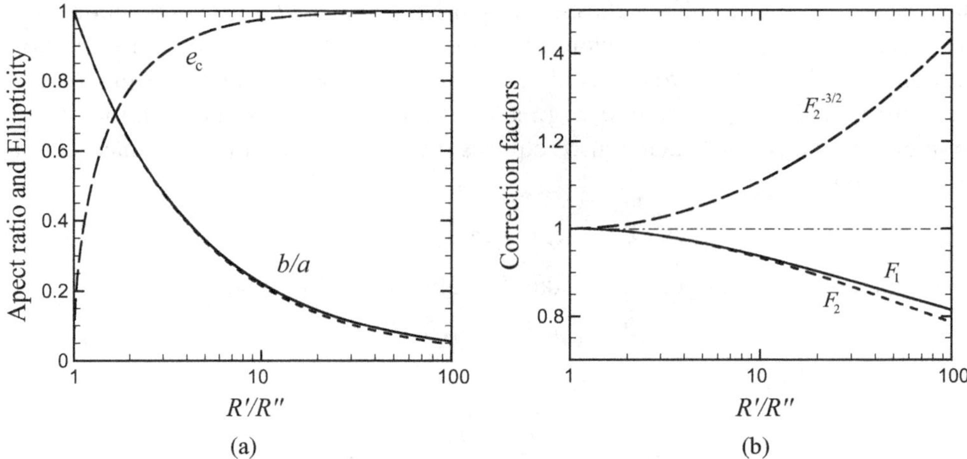

Figure 7.7. Plots showing characteristics of contact between ellipsoidal particles as a function of the ratio R'/R'' of the maximum and minimum principal relative radii of curvature. (a) Ellipticity e_c (long dashed line) and aspect ratio b/a of the contact region (solid line). Also shown as a dashed line is the approximation (7.4.22) for the aspect ratio. (b) Correction factors $F_1(e_c)$ (solid line) and $F_2(e_c)$ (dashed line) for the mean contact region radius and the normal overlap. Also shown is the correction $[F_2(e_c)]^{-3/2}$ for the stiffness coefficient (long dashed line).

Figure 7.7 plots various terms and corrections for collision of ellipsoidal particles as a function of the ratio R'/R'' of the maximum and minimum principal relative radii of curvature. The ellipticity e_c is solved iteratively from (7.4.12) and plotted as a function of R'/R'', along with the corresponding contact region aspect ratio b/a. The ratio a/b for the contact region is observed to be significantly less than the corresponding ratio R'/R'' based on the maximum and minimum curvatures of the ellipsoids as a whole. Also plotted is an approximation due to Johnson (1985) for contact region aspect ratio, given by

$$b/a = (R'/R'')^{-2/3}, \qquad (7.4.22)$$

which is found to be in excellent agreement with the exact solution. The correction factors $F_1(e_c)$ and $F_2(e_c)$ for the effective contact region radius a_e and the normal overlap δ_N are plotted in Figure 7.7b as functions of R'/R''. Also shown is the correction factor $[F_2(e_c)]^{-3/2}$ for the stiffness coefficient K_H shown in (7.4.21). As each of these correction factors approaches unity, the results approach those for spherical particles. It is clear from Figure 7.7b that for ellipsoids with modest aspect ratio, there is a fairly small difference in the normal force between the exact solution and that obtained for spherical particles with the same effective radii and normal overlap. Indeed, even for R'/R'' as high as 10, the relative error in normal force would only have been about 11% had we simply used the original Hertz formula (3.2.8) for spherical particles. Most of the other contact forces can similarly be approximated using the expressions for equivalent spherical particles to within reasonable accuracy, particularly if the particles are not too elongated.

Many investigators using the discrete element model with ellipsoidal particles simply use the corresponding contact force expressions for spherical particles, unless the particles are extremely elongated into needle-like or disk-like shapes. The major differences in the dynamics of spherical and ellipsoidal particles are that (1) the torque associated with the normal force causes a rotational motion for ellipsoidal particles when a normal collision occurs, which is not present for spherical particles; and (2) it is more difficult for ellipsoidal particles to roll for an appreciable distance along a flat surface because there is a potential energy hill that must be overcome with each half-rotation of the particle that is not present for spherical particles. Many practitioners feel that simulations with spherical particles allow too much rolling motion compared to "real" (generally nonspherical) particles, and for this reason ellipsoidal particles are often a desirable choice to obtain more realistic levels of particle rolling.

REFERENCES

Alfano S, Greer ML. Determining if two solid ellipsoids intersect. *Journal of Guidance, Control, and Dynamics* **26**, 106–110 (2003).

Brenner H. The Stokes resistance of an arbitrary particle. IV. Arbitrary fields of flow. *Chemical Engineering Science* **19**(10), 703–727 (1964).

Camp PJ, Allen MP. Hard ellipsoid rod-plate mixtures: Onsager theory and computer simulations. *Physica A* **229**, 410–427 (1996).

Campbell CS. Elastic granular flows of ellipsoidal particles. *Physics of Fluids* **23**, 013306 (2011).

Chan K. A simple mathematical approach for determining intersection of quadratic surfaces. *Proceedings of the American Astronautical Society*, AAS Paper 01–358, July–Aug. (2001).

Chesnutt JKW, Marshall JS. Blood cell transport and aggregation using discrete ellipsoidal particles. *Computers & Fluids* **38**, 1782–1794 (2009).

Chesnutt JKW, Marshall JS. Structural analysis of red blood cell aggregates under shear flow. *Annals of Biomedical Engineering* **38**(3), 714–728 (2010).

Chhabra RP, Agarwal L, Sinha NK. Drag on non-spherical particles: An evaluation of available methods. *Powder Technology* **101**, 288–295 (1999).

Dahneke BE. Slip correction factors for non-spherical bodies. III. The form of the general law. *Journal of Aerosol Science* **4**(2), 163–170 (1973).

Džiugys A, Peters B. An approach to simulate the motion of spherical and non-spherical fuel particles in combustion chambers. *Granular Matter* **3**, 231–265 (2001).

Fan FG, Ahmadi G. Wall deposition of small ellipsoids from turbulent air flows: A Brownian dynamics simulation. *Journal of Aerosol Science* **31**, 1205–1229 (2000).

Gallily I, Cohen AH. On the orderly nature of the motion of nonspherical aerosol particles. *Journal of Colloid and Interface Science* **68**, 338–356 (1979).

Happel J, Brenner H. *Low Reynolds Number Hydrodynamics*. Martinus Nijhoff Publishers, The Hague (1963).

Harper EY, Chang I-D. Maximum dissipation resulting from lift in a slow viscous shear flow. *Journal of Fluid Mechanics* **33**(2), 209–225 (1968).

Hogue C, Newland D. Efficient computer simulation of moving granular particles. *Powder Technology* **78**, 51–66 (1994).

Hughes PC. *Spacecraft Attitude Dynamics*. John Wiley & Sons, Inc., New York (1986).

Jeffery GB. The motion of ellipsoidal particles immersed in a viscous fluid. *Proceedings of the Royal Society of London A* **102**, 161–179 (1922).

Johnson, K.L., *Contact Mechanics*, Cambridge University Press (1985).

Knuth MA, Johnson JB, Hopkins MA, Sullivan RJ, Moore JM. Discrete element modeling of a Mars Exploration Rover wheel in granular material. *Journal of Terramechanics* **49**, 27–36 (2012).

Lin X, Ng T-T. Contact detection algorithms for three-dimensional ellipsoids in discrete element modelling. *International Journal for Numerical and Analytical Methoids in Geomechanics* **19**, 653–659 (1995).

Oberbeck A. Ueber stationäre Füssigkeitsbewegungen mit Berücksichtigung der inneren Reibung. *Journal für die reine und angewandte Mathematik* **81**, 62–80 (1876).

Perram JW, Wertheim MS, Lebowitz JL, Williams GO. Monte Carlo simulations of hard spheroids. *Chemical Physics Letters* **105**(3), 277–280 (1984).

Schneider PJ, Eberly DH. *Geometric Tools for Computer Graphics*. Morgan Kaufmann Publishers, San Francisco (2003).

Solis KJ, Martin JE. Isothermal magnetic advection: Creating functional fluid flows for heat and mass transfer. *Applied Physics Letters* **97**, 034101 (2010).

Ting JM. A robust algorithm for ellipse-based discrete element modeling for granular materials. *Computers and Geotechnics* **13**(3), 175–186 (1992).

Yin C, Rosendahl L, Kaer SK, Sørensen H. Modelling the motion of cylindrical particles in a nonuniform flow. *Chemical Engineering Science* **58**, 3489–3498 (2003).

8 Particle Interactions with Electric and Magnetic Fields

Particle interactions with electric and magnetic fields are important for a wide range of industrial applications, including mixing and separation processes in microfluidic flows and biological assay processes, electrospray coating processes, particle separation devices such as electrostatic precipitators, nanoparticle dispersion and manufacturing processes, and electrostatic classification processes for aerosol-generation or particle-sizing applications. In these processes, as well as in many others, particles are significantly influenced by forces arising from particles traveling in electric or magnetic fields, as well as from charged or magnetic particles interacting with each other. Interactions of charged particles immersed in an electrolytic solution result in short-range interaction force due to ionic shielding, as discussed in Chapter 4. However, in an aerosol there are no ions available, and the resulting electric and magnetic forces between particles decay slowly in space.

This chapter discusses the physics of electric and magnetic forces on particles, as well as computational methods for simulation of the electric field that can be used in conjunction with DEM computations. A description of the forces and torques imposed on a particle in both DC and AC electric fields is presented in the first section, including the phenomenon of dielectrophoresis. The second section discusses different methods by which particles in electric fields become charged, including contact electrification and de-electrification. The third section discusses forces and torques on particles in a magnetic field, including the phenomenon of magnetophoresis. Review of particle physics concludes in the sixth section, which examines the agglomerate chain structure of particles in an electric field.

Methods for computation of the electric field are the focus of the fourth and fifth sections, including the boundary element method (BEM) and the fast multipole method (FMM). The FMM provides a method for rapid computation of the electric field induced by charged particles. The BEM can be used to solve for the electrostatic field induced by macroscopic bodies immersed in the flow, such as electrodes, or the modification of an external electric field by dielectric surfaces. The bodies considered with the BEM are generally much larger than the particles. The total electric field to which the particles are exposed is the sum of an external electric field (typically prescribed), the electric field induced by other particles (obtained by FMM), and the electric field induced by macroscopic bodies in the flow (obtained by BEM), as illustrated in Figure 8.1. Summing all of these different fields yields the total electric

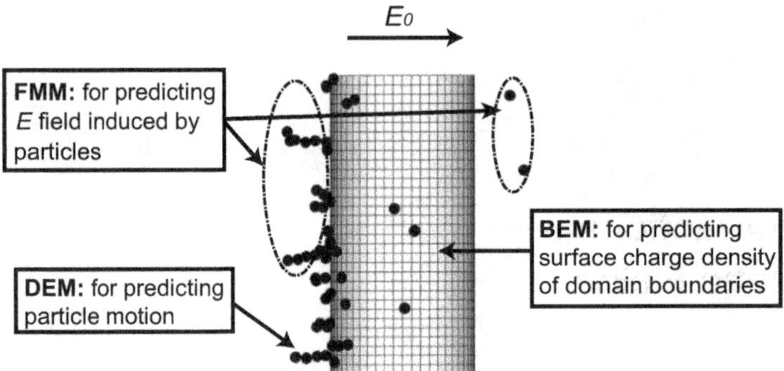

Figure 8.1. Illustration of the roles of DEM, FMM, and BEM for prediction of the dynamics and adhesion of particles near a body in the presence of an electric field.

field vector, which is used in DEM to compute the forces and torques on the particles and to move the particles in time. In the fourth and fifth sections of this chapter, we present general computational approaches for FMM and BEM applicable to systems consisting of a large number of particles immersed in a fluid flow in the presence of conducting or dielectric "macroscopic" bodies of arbitrary shape.

8.1. Electric Field Forces and Torques

8.1.1. Coulomb Force and Dielectrophoresis

A particle with charge q placed in an electric field $\mathbf{E}$ is subject to the Coulomb force

$$\mathbf{F}_{CL} = q\,\mathbf{E}. \tag{8.1.1}$$

An electric dipole is formed of positive and negative charges $\pm q$ separated by a displacement $\mathbf{d}$ in the limit as $|\mathbf{d}| \to 0$ and $q \to \infty$, as shown in Figure 8.2. If the negative charge is located at a position $\mathbf{x}$ and the positive charge is located at a position $\mathbf{x} + \mathbf{d}$, the resulting force on the dipole is given by

$$\mathbf{F}_{dipole} = \lim_{d \to 0}[-q\mathbf{E}(\mathbf{x}) + q\mathbf{E}(\mathbf{x} + \mathbf{d})]. \tag{8.1.2}$$

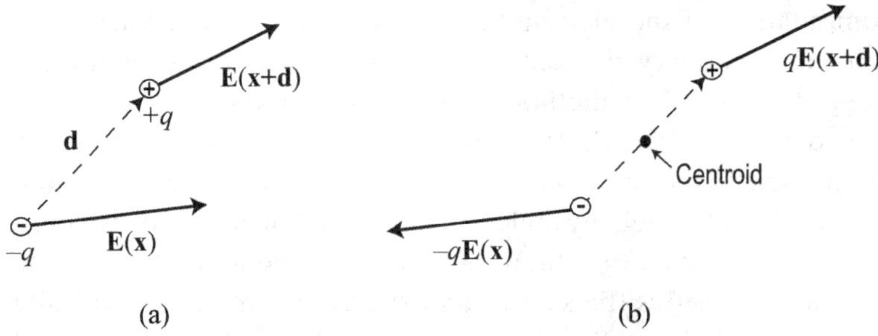

Figure 8.2. Illustration of a particle with negative charge $-q$ and a particle with positive charge $+q$ in an electric field, separated by a displacement $\mathbf{d}$. (a) Electric field on the particles; (b) Coulomb forces acting on the particles.

Expanding $\mathbf{E}(\mathbf{x} + \mathbf{d})$ in a Taylor series about $\mathbf{x}$ gives

$$\mathbf{E}(\mathbf{x} + \mathbf{d}) = \mathbf{E}(\mathbf{x}) + \mathbf{d} \cdot \nabla \mathbf{E} + \cdots . \tag{8.1.3}$$

Substituting (8.1.3) into (8.1.2) gives

$$\mathbf{F}_{dipole} = \lim_{d \to 0} q\mathbf{d} \cdot \nabla \mathbf{E}. \tag{8.1.4}$$

For a dipole, we require that $|\mathbf{d}| \to 0$ and the point charge $q \to \infty$ in such a manner that the product $q\mathbf{d}$ approaches a constant $\mathbf{p}$, called the *dipole moment*. The resulting force on the dipole is obtained from (8.1.4) as

$$\mathbf{F}_{dipole} = \mathbf{p} \cdot \nabla \mathbf{E}. \tag{8.1.5}$$

There is also a moment exerted on the dipole about the centroid point, located at the midpoint of the line segment connecting the two charge points. The moment $\mathbf{M}_{dipole}$ is obtained by summing the moment exerted by the Coulomb force on the two charge points making up the dipole about the centroid, giving

$$\mathbf{M}_{dipole} = \tfrac{1}{2}\mathbf{d} \times q\mathbf{E}(\mathbf{x} + \mathbf{d}) + \left(-\tfrac{1}{2}\mathbf{d}\right) \times (-q)\mathbf{E}(\mathbf{x}). \tag{8.1.6}$$

The lever arm from the centroid is $\mathbf{d}/2$ for the positively charged particle and $-\mathbf{d}/2$ for the negatively charged particle. Taking the limit $q\mathbf{d} \to \mathbf{p}$ while $\mathbf{d} \to 0$, (8.1.6) becomes

$$\mathbf{M}_{dipole} = \mathbf{p} \times \mathbf{E}. \tag{8.1.7}$$

The dipole moment of a conductive particle is caused by migration of charge on the particle surface, whereas for a dielectric particle, the dipole moment is generated by charge migration within the particle volume. Particles can possess two different kinds of dipole moments. The first type is a *permanent dipole*, for which the dipole exists within the material independent of other fields. An example of a permanent dipole is a polarized molecule, such as a water molecule. A second example is a material called an *electret*, which can be formed from certain dielectric materials that are first heated above their melting temperature and then cooled while exposed to a strong electric field.

The second type of dipole moment is an *induced dipole*, for which the polarization arises as a result of exposure of the particle to an electric field, but vanishes when the electric field is removed. Polarization occurs within dielectric materials due to displacement of positive charges in the "upstream" direction relative to the streamlines of the electric field vector, and displacement of negative charges in the opposite direction (Figure 8.3). The value of the induced dipole moment for a dielectric spherical particle of radius r_p immersed in a constant (DC) electric field $\mathbf{E} = E_0\mathbf{e}_x$ can be determined by recalling the solution of the Laplace equation $\nabla^2\Phi = 0$ governing the electrostatic potential Φ, where $\mathbf{E} = -\nabla\Phi$, both inside and outside of a sphere, given in spherical polar coordinates by

$$\Phi = \begin{cases} -E_0 r \cos\theta + (C_1/r^2)\cos\theta & \text{for } r > r_p \\ -C_2 r \cos\theta & \text{for } r \leq r_p \end{cases} . \tag{8.1.8}$$

The constants of integration C_1 and C_2 can be determined from the boundary conditions on the sphere surface. If the sphere has permittivity ε_p and the surrounding fluid medium has permittivity ε_f, then continuity of the tangential component of the

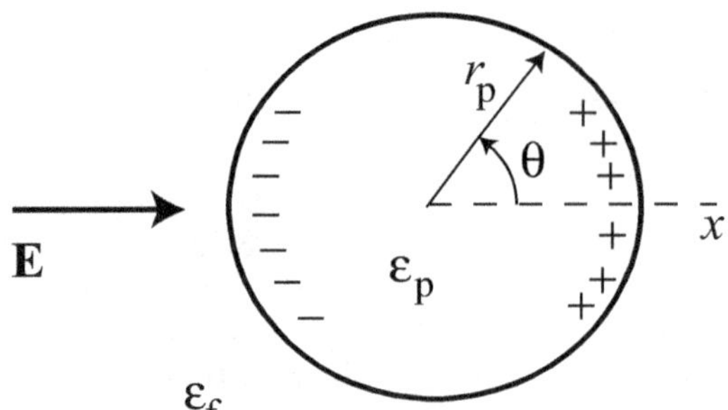

Figure 8.3. Dielectric sphere with permittivity ε_p immersed in a uniform electric field $\mathbf{E}$ in a medium with permittivity ε_f.

electric field vector implies that Φ must be continuous across the sphere surface, or

$$\Phi_1(r_p, \theta) = \Phi_2(r_p, \theta), \tag{8.1.9}$$

where the subscript 1 denotes the fluid side of the interface and the subscript 2 denotes the particle side. The second boundary condition requires continuity of the normal component of the electric displacement vector $\mathbf{D} = \varepsilon\, \mathbf{E}$, so that

$$\varepsilon_f \frac{\partial \Phi_1}{\partial r}(r_p, \theta) = \varepsilon_p \frac{\partial \Phi_2}{\partial r}(r_p, \theta). \tag{8.1.10}$$

Substituting (8.1.8) into (8.1.9) and (8.1.10) yields two equations for the coefficients C_1 and C_2 as

$$C_1 = \frac{\varepsilon_p - \varepsilon_f}{\varepsilon_p + 2\varepsilon_f} r_p^3 E_0, \quad C_2 = \frac{3\varepsilon_f}{\varepsilon_p + 2\varepsilon_f} E_0. \tag{8.1.11}$$

The first term in the outer $(r > r_p)$ solution (8.1.8) for Φ_1 represents the uniform electrostatic field and the second term has the same form as the potential function for an electric dipole with dipole moment $\mathbf{p} = 4\pi\varepsilon_f C_1 \mathbf{e}_x$. Substituting (8.1.11) for C_1 into this equation, we find that in the outside region $r > r_p$ the electric field has the form of a uniform electric field vector plus a point dipole with induced dipole moment

$$\mathbf{p} = 4\pi\varepsilon_f K_{CM} r_p^3 \mathbf{E}, \tag{8.1.12}$$

where the coefficient K_{CM}, known as the *Clausius-Mossotti function*, is given by

$$K_{CM} = \frac{\varepsilon_p - \varepsilon_f}{\varepsilon_p + 2\varepsilon_f}. \tag{8.1.13}$$

The value of K_{CM} is limited by (8.1.13) to the range $-0.5 \le K_{CM} \le 1$. The value of the vacuum permittivity ε_0 is 8.85×10^{-12} C^2/N m^2. The relative permittivity ε is commonly used to reflect the strength of the electrostatic field produced in different materials by a fixed potential relative to that produced in a vacuum under identical conditions. For most materials, relative permittivity falls in the range $1 < \varepsilon < 10$; for example, ε is 1.00059 for air, 2.1 for Teflon, 2.4–2.7 for polystyrene, and 3.9–4.3 for quartz. Some materials exhibit much higher values of relative permittivity; for example, ε is equal to 80 for pure water at room temperature and 86–173 for titania.

The process by which an induced dipole moment forms in a dielectric particle exposed to an electric field is called *dielectrophoresis*, and the force resulting from substitution of (8.1.12) into (8.1.5) is called the *dielectrophoretic (DEP) force*. Because the induced dipole moment in (8.1.12) is proportional to **E**, the resulting DEP force can be written as

$$\mathbf{F}_{DEP} = 2\pi \varepsilon_f K_{CM} r_p^3 \nabla E^2, \tag{8.1.14}$$

where E is the magnitude of the electric field vector **E**. The result (8.1.14) implies that particles with $K_{CM} < 0$ are repelled from regions with high electric field strength and particles with $K_{CM} > 0$ are attracted to regions with high electric field strength.

A similar derivation can be performed for a conducting particle immersed in a fluid with a constant (DC) electrostatic field. In this case the boundary condition (8.1.10) is replaced by the condition that the normal component of the current density vector $\mathbf{J} = \sigma \mathbf{E}$ is continuous at the sphere surface, or

$$\sigma_f \frac{\partial \Phi_1}{\partial r}(r_p, \theta) = \sigma_p \frac{\partial \Phi_2}{\partial r}(r_p, \theta), \tag{8.1.15}$$

where σ_p and σ_f are the electrical conductivities of the particle and fluid, respectively. Solving for C_1 from (8.1.9) and (8.1.15) and substituting into (8.1.8) gives the Clausius-Mossotti function for a conducting sphere as

$$K_{CM} = \frac{\sigma_p - \sigma_f}{\sigma_p + 2\sigma_f}. \tag{8.1.16}$$

The total electric force on a particle is given by $\mathbf{F}_E = \mathbf{F}_{CL} + \mathbf{F}_{DEP}$. For spherical particles with no permanent dipole and with induced dipole moment aligned with the electric field, as in (8.1.12), there is no electric torque on the particle ($\mathbf{M}_E = 0$). Particles respond to any electric field to which they are exposed. This electric field could originate from an electrode immersed in the fluid, or it could come from the electric field emitted by other nearby particles. The electric field emitted by nearby particles could in turn be associated with particle charge, or it could itself be induced by response to an electric field generated elsewhere. A further complication is that the presence of nonuniformity in the incident electric field causes the induced field to be more complex than a simple dipole, but instead to consist of the sum of a dipole, a quadrupole, and higher order multipoles. As a consequence, the problem of a cloud of particles immersed in a fluid in the presence of an electrode is highly nonlinear, with the electric field induced by polarization of each particle influencing that induced by every other particle. Nevertheless, this problem can often be simplified with reasonable accuracy by treating each particle as a monopole (for a charged particle) plus a dipole, ignoring higher-order multipoles, and lagging the interaction terms between the particles in time during the computation. An exception occurs for cases where the particles are actually touching each other, which is discussed in Section 8.5.

8.1.2. Dielectrophoresis in an AC Electric Field

In an AC electric field, the presence of ohmic energy loss due to oscillation of the ambient electric field introduces some important changes in the nature of the induced dipole moment. Let us reconsider the problem of a uniform electric field past a sphere considered in the previous section, but this time with an oscillating electric field vector given by

$$\mathbf{E}(t) = \text{Re}[E_0 \exp(j\omega t)]\,\mathbf{e}_x, \tag{8.1.17}$$

where $j \equiv \sqrt{-1}$. In an oscillating electric field, the boundary condition (8.1.10) is replaced by

$$J_{r1} - J_{r2} + \frac{\partial q_s}{\partial t} = 0 \quad \text{at } r = r_p, \tag{8.1.18}$$

where J_r is the radial component of the current density vector $\mathbf{J} = \sigma\mathbf{E}$ and q_s is the free electric surface charge, defined by

$$q_s = \varepsilon_f E_{r1} - \varepsilon_p E_{r2} \quad \text{at } r = r_p. \tag{8.1.19}$$

Assuming that all variables oscillate harmonically in time in proportion to $\exp(j\omega t)$ and substituting (8.1.19) into (8.1.18), the normal flux boundary condition at $r = r_p$ can be written as

$$\varepsilon_f^* E_{r1}(r_p, \theta) = \varepsilon_p^* E_{r2}(r_p, \theta), \tag{8.1.20}$$

where the complex dielectric constants ε_f^* and ε_p^* are defined by

$$\varepsilon_f^* = \varepsilon_f - j\sigma_f/\omega, \quad \varepsilon_p^* = \varepsilon_p - j\sigma_p/\omega. \tag{8.1.21}$$

Equation (8.1.20) has the same form as (8.1.10) with the replacement of the permittivities by the complex dielectric constants. Substituting the inner and outer solutions (8.1.8) into the boundary conditions (8.1.9) and (8.1.20) gives a solution for the coefficient C_1 in (8.1.8) as

$$C_1 = \frac{\varepsilon_p^* - \varepsilon_f^*}{\varepsilon_p^* + 2\varepsilon_f^*} r_p^3 E_0. \tag{8.1.22}$$

The second term in the outer flow solution of (8.1.8) has the form of an induced dipole with a complex-valued effective dipole moment

$$\mathbf{p}_{eff}^*(t) = 4\pi\varepsilon_f K_{CM}^* r_p^3 \mathbf{E}(t), \tag{8.1.23}$$

where K_{CM}^* is the complex Clausius-Mossotti function

$$K_{CM}^* = \frac{\varepsilon_p^* - \varepsilon_f^*}{\varepsilon_p^* + 2\varepsilon_f^*}. \tag{8.1.24}$$

A real-valued effective dipole moment can be defined by

$$\mathbf{p}_{eff}(t) = \text{Re}[\mathbf{p}_{eff}^*(t)\exp(j\omega t)] = 4\pi\varepsilon_f \text{Re}[K_{CM}^*]\,r_p^3 \mathbf{E}. \tag{8.1.25}$$

Equations (8.1.17) and (8.1.25) give expressions for the instantaneous electric field $\mathbf{E}(t)$ and effective dipole moment $\mathbf{p}_{eff}(t)$, both of which oscillate with the driving frequency ω of the AC electric field. The instantaneous DEP force and torque are given by

$$\mathbf{F}_{DEP}(t) = \mathbf{p}_{eff}(t) \cdot \nabla\mathbf{E}, \tag{8.1.26a}$$

$$\mathbf{M}_{DEP}(t) = \mathbf{p}_{eff}(t) \times \mathbf{E}. \tag{8.1.26b}$$

Substituting (8.1.25), the time-averaged DEP force can be written as

$$\langle \mathbf{F}_{DEP} \rangle = 2\pi \varepsilon_f r_p^3 \mathrm{Re}[K_{CM}^*(\omega)] \, \nabla E_{rms}^2, \tag{8.1.27}$$

where $E_{rms} = \langle \mathbf{E} \cdot \mathbf{E} \rangle^{1/2}$ is the root-mean-square electric field strength.

The torque vanishes for the case of a unidirectional electric field, as given by (8.1.17). However, many applications involve electric fields that are circularly polarized, in which the magnitude of $\mathbf{E}$ is constant in time but the vector direction rotates in a circular manner. For a right-polarized electric field, (8.1.17) is replaced by

$$\mathbf{E}(t) = \mathrm{Re}[E_0(\mathbf{e}_x - j\mathbf{e}_y)\exp(j\omega t)]. \tag{8.1.28}$$

If $\mathbf{E}^*(t)$ is the complex electric field vector, given by the term in brackets in (8.1.28), and $\mathbf{p}_{eff}^*(t)$ is the complex dipole moment defined by (8.1.23), then the DEP torque on the particle is given by

$$\mathbf{M}_{DEP}(t) = \mathrm{Re}[\mathbf{p}_{eff}^*(t)] \times \mathrm{Re}[\mathbf{E}^*(t)], \tag{8.1.29}$$

where for the right-polarized electric field, (8.1.23) gives

$$\mathrm{Re}[\mathbf{p}_{eff}^*(t)] = 4\pi \varepsilon_f r_p^3 \left(\mathrm{Re}[K_{CM}^*]\mathrm{Re}[\mathbf{E}^*] - \mathrm{Im}[K_{CM}^*]\mathrm{Im}[\mathbf{E}^*] \right). \tag{8.1.30}$$

Substituting (8.1.30) into (8.1.29) gives

$$\mathbf{M}_{DEP} = -4\pi \varepsilon_f r_p^3 \mathrm{Im}[K_{CM}^*] \left(\mathrm{Im}[\mathbf{E}^*] \times \mathrm{Re}[\mathbf{E}^*] \right). \tag{8.1.31}$$

From (8.1.28), $\mathrm{Im}[\mathbf{E}^*] \times \mathrm{Re}[\mathbf{E}^*] = E_0^2 \mathbf{e}_z$, so the DEP torque on the particle for this right-polarized field is obtained from (8.1.31) as

$$\mathbf{M}_{DEP} = -4\pi \varepsilon_f r_p^3 \, \mathrm{Im}[K_{CM}^*] \, E_0^2 \, \mathbf{e}_z, \tag{8.1.32}$$

where $\mathbf{e}_z$ is a unit vector along the axis of polarization. Although the electric field varies in time, the time dependence cancels out from the cross-product of the imaginary and real components of $\mathbf{E}^*(t)$ due to the identity $\sin^2(\omega t) + \cos^2(\omega t) = 1$.

8.1.3. Application to Particle Separation and Focusing

The complex Clausius-Mossotti function given in (8.1.24) can be rearranged as

$$K_{CM}^* = \left(\frac{\sigma_p - \sigma_f}{\sigma_p + 2\sigma_f} \right) \left[\frac{j\omega\tau_0 + 1}{j\omega\tau_{MW} + 1} \right], \tag{8.1.33}$$

where

$$\tau_{MW} = \frac{\varepsilon_p + 2\varepsilon_f}{\sigma_p + 2\sigma_f}, \quad \tau_0 = \frac{\varepsilon_p - \varepsilon_f}{\sigma_p - \sigma_f}. \tag{8.1.34}$$

The parameter τ_{MW}, called the *Maxwell-Wagner relaxation time*, characterizes the decay of free charge at the sphere surface in response to the AC electric field. Taking

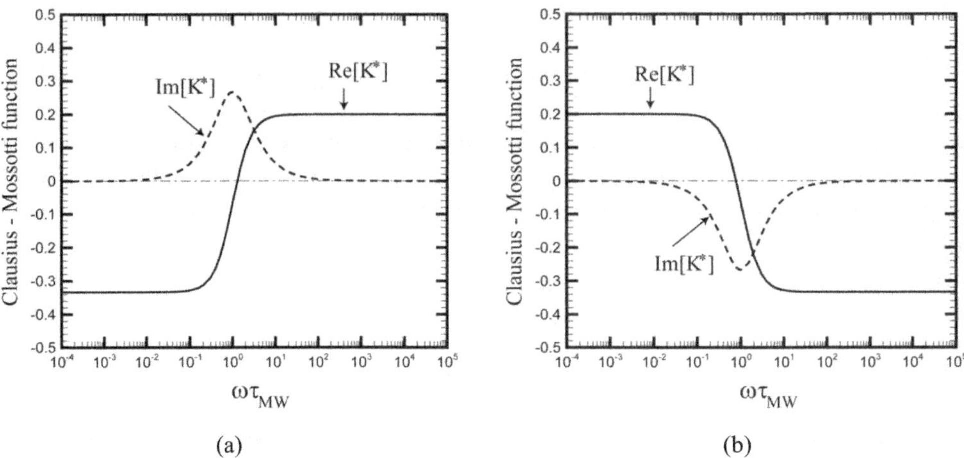

(a)   (b)

Figure 8.4. Plots showing the real and imaginary parts of the Clausius-Mossotti functions as a function of the dimensionless parameter $\omega\tau_{MW}$ for cases with (a) $E_\varepsilon = 1/2$ and $E_\sigma = -1/3$ and (b) $E_\varepsilon = -1/3$ and $E_\sigma = 1/2$.

the real and imaginary parts of K_{CM}^*, we find

$$\mathrm{Re}[K_{CM}^*(\omega)] = \left(\frac{\sigma_p - \sigma_f}{\sigma_p + 2\sigma_f}\right)\frac{1}{1 + \omega^2\tau_{MW}^2} + \left(\frac{\varepsilon_p - \varepsilon_f}{\varepsilon_p + 2\varepsilon_f}\right)\frac{\omega^2\tau_{MW}^2}{1 + \omega^2\tau_{MW}^2}, \qquad (8.1.35a)$$

$$\mathrm{Im}[K_{CM}^*(\omega)] = \left(\frac{\varepsilon_p - \varepsilon_f}{\varepsilon_p + 2\varepsilon_f} - \frac{\sigma_p - \sigma_f}{\sigma_p + 2\sigma_f}\right)\frac{\omega\tau_{MW}}{1 + \omega^2\tau_{MW}^2}. \qquad (8.1.35b)$$

As $\omega\tau_{MW} \to 0$, the real part of the Clausius-Mossotti function approaches the value $E_\sigma \equiv (\sigma_p - \sigma_f)/(\sigma_p + 2\sigma_f)$ given in (8.1.16) for a conducting particle in a DC field and the imaginary part vanishes. As $\omega\tau_{MW} \to \infty$, the real part of the Clausius-Mossotti function approaches the value $E_\varepsilon \equiv (\varepsilon_p - \varepsilon_f)/(\varepsilon_p + 2\varepsilon_f)$ given in (8.1.13) for a dielectric particle in a DC field and the imaginary part again vanishes. Examples showing variation of both the real and imaginary parts of $K_{CM}^*(\omega)$ as a function of frequency are given in Figure 8.4. For the case shown in Figure 8.4a where $E_\varepsilon > 0$ and $E_\sigma < 0$, the value of the real part of $K_{CM}^*(\omega)$ is negative for $\omega\tau_{MW} \ll 1$ and positive for $\omega\tau_{MW} \gg 1$ and the imaginary part of $K_{CM}^*(\omega)$ is everywhere positive. The opposite occurs when $E_\varepsilon < 0$ and $E_\sigma > 0$, as shown in Figure 8.4b. Most of the variation in the real and imaginary parts of $K_{CM}^*(\omega)$ occurs in the interval $0.1 < \omega\tau_{MW} < 10$; outside of this interval the imaginary part is close to zero and the real part is nearly constant.

Cases such as those shown in Figure 8.4, where $\varepsilon_p - \varepsilon_f$ and $\sigma_p - \sigma_f$ are of opposite signs, are of particular interest in applications, as in such cases the sign of $\mathrm{Re}[K_{CM}^*(\omega)]$ changes as the frequency is varied from a low to a high value. By adjusting the permittivity and conductivity of the fluid, the DEP force can be used to induce a differential drift on particles with different values of ε_p and σ_p, including in some cases a drift velocity in different directions. This observation is the basis of the dielectrophoretic method used by several researchers for separation of small particles of different sizes or different materials from each other, or even for separation of biological cells of different types (Fiedler et al., 1998; Chen and Du,

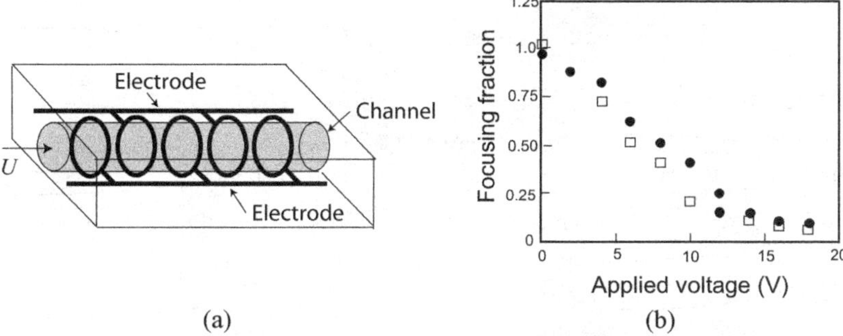

(a)   (b)

Figure 8.5. (a) Dielectrophoretic particle focusing system used to transport particles in a microchannel into a thin stream by Yu et al. (2005). (b) Results with latex particles for the focusing fraction, defined as the final width of the particle stream divided by the tube width, as a function of applied voltage at frequency 10 kHz (filled circles) and 30 kHz (open squares).

2007; Zhang et al., 2006). In a related application, dielectrophoresis has been used for particle focusing, in which particles that are initially dispersed in a microchannel flow are maneuvered into a concentrated stream, typically at the center of the channel (Morgan et al., 2003; Yu et al., 2005). Focusing is typically achieved by wrapping electrodes around the outside of the microchannel and oscillating the potential of the electrodes at a set frequency. For example, the particle focusing system proposed by Yu et al. (2005), shown in Figure 8.5a, consists of an elliptical microchannel surrounded at regular intervals by ring-like elliptical electrodes. The electrodes are attached to conductors running along the length of the channel in an alternating manner so as to produce a two-phase AC electric field. The DEP force for cases with $\mathrm{Re}[K^*_{CM}(\omega)] < 0$, or so-called negative DEP, acts to move particles away from the regions near the electrodes with high electric field strength toward the region in the center, at which the electric field strength is at a minimum. The negative DEP phenomenon thus produces a thin stream of particles at the channel center, the thickness of which is limited by diffusion related to particle collision. Figure 8.5b plots the ratio of the width of the particle stream after focusing to the tube width, which we call the focusing fraction, as a function of peak-to-peak applied voltage based on data from Yu et al. (2005). Cases with different driving frequencies appear to give similar results, although the time required to achieve focusing depends on frequency. These results illustrate the increase in DEP-induced particle drift as the electric field strength increases.

8.2. Mechanisms of Particle Charging

Determination of particle charge is essential for accurately predicting the electric force on particles. The principal mechanisms by which a solid particle acquires charge in an aerosol include field charging, diffusion charging, contact electrification (tribocharging), and thermionic emission (known as the Edison effect). Of particular interest for many manufacturing processes are liquid droplet aerosols, for which spray charging and electrolytic charging during the processes of atomization/nebulization are of most importance, along with electrokinetic streaming due to the relative motion between solid and fluid phases. Among these mechanisms,

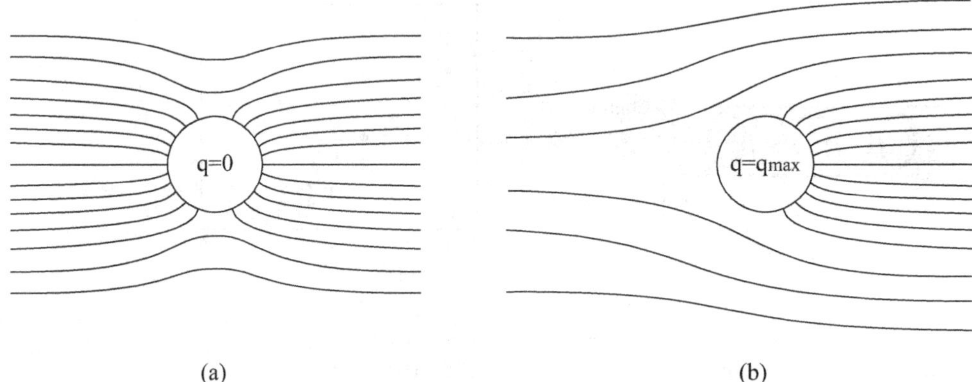

Figure 8.6. Schematic diagrams of the electric field lines near (a) an uncharged particle and (b) a particle at saturation charge.

the field and diffusion charging mechanisms are feasible and easily operational for production of highly charged aerosols, either of solid or liquid phases. These two mechanisms both depend on the collisions of particles with gaseous ions, which are generated by unipolar or bipolar chargers. The difference lies in that field charging is due to the particle-ion collisions influenced by an applied external field, whereas diffusion charging is due to collisions driven by random Brownian motion of ions. Also of importance for problems with frequent particle collisions is the phenomenon of contact electrification, which results in a transfer of electrons from one material to another upon contact. The current section discusses and compares these different methods for particle charging in aerosol flows.

8.2.1. Field Charging

Field charging occurs when gas ions are driven by an external electric field to collide with and adhere to the particle surface. This method of charging is dominant for particles larger than about 1 μm. For a spherical particle with radius r_p and with a charge q placed in a uniform external electric field $\mathbf{E} = E_0 \mathbf{e}_x$, the resultant electric field around the particle is a summation of the external field $\mathbf{E}$, the field $\mathbf{E}_p$ due to the effect of induced polarization of the particle, and the field $\mathbf{E}_q$ due to the particle charge, so the total electric field vector is given by $\mathbf{E}_T = \mathbf{E} + \mathbf{E}_p + \mathbf{E}_q$. The field at a point $\mathbf{x}$ due to a dipole of strength $\mathbf{p}$ at $\mathbf{x}_0$, with $r = |\mathbf{x} - \mathbf{x}_0|$, is given by (Jackson, 1962)

$$\mathbf{E}_p = \frac{3\mathbf{n}(\mathbf{p} \cdot \mathbf{n}) - \mathbf{p}}{4\pi\varepsilon_f r^3}. \tag{8.2.1}$$

Substituting the expression (8.1.12) for the induced dipole moment strength into (8.2.1) gives

$$\mathbf{E}_T = \mathbf{E} + K_{CM}\frac{r_p^3}{r^3}[3(\mathbf{E} \cdot \mathbf{n})\mathbf{n} - \mathbf{E}] + \frac{q}{4\pi\varepsilon_f r^2}\mathbf{n}, \tag{8.2.2}$$

where $\mathbf{n}$ is a unit vector defined by $\mathbf{n} = (\mathbf{x} - \mathbf{x}_0)/|\mathbf{x} - \mathbf{x}_0|$. Figure 8.6 illustrates the electric field lines around a particle both with no charge and with the saturation charge. Because the ions in the air migrate along the electric field lines in the direction of $\mathbf{E}_T$, the ions only reach the particle surface (and hence charge the particle) if the electric field lines terminate on the particle with negative value of the radial component $\mathbf{E}_T \cdot \mathbf{n}$ on the particle surface. Using (8.2.2) together with $E_0 = |\mathbf{E}|$, the radial component E_r of the electric field at the particle surface ($r = r_p$) can be calculated as

$$E_r|_{r=r_p} = E_0 \cos\theta\,(2K_{CM} + 1) + \frac{q}{4\pi\varepsilon_f r_p^2},\qquad(8.2.3)$$

where θ is the angle between $\mathbf{E}$ and the radial direction $\mathbf{n}$. If $E_r \geq 0$ for all θ, no electric field line can reach the particle surface, implying that no ions can impact with the particle. Using this condition, the particle saturation charge by the field charging method can be estimated by setting $\cos\theta = -1$ in (8.2.3) to obtain

$$q_{\max} = 4\pi\varepsilon_f r_p^2 E_0(2K_{CM} + 1) = 4\pi\varepsilon_f r_p^2 E_0 \frac{3\varepsilon_p}{\varepsilon_p + 2\varepsilon_f}.\qquad(8.2.4)$$

The time variation of the particle charge via the field-charging method is commonly approximated by the equation

$$q_f(t) = q_{\max}\frac{t}{t + \tau} = q_{\max}\frac{t}{t + 4\dfrac{\varepsilon_f}{eN_0 Z}},\qquad(8.2.5)$$

where the characteristic field-charging time τ is dependent on the gas phase permittivity ε_f, the ion concentration N_0 (with a magnitude of $10^{13}/\text{m}^3$ or greater), and the electric mobility of the ion Z (with a value of $10^{-4} - 10^{-3}\ \text{m}^2/\text{V}\cdot\text{s}$). Equation (8.2.5), is known as the *Pauthenier and Moreau-Hanot equation*, was developed to model charging of dust particles in the continuum regime (Pauthenier and Moreau-Hanot, 1937).

8.2.2. Diffusion Charging

Diffusion charging occurs when ions subject to a random Brown motion collide with and adhere to a particle surface. The diffusion-charging mechanism is dominant for ultra-fine particles, where the continuum approximation breaks down and the free-molecular regime is possible (see Chapter 9). White (1951) derived a simplified diffusion charging model for particles in the free-molecular regime. Under equilibrium diffusion charging, a particle of diameter r_p possessing n elementary charges (i.e., $q_d = ne$) has an electrostatic surface potential $\Phi = ne/4\pi\varepsilon_f r_p$. The number concentration of ions in the gas, N_i, satisfies the Boltzmann distribution $N_i = N_0 \exp(-e\Phi/k_B T)$, where k_B is the Boltzmann constant and T is absolute temperature. The mean thermal velocity of ions of mass m_i is given by $\bar{v}_i = \sqrt{8k_B T/\pi m_i}$. From molecular collision theory, the rate of ion collisions on the particle can be estimated as $dn/dt = \pi r_p^2 N_i \bar{v}_i$, provided the ion size is much smaller than the particle size. Substituting for N_i and $\bar{v}_i$ gives

$$\frac{dn}{dt} = \pi\,r_p^2 N_0 \exp\left(-\frac{ne^2}{4\pi\varepsilon_f r_p k_B T}\right)\left(\frac{8k_B T}{\pi m_i}\right)^{1/2}.\qquad(8.2.6)$$

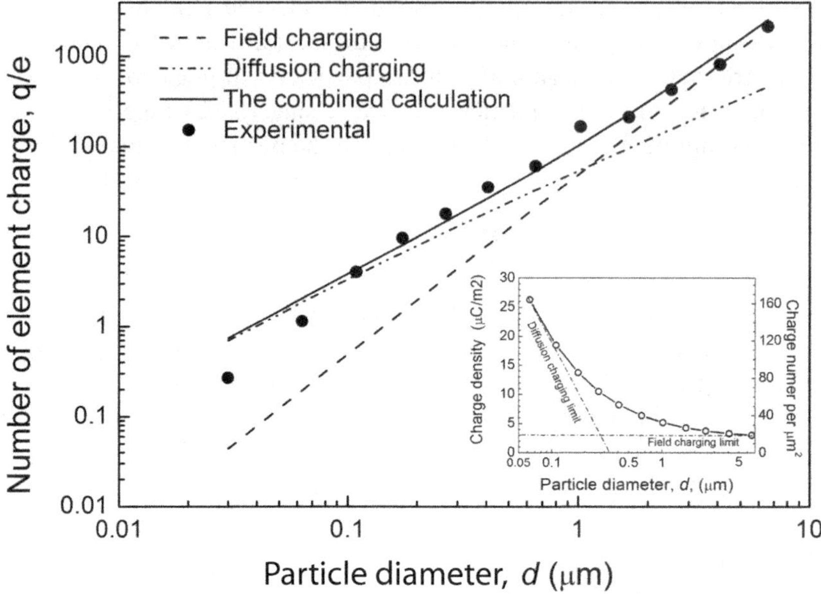

Figure 8.7. Number of elementary charges on dust particles in an electrostatic precipitator with external electric field $E = 1.5 \times 10^5$ V/m, ion concentration $N_i = 10^{13}$ /m^3, $t = 1$s, and ion mean velocity $\bar{v}_i = 240$ m/s. The experimental data from Jin (2013) was measured using an electric low pressure impactor (ELPI) device, while the calculation is based on the Pauthenier and Moreau-Hanot equation for field charging and the White equation for diffusion charging.

Integrating this equation in time, with initial condition $n(0) = 0$, gives the particle charge due to diffusion charging as

$$q_d(t) = \frac{4\pi \varepsilon_f r_p k_B T}{e} \ \ln \left(1 + \frac{\bar{v}_i r_p e^2 N_0 t}{4\varepsilon_f k_B T} \right). \tag{8.2.7}$$

This equation was first derived by White (1951). In contrast to the relationship $q_f \propto d^2$ for field charging, (8.2.7) yields a relationship $q_d \propto d$ for diffusion charging. The total charge of a particle in an electric field is the sum of both field and diffusion charging.

The number of elementary charges on a particle, q/e, can be measured as a function of particle size in an electrostatic precipitator by using an electrical low-pressure impactor (ELPI) device (Glover and Chan, 2004). Jin (2013) reported experimental data for particle charging with electric field $E = 1.5 \times 10^5$ V/m, ion concentration $N_i = 10^{13}/$m^3, $t = 1$s, $\varepsilon_f = 8.854 \times 10^{-12}$ F·m^{-1}, and mean ion velocity $\bar{v}_i = 240$ m/s. As shown in Figure 8.7, the sum of White's formula (8.2.7) for diffusion charging and the Pauthenier and Moreau-Hanot formula (8.2.5) for field charging can accurately predict the experimental data of Jin (2013) for the combined charge of particles over a wide range of sizes. For instance, for 1 μm diameter particles, Figure 8.7 indicates that the particle charge will be approximately 110 e, which corresponds to a surface charge density of ~35 charges/μm^2, or 5~6 μC/m^2. Because the particle charge for field charging varies in proportion with the surface area, both of which increase as ~d^2, it follows that the surface charge density for particles under field charging

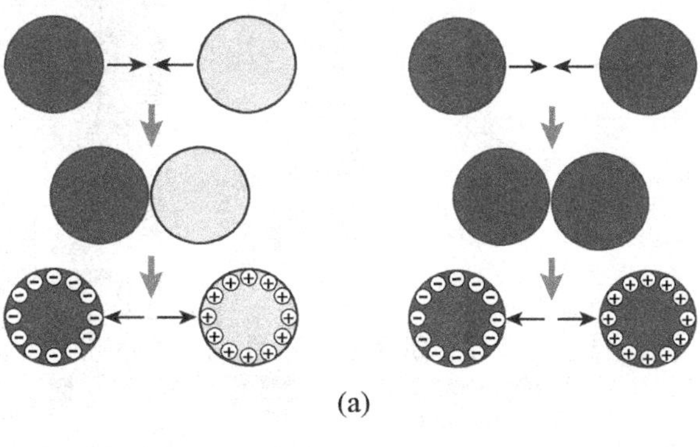

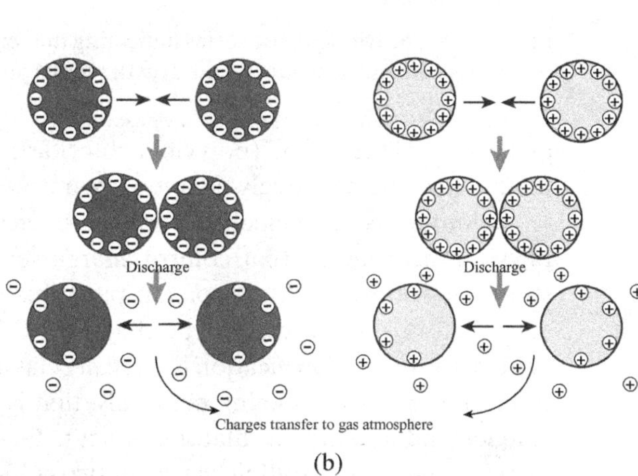

(b)

Figure 8.8. Mechanisms for (a) contact electrification and (b) contact de-electrification between two spheres of both different and identical materials (adapted from Soh et al., 2012).

is independent of particle diameter. Thus, this value of 35 charges/μm^2 can help to approximately determine the magnitude of the field charge for particles with various sizes. Brocilo et al. (2003) made a thorough review of the various kinds of field and diffusion charging models across the broad range of Knudsen numbers for free-molecular, transition, and continuum regimes. The results of this study confirm the validity of combining the White and Pauthenier-Moreau-Hanot equations, though other more detailed models are also available.

8.2.3. Contact Electrification

When two particles, made of different materials, are brought into contact and then separated, an electric charge is usually transferred from one particle to the other across the contact region, as illustrated in Figure 8.8a. This phenomenon is called *contact electrification* or *contact charging*. It is termed as *impact charging* when the contact is fast and in a normal direction, as in a particle collision, and it is termed as *tribocharging*, or frictional electrification, when the surfaces of two materials are rubbed relative to each other, as when a particle rolls or slides along a surface. Figure 8.9 provides a chart, known as the triboelectric series, which illustrates the tendency of different kinds of materials to acquire a positive or negative charge during contact electrification. As can be seen from this figure, polar materials such as glass, mica, fur, and silica easily become positively charged, whereas nonpolar materials such

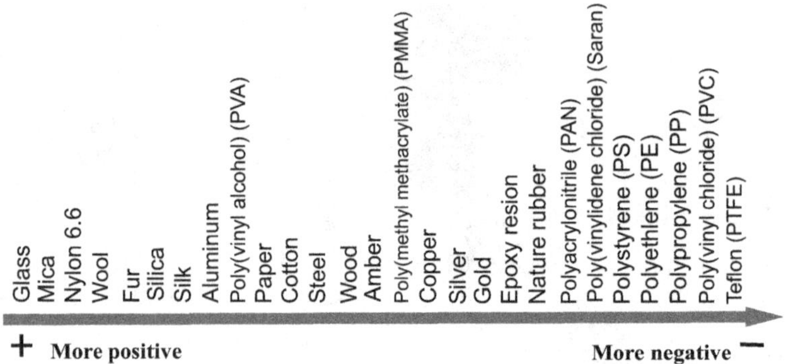

Figure 8.9. The triboelectric series indicating the tendency of materials for acquiring a positive or negative charge on contact, based on the series proposed by McCarty and Whitesides (2008).

as Teflon (PTFE), PVC (polyvinyl chloride), PS (polystyrene), and other polymers tend to become negatively charged upon contact (McCarty and Whitesides, 2008).

Although contact electrification has been studied since antiquity, the underlying physical mechanisms that control charge separation during particle collisions are still topics of active discussion. In part, this is because multiple mechanisms may contribute to the movement of charge in different systems, or even in the same system. Contact electrification is typically classified into three categories based on the properties of the contacting materials, that is, metal-metal contact, metal-insulator contact, and insulator-insulator contact (Matsusaka and Masuda, 2003). For metal-metal contact, the resulting charge of two colliding particles depends on the electron transfer from one metal to another. The charge change after separation, Δq_c, is approximated by the following simple equation

$$\Delta q_c = C_p \frac{W_2 - W_1}{e}, \tag{8.2.8}$$

where $W_{1,2}$ are the work functions of metals 1 and 2, respectively, and C_p is the capacitance between the two particle surfaces at the equilibrium contact region separation distance δ and contact area S, given by $C_p = \varepsilon_f S/\delta$. For instance, a 4 mm chromium sphere colliding with Ag or Rh particles produces a charge density of about 0.06 charges/μm^2, or 0.01 μC/m^2, which is so small as to be almost unnoticeable. For metal-insulator contacts, a similar relationship between Δq_c and the work function difference between metal and insulator, $(W_M - W_I)$, can be assumed, and interpreted to suggest that the charge carrier is also an electron (Soh et al., 2012). However, the mechanism appears not to be universal for all types of insulators.

Tribocharging of insulator-insulator contacts is of considerable importance in applications such as the flow of granular matter, the processing of pharmaceutical powders, the movements of small particles in dust storms and dust clouds, and so on. Though the separation of charges by the transfer of electrons plays an important role in metal-metal contacts, the contact electrification of insulators does not correlate with bulk electronic properties. Horn et al. (1993) noted a correlation between contact electrification of insulators and the pH of insulating materials. Whitesides and co-workers confirmed that the charge carriers in insulator-insulator contacts are the mobile counterions (McCarty and Whitesides, 2008; Thomas et al. 2008; Apodaca et al., 2010). They introduced a "rolling sphere tool," in which a rotating bar magnet causes a ferromagnetic steel sphere to roll along a circular path on an insulator surface. The charging density that was achieved by the sphere in this experiment was about 200 charges/μm^2, or 32 $\mu C/m^2$, which is about five times larger than the maximum charge obtained by field and diffusion charging in an electrostatic field (Thomas et al. 2008).

8.2.4. Contact De-electrification

The amount of contact charging achievable in a system of colliding particles eventually reaches an asymptotic limit, which implies the possibility of a discharging (or de-electrification) process that balances the charging process (Soh et al., 2012). This discharging process is illustrated in Figure 8.8b, where the contact of two previously charged insulator particles with the same polarity results in the discharging of both particles. Contact de-electrification occurs when the electric field strength in the gap within the contact region between two colliding charged particles exceeds the breakdown voltage of the gas (called the Paschen limit). Above the breakdown limit, the gas molecules ionize and generate ions (either positive or negative) and electrons. Depending on the electrical polarity of the particle surfaces, either the positive ions or the electrons are attracted to the particle surface while the other ones are repelled from the surface. Attraction of either positive ions or electrons to the particle surface causes the particles to discharge. Soh et al. (2012) reported that both Nylon and Teflon particles lost approximately 30% of their initial charge after contact, whereas without particle contact the charge of individual Nylon or Teflon particles only decreases by about 12% over a 10 hour period. Continuation of research to clarify physical mechanisms and validate prediction models for contact electrification and de-electrification is important in order to better account for the significant role of these phenomena in many particle flow systems.

8.3. Magnetic Field Forces

A particle with charge q traveling with velocity $\mathbf{v}$ in a space with electric field $\mathbf{E}$ and magnetic flux density $\mathbf{B}$ experiences the *Lorentz force*

$$\mathbf{F}_{LOR} = q(\mathbf{E} + \mathbf{v} \times \mathbf{B}), \tag{8.3.1}$$

The first term within parentheses is simply the Coulomb force given in (8.1.1), and the second force arises from the motion of a charged particle within a magnetic field. This force underlies the operation of electric motors and generators. The magnetic flux density is related to the volume magnetization vector $\mathbf{M}$ and the magnetic field intensity vector $\mathbf{H}$ by

$$\mathbf{B} = \mu_0(\mathbf{H} + \mathbf{M}), \tag{8.3.2}$$

where μ_0 is the permeability of free space and has the value $\mu_0 = 4\pi \cdot 10^{-7}\,\mathrm{H/m}$. In a linear magnetic material $\mathbf{M}$ and $\mathbf{H}$ are related by $\mathbf{M} = \chi_M \mathbf{H}$, where $\chi_M = (\mu_M/\mu_0) - 1$ is the magnetic susceptibility and μ_M is the magnetic permeability. Substituting this relationship into (8.3.2) gives

$$\mathbf{B} = \mu_M \mathbf{H}. \tag{8.3.3}$$

A noncharged particle in a magnetic field can also exhibit a nonzero force due to the phenomenon of *magnetophoresis*, or MAP. We consider a spherical particle of linear magnetic material with radius r_p immersed in a constant external magnetic field vector $\mathbf{H}_0 = H_0 \mathbf{e}_x$. The vector $\mathbf{H}$ has zero curl both inside and outside the sphere, so a potential ψ can be defined such that $\mathbf{H} = -\nabla \psi$. Because $\mathbf{H}$ has zero divergence, it follows that ψ satisfies the Laplace equation $\nabla^2 \psi = 0$ everywhere throughout the space. The Laplace equation has solutions inside and outside the sphere, given in spherical polar coordinates by

$$\psi = \begin{cases} -H_0 r \cos\theta + (A_1/r^2)\cos\theta & \text{for } r > r_p \\ -A_2 r \cos\theta & \text{for } r \le r_p \end{cases}. \tag{8.3.4}$$

If the sphere has magnetic permeability μ_p and the surrounding medium has magnetic permeability μ_f, then continuity of the tangential component of the magnetic field intensity vector $\mathbf{H}$ implies that ψ must be continuous across the sphere surface, or

$$\psi_1(r_p, \theta) = \psi_2(r_p, \theta). \tag{8.3.5}$$

The second boundary condition requires continuity of the normal component of the magnetic flux density vector $\mathbf{B} = \mu_M \mathbf{H}$, so that

$$\mu_f \frac{\partial \psi_1}{\partial r}(r_p, \theta) = \mu_p \frac{\partial \psi_2}{\partial r}(r_p, \theta). \tag{8.3.6}$$

These two boundary conditions yield solutions for the two coefficients A_1 and A_2 as

$$A_1 = \frac{\mu_p - \mu_f}{\mu_p + 2\mu_f} r_p^3 H_0, \quad A_2 = \frac{3\mu_f}{\mu_p + 2\mu_f} H_0. \tag{8.3.7}$$

In the region outside of the sphere, the magnetic field vector has the form of a uniform magnetic field $\mathbf{H}_0$ plus a magnetic dipole with induced dipole moment

$$\mathbf{m} = 4\pi \mu_f K_M r_p^3 \mathbf{H}_0. \tag{8.3.8}$$

The coefficient K_M is the magnetic *Clausius-Mossotti function*, given by

$$K_M = \frac{\mu_p - \mu_f}{\mu_p + 2\mu_f}. \tag{8.3.9}$$

The derivation has a similar form to that given for the induced dipole moment of a sphere in a DC electric field, with the magnetic permeabilities replacing the electric permittivities.

The force on a magnetic dipole is given by

$$\mathbf{F}_{dipole} = \mu_f \mathbf{m} \cdot \nabla \mathbf{H}. \tag{8.3.10}$$

Substituting the effective dipole moment given in (8.3.8) yields the magnetophoretic (MAP) force as

$$\mathbf{F}_{MAP} = 4\pi \mu_f K_M r_p^3 \mathbf{H}_0 \cdot \nabla \mathbf{H}_0 = 2\pi \mu_f K_M r_p^3 \nabla H_0^2. \tag{8.3.11}$$

As we see from the discussion presented here, under the assumption of a linear magnetic material, such that $\mathbf{M} = \chi_M \mathbf{H}$, the influence of constant magnetic field on particles closely parallels that of a DC electric field. However, many materials of

Figure 8.10. A body with bounding surface S_B immersed in an external electric field $\mathbf{E}_0$. The potential field and permittivity are denoted, respectively, by Φ_I and ε_I interior to the body and by Φ_E and ε_E exterior to the body.

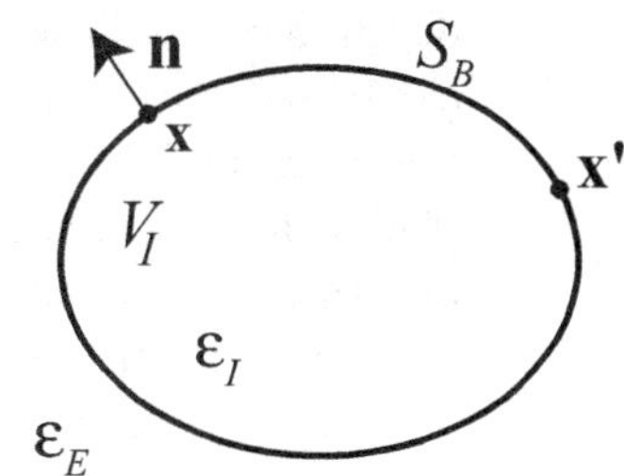

interest either behave in a nonlinear manner, where the volume magnetization vector **M** is not proportional to **H**, or else exhibit eddy currents as a response to fluctuating magnetic fields, and this analogy breaks down. In general, the problem of particles in electric and magnetic fields exhibits a rich physics, particularly when consideration is given to nonspherical or nonhomogeneous particles, but such topics are outside the scope of the current book. More extensive consideration of the electromechanics of particles of different types exposed to electric and magnetic fields can be found in Jones (1995).

8.4. Boundary Element Method

Section 8.1 discusses how particles, even if they carry no charge, are sensitive to the electric field vector and its gradient. In conducting a DEM simulation with charged particles, it is therefore necessary to compute the electric field to which the particles are exposed. This electric field is influenced by two features of the flow field – the domain boundaries and the presence of other particles. Domain boundaries may include electrode surfaces that emit an electric field into the flow, or dielectric surfaces that nevertheless alter the electric field within the domain. An electric field may also be introduced by the presence of charged particles, each of which generates an electric field that decays slowly with distance away from the particle. Even if the other particles are uncharged, they will exhibit an induced dipole in the presence of an electric field that will influence forces and torques on nearby particles. The electric field induced by charged particles is particularly challenging for DEM simulation because the decay rate of the electrostatic potential is sufficiently slow that the force induced by a single particle will have significant influence throughout the flow field, thereby considerably slowing the computation of the electric field in the presence of many particles. The next section discusses a computational method for accelerating the computation in the presence of charged particles. The current section is concerned with use of the boundary element method (BEM) for simulation of the influence of domain boundaries on the electric field. After a general introduction to BEM, we discuss two particular problems that occur when using BEM together with DEM for a bounded particulate flow in an electric field, as well as methods by which these problems might be resolved.

8.4.1. General Boundary Element Method

We consider a flow within a three-dimensional space V_I with boundary surface S_B having outward unit normal **n**, as shown in Figure 8.10. The region exterior to S_B is denoted by V_E. When the electrostatic assumption applies, the electric field vector $\mathbf{E}$ can be written in terms of a potential Φ as $\mathbf{E} = -\nabla\Phi$, where Φ satisfies $\nabla^2\Phi = 0$. The potential field $\Phi(\mathbf{x})$ at any point $\mathbf{x}$ within the interior domain V_I can be generated by placing a sheet charge of strength $q(\mathbf{x})$ along the boundary S_B, which can be written in terms of the local surface charge density $q_s(\mathbf{x})$ on S_B as $q(\mathbf{x}) = -q_s(\mathbf{x})/\varepsilon_E$, where ε_E is the permittivity of the external medium. Using the method of Green's functions (Marshall, 2001, chapter 12) yields

$$\Phi(\mathbf{x}) - \Phi_0(\mathbf{x}) = \int_{S_B} q(\mathbf{x}') \, G(\mathbf{x} - \mathbf{x}') \, da', \tag{8.4.1}$$

where $\Phi_0(\mathbf{x})$ is the potential generated by a prescribed external field $\mathbf{E}_0 = -\nabla\Phi_0$ and $G = G(\mathbf{x} - \mathbf{x}')$ is the Green's function of the Poisson equation in three-dimensional space, given by

$$G(\mathbf{x} - \mathbf{x}') = -\frac{1}{4\pi \, |\mathbf{x} - \mathbf{x}'|}. \tag{8.4.2}$$

The source sheet strength is related to the jump in the gradient of the potential across S_B, and can be written as

$$q(\mathbf{x}) \equiv \partial(\Phi_E - \Phi_I)/\partial n, \tag{8.4.3}$$

where Φ_I and Φ_E are values of the electrostatic potential just inside and just outside S_B, respectively, as shown in Figure 8.10. Taking the normal derivative of (8.4.1) and evaluating this equation at a point $\mathbf{x}$ on the surface S_B yields an integral equation for $q(\mathbf{x})$ as

$$\frac{1}{2}(\partial\Phi_E/\partial n + \partial\Phi_I/\partial n) - \nabla\Phi_0 \cdot \mathbf{n} = \int_{S_B} q(\mathbf{x}')\frac{\partial G}{\partial n}\, da', \tag{8.4.4}$$

where $\partial\Phi/\partial n$ on the boundary S_B is set equal to the average of its value on either side of the boundary. However, this problem specification is not complete, because as shown in (8.4.3), $q(\mathbf{x})$ is related to the normal gradients of Φ_I and Φ_E.

The next step in the formulation of the problem depends on whether the material in region V_I is a dielectric or a conductor. For convenience, in the following discussion we assume that the material in the exterior region V_E is a dielectric. As noted in Section 8.1 the boundary conditions for conducting and dielectric bodies are different. In particular, the surface of a conducting body is everywhere equipotential, such that for any point $\mathbf{x}$ on the bounding surface S_B of a conducting body the potential satisfies the Dirichlet boundary condition $\Phi = \Phi_B$, where Φ_B is a constant value. Consequently, when evaluated at a point $\mathbf{x} \in S_B$ on the surface of a conducting body, (8.4.1) becomes

$$\Phi_B - \Phi_0(\mathbf{x}) = \int_{S_B} q(\mathbf{x}')G(\mathbf{x} - \mathbf{x}') \, da'. \tag{8.4.5}$$

Employing the three-dimensional Green's function and using $q(\mathbf{x}) = -q_s(\mathbf{x})/\varepsilon_E$, (8.4.5) becomes

$$\Phi_B - \Phi_0(\mathbf{x}) = \int_{S_B} \frac{q_s(\mathbf{x}')}{4\pi\,\varepsilon_E r} \, da', \tag{8.4.6}$$

where $r = |\mathbf{x} - \mathbf{x}'|$. Equation (8.4.6) is a Fredholm integral equation of the first kind for $q_s(\mathbf{x})$. For problems such as determination of the potential field around an electrode, it can be assumed that the value of surface potential Φ_B is known as part of the problem specification. However, in some situations the value of Φ_B might not be known *a priori*, such as for the case of an isolated conductor. In such cases, we might instead know the total charge q_B on the body surface S_B. This total charge can be used to pose an additional constraint as

$$q_B = \int_{S_B} q_s(\mathbf{x}')\, ds', \qquad (8.4.7)$$

which is sufficient to make the solution of (8.4.6) determinate.

By contrast, the potential function on the interface between two dielectric bodies, in the absence of a permanent surface charge, satisfies a Neumann boundary condition of the form

$$\varepsilon_E \partial \Phi_E / \partial n - \varepsilon_I \partial \Phi_I / \partial n = 0, \qquad (8.4.8)$$

which is a generalization to arbitrary bodies of the boundary condition given in (8.1.10). Substituting this boundary condition into (8.4.3), the surface sheet strength can be written as

$$q = [(\varepsilon_I - \varepsilon_E)/\varepsilon_I] \partial \Phi_E / \partial n. \qquad (8.4.9)$$

Substituting (8.4.9) into (8.4.4) gives a Fredholm integral equation of the second kind for $q(\mathbf{x})$ as

$$-\frac{1}{2}\frac{\varepsilon_E + \varepsilon_I}{\varepsilon_E - \varepsilon_I} q(\mathbf{x}) - \nabla \Phi_0 \cdot \mathbf{n} = \int_{S_B} q(\mathbf{x}') \frac{\partial G}{\partial n}\, ds'. \qquad (8.4.10)$$

The integral equations (8.4.6) for a conducting body and (8.4.10) for a dielectric body must be solved numerically to obtain the surface sheet strength $q(\mathbf{x})$, or equivalently the surface charge density $q_s(\mathbf{x})$. Equation (8.4.6) for conducting bodies and equation (8.4.10) for dielectric bodies can be discretized using flat panels on the body surface, using a procedure analogous to that discussed by Hess and Smith (1967) for aerodynamic applications. This discretization results in a matrix equation of the form $\mathbf{Aq} = \mathbf{b}$, where

$$A_{ij} = \int_{\Delta S_{B,j}} G(\mathbf{x}_i - \mathbf{x}_j)\, da, \quad b_i = \Phi_B - \Phi_0(\mathbf{x}_i) \text{ for conductors}, \qquad (8.4.11a)$$

$$A_{ij} = \int_{\Delta S_{B,j}} \left[\frac{\partial G}{\partial n}(\mathbf{x}_i - \mathbf{x}_j) + \frac{1}{2}\frac{\varepsilon_E + \varepsilon_I}{\varepsilon_E - \varepsilon_I}\delta_{ij} \right] da, \quad b_i = -\nabla \Phi_0 \cdot \mathbf{n} \text{ for dielectrics}.$$

$$(8.4.11b)$$

In this equation, δ_{ij} is the Kronecker delta, which is equal to 0 if $i \neq j$ and 1 if $i = j$. After $\mathbf{q}$ is obtained by solution of the matrix equation, the potential field at any point $\mathbf{x} \in V_I$ is obtained by substitution into (8.4.1) and evaluating the resulting integral using the panel discretization. The matrix $\mathbf{A}$ is dependent only on the relative location of the body panels. The electric field generated by charged particles or by the induced field of particle dipoles is accounted for in the earlier equations by the prescribed field Φ_0, which enters into the vector $\mathbf{b}$. Consequently, when BEM is applied to nondeforming bodies using the LU-decomposition method, the matrix $\mathbf{A}$ needs to be generated and decomposed only once at the beginning of the calculation.

8.4.2. Pseudoimage Method for Particles Near an Electrode Surface

When the BEM approach for computation of the electric field is used together with a computational method such as DEM for the particles, a number of problems with the combined computational approaches can occur that are related to the relative sizes of the particles and the panels used to discretize the surface S_B. These problems in combining BEM and DEM were examined by Liu et al. (2010), who also recommended a number of effective approaches for resolving these problems without significantly increasing the computational time or memory requirements.

The first problem arises from the fact that a particle located near a surface admits an image field over the surface, where this image field has a length scale on the order of the distance between the particle and the surface. As the particle approaches the surface, the length scale associated with this image field becomes progressively smaller, decreasing to a length on the order of the particle diameter for a particle touching the surface. If the particle diameter is much smaller than the size of the panels on the bounding surface S_B, as would typically be the case, the inability of the panel discretization to resolve the small length scale of the image field results in a considerable loss in accuracy of the electric field computation.

The resolution to this problem proposed by Liu et al. (2010) is based on the observation that if ℓ_{image} is the length scale associated with the image field, such that $\ell_{image} = O(d)$ as the particle approaches the surface, and κ is the mean curvature at some point on the surface S_B, then $\kappa \ell_{image} = O(\kappa d)$. Assuming that the particle size is sufficiently small that $\kappa d \ll 1$, we find that as the particle approaches the surface S_B, its image field becomes increasingly similar to the image of a particle over a flat surface. Put another way, if $\mathbf{E}_{image}$ is the electric field associated with the exact image field and $\mathbf{E}_{flat}$ is the image field associated with the same particle located at the same distance from a flat surface, then the characteristic length scale ℓ_{diff} of the difference field $\mathbf{E}_{image} - \mathbf{E}_{flat}$ must satisfy $\ell_{diff}/\ell_{image} = O(\kappa d)^{-1} \gg 1$ as the particle approaches the surface S_B.

The resolution for this problem proposed by Liu et al. (2010) is to add a pseudoimage field $\mathbf{E}_{flat}$ for a flat surface to the electric field $\mathbf{E}$ for all particles that lie within some distance $\ell_{crit} = O(\kappa^{-1})$ of S_B. For instance, for a flat conducting surface the image of a point charge is given simply by a charge of the opposite sign located at the reflection of the original charge (Jackson, 1962, 27). The problem of a point charge image over the boundary between two dielectrics was discussed by Sometani (2000). The difference between the exact image field $\mathbf{E}_{image}$ over the curved boundary S_B and the approximate image field $\mathbf{E}_{flat}$ for a flat surface is computed using the boundary element method, which is achieved simply by including the pseudoimage field in the prescribed potential Φ_0. Because the difference field has characteristic length scale $\ell_{diff} = O(\kappa^{-1})$, it can be well resolved by the surface panel discretization even as the particle moves very close to the body surface.

The utility of this pseudoimage method can be demonstrated by comparing the classic analytical solution for a point charge Q outside of a grounded conducting sphere with radius r_p (Figure 8.11a) to the numerically computed result obtained using BEM. If the point charge is located at a point $(x, y) = (-x_0, 0)$ and the sphere is centered at the origin, the exact image of Q has strength $Q_{I,exact} = r_p Q/x_0$ and is located at $(x_{I,exact}, y_{I,exact}) = (-r_p^2/x_0, 0)$. The distance between the point charge and

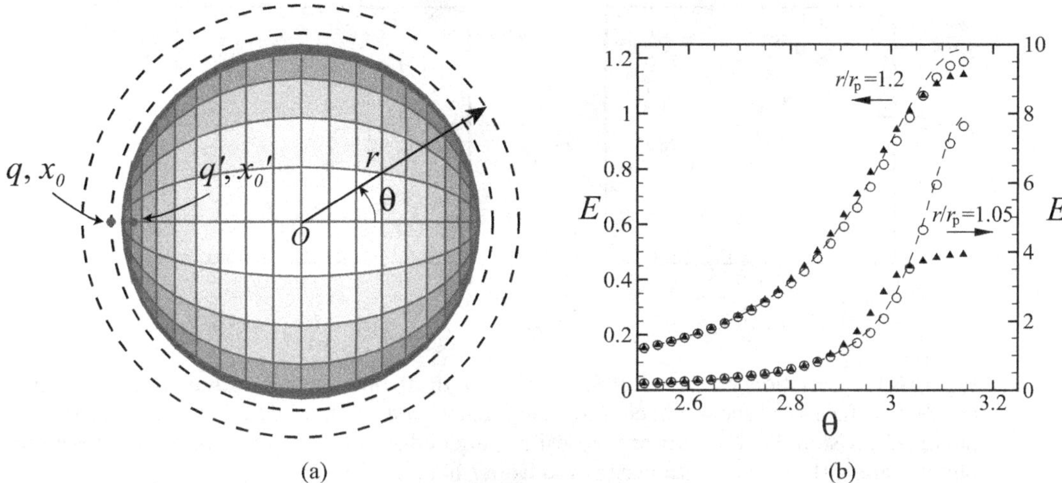

Figure 8.11. Results for a point charge q outside of a grounded conducting sphere of radius r_p: (a) two circular arcs used for validation of the pseudoimage method in the x-y plane; (b) comparison of the electric field magnitude on the two arcs computed both with the pseudoimage method (open circles) and with the standard BEM (filled triangles). The exact solution is indicated by a dashed line. [Reprinted with permission from Liu et al., 2010.]

the sphere surface is $b = x_0 - r_p$, where it is assumed that $b \ll r_p$. The electrostatic field is computed in three different ways: (a) analytically using the exact image field, (b) numerically using the standard boundary element method with no prescribed images, and (c) numerically using BEM with the pseudoimage approach. For the two numerical solutions, the sphere surface is approximated by 400 quadrilateral panels, with $\Phi_B = 0$. For standard BEM calculations, $\Phi_0(\mathbf{x})$ is set equal to the potential field of the point charge Q. For the pseudoimage method, $\Phi_0(\mathbf{x})$ is set equal to the sum of the potential field of the point charge Q plus an approximate image with strength $Q_I = -Q$ located at $x_I = -x_0 + 2b$.

The magnitude of the electric field is compared in Figure 8.11b for the three solutions described earlier as functions of angle θ along circular arcs with radii $r/r_p = 1.05$ and 1.2, where the point charge is located at angle $\theta = \pi$. For the arc at $r/r_p = 1.2$, both the standard BEM solution and the solution using the pseudoimage BEM are close to the exact solution. However, for the closer arc at $r/r_p = 1.05$, the standard BEM solution exhibits significant deviation from the exact result, giving a value of E at $\theta = \pi$ of approximately half the exact solution. By contrast, the pseudoimage method maintains excellent accuracy for all values of θ, even as we approach the point charge position at $\theta = \pi$. Even though the approximate source image location and strength do not correspond to the exact image location or strength, the difference between the electric fields induced by the exact and approximate images has a length scale that is on the order of the sphere radius, which is large compared to the panel size, and hence this difference field can be well approximated by the BEM approach.

8.4.3. Problems with DEP Force Near Panel Edges

An additional problem occurs as particles move close to panels placed on the surface of a curved body due to variation of the electric field along the panel surface,

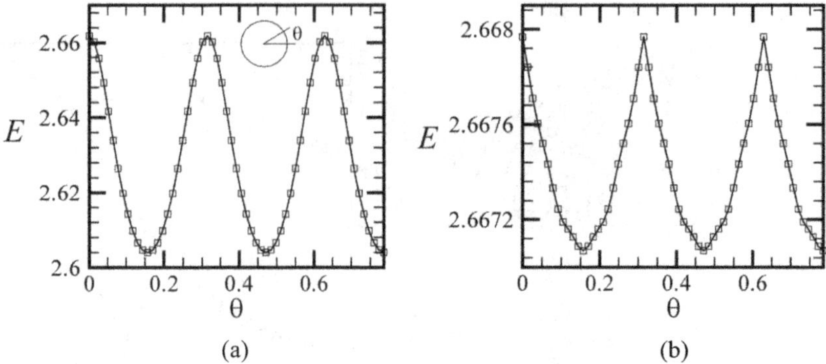

(a)   (b)

Figure 8.12. Variation of the electric field strength at one particle radius away from the cylinder surface as a function of circumferential direction, computed (a) without and (b) with panel subdivision. Peak points are closest to panel edges while trough points are closest to panel centers. [Reprinted with permission from Liu et al. (2010).]

particularly near the panel edges. This variation of the electric field due to panel discretization is illustrated in Figure 8.12a, which shows the magnitude E of the electric field vector $\mathbf{E}$ for the problem of a charged cylinder of unit radius with a constant surface potential in an unbounded domain. There is no imposed electric field far away from the cylinder, so the resulting value of E should be independent of angle around the cylinder. The cylinder surface is discretized in the circumferential direction using 20 panels with uniform size, with panel edges located at angles $\theta_n = n\pi/10$, where $n = 0, 1, \ldots, 19$. The electric field is calculated on a circle with radius $r = 1.05$, and the resulting value of E is plotted against θ in Figure 8.12a. Instead of being constant, the value of E is observed to change periodically with distance over the surface, with maximum values coinciding with positions of the panel edges and minimum values coinciding with the panel centers. Particles placed in this electric field would tend to move toward the panel edges for cases with positive DEP and toward the panel centers for cases with negative DEP. Of course, particle drift along the panels is mitigated by the particle inertia and, for adhesive particles, by the particle adhesive forces that bind particles to the panel surfaces. In the case of positive DEP, the particle is also attracted to the dipole and, for a charged particle, to the monopole image over the panel surface, which may be sufficiently strong to overcome the tendency to drift along the panel. The percentage variation in electric field strength along the panel decreases with decrease in the product $\ell_{panel}\kappa$, where ℓ_{panel} is the arc length of the panels and κ is the surface curvature. This electric field fluctuation decays with distance on the order of ℓ_{panel} from the body surface, so this error is primarily important for particles colliding with the body.

The fluctuations in electric field strength are caused by the fact that the panels do not exactly lie on the surface of the body, but instead only the panel vertices are on the body surface. This edge effect can be reduced by increasing the number of panels, which has the effect of reducing the change in angle between panels. Of course, since the computational cost of the BEM matrix inversion with N panels varies with $O(N^3)$, assuming direct matrix inversion, increase in the number of panels rapidly increases computational time and memory requirements. Rather than increase panel numbers, Liu et al. (2010) proposed a *panel subdivision method* that resolves the problem with negligible additional computational cost. In this method, the original panels along the body surface are used to compute the electric field on particles that are not close to the body. For near-surface particles, the panels within the region of the surface closest to each near-surface particle (called the "near-surface region") are identified. These "parent" panels are each subdivided into "child" panels, each of which has the same strength as the corresponding parent panel. However, instead of lying on the plane of the parent panels, the vertices of the child panels are moved to lie on the body surface S_B. The electric field on each near-surface particle is calculated using the child panels for panels within the near-surface region and the original parent panels for panels outside of this region. This subdivision process significantly reduces the panel edge error by bringing the location of the panel points closer to the body surface, without the expense of recomputing the panel strengths for the subpanels.

Figure 8.12 shows the variation in E with θ for cases both with and without subdivision of the near-surface panels for the constant-potential cylinder problem, where five child panels are used in the circumferential direction for each parent panel. Comparison of Figure 8.12a and 8.12b shows that the panel subdivision procedure reduced the peak-to-peak variation of the electric field magnitude by a factor of nearly sixty compared to the case with no subdivision.

8.5. Fast Multipole Method for Long-Range Forces

As noted in the previous section, the presence of charged particles within the computational domain induces an electric field that decays slowly with distance away from each particle. The electric field induced by a charged particle subject to an external electric field can be approximated using the effective moment method by the sum of a monopole and dipole field, given respectively by $\mathbf{E}_{monopole} = q\mathbf{r}/4\pi\varepsilon r^3$ and $\mathbf{E}_{dipole} = -\nabla(\mathbf{p}\cdot\mathbf{r}/4\pi\varepsilon r^3)$. Here q and $\mathbf{p}$ denote the particle charge and dipole moment, respectively, $\mathbf{r}$ is the vector from the particle centroid to a space point $\mathbf{x}$, and $r = |\mathbf{r}|$ is the distance from the particle centroid to the space point $\mathbf{x}$. The magnitude of the electric field vector induced by a dipole decays as $O(r^{-3})$, which is sufficiently fast that dipoles beyond a relatively small distance from the space point $\mathbf{x}$ can be ignored. The electric field magnitude for an electrostatic source (monopole) decays with distance as $O(r^{-2})$, which is slow enough that a large group of charged particles will generate a significant electric field even at fairly large distances. For a system with N particles, the cost of computing the pair-wise electrostatic interactions to determine the electric field strength at all particle locations varies in proportion to $O(N^2)$, which for large N can easily become the limiting factor in determining the computation speed. This section introduces the *fast multipole method* (FMM), which is a popular and effective approach for accelerating the computation of the electric field induced by charged particles. This method is also used in other fields governed by a convolution integral with a kernel that decreases slowly in space, such as computation of gravitational interactions of massive bodies or the computation of fluid flows using vortex methods.

The objective of the fast multipole method is to obtain an approximation for the electric field induced by a group of particles (the source) at a "target" point $\mathbf{x}$ located sufficiently far from the source charges. Let us assume that the various charged particles within the flow field are sorted into some box structure, so that at a given level of the box structure all particles are contained in exactly one box.

Let us consider a box numbered ℓ within this box structure that contains N_ℓ charged particles and has centroid located at $\hat{\mathbf{x}}_\ell$. If $q(\mathbf{x}_i)$ is the charge of a particle located at position $\mathbf{x}_i$ within box ℓ and the interaction kernel is denoted by $\mathbf{K}(\mathbf{r}_\ell)$ $(= \mathbf{r}_\ell/4\pi\varepsilon\, r_\ell^3)$, where $\mathbf{r}_\ell \equiv \mathbf{x} - \hat{\mathbf{x}}_\ell$ and $r_\ell = |\mathbf{r}_\ell|$, the electric field $\mathbf{E}_\ell(\mathbf{x})$ generated by the N_ℓ source particles at a target point $\mathbf{x}$ can be expressed in terms of a multipole expansion as

$$\mathbf{E}_\ell(\mathbf{x}) = \sum_{m=0}^{+\infty}\sum_{n=0}^{+\infty}\sum_{k=0}^{+\infty} \frac{(-1)^{m+n+k}}{m!\,n!\,k!} I_{\ell,mnk} \frac{\partial^{m+n+k}}{\partial x^m \partial y^n \partial z^k} \mathbf{K}(\mathbf{x} - \hat{\mathbf{x}}_\ell). \tag{8.5.1}$$

The box moments $I_{\ell,mnk}$ of a box are defined by

$$I_{\ell,mnk} = \sum_{i=1}^{N_\ell} Q_i(x_i - \hat{x}_\ell)^m (y_i - \hat{y}_\ell)^n (z_i - \hat{z}_\ell)^k. \tag{8.5.2}$$

The box moments are independent of the target point location, and can thus be evaluated once per time step for each box. On the other hand, the interaction kernel $\mathbf{K}(\mathbf{x} - \hat{\mathbf{x}}_\ell)$ depends only on the difference between the box centroid location $\hat{\mathbf{x}}_\ell$ and the target point location $\mathbf{x}$, and it is independent of the number of particles in the box.

The speed-up from the multipole acceleration approach comes about from the fact that there are many fewer boxes than there are charged particles. For instance, suppose there are a total of N particles contained in M boxes at a given level of the box structure, where on average $\bar{N}_\ell = N/M$. For each time step we desire to find the value of $\mathbf{E}$ on each particle induced by all of the other particles. Using the multipole acceleration procedure, the first step is to determine the box number in which each particle is contained. There are several ways to do this sorting procedure, depending on the box structure, but with an efficient algorithm and a standard box structure, it usually can be accomplished in $O(N)$ computations. Second, we compute the box moments $I_{\ell,mnk}$ for each box, which also requires $O(N)$ computations. Third, we compute and store the derivatives of the interaction kernel $\mathbf{K}(\mathbf{x} - \hat{\mathbf{x}}_\ell)$ that appear in (8.5.1) for each box. Although the range of the indices m, n, and k in (8.5.1) is unbounded, in a computation these indices would be limited such that $m + n + k \leq H$, where a small value of H is typically assumed so that only a few terms of the multipole expansion are used. Therefore omitting the factor of H, computation of the derivatives requires $O(M)$ computations for each target point. A tree-code boxing structure was proposed by Barnes and Hut (1986) in which the value of M increases in proportion to $\log N$, and the resulting fast multipole method therefore requires a computation count of $O(N \log N)$. Of course, the FMM has a much higher overhead than does the direct calculation $O(N^2)$ approach, and so we typically find that the method is only worthwhile compared with direct calculation if N is sufficiently large.

In the FMM presented in this section the electric field is computed at each particle. As shown by Greengard and Rokhlin (1987), additional speed-up can be gained for large numbers of particles by also using a local expansion procedure, in which the electric field is computed on a grid covering the flow and then interpolated onto the particles via a local Taylor expansion. This local expansion procedure reduces the operational count for the FMM to $O(N)$, but unless N is very large the speed-up gained with the local expansion approach may not be worth the additional overhead and coding complexity.

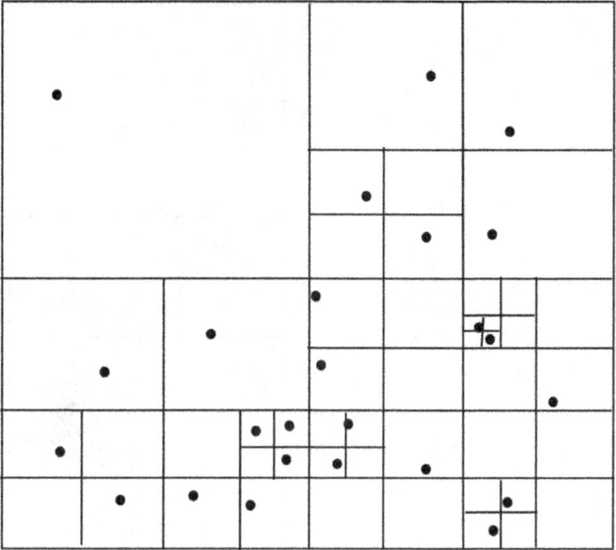

Figure 8.13. Illustration of the Barnes-Hut box family with $n_p = 1$.

The fast multipole method can be optimized by using a theoretical error bound developed by Salmon and Warren (1994) for an H-order multipole expansion, in which the absolute error Δe_H in the expansion is found to satisfy the bound

$$\Delta e_H(\mathbf{x}) \leq \frac{1}{(d_\ell - b_{\max})^2} \left[(H+2)\frac{B_{H+1}}{d_\ell^{H+1}} - (H+1)\frac{B_{H+2}}{d_\ell^{H+2}} \right], \qquad (8.5.3)$$

where $d_\ell \equiv |\mathbf{x} - \mathbf{x}_l|$ is the distance between the target point and the centroid of box ℓ, $b_{\max}$ is the maximum distance between the box centroid and any other point in the box, and B_H is defined by

$$B_H \equiv \sum_{i=1}^{N_l} q_i |\mathbf{x}_i - \mathbf{x}_l|^H. \qquad (8.5.4)$$

By specifying the computational precision $\Delta e_H(\mathbf{x})$ and the multipole order H, (8.5.3) can be used to determine the lower limit of d_ℓ for which to apply an H-order multipole expansion and the maximum size of the box that can be used to achieve the stated precision.

Prior to implementation of the fast multipole method, it is necessary to form a family of boxes, such that all particles are contained within at most one box at each level of the box family. Many different types of box families are used by different investigators, but two primary examples are discussed here. An illustration of the Barnes-Hut (1986) algorithm is shown in two dimensions in Figure 8.13. This algorithm starts by placing a single box over the flow field, which constitutes the first level of the box tree-structure. At the next level, the box is divided into four equal-area parts in two dimensions, or eight equal-volume parts in three dimensions. We call the original box the "parent" box and the different smaller boxes after subdividing the "child" boxes. The number of particles in each child box is now counted, and any child box that contains more than a prescribed number n_p particles is again divided into four parts. The procedure is repeated until all boxes at the

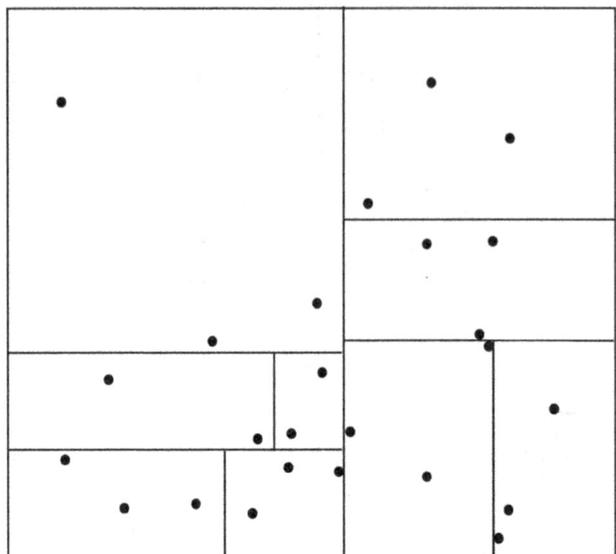

Figure 8.14. Illustration of the Clarke-Tutty box family, with the same points as in Figure 8.13 and $n_p = 3$.

smallest level contain n_p particles or less. Barnes and Hut set $n_p = 1$, but it is often convenient to use a higher value, say, $n_p = 20$–100, so as to reduce the number of boxes. Structured box families such as in this example have the advantage that they are easy to construct. The disadvantage of this box family is that the number of particles per box can vary widely, including generation of some boxes with no particles. This can particularly become a problem for highly clustered particles, as is often observed in the presence of particle adhesive force.

An alternative example is the Clarke-Tutty (1994) box family, shown in Figure 8.14, which uses a tree structure that is designed to maintain approximately the same number of particles in each box at each level of the structure. The Clarke-Tutty box family can be generated by a recursive approach in which the particles within the flow field are initially contained within a single box, which constitutes the first level of the box structure. This box is then divided along the largest dimension into two "child" boxes, in such a manner that each child box contains approximately an equal number of particles. The procedure is repeated until the number of particles in each box is approximately equal to a prescribed value n_p.

Although the Clarke-Tutty box family requires a slightly more complex algorithm to form, the resulting box family is well suited for parallel processing applications and for use with optimized fast multipole methods due to the fact that it possesses the minimal number of boxes for a give value of n_p and the property that each box at each level has approximately the same number of particles. For example, the set of points shown in Figure 8.13 requires 49 boxes using a Barnes-Hut box structure with $n_p = 1$, of which 19 are empty. The same set of points in Figure 8.14 with a Clarke-Tutty box structure and $n_p = 3$ requires nine boxes, where at the smallest level each box contains either two or three particles. Had we conducted the Barnes-Hut division in Figure 8.13 with $n_p = 3$, we would have required 16 boxes. Although these differences might not seem large for a small number of particles, the differences in number of boxes can become quite significant with a large number of particles, particularly in three dimensions with highly clustered particles.

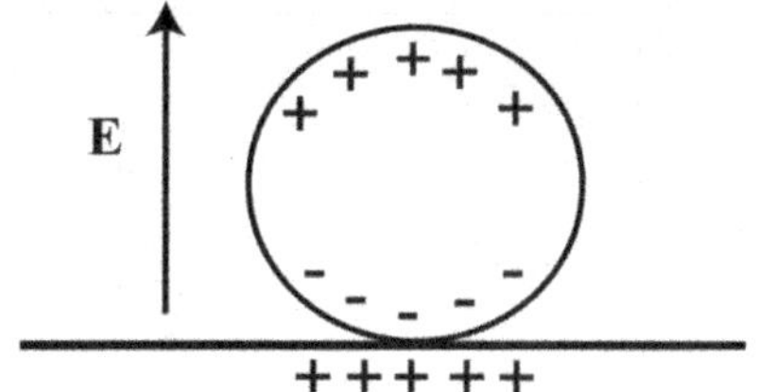

Figure 8.15. Schematic diagram of a particle attached to a planar surface with an upward electric field.

8.6. Electrostatic Agglomeration Processes

Electric field forces between oppositely charged particles will move the particles toward one another. Even for uncharged particles placed in an electric field, DEP force between a particle and surrounding particles will make the particles adhere together. A similar phenomenon occurs with DEP force between a particle and its image over a conducting or dielectric surface. This electrostatic adhesion force acts in concert with the van der Waals adhesion force discussed in Chapter 4; however, the electrostatic force has a very long range compared to the van der Waals force.

8.6.1. Relative Importance of Electrostatic and van der Waals Adhesion Forces

The question of the relative importance of electrostatic and van der Waals forces on particle adhesion on a substrate has long been an issue of debate in the literature. Although it might sound academic, the problem is of a great deal of importance for processes that require control of particle adhesion. For instance, it is necessary to remove charged particles from the surface of semiconductor wafers during the manufacturing process so as to avoid unwanted electrostatic discharge. Electrostatic removal processes would be effective for cleaning particles from the wafer if electrostatic forces dominate particle adhesion, but these same measures would make adhesion by van der Waals forces much worse. Similar control problems arise in electrophotographic printing processes, in which it is necessary to electrostatically deposit and remove particles from the paper during different parts of the process (Rimai et al., 2006).

Different studies examining the problem of electrostatic particle attachment to a plane surface have been performed using different techniques to measure the adhesion force, including atomic force microscopy (Zhou et al., 2003). Mathematical models for electrostatic adhesion processes (Rimai and Quesnel, 2002; Feng and Hays, 2003; Tang et al., 2006) typically include three types of electrostatic forces involved in particle adhesion: (1) the Coulomb force, (2) the attraction between a charged particle and its image charge (of the opposite sign) over the surface, and (3) DEP attraction between the induced dipole on the particle and its image over the surface. In the presence of an upward electric field, as shown in Figure 8.15, expressions for these three terms can be written as (Feng and Hays, 2003)

$$F_{E,adh} = -\frac{\alpha_1 q^2}{4\pi\varepsilon d^2} + \alpha_2 qE - \frac{1}{4}\alpha_3 \pi\varepsilon d^2 E^2, \tag{8.6.1}$$

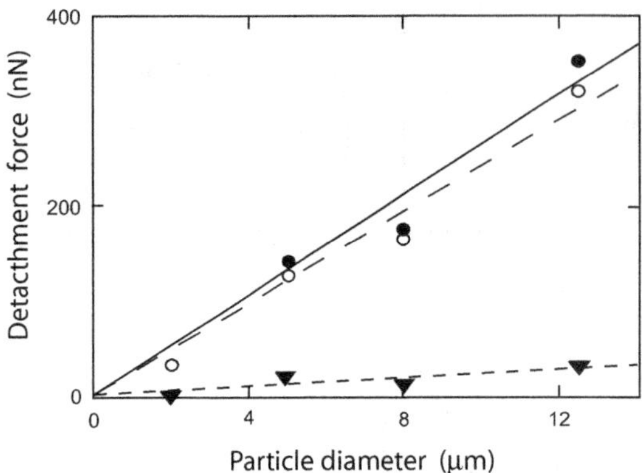

Figure 8.16. Plot of detachment force for a particle from a plane surface as a function of particle diameter, showing total detachment force (filled circles, solid line), van der Waals force (open circles, long dashed line), and electrostatic force (filled triangles, short dashed line). For $d = 2\,\mu$m, the electrostatic force was so small that the total and van der Waals forces are both shown as a single open circle. [Reprinted with permission from Rimai and Quesnel (2002).]

where the coefficients α_1, α_2, and α_3 are functions of the dielectric coefficients of the two media. A negative term in (8.6.1) is attractive and a positive term is repulsive. The critical van der Waals force for detachment of a particle from a surface is given by JKR theory by (4.2.37), or

$$F_C = \tfrac{3}{2}\pi \gamma\, d. \tag{8.6.2}$$

If the particle surface charge density q_s is assumed to be constant, then the particle charge q can be written as a function of diameter d as

$$q = \pi\, q_s d^2. \tag{8.6.3}$$

Substituting (8.6.3) into (8.6.1), all three terms of the electrostatic force are found to vary in proportion to d^2. By contrast, the critical van der Waals force in (8.6.2) varies linearly in d.

As a consequence of the linear dependence of van der Waals force and the quadratic dependence of electrostatic force on d, it is expected that the van der Waals force would dominate for sufficiently small particles and the electrostatic force would dominate for sufficiently large particles. A study of the dependence of the detachment force of charged particles from a surface in an electric field on the particle diameter was reported by Rimai and Quesnel (2002). Data are shown in Figure 8.16 for the total detachment force, the van der Waals force, and the electrostatic force as a function of particle diameter in the range $2\,\mu$m $\leq d \leq 13\,\mu$m. The total detachment force in this range is found to vary approximately linearly with d and to be primarily due to the van der Waals force. Other studies, such as Zhou et al. (2003), noted that the relative importance of van der Waals and electrostatic force for adhesion depends critically on factors such as surface roughness, where rough surfaces can have significantly lower van der Waals forces than smooth surfaces without significant change in electrostatic force.

8.6.2. Particle Chain Formation

Uncharged particles immersed in an electric field exhibit an induced dipole moment that arises from charge polarization, as discussed in Section 8.1. The dipole moments from one particle can interact strongly with the induced dipole moments from surrounding particles to influence the structure of agglomerates formed of the particles. The net effect of this interaction is to cause the particles to align into a chain-like

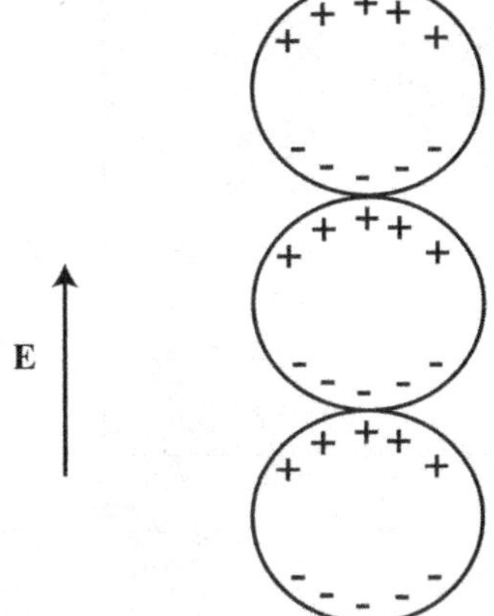

Figure 8.17. Schematic diagram of chain-like agglomerate formed of uncharged particles in an electric field **E**, with particles lined up such that the induced dipoles of neighboring particles are aligned.

structure, in which the positively charged side of one induced dipole is attracted to the negatively charged side of the induced dipole on a neighboring particle, and vice versa, as illustrated in Figure 8.17. The electric field strength is high in the regions between the particles where the negative and positive sides of the dipoles meet, and it is low along the sides of the particles.

If the dipole moment on a given particle (labeled particle A in Figure 8.18a) is given by $\mathbf{p} = p\mathbf{e}_y$, the electrostatic potential field generated by the dipole can be written in polar coordinates as

$$\Phi = \frac{p\cos\theta}{4\pi\varepsilon_f r^2}.\tag{8.6.4}$$

The electrostatic force **F** acting on a second particle (labeled particle B in Figure 8.18a) located at position (r, θ) has components

$$F_r = -\frac{3p^2}{4\pi\varepsilon_0 r^4}(3\cos^2\theta - 1), \qquad F_\theta = -\frac{6p^2}{4\pi\varepsilon_0 r^4}\sin\theta\cos\theta.\tag{8.6.5}$$

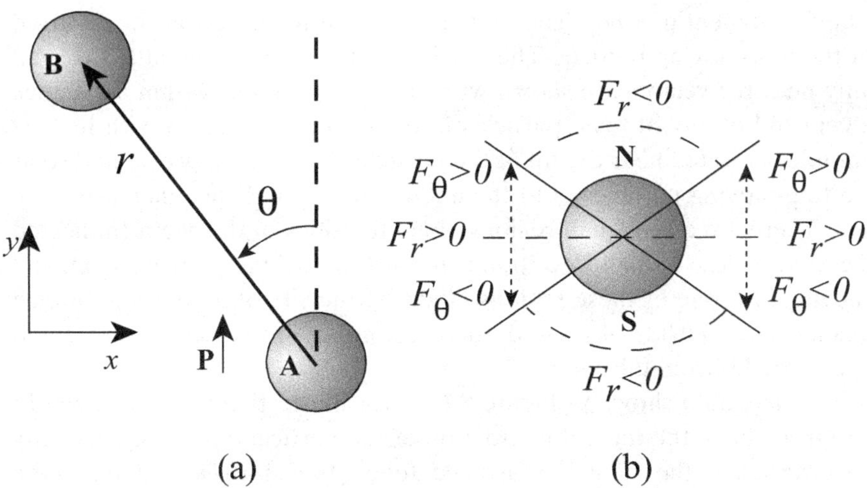

Figure 8.18. Schematic illustrating the sign of the interaction force between two polarized particles with the same dipole moment $\mathbf{p} = p\mathbf{e}_y$: (a) coordinate system; (b) regions with positive and negative signs of the radial force F_r.

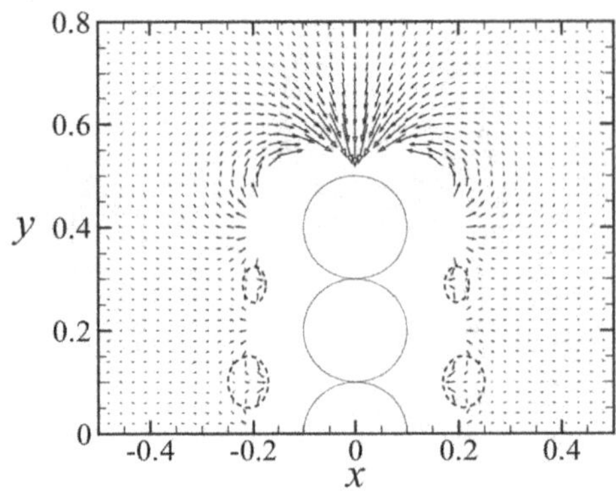

Figure 8.19. Numerical computation of the force vector that would be imposed on a particle, with force vectors drawn at the particle centroid location, near the tip of a particle chain. [Reprinted with permission from Liu et al. (2010).]

The radial component F_r is negative within a conical region with boundary $\cos\theta \sim \pm \sqrt{1/3}$ located both above and below the particle, labeled N and S in Figure 8.18b. In this figure, $\theta = 0$ at the N polar axis. Outside of these two cones, F_r is positive. A second particle B lying in either of the regions N or S will be attracted toward the original particle A, whereas a particle B lying outside of these conical regions will be repelled from particle A. The tangential force component F_θ is oriented so as to always push particle B toward the equatorial plane of particle A. If A and B collide as particle B is drawn inward at a poles of A, the van der Waals adhesion force between the two particles can be sufficient to overcome the tangential electrostatic force. It is noted that this simple discussion does not account for the fact that the induced dipoles of the two particles will also be influenced by the mutual interaction of the particles, and hence will not point exactly in the y-direction. Nevertheless, this simple thought experiment illustrates how one particle might be drawn toward the poles of a second particle and be repelled from the equatorial region of the particle.

The electrostatic force $\mathbf{F}(x, y)$ that would be exerted on a free-moving uncharged particle traveling near the tip of a particle chain is plotted in Figure 8.19, from Liu et al. (2010). The particle chain in this example is forced from five particles aligned with the direction of an external electric field, and all particles are required to have the same dipole moment $\mathbf{p} = p\mathbf{e}_y$. The plot shows the force vector at the centroid position of the free-moving particle. The particle radius r_p is set equal to 0.1, and consequently no force vectors are shown with centroid positions within a distance $2r_p$ of the centroid of any of fixed particles in the chain. The region with highest force magnitude is located near the tip of the particle chain, and it is oriented so as to draw the free-moving particles onto the end of the chain, hence increasing the chain length. There are a few small regions along the sides of the chain (indicated by dashed curves) where a small force points inward toward the particle chain, and particles moving into one of these regions might be the site of a dendritic branch from the main chain. Outside of these dashed regions, the force vector is oriented so as to repel particles from the chain.

The force calculation shown in Figure 8.19 is for illustrative purposes only. In an actual particle chain, the mutual interaction of the particle dipoles significantly alters the electric field of the chain. For instance, Jones (1995, 146) estimated that the effective dipole moment of each particle in a chain of length N increases by a factor of 2.4 for $N = 2$, 4.0 for $N = 3$, and 5.95 for $N = 4$ compared with what it would have been for a particle exposed only to the ambient electric field. Of course, this

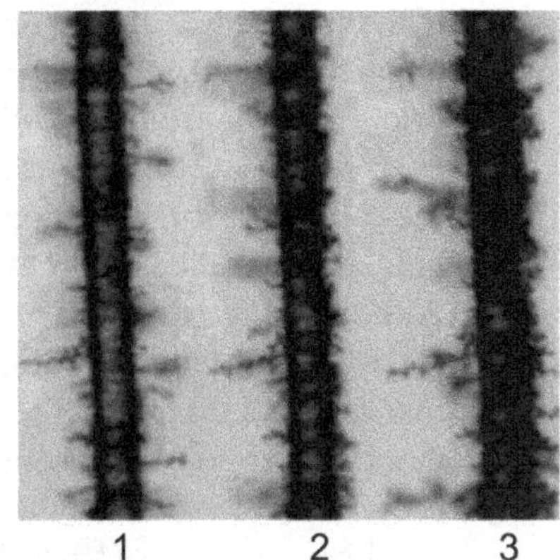
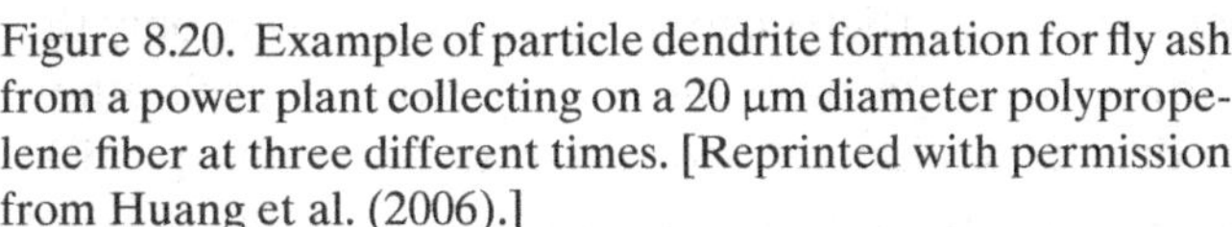

Figure 8.20. Example of particle dendrite formation for fly ash from a power plant collecting on a 20 μm diameter polypropelene fiber at three different times. [Reprinted with permission from Huang et al. (2006).]

mutual interaction effect is included in most DEM simulations, because as discussed previously in this chapter the induced field of each particle is computed based on the sum the electric field generated by the electrodes and that generated by the other particles, where the latter includes both monopole sources from charged particles and induced dipole sources.

Because the chain-like particle agglomerates are observed to form with either single-chain or branched structures, they are sometimes referred to as *pearl chains* or *dendrites*, respectively. An example of particle chain formation is shown in Figure 8.20 for the problem of collection of polarized fly ash particles in air on a cylindrical fiber oriented orthogonal to the ambient velocity and electric field vectors (Huang et al., 2006). The particles are observed to form dendritic structures, which primarily extend outward from the cylinder but also exhibit multiple branches as they extend away from the fiber. Examples of particle chain formation have also been observed in DEM simulations. For instance, particle capture by a cylinder in the presence of an external electric field has been examined using DEM by Park and Park (2005) and Liu et al. (2010). Simulation results of electrostatic particle capture by a cylinder from Liu et al. (2010) are shown in Figure 8.21. These simulations employ a combination of the pseudoimage BEM approach for the electric field,

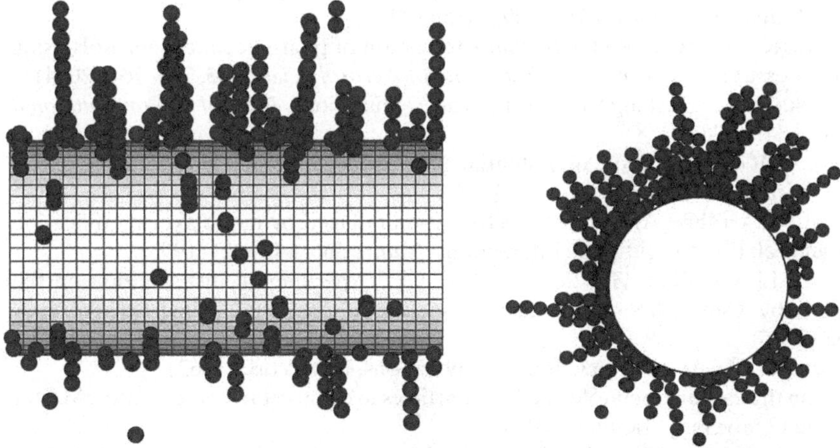

Figure 8.21. Results of DEM simulations of Liu et al. (2010) for chain formation of uncharged particles in an air flow past a cylinder subject to a uniform external electric field. [Reprinted with permission from Liu et al. (2010).]

including particle images and particle mutual electrostatic interactions. The presence of particle chain formation is quite noticeable in the figure, with preferential direction oriented away from the cylinder.

Finally, we would be remiss if we did not note that the simple assumption that the electrostatic interaction between the particles can be represented using only the induced dipole fields breaks down as the particles move very close to each other. Although this assumption produced qualitatively correct results for the particle electrostatic agglomeration, as we have seen, many investigators have pointed out that the induced field for touching particles is not accurately represented by a simple monopole-dipole representation, but instead higher-order multipoles must also be used. Details of the higher-order calculation can be found in references such as Washizu and Jones (1996a,b), Techaumnat and Takuma (2003), and Nakajima and Matsuyama (2002). Collision of electrically charged particles in a surrounding gas or liquid medium can also exhibit complications due to the electrical breakdown of the fluid if the electric field strength between the particles exceeds the Paschen limit, resulting in formation of an arc between the particles (Jones and Saha, 1990). However, such phenomena are outside the scope of the present text.

REFERENCES

Apodaca MM, Wesson PJ, Bishop KJM, Ratner MA, Grzybowski BA. Contact electrification between identical materials. *Angewandte Chemie-International Edition* **49**, 946–949 (2010).
Barnes JE, Hut P. A hierarchical O(NlogN) force calculation algorithm. *Nature* **324**(4), 446–449 (1986).
Brocilo D, Eng B, Eng M. Electrode geometry effects on the collection efficiency of submicron and ultrafine dust particles in wire-plate electrostatic precipitators. PhD dissertation, McMaster University, pp. 41–61 (2003).
Chen D, Du H. A dielectrophoretic barrier-based microsystem for separation of microparticles. *Microfluidics and Nanofluidics* **3**, 603–610 (2007).
Clarke NR, Tutty OR. Construction and validation of a discrete vortex method for the two-dimensional incompressible Navier-Stokes equations. *Computers Fluids* **23**(6), 751–783 (1994).
Feng JQ, Hays DA. Relative importance of electrostatic forces on powder particles. *Powder Technology* **135**–136, 65–75 (2003).
Fiedler S, Shirley SG, Schnelle T, Fuhr G. Dielectrophoretic sorting of particles and cells in a microsystem. *Analytical Chemistry* **70**, 1909–1915 (1998).
Glover W, Chan HK. Electrostatic charge characterization of pharmaceutical aerosols using electrical low-pressure impaction (ELPI). *Journal of Aerosol Science* **35**, 755–764 (2004).
Greengard L, Rokhlin V. A fast algorithm for particle simulations. *Journal of Computational Physics* **73**, 325–348 (1987).
Hess JL, Smith AMO. Calculation of potential flow about arbitrary bodies. *Progress in Aerospace Sciences* **8**, 1–138 (1967).
Horn RG, Smith DT, Grabbe A. Contact electrification induced by monolayer modification of a surface and relation to acid–base interactions. *Nature* **366**, 442–443 (1993).
Huang B, Yao Q, Li SQ, Zhao HL, Song Q, You CF. Experimental investigation on the particle capture by a single fiber using microscopic image technique. *Powder Technology* **163**, 125–133 (2006).
Jackson JD. *Classical Electrodynamics*. John Wiley & Sons, New York (1962).
Jin X. Research on the capture technology of fine particles in electrostatic precipitator. Master Thesis, Tsinghua University, Beijing (2013).
Jones TB. *Electromechanics of Particles*. Cambridge University Press, Cambridge UK (1995).
Jones TB, Saha B. Nonlinear interactions of particles in chains. *Journal of Applied Physics* **68**(2), 404–410 (1990).

Liu G, Marshall JS, Li S-Q, Yao Q. Discrete-element method for particle capture by a body in an electrostatic field. *International Journal for Numerical Methods in Engineering* **84**(13), 1589–1612 (2010).

Marshall JS. *Inviscid Incompressible Flow*, John Wiley & Sons, New York (2001).

Matsusaka S, Masuda H. Electrostatics of particles. *Advanced Powder Technology* **14**(2), 143–166 (2003).

McCarty LS, Whitesides GM. Electrostatic charging due to separation of ions at interfaces: Contact electrification of ionic electrets. *Angewandte Chemie-International Edition* **47**, 2188–2207 (2008).

Morgan H, Holmes D, Green NG. 3D focusing of nanoparticles in microfluidic channels. *IEE Proceedings – Nanobiotechnology* **150**(2), 76–81 (2003).

Nakajima Y, Matsuyama T. Electrostatic field and force calculation for a chain of identical dielectric spheres aligned parallel to uniformly applied electric field. *Journal of Electrostatics* **55**, 203–221 (2002).

Park HS, Park YO. Simulation of particle deposition on filter fiber in an external electric field. *Korean Journal of Chemical Engineering* **22**(2), 303–314 (2005).

Pauthenier MM, Moreau-Hanot M. Generateurs de haute tension a courants gazeux. *Journal De Physique* **5**, 193–196 (1937).

Rimai DS, Quesnel DJ. The adhesion of spherical particles: Contributions of van der Waals and electrostatic interactions. *The Journal of Adhesion* **78**, 413–429 (2002).

Rimai DS, Weiss DS, de Jesus MC, Quesnel DJ. Electrophotography as a means of microfabrication: The role of electrodynamic and electrostatic forces. *C. R. Chimie* **9**, 3–12 (2006).

Salmon JK, Warren MS. Skeletons from the treecode closet. *Journal of Computational Physics* **111**, 136–155 (1994).

Soh SL, Kwok SW, Liu HL, Whitesides GM. Contact de-electrification of electrostatically charged polymers. *Journal of the America Chemical Society* **134**, 20151–20159 (2012).

Sometani T. Image method for a dielectric plate and a point charge. *European Journal of Physics* **21**(6), 549–554 (2000).

Tang T, Hui C-Y, Jagota A. Adhesive contact driven by electrostatic forces. *Journal of Applied Physics* **99**, 054906 (2006).

Techaumnat B, Takuma T. Calculation of the electric field for lined-up spherical dielectric particles. *IEEE Transactions on Dielectrics and Electrical Insulation* **10**(4), 623–633 (2003).

Thomas SW, Vella SL, Kaufman GK, Whitesides GM. Patterns of electrostatic charge and discharge in contact electrification *Angewandte Chemie-International Edition* **47**, 6654–6656 (2008).

Washizu M, Jones T. Dielectrophoretic interaction of two spherical particles calculated by elquivalent multipole-moment method. *IEEE Transactions on Industry Applications* **32**(2), 233–242 (1996a).

Washizu M, Jones T. Generalized multipolar dielectrophoretic force and electrorotational torque calculation. *Journal of Electrostatics* **38**, 199–211 (1996b).

White HJ. Particle charging in electrostatic precipitation. *AIEE Transactions* **70**, 1186–1191 (1951).

Yu C, Vykoukal J, Vykoukal DM, Schwartz JA, Shi L, Gascoyne PRC. A three-dimensional dielectrophoretic particle focusing channel for microcytometry applications. *Journal of Microelectromechanical Systems* **14**(3), 480–487 (2005).

Zhang L, Tatar F, Turmezei P, Bastemeijer J, Mollinger JR, Piciu O, Bossche A. Continuous electrodeless dielectrophoretic separation in a circular channel. *Journal of Physics: Conference Series* **34**, 527–532 (2006).

Zhou H, Götzinger M, Peukert W. The influence of particle charge and roughness on particle-substrate adhesion. *Powder Technology* **135–136**, 82–91 (2003).

9 Nanoscale Particle Dynamics

Nanotechnology is expected to have major impacts in medicine, catalysis, electronics, and materials science. The basic science and engineering of nanoparticles increasingly play a central role in many engineering fields, including the environment (air pollution, climate change), energy utilization (nanoenergetic or nanocatalytic materials, fossil fuel combustion, and fly ash formation and capture), and biotechnology (medical diagnostics and drug delivery). Despite the fact that selective use of nanoparticles dates back to ancient times, the term *nanoparticle* appears to have first been used at the international symposium on small particles and inorganic clusters in the 1980s (Goesmann and Feldmann, 2010). In certain fields, nanoparticles are also called *ultrafine particles*, particularly in the older literature in energy and environmental engineering (Madler and Friedlander, 2007). Typically, a nanoparticle has a diameter ranging in size from about 1 to 100 nm, which is several orders of magnitude smaller than the micrometer-size particles that have been the focus of the majority of the book. This small size gives nanoparticles a number of unique properties and it significantly modifies some of the computational approaches that must be used to simulate nanoparticle systems. Modeling of nanoparticle interaction with a surrounding fluid is particularly influenced in cases where the particle size is of the same order of magnitude as, or smaller than, the mean free path of the fluid, at which point noncontinuum effects become important.

A topic of significant interest in the nanoparticle literature deals with methods for synthesis of nanoparticles with specific desired properties. There is a large literature on methods to produce particles with desired coatings, electric and magnetic properties, hollow and solid cores, various crystalline structures, and so on. A second area of great interest has to do with the self-assembly of nanoparticles into larger assemblages or agglomerates, as well as ways to control these self-assembly processes. Related to this area is the problem of how nanoparticles can be dispersed over a broad region during a manufacturing process with a desired agglomerate size and structure. Meanwhile, the rapid development of nanotechnology requires us to develop more strict control devices for high-efficiency capture of nanoparticles, which is important considering their serious environmental and health effects.

All of these issues benefit from the ability to accurately model and simulate nanoparticle motion and agglomeration under different flow fields. This chapter describes various types of interparticle interactions (van der Waals, electrostatic, etc.) and transport phenomena (drag, diffusion, thermophoresis, etc.) at the nanoscale, which are necessary for accurate modeling of nanoparticle flows at a particle scale. The focus of the chapter is on nanoparticle aerosols. This choice is made in part because the mean free path of many gases is on the same scale as the nanoparticle diameter, whereas that of a liquid is of the scale of the molecular radius, which is of the order of 0.1 nm. Consequently, nanoparticles in a liquid can generally be modeled using standard continuum theory, whereas nanoparticles in a gas require special treatment to account for noncontinuum effects. Another reason to focus on nanoparticle aerosols is that many of the most interesting nanoparticle transport problems occur in aerosols. In particular, many commodity nanoparticle products are made by flame aerosol processes including carbon blacks, fumed silica, pigmentary titania, and optical fibers. Compared with wet chemistry approaches (e.g., sol–gel), the aerosol technology has so far proved advantageous for manufacture of commercial quantities of nanoparticles due to the high-throughput production, the short processing time, and the simplicity of the one-step process (Madler and Friedlander, 2007). Nanoparticles with complex composition and most mixed oxide ceramics can be manufactured more cheaply and with less environmental harm by aerosol technology than by wet-chemistry processes (Feng et al., 2006).

The first section of this chapter discusses modification to the fluid force on nanoparticles due to noncontinuum effects, including drag, diffusion, and thermophoretic effects. The second section discusses nanoparticle interaction problems, including evaluation of adhesive contact models and electrostatic interaction for nanoparticle systems. The third section discusses Brownian coagulation of nanoparticles, including particle sintering following adhesive contact.

9.1. Continuum and Free-Molecular Regimes

Equation (5.1.15) defines the Knudsen number as the ratio of the fluid mean free path to the particle diameter, or $\mathrm{Kn} \equiv \lambda/d$. The Knudsen number serves as a measure for the accuracy of the assumptions of continuum theory. The value of λ can be obtained from (5.1.11), and has a value of about 68 nm for air at room temperature. Nanoparticles therefore possess a value of Knudsen number that is of order unity for large nanoparticles, or significantly larger than unity for small nanoparticles. The case with $\mathrm{Kn} \gg 1$ is called the *free-molecular regime*, for which case the flow about the particle differs significantly from that obtained by continuum theory. The case with $\mathrm{Kn} \sim 1$ is called the *transition regime*, and for this case the continuum theory still approximately holds but it is subject to a number of modifications having to do with breakdown of the no-slip boundary condition and the presence of Brownian motion.

An estimate of scaling parameters for particles with diameter ranging from 10 nm to 1 μm is given in Table 9.1. This estimate is obtained using fluid length and velocity scales typical of a suspended particle in the atmosphere. For small particle sizes, the fluid scaling might be set equal to that of the Kolmogorov (dissipation) scale eddies of the atmospheric turbulent motion, for which the length and time scales are given by $\eta = (v^3/\varepsilon)^{1/4}$ and $\tau_K = (v/\varepsilon)^{1/2}$, where v is the fluid kinematic viscosity and ε is the turbulence energy dissipation rate per unit mass. In atmospheric turbulence, the energy dissipation varies in the range $\varepsilon = 10^{-3} - 10^{-1}$ m²/s³, depending on height and atmospheric conditions. Assuming an average value of $\varepsilon = 0.01$ m²/s³

Table 9.1. *Values of the particle Peclet number, slip correction factor, and particle time scale for nanoparticles with specific gravity S = 2.5 in air at standard temperature and pressure. The mean free path of air is $\lambda = 68$ nm, and the fluid convective time scale based on the typical Kolmogorov scale of atmospheric turbulence is $\tau_f = 38$ ms. The Tabor parameter is computed for quartz particles with adhesion length scale $\delta_A = 0.165$ nm*

Particle diameter, d (nm)	Particle time scale, τ_p (ns)	$Pe_p^{-1/2}$	Knudsen number, Kn	Slip correction factor, C_C	Tabor parameter, λ_T
10	0.77	24,520	6.8	0.044	0.12
20	3.1	6130	3.4	0.089	0.15
50	19	981	1.4	0.21	0.21
100	77	245	0.68	0.37	0.27
200	309	61	0.34	0.54	0.33
500	1929	9.8	0.14	0.75	0.45
1000	7716	2.5	0.068	0.86	0.57

gives $L = 0.76$ mm and $U = 20$ mm/s. In order to estimate the Tabor parameter in Table 9.1, the relationship $\gamma = A/24\pi\delta^2$ between surface energy density γ and adhesion length scale δ is used, where A is the Hamaker constant for the material. The value of the Tabor parameter is computed for quartz particles with $\delta = 0.165$ nm, elastic modulus $E_p = 72$ GPa, and Hamaker coefficient $A = 6.5 \times 10^{-20}$ J.

The particle time scale listed in Table 9.1 is given by $\tau_p = m/3\pi\,d\mu$. The listed values can be compared to the fluid time scale for this problem, $\tau_f = L/U = 0.04$ s, resulting in Stokes numbers St_K ranging from 2×10^{-8} for 10 nm particles to 2×10^{-4} for 1 μm particles. The Knudsen number ranges from 0.07 to about 7 over this range of particle diameters, with the resulting Cunningham slip correction factors C_C ranging from 0.86 for 1 μm particles to 0.04 for 10 nm particles. It was observed in Section 5.5 that the ratio of Brownian force to drag force is proportional to $\mathrm{Pe}_p^{-1/2}$, where Pe_p is the particle Peclet number. Values of $\mathrm{Pe}_p^{-1/2}$ listed for this problem in Table 9.1 range from order unity for 1 μm particles to much greater than unity for 10 nm particles. These scaling results indicate that both surface slip and Brownian motion have a strong influence on the transport of nanoparticles, with the result that the fluid forces acting on nanoparticles are much different from those for micrometer-size particles, for which these effects are generally of small importance. The collision and adhesion forces between particles are characterized by the Tabor parameter λ_T, as discussed in Chapter 4. Estimated values for the Tabor parameter listed in Table 9.1 vary from about 0.1 to 0.6 as particle diameter varies from 10 to 1000 nm. These values lie between the regions covered by the DMT and JKR models, but they are still well above the range of Tabor parameter values for which the more extreme assumptions of the Bradley model would be appropriate. More discussion of the effect of small size on nanoparticle collision and adhesion forces is given in the remaining sections of the chapter.

9.1.1. Drag Force

One of the early consequences of the breakdown of the continuum assumption at finite values of the Knudsen number is the development of slip on the interface between the particle and the surrounding gas. The effects of surface slip were discussed in Section 5.1.3, where we showed that the effect of surface slip on the drag force can be accounted for using slip correction factor C_C (also known as the Cunningham correction factor). Using the empirical Millikan expression for C_C, given in (5.1.22), the drag force becomes

$$\mathbf{F}_d = -3\pi d\mu(\mathbf{v} - \mathbf{u})C_C = \frac{-3\pi d\mu(\mathbf{v} - \mathbf{u})}{1 + \mathrm{Kn}[A_1 + A_2 \exp(-A_3/\mathrm{Kn})]}. \tag{9.1.1}$$

The selection of the parameters A_1, A_2, and A_3 has been reviewed by Li and Wang (2003a,b). Equation (9.1.1) simplifies to the original Stokes equation for the continuum region when $\mathrm{Kn} \to 0$, for which the drag force is proportional to particle diameter d. In the free-molecule regime ($\mathrm{Kn} \gg 1$), this equation asymptotically approaches

$$\mathbf{F}_d = \frac{-3\pi d^2\mu(\mathbf{v} - \mathbf{u})}{\lambda(A_1 + A_2)}. \tag{9.1.2}$$

In this large Knudsen number limit, the drag force is proportional to d^2, rather than d.

Using kinetic theory, Epstein (1924) argued that the drag force on an ultrafine particle in the free-molecule regime can be expressed as a linear combination of terms resulting from diffuse and specular scattering effects. Introducing an accommodation coefficient α to represent the fraction of gas molecules that are reflected diffusively from the particle, the drag force on a particle in the free-molecular regime can be written as

$$\mathbf{F}_d = -\frac{2}{3}\left(1 + \frac{\pi\alpha}{8}\right) d^2\rho_f\sqrt{\frac{2\pi k_B T}{M_g}}(\mathbf{v} - \mathbf{u}), \tag{9.1.3}$$

where M_g is the mass of a gas molecule. The coefficient α can be evaluated experimentally, but it is usually near 0.9 for momentum transfer (Friedlander, 2000).

In Section 5.1.3, it is noted that kinetic theory also provides a relationship (5.1.12) between the molecular diameter σ_M and the fluid viscosity μ. Substituting this relationship into (9.1.2), with the mean free path given by (5.1.11), or $\lambda = k_B T/\sqrt{2}\pi\sigma^2 P$, and $P = N k_B T = \rho_f k_B T/M_g$, where N is the number density of gas molecules and ρ_f is the fluid mass density, the Stokes-Cunningham drag force for a nanoparticle in the free-molecular regime can be written as

$$\mathbf{F}_d = -\frac{15\pi}{16(A_1 + A_2)} d^2\rho_f\sqrt{\frac{2\pi k_B T}{M_g}}(\mathbf{v} - \mathbf{u}). \tag{9.1.4}$$

Comparison of (9.1.3) and (9.1.4) yields an approximation for the sum $A_1 + A_2$ as $A_1 + A_2 = 45\pi / [4(8 + \pi\alpha)] \approx 3.274$. This approximation is compared to values of the sum $A_1 + A_2$ from various references in Table 9.2, and it is found to be within about 1.5% of the empirically obtained values.

Using kinetic theory and considering the influence of van der Waals interactions (via the Lennard-Jones potential) on momentum transfer during collision of a gas molecule with a particle in the specular and diffuse scattering limits, a rigorous generalized expression for drag force of spherical particles in the entire range of the Knudsen number was derived by Li and Wang (2003a,b). Their theoretical expression agrees well with the empirical correlation (9.1.1) using the parameter values proposed by Allen and Raabe (1982), given in Table 9.2. We recommend

Table 9.2. *Comparison of theoretical approximation for $A_1 + A_2$ obtained from Equations (9.1.3) and (9.1.4) with empirical values obtained in the literature*

Source	A_1	A_2	A_3	$A_1 + A_2$
Theory				3.274
Davies (1945)	2.514	0.80	0.55	3.314
Fuchs (1964)	2.49	0.84	0.435	3.33
Allen and Raabe (1982)	2.310	0.942	0.298	3.252

that (9.1.1) with the Allen-Raabe coefficients be used for numerical computations, both because of its ease of implementation and because of its close agreement with theoretical and experimental results. Figure 9.1 plots the slip correction factor C_C as a function of the particle diameter in air in a range from 1 nm to 100 μm, for which the Knudsen number reduces from 70 to 7×10^{-4}, respectively. In particular, for nanoparticles with diameter less than 50 nm, the Epstein theory (9.1.3) predicts the drag with less than a 20% difference compared to the empirical Stokes-Cunningham formula (9.1.1) in the free-molecular regime. Using (9.1.3) and (5.1.12), an expression for C_C for nanoparticles in the free-molecular regime is obtained as

$$C_C = \frac{32\sqrt{2}\,\sigma_M^2 \rho_f}{45 M_g} \left(1 + \frac{\pi\alpha}{8}\right) d. \tag{9.1.5}$$

9.1.2. Brownian Force

In Section 5.4, the particle diffusion coefficient under Brownian motion is given by the Stokes-Einstein expression $D_b = k_B T / 3\pi\mu d$. A generalized expression for

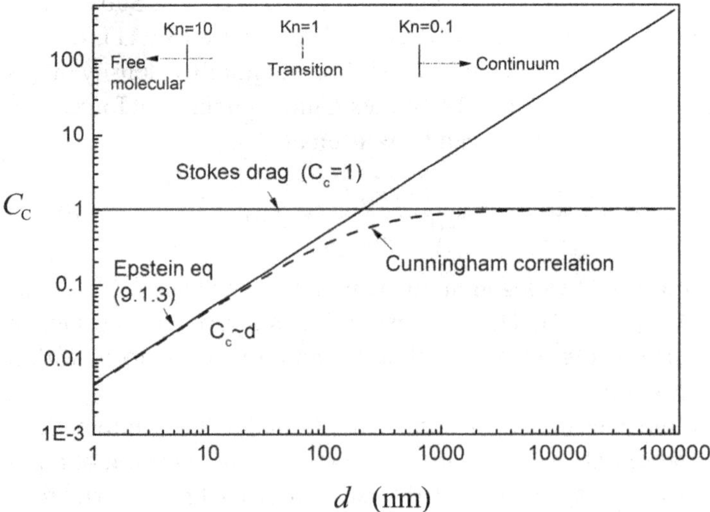

Figure 9.1. The slip correction factor C_C as a function of particle diameter for spherical particles in both the free molecular regime and the continuum regime (with $T = 293.13$ K, $\rho_f = 1.204$ kg/m^3, $M_g = 4.81 \times 10^{-26}$ kg, $\sigma_M = 0.37$ nm).

Brownian diffusion coefficient that accounts for particle surface slip can be written as

$$D_b = \frac{k_B T}{3\pi \mu d C_C}.$$ (9.1.6)

Using the expression (9.1.5) for the slip correction factor in the free-molecular regime, (9.1.6) becomes

$$D_b = \frac{1}{\frac{2}{3}\left(1 + \frac{\pi \alpha}{8}\right)\sqrt{2\pi}} \frac{\sqrt{k_B T M_g}}{\rho_f} d^{-2}.$$ (9.1.7)

Assuming that the absolute pressure P and temperature T are related to ρ_f by the ideal gas law, the Brownian diffusion coefficient in (9.1.7) satisfies $D_b \propto P^{-1} T^{3/2} d^{-2}$, which is similar to the diffusion coefficient of a molecule. For a larger particle in the continuum regime, we would instead obtain $D_b \propto T^{1/2} d^{-1}$. This relationship suggests that a nanoparticle in the free-molecular regime behaves much like an individual molecule, in the sense that the kinetic energy of the particle is dominated by the thermal energy. Knowing the particle diffusion coefficient D_b, the mean-square displacement (MSD) for Brownian motion of the nanoparticles can be predicted using (5.4.12).

9.1.3. Mean Free Path of Nanoparticles

When a fluid molecule hits a particle, the particle velocity and direction persists to a large extent due to the much larger inertia of the particle. This is in contrast to the case where a molecule collides which another molecule of a similar type, which often results in reversal of molecule travel direction with a single collision. As a consequence, it requires collisions with a large number of fluid molecules to change the direction of a nanoparticle. The collision of molecules with a nanoparticle occurs quite frequently, for example, about 10^{12} collisions per second for a 10 nm particle, so the change in direction of a nanoparticle can occur over a relatively short length scale. The average distance a particle travels before it changes direction is defined as the *particle mean free path* (Fuchs, 1964). The root-mean-square velocity of a nanoparticle of mass m due to thermal motion in one direction is given by

$$\bar{v}_{p,rms,r} = \sqrt{\frac{k_B T}{m}}.$$ (9.1.8)

The response time of a particle in fluid is given by $\tau_p = m/(3\pi \mu d C_C)$, with C_c given by (9.1.5) for the free-molecular regime. The particle mean free path λ_p is equal to the product of the root-mean-square velocity and the response time, giving

$$\lambda_p = \bar{v}_{p,rms,r} \cdot \tau_p = \frac{\sqrt{m k_B T}}{3\pi \mu d C_C}.$$ (9.1.9)

The particle mean free path plays an important role in particle deposition processes. When λ_p is larger than d, particle diffusion may exceed inertial impact to become the dominant mechanism for deposition as well as interparticle collision (Madler et al., 2006). Figure 9.2 illustrates the ratio of the particle mean free path to the particle diameter over a range of particle sizes from 1 nm to 1 μm with an assumed

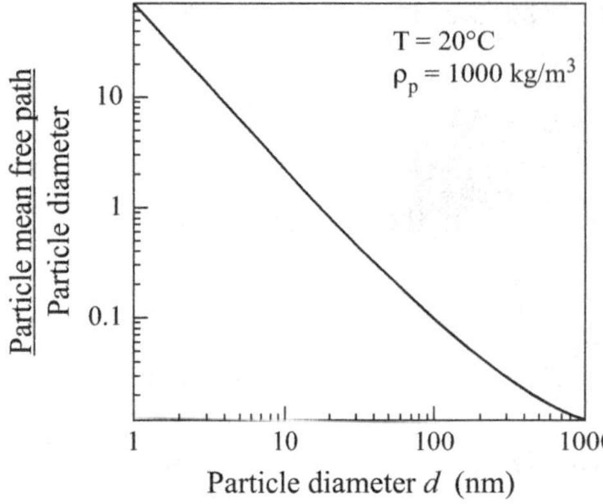

Figure 9.2. Ratio of particle mean free path to particle diameter ranging from 1 nm to 1 μm. Particles smaller than about 20 nm travel an average distance greater than their own diameter before changing directions due to random collisions with gas.

particle density of 1000 kg/m³. The λ_p/d ratio decreases with increasing particle size. Nanoparticles with diameter less than about 20 nm travel an average distance that is more than their own diameter under Brownian motion before changing direction.

9.1.4. Thermophoretic Force

Thermophoresis describes particle drift in the presence of a temperature gradient resulting from the gradient in root-mean-square velocity of the surrounding gas molecules. The molecular root-mean-square velocity increases with temperature in proportion to $T^{1/2}$, so that molecular collisions on the "hot" side of the particle import more momentum than do collisions with the slower-moving molecules on the "cold" side of the particle. As illustrated in Figure 9.3, the molecular collisions on the hot side push the particle toward the cold side and those on the cold side push the particle toward the hot side, with the result that the net thermophoretic force on the particle is oriented opposite to the direction of the temperature gradient. This effect was first reported by Tyndall (1870), who noticed a particle-free region around a heated metal wire that was visible as a dark space in the bright background of scattered light from the particles. Thermophoresis is of major importance in a variety of nanoengineering

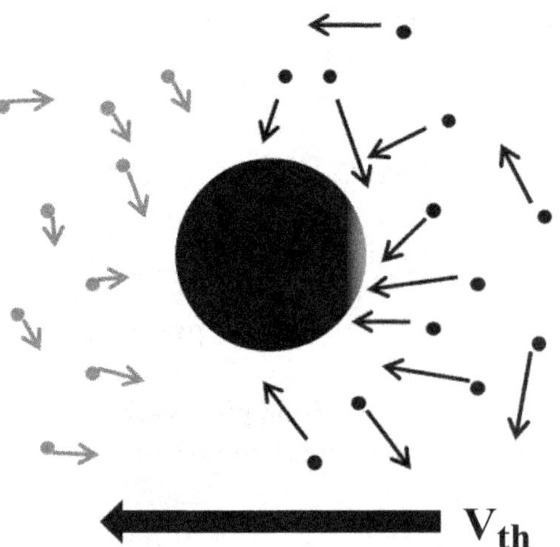

Figure 9.3. Schematic diagram of a particle (large center sphere) immersed in a gas. The small dots represent the gas molecules that are randomly fluctuating. The gradient in the mean molecular velocities produces a force on the particle that pushes the particle against the temperature gradient toward the colder region.

applications. For instance, in the flame synthesis process for nanoparticle films, thermophoresis is used as a deposition method for porous thin-film manufacture (Tolmachoff et al., 2009; Zhang et al., 2012c). A modified chemical vapor deposition process has been used to manufacture optical fiber by deposition of particles on the tube wall (Park et al., 1999). The semiconductor industry uses thermophoresis to avoid "light point defects" on wafers and lithographic masks by particle deposition (Stratmann et al., 1988; Ye et al., 1991). Thermophoresis can also be used as a high-efficiency capture mechanism to remove nanoparticles from high-temperature gas streams, where the collecting medium is cooler than the surrounding gas.

Similar to the particle drag force, the thermophoretic force is strongly dependent on the Knudsen number. In the free-molecule regime ($Kn \gg 1$), Waldmann (1959) proposed an expression for thermophoretic force as

$$\mathbf{F}_{th} = -\frac{8}{15}\frac{\kappa_{tr}d^2}{\bar{c}}\nabla T = -\frac{2}{15}\sqrt{\frac{2\pi M_g}{k_B T}}\kappa_{tr}d^2\nabla T, \qquad (9.1.10)$$

where the mean speed $\bar{c}$ of gas molecules is given by $\bar{c} = \sqrt{8k_B T/\pi M_g}$ and the gas thermal conductivity κ_{tr} pertains to the translational part of the thermal energy. κ_{tr} is linearly related to the fluid viscosity μ as $\kappa_{tr} = 15k_B\mu/(4M_g)$. The thermophoretic force is often counteracted by the fluid drag on the particle, and under equilibrium conditions the combination of these two forces leads to a terminal particle drift velocity that can be derived by equating (9.1.10) and (9.1.3) to obtain

$$\mathbf{v}_{th} = -\frac{\kappa_{tr}}{5(1+\frac{\pi\alpha}{8})k_B N}\frac{\nabla T}{T} = -\frac{3\mu}{4\rho_f(1+\frac{\pi\alpha}{8})}\frac{\nabla T}{T}. \qquad (9.1.11)$$

From the equation derived using the Waldmann theory, the thermophoretic drift velocity for particles in the free-molecular regime is observed to be independent of the particle size. However, the Waldmann analysis employs an assumption that the particles collide as rigid bodies, which is of questionable validity for nanoparticles. Li and Wang (2004) examined the influence of nonrigid body collision by accounting for van der Waals or other long-range forces between the particles and surrounding gas molecules, and these authors find that long-range forces play a significant role for particles with diameter less than about 10 nm. This influence was expressed in terms of reduced collision integrals in the kinetic theory approach. The resulting expression for the thermophoretic force obtained with this modified theory is

$$\mathbf{F}_{th} = -\frac{2}{15}\sqrt{\frac{2\pi m_r}{k_B T}}\kappa_{tr}d^2\nabla T \left|5\Omega_{avg}^{(1,1)}-6\Omega_{avg}^{(1,2)}\right|, \qquad (9.1.12)$$

where $m_r \equiv m_p M_g/(m_p + M_g)$ is the reduced mass of the gas molecule and a particle and $\Omega_{avg}^{(1,1)}$ and $\Omega_{avg}^{(1,2)}$ are the average reduced collision integrals for specular and diffuse scattering, respectively (Li and Wang, 2004; Li, 2005). The corresponding thermophoretic velocity is given by

$$\mathbf{v}_{th} = -\left(1-\frac{6\Omega_{avg}^{(1,2)}}{5\Omega_{avg}^{(1,1)}}\right)\frac{\kappa_{tr}}{Nk_B}\frac{\nabla T}{T}. \qquad (9.1.13)$$

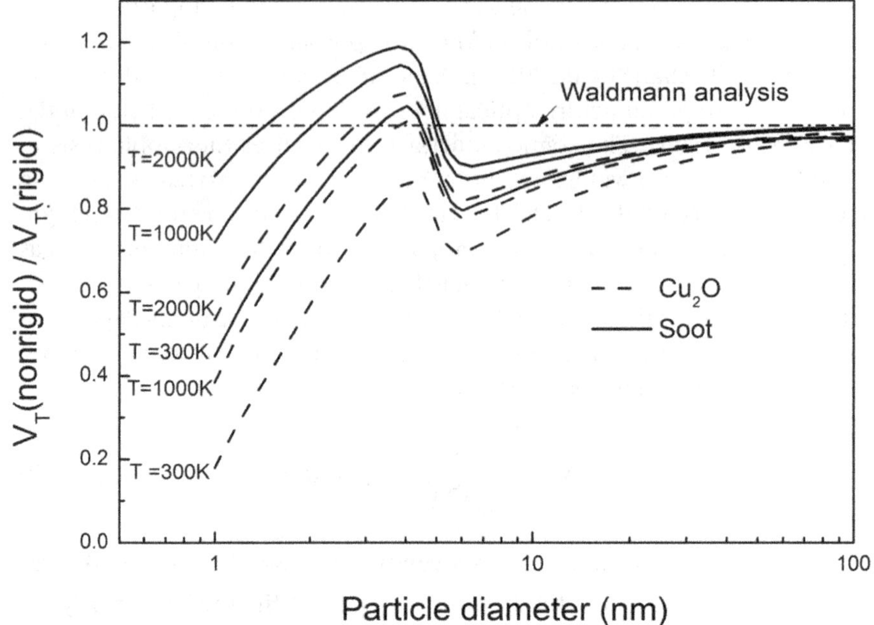

Figure 9.4. Ratio of thermophoretic drift velocity with nonrigid collisions (i.e., van der Waals interaction) to that with rigid body collisions for Cu_2O and soot particles. (Based on calculations of Li and Wang, 2004.)

In the limit of rigid body collision, both $\Omega_{avg}^{(1,1)}$ and $\Omega_{avg}^{(1,2)}$ are equal to unity and m_r approaches the molecular mass M_g, so that (9.1.12) reduces to the Waldmann equation (9.1.10) and (9.1.13) reduces to (9.1.11).

Figure 9.4 shows the ratio of v_{th} with nonrigid body collisions to that with rigid body collisions for Cu_2O and soot particles, obtained using (9.1.11) and (9.1.13). Soot is assumed to be composed of pure carbon and its density is chosen to be 1,800 kg/m^3. The coefficient α in the Waldmann formula is chosen to be 0.9 for both Cu_2O and soot, but the corresponding coefficient in Li and Wang's formula is varied. For ultrafine particles, with diameter around 1 nm, the value of the thermophoretic velocity obtained from the Waldmann theory is significantly larger than that obtained from the more refined Li-Wang analysis. For particles with diameter larger than 30 nm, the predictions of the two theories are within 10% of each other.

Compared with the free-molecular regime, the analysis of thermophoresis for intermediate Knudsen numbers (in the transition regime) is relatively complicated. Most theories for thermophoretic force with arbitrary Kn are derived by solution of the Boltzmann equation, or some approximation of the Boltzmann equation. Similar to (9.1.10), a general form for thermophoretic force at intermediate Knudsen numbers can be written as

$$\mathbf{F}_{th} = -\frac{\sqrt{\pi}}{8}h(\mathrm{Kn}, \kappa_{21})\frac{\kappa_{tr}d^2}{\bar{c}}\nabla T, \qquad (9.1.14)$$

where $h(\mathrm{Kn}, \kappa_{21})$, sometimes called the dimensionless thermophoretic force, is a function of the Knudsen number Kn and the gas/particle thermal conductivity ratio κ_{21}, and $\bar{c} = \sqrt{8k_B T/\pi M_g}$. Combining (9.1.14) with the expression (9.1.1) for

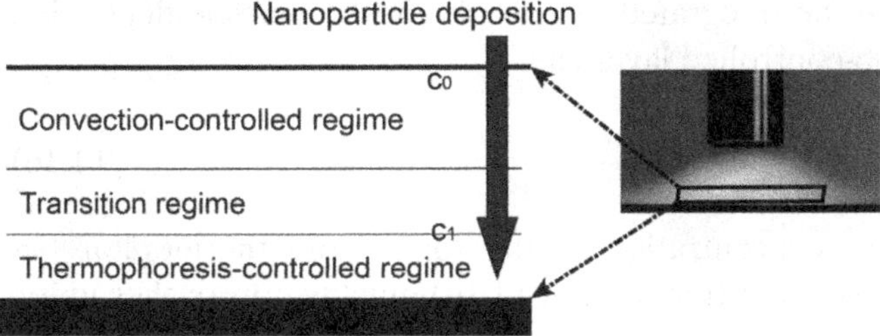

Figure 9.5. Schematic illustration of nanoparticle deposition adjacent to the substrate during thin film production.

drag force, the equilibrium thermophoretic velocity on a particle with broad Kn range can be predicted. Expressions for $h(\text{Kn}, \kappa_{21})$ were proposed by Sone and Aoki (1983), Yamamoto and Ishihara (1988), and Loyalka (1992) on the basis of the linearized Boltzmann equation. Talbot and co-workers (Talbot et al., 1980; Talbot, 1981) presented an empirical formula for $h(\text{Kn}, \kappa_{21})$ based on the assumption of near-continuum conditions. Comparisons between different expressions for thermophoretic force were given by Talbot (1981), Davis and Schweiger (2002), Li (2005), and Madler and Friedlander (2007).

9.1.5. Competition between Diffusion and Thermophoresis during Deposition

We consider particle deposition in an axisymmetric stagnation-point flow with strain rate $s = U_0/2h$, where U_0 is the characteristic velocity along the symmetry axis at a distance h from the impingement surface. The axisymmetric Heimenz flow admits a boundary layer with characteristic thickness $\delta_{BL} = 2.8\sqrt{v/s}$, where v is the fluid kinematic viscosity (White, 2006). As shown in Figure 9.5, the boundary layer can be divided into three layers based on comparison of the particle thermophoretic velocity and the axial gas velocity. The convection-controlled layer is the outermost layer of the boundary layer, in which the axial gas velocity dominates the particle transport. The thermophoresis-controlled layer is the layer closest to the impingement surface, in which the gas velocity is small compared with the particle thermophoretic velocity. This inner layer has a thickness of between 2 and 10% of the total boundary layer thickness, depending on the temperature gradient. Between these two regions is the transition layer, in which the gas velocity is of the same order of magnitude as the thermophoretic velocity.

Assuming that the particle concentration is a function only of the axial coordinate x, the equation for particle transport via convection, thermophoresis, and diffusion in the boundary layer is given by

$$(u + v_{th})\frac{d\phi_m}{dx} = \frac{d}{dx}\left(D_b \frac{d\phi_m}{dx}\right), \qquad (9.1.15)$$

where ϕ_m is particle mass concentration and v_{th} is the particle thermophoretic velocity. Particle deposition on the wall is controlled by the thermophoretic-controlled layer within which $u \ll v_{th}$, where v_{th} is assumed to be approximately constant within this inner layer. Defining a Peclet number based on the thermophoretic velocity as

$\text{Pe}_{th} = v_{th}\delta_{BL}/D_b$, (9.1.15) can be integrated in x to obtain the particle deposition flux $\dot{m}''_p$ in the thermophoretic-controlled layer as

$$\dot{m}''_p = \phi_{m1}v_{th}\frac{e^{\text{Pe}_{th}}}{e^{\text{Pe}_{th}} - 1}, \qquad (9.1.16)$$

where ϕ_{m1} denotes the mass concentration at the top of the thermophoretic-controlled layer. When $\text{Pe}_{th} \gg 1$, the fraction in (9.1.16) quickly approaches unity, giving $\dot{m}''_p = \phi_{m1}v_{th}$, so that the particle mass flux is controlled entirely by thermophoretic transport. For $\text{Pe}_{th} < O(1)$, the deposition flux is influenced both by the thermophoretic force and by Brownian diffusion. The fluid convection is important for determining the coefficient ϕ_{m1} in the equation.

9.2. Nanoparticle Interactions

When two nanoparticles move close to each other, they experience a number of different types of interaction forces. For particles on the larger end of the nanoparticle size spectrum, the length scale of adhesive forces remains small compared with the particle diameter and it is reasonable to use the adhesive contact models discussed in Chapter 4, which assume that adhesive forces have no influence until the particles collide. However, for particles on the smaller end of the nanoparticle size spectrum, the length scale of adhesive forces is a significant fraction of the particle diameter and action of adhesive forces prior to collision becomes important.

This section first discusses the collision and adhesion forces between nanoparticles in the large end of the size spectrum, followed by a discussion of the influence of long-range effects of adhesive force for very small nanoparticles. Here, we refer to "long range" as having a length scale on the order of the particle diameter. The final two parts of this section discuss electrostatic Coulomb and dipolar interactions between nanoparticles, which play a major role in the synthesis, design, and manipulation of nanostructured materials.

9.2.1. Collision of Large Nanoparticles

The accuracy of the continuum approximation for nanoscale collision problems with particles in the large end of the nanoparticle size spectrum, given approximately by $d > 20$ nm, has been examined by a number of investigators using molecular dynamics (MD) to simulate the atomic interactions in nanoparticle collision problems (Wenning and Müser, 2001; Luan and Robbins, 2005). Luan and Robbins (2005) examined both normal and tangential (frictional) forces for contact between a particle (with either a spherical or cylindrical shape) and a flat substrate. Results are shown in Figure 9.6 for a cylinder with radius of 100 atomic diameters (approximately 30 nm) colliding on a flat substrate. The MD simulation results are compared with continuum theory predictions obtained using the Maugis-Dugdale adhesive contact model. Use of this model is consistent with the range of Tabor parameter values for nanoparticles in this size range listed in Table 9.1. Figure 9.6 compares the MD simulations with continuum theory for normal displacement, contact region radius, and tangential friction force as a function of normal force. The normal displacement results exhibit excellent agreement between the MD simulations and continuum

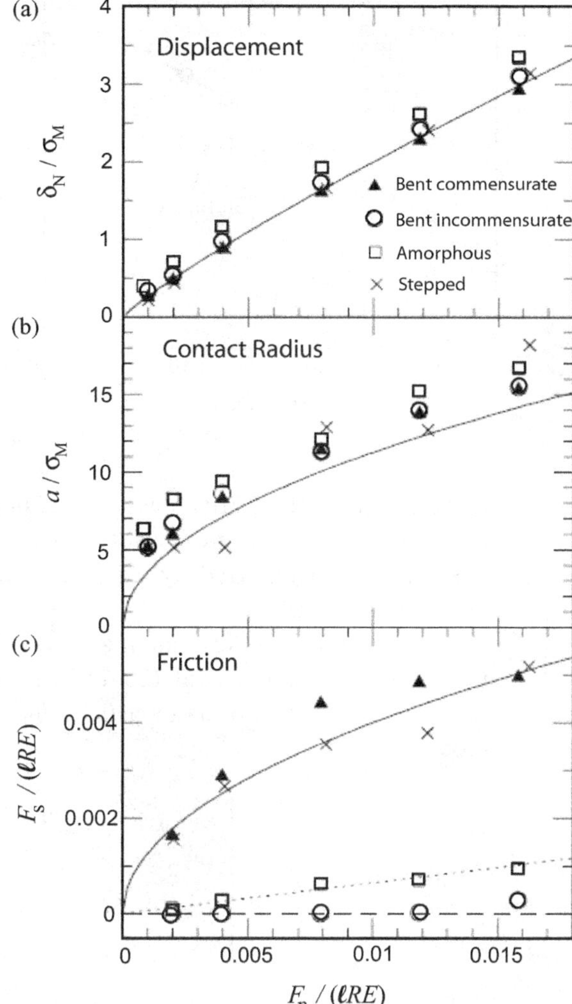

Figure 9.6. Comparison of the predictions of the Maugis-Dugdale continuum theory (solid lines) and molecular dynamics simulations (symbols) for different atomic arrangements with a 30 nm diameter cylinder colliding with a flat substrate, for (a) normal displacement; (b) contact region radius; and (c) frictional force as a function of the dimensionless normal force. The dashed and dotted lines in (c) are best fits to the data. [Reprinted with permission from Luan and Robbins (2005).]

theory. The contact region radius obtained from MD solutions is about 10–20% higher than the continuum theory predictions, but the trend is the same between the two. It is noted that the Maugis-Dugdale model is approximate, so this modest disagreement might well be due to inaccuracy in solution of the continuum problem.

Whereas the normal force solutions agree well between MD simulation and continuum theory, the frictional sliding force obtained from MD is highly sensitive to the atomic arrangement within the nanoparticle, as shown in Figure 9.6c. Although the MD simulations for materials with crystalline atomic structure agree reasonably well with the continuum predictions, the frictional force is much smaller for materials with random amorphous atomic structure. The study by Wenning and Müser (2001) used MD simulations to examine the relationship between frictional force F_s and normal load F_n under static friction conditions for contacts with radius of curvature ranging between 1 and 15 nm. Whereas the Amonton theory used in continuum mechanics descriptions assumes that friction force is proportional to normal load, Wenning and Müser only observe linear behavior for crystalline (commensurate) atomic arrangements. For amorphous atomic arrangements, the friction force varies

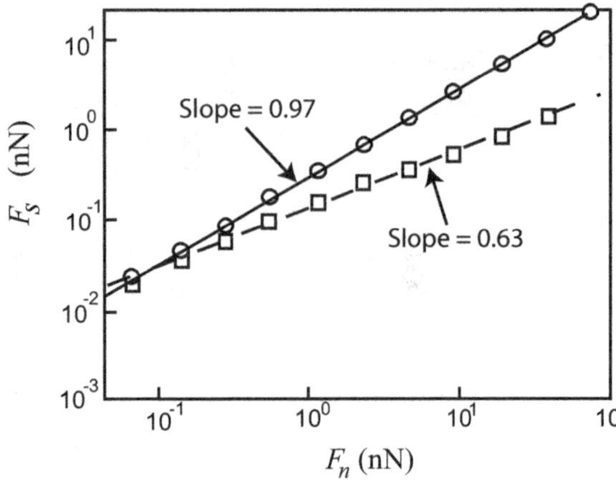

Figure 9.7. Plot showing normalized sliding force F_s versus normal load F_n (in nano-Newtons) for nanoparticles with commensurate structure (circles, solid line) and amorphous structure (squares, dashed line), along with best-fit lines, based on the MD simulation results of Wenning and Müsser (2001). The upper (solid) line has slope 0.97 and the lower (dashed) line has slope 0.63.

more closely to the 2/3rds power of the normal load (Figure 9.7), and yields results for friction force that are considerably smaller than for crystalline materials. This observation is supported by experimental results with friction force microscopes (Schwarz et al., 1997).

Normal force models similar to the DMT model were used by Eggersdorfer et al. (2010) and Higashitani et al. (2001) for simulation of the restructuring and break-up of soft agglomerates of nanoparticles under a shear flow. The normal force in this model is given by

$$
F_n = \begin{cases} k_N\delta_N + \eta_N(\mathbf{v}_R \cdot \mathbf{n}) - \dfrac{A}{6\delta^2}R & \delta_N > 0 \\[2ex] -\dfrac{A}{6(\delta + \delta_N)^2}R & -\delta_C \le \delta_N \le 0 \\[2ex] 0 & \delta < -\delta_C \end{cases} \tag{9.2.1}
$$

where the stiffness k_N can be written as $k_N = 4E\,a/3$ using the Hertz contact result (3.2.11), the normal dissipation coefficient η_N is determined by the TTI model given in (3.3.12), and δ is the minimum separation distance within the contact region, with value ranging between 0.15 and 0.40 nm. For convenience in simulation, Higashitani et al. (2001) proposed a linear empirical relationship of $k_N = \pi r_i E_i$, where $E = E_i/2(1 - v_i^2)$. The first line in Equation (9.2.1), valid for $\delta_N > 0$, is just the DMT model (4.2.15), modified by addition of the normal damping term. The simulation of Eggersdorfer et al. (2010) indicated that the variation of elastic moduli has no influence on the agglomerate breakage time or general behavior. This observation is in agreement with other studies of DEM of adhesive particles. For instance, for the problem of particle capture by a circular cylinder, Li and Marshall (2007) observed that the particle dynamics is insensitive to the value of the elasticity parameter Γ ($\equiv E/\rho_p U_0^2$), defined as the ratio of elastic force to particle inertia. The second line in (9.2.1), valid for $-\delta_C \le \delta_N \le 0$, is an approximate expression of the van der Waals force obtained by applying Hamaker theory for two spheres separated by a small distance. The critical pull-off distance δ_C is arbitrarily set as 10 nm in these papers, where van der Waals force drops to less than 1% of the maximum attractive force.

Tangential forces play a significant role in the formation of particle agglomerates. Pantina and Furst (2005) experimentally demonstrated the existence of tangential forces between bonded nanoparticles and the capability of these bonds to supporting bending moments, and then presented a possible explanation for the tangential forces in terms of the JKR theory for adhesive surfaces. They related the single bond rigidity to the work of adhesion derived from the JKR theory and to the elastic modulus of the particles. Becker and Briesen (2008) developed tangential force models for DEM simulations including both rolling and twisting frictions, in which several model parameters were determined by the experiments of Pantina and Furst (2005). The more detailed effects of adhesion on the sliding, twisting, and rolling frictions are summarized in Sections 4.2.3 and 4.2.4.

9.2.2. Collision of Small Nanoparticles

As particles get increasingly smaller within the nanoparticle size range, the models used for adhesive contact forces and torques begin to break down. As discussed earlier, this occurs first with the models for tangential resistance, as indicated by observations that the atomic arrangement within the nanoparticle has a significant influence on the sliding resistance. At very small particle sizes, the models for normal force begin to break down in cases where the particle diameter is about an order of magnitude larger than the length scale of the van der Waals force, characterized by the equilibrium gap thickness δ. For such cases, the primary difference with the theory discussed in Chapter 4 is the fact that the van der Waals force begins to have a significant influence on the particle dynamics prior to collision of the particles.

For nanoparticles with diameters smaller than about 10 nm, the Tabor parameter is sufficiently small (Table 9.1) that we can adopt the assumption employed in the DMT model and consider the van der Waals force and the repulsive elastic force separately, so that the particle deformation is assumed to have negligible effect on the van der Waals force. The interaction energy $w_{s-s}(h)$ between two spherical particles with radii r_1 and r_2 and separation distance h is given by

$$w_{s-s} = -\frac{A_{12}}{6}\left[\frac{2r_1r_2}{(r_1+r_2+h)^2-(r_1+r_2)^2} + \frac{2r_1r_2}{(r_1+r_2+h)^2-(r_1-r_2)^2}\right.$$
$$\left. + \ln\left(\frac{(r_1+r_2+h)^2-(r_1+r_2)^2}{(r_1+r_2+h)^2-(r_1-r_2)^2}\right)\right], \tag{9.2.2}$$

where the Hamaker coefficient A_{12} between media 1 and 2 can be determined by (4.1.7). For two equal-size particles ($r_1 = r_2 = d/2$), this equation reduces to

$$w_{s-s} = -\frac{A}{6}\left[\frac{d^2}{2(d+h)^2-2d^2} + \frac{d^2}{2(d+h)^2} + \ln\left(1 - \frac{d^2}{(d+h)^2}\right)\right]. \tag{9.2.3}$$

The van der Waals force between the spheres is given by

$$F_{vdW} = -\frac{dw_{s-s}}{dh} = -\frac{A}{6d}\frac{1}{x^2(x+1)^3(x+2)^2}, \tag{9.2.4}$$

where $x = h/d$ is a dimensionless separation distance. Equation (9.2.4) indicates that the van der Waals force at a certain separation distance (relative to particle

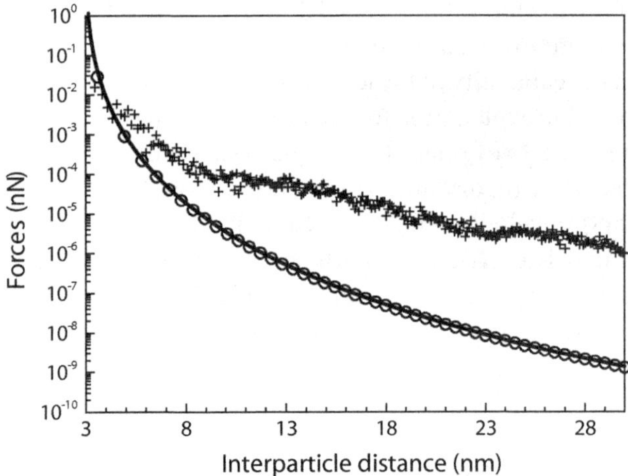

Figure 9.8. Interaction forces between two co-oriented 3 nm TiO_2 nanoparticles at different separation distances. Data are for MD simulations of the dipolar force (+ symbols) and the van der Waals force (O symbols), and for prediction of the van der Waals force from Equation (9.2.3) (solid line). [Reprinted with permission from Yan et al. (2010).]

diameter) scales inversely with particle diameter. Figure 9.8 shows the interaction forces between two approaching 3 nm charge-neutral anatase TiO_2 nanoparticles at 273 K as a function of the interparticle center-to-center distance $(d + h)$, calculated by a molecular dynamics simulation with a Matsui-Akaogi potential that includes electrostatic, van der Waals, and repulsive interactions (Matsui and Akaogi, 1991). The van der Waals force predicted by (9.2.4) agrees well with that obtained by the MD simulation. At sufficiently large interparticle separation distance, the adhesive force is dominated by the dipole-dipole interaction of the nanoparticles, which is discussed in the next section.

The significance of the long-range effect of van der Waals force can be evaluated by comparing the van der Waals interaction given in (9.2.3) with the particle kinetic energy. As noted at the beginning of this chapter, nanoparticle dynamics prior to particle collision is dominated by a balance between the drag force and the Brownian force. For very large nanoparticles with diameter of roughly 1,000 nm, the particle inertial force is of a similar order of magnitude to the drag and Brownian forces, but for much smaller nanoparticles the particle inertia is significantly smaller than drag or Brownian force. Taking the ratio of $|w_{s-s}|$ from (9.2.3) with the energy associated with Brownian motion, given by $(3/2)k_B T$, gives the ratio of the adhesive energy to the particle kinetic energy as

$$\frac{W_A}{W_B} = \frac{A}{9k_\mathrm{B}T}\left[\frac{1}{2(1+x)^2 - 2} + \frac{1}{2(1+x)^2} + \ln\left(1 - \frac{1}{(1+x)^2}\right)\right]. \quad (9.2.5)$$

This ratio is plotted as a function of dimensionless separation distance $x = h/d$ in Figure 9.9, assuming room temperature $(T = 300\mathrm{K})$ and values of the Hamaker constant for particles interacting in air of $A = 4 \times 10^{-20}\mathrm{J}$ and $4 \times 10^{-19}\mathrm{J}$, where the latter value applies primarily to metallic particles. The energy ratio is equal to unity for $x_{crit} = 0.1$ with the smaller value of Hamaker coefficient, with Brownian motion dominating adhesion for values of $x > x_{crit}$. For the larger value of Hamaker coefficient, the critical separation distance increases to $x_{crit} = 0.34$.

Fanelli et al. (2006) developed a model based on DEM to simulate the agglomerate behavior of nanoparticles with size below 4 nm. They used a Born repulsive force model for the elastic force instead of the elastic force expression obtained

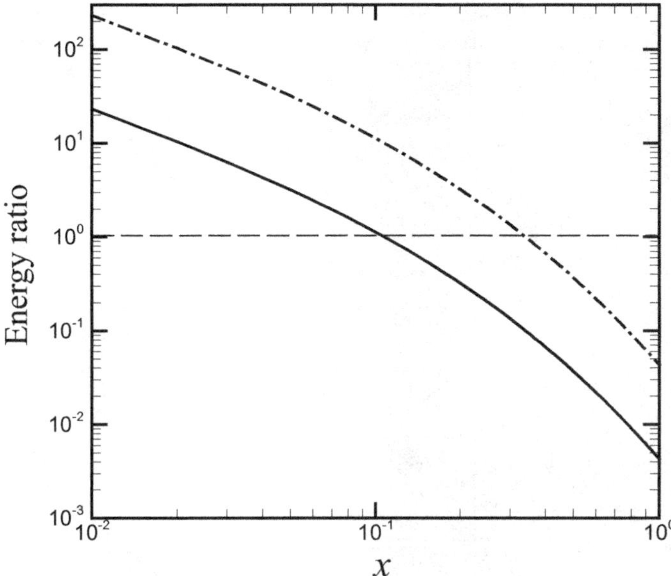

Figure 9.9. Ratio of the adhesive energy to the Brownian thermal energy of a nanoparticle as a function of the dimensionless particle separation distance $x = h/d$. Plots are given both for Hamaker constants of 4×10^{-20}J (solid line) and of 4×10^{-19}J (dashed-dotted line). The point where the two energies are equal is indicated by a dashed line.

from Hertz theory. Results of MD simulations for the normal contact force of two identical nanoparticles were reported by Sun et al. (2013) using particle diameters of 1.95, 4.01, 6.13, or 8.24 nm. Results of these simulations comparing the MD results with predictions of the JKR and DMT models are shown in Figure 9.10. The DMT predictions agree reasonably well with the MD predictions, but the JKR predictions indicate that the normal force is too low. This difference between DMT and JKR is consistent with what would be expected from consideration of the Tabor parameter values for particles of this small size. This work focuses on the initial stage of particle contact. At later times additional bonding mechanisms can result in sintering of the particles, as discussed in Section 9.3.

9.2.3. Long-Range Interparticle Electrostatic Forces

A theoretical calculation of nanoparticle charging from Sahu et al. (2012) based on the attachment of ions to particles suggests that particles smaller than 10 nm are charged primarily by diffusion charging, whereas the experimental charge measurements shown in Figure 9.11 confirm that particles smaller than 7 nm carry no charge. Particles smaller than 25 nm are primarily observed to carry no more than one elementary unit of charge. In general, it is very difficult to charge nanoparticles by field charging mechanisms when the saturation charge is less than one elementary charge.

Even without a net charge or the presence of an external electric field, the nanoparticles themselves can possess permanent dipoles whose interaction leads to important forces, particularly during particle collisions, as seen in Figure 9.8. Some dipoles arise from the asymmetry of wurzite crystalline structure of the nanoparticles, such as in CdSe nanocrystals. Blanton et al. (1997) reported values of dipole

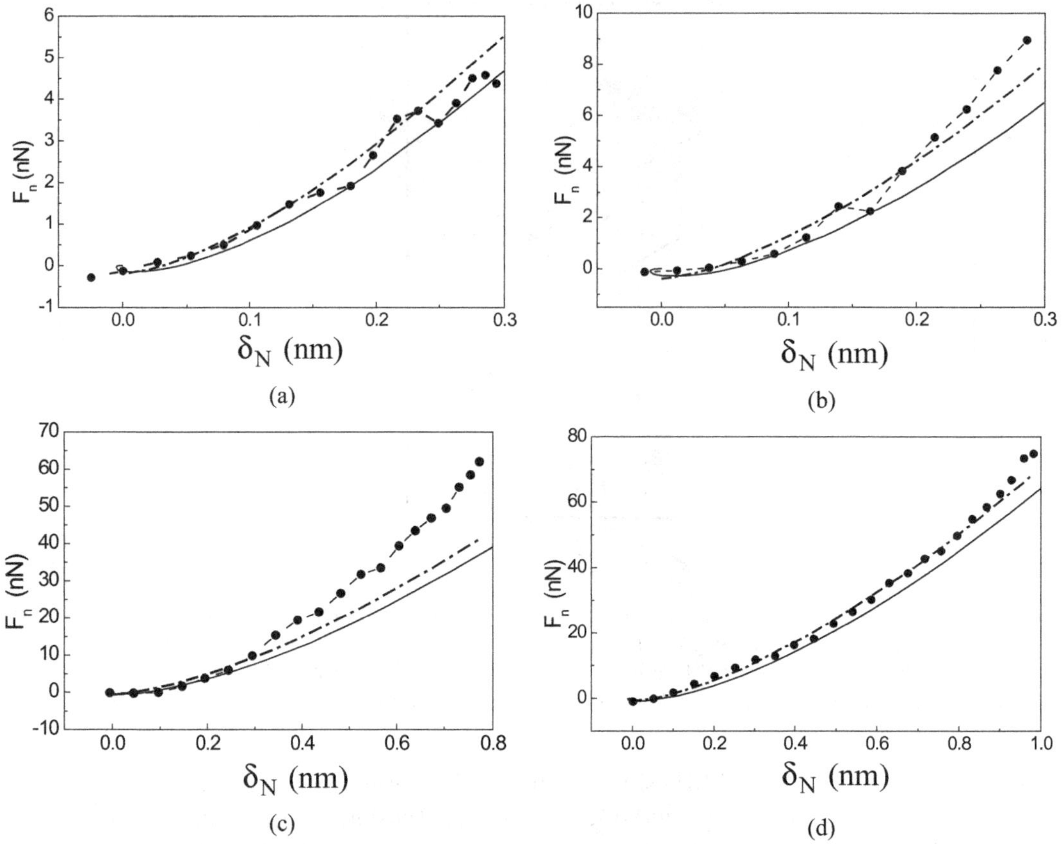

Figure 9.10 Comparison of the normal forces F_n between silica nanospheres of different diameter of (a) 1.95; (b) 4.01; (c) 6.13; to (d) 8.24 nm obtained from the MD simulations of Sun et al. (2013) (symbols) with those predicted by JKR model (solid line) and DMT model (dashed line).

moments of 3.4 and 4.6 nm CdSe nanocrystals as 25 D (Debye) and 47 D, respectively. For some nanocrystals with centrosymmetric lattice, for example, ZnSe and TiO_2, the asymmetrical distribution of ions at the surface could also create permanent dipoles, verified by both experiments (Shim and Guyot-Sionnest, 1999) and more recently by molecular dynamics (MD) simulation results (Yan et al., 2010). Figure 9.8

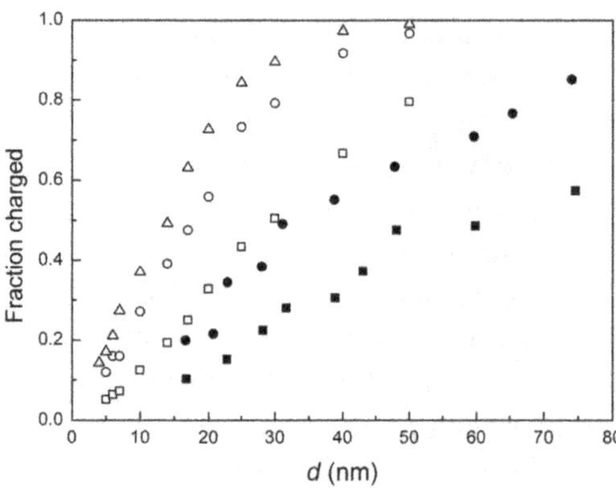

Figure 9.11. Fraction of particles charged after exposure to a unipolar ion source, with ion concentration n and charging time t. Open symbols are used for the data reported by Pui et al. (1988) for NaCl with $nt = 3.0 \times 10^6$, 7.2×10^6, and 10.3×10^6 s/cm^3 indicated by squares, circles, and triangles, respectively. Filled symbols are used for the data reported by Adachi et al. (1985) for $ZnCl_2$ with $nt = 0.58 \times 10^6$ and 1.2×10^6 s/cm^3 indicated by squares and circles, respectively.

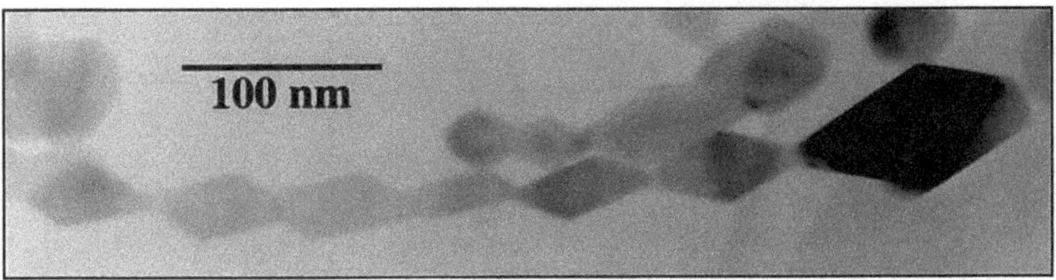

Figure 9.12. Self-assembly of TiO_2 nanoparticles under dipole-dipole interaction. [Reprinted with permission from Penn and Banfield, 1999.]

shows the interaction forces between two approaching charge-neutral anatase TiO_2 nanoparticles with diameter 3 nm at 273 K as a function of the interparticle center-to-center distance. The dipole-dipole interaction energy between two nanoparticles can be expressed by

$$w_{12}^{d-d} = \frac{1}{4\pi\varepsilon_0} \frac{\mathbf{p_1} \cdot \mathbf{p_2} - 3(\mathbf{n} \cdot \mathbf{p_1})(\mathbf{n} \cdot \mathbf{p_2})}{(r_1 + r_2 + h)^3}, \tag{9.2.6}$$

where $\mathbf{p_1}$ and $\mathbf{p_2}$ are the permanent dipole moments of the two nanoparticles, r_1 and r_2 are the radii, and $\mathbf{n}$ is the unit vector between the particle centroids. Equation (9.2.6) is a close approximation to the MD simulation predictions by Yan et al. (2010) and Zhang et al. (2011). As seen in Figure 9.8, the electrostatic force resulting from the dipole-dipole interaction is several orders of magnitude larger than the van der Waals force at large separations (>10 nm). The additional long-range attraction provided by dipole-dipole interaction enlarges the effective particle collision radius, which serves to shorten the collision time scale.

Unlike the isotropic nature of the van der Waals force, the dipole-dipole interaction acts to align particles as they collide, which is important for certain nanoparticle self-assembly processes. Figure 9.12 shows an electron microscope photo of a self-assembled structure of TiO_2 nanoparticles from Penn and Banfield (1999). The particles in this structure are aligned end-to-end due to dipole-dipole interaction. Nanoparticles are aligned as they approach each other, resulting in preferential orientation at the time of attachment. Similar observations of nanoparticle alignment during self-assembly were reported for CdTe nanoparticles by Tang et al. (2002). The relative strength of the interaction potential energy compared with the energy of Brownian fluctuations is an important factor in determining whether this self-assembled structure will be achieved. A detailed discussion of the relationship between nanoparticle self-assembly, Brownian motion, and interparticle forces was given in the review by Bishop et al. (2009).

As the particle scale reduces, thermal fluctuations play an increasingly important role in the particle dynamics, especially at high temperatures. Figure 9.13 shows the effect of temperature on the dipole moment of TiO_2 nanoparticles, obtained by molecular dynamics simulation. As the temperature increases from 273 K to 1673 K, the magnitude of the time-averaged dipole moment dramatically decreases, from 60.1 D to 2.1 D. As the temperature further increases, the dipole moment remains below 4.0 D with relatively small deviation. The decrease in magnitude of the time-averaged dipole moment caused by an increase in temperature can be attributed to the increasing fluctuation of the instantaneous dipole direction. Inset A of Figure 9.13

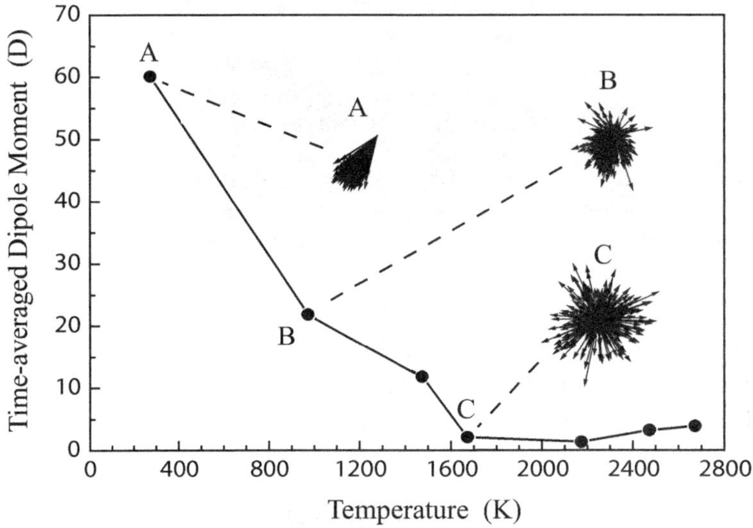

Figure 9.13. Time-averaged dipole moments of a TiO_2 nanoparticle at different temperatures. Insets A, B, and C show the distribution of dipole vectors projected onto the x–y plane along the time period at different temperatures. [Reprinted with permission from Yan et al. (2010).]

shows the distribution of dipole vectors projected onto the x–y plane at 273 K. The distribution of the fluctuating dipole directions is relatively narrow and is constrained to angles close to $\pi/4$. As the temperature increases to 973 K (Inset B), the direction of the dipole vector becomes more widely spread out, but the distribution is still obviously asymmetrical, resulting in a dipole moment value of 21.8 D. When the temperature further increases to 1,673 K (Inset C), the dipole vectors have widely different orientations spreading over the entire angular range, with the result that the time-averaged dipole moment is very small. This phenomenon is caused by the fact that Brownian motion becomes progressively more important as the temperature increases, eventually outweighing the energy associated with dipole-dipole interaction, leading to random nanoparticle aggregates rather than an ordered assembly.

9.3. Time Scales of Nanoparticle Collision-Coalescence Mechanism

A great deal has been learned about the formation of solid nanoparticles using inorganic metal oxides in flame aerosol processes, as well as other kinds of high-temperature vapor-phase methods, based on the *collision-coalescence mechanism*. As illustrated in Figure 9.14, this mechanism is a limiting case in which the chemical reactions that lead to particle formation are very rapid, so that all particles stick to each other when they collide (i.e., zero activation energy for nucleation). Therefore, the competition between the time scales for particle–particle collisions (coagulation) and coalescence (sintering) eventually determines the morphology of nanomaterials (Xing et al., 1996; Hawa and Zachariah, 2004). If the characteristic sintering time is significantly smaller than the characteristic collision time, colliding particles will merge into a single sphere before another collision event occurs, resulting in the formation of nearly monodispersed spherical particles. On the other hand, if the collision time is much smaller than the characteristic sintering time, particle chain

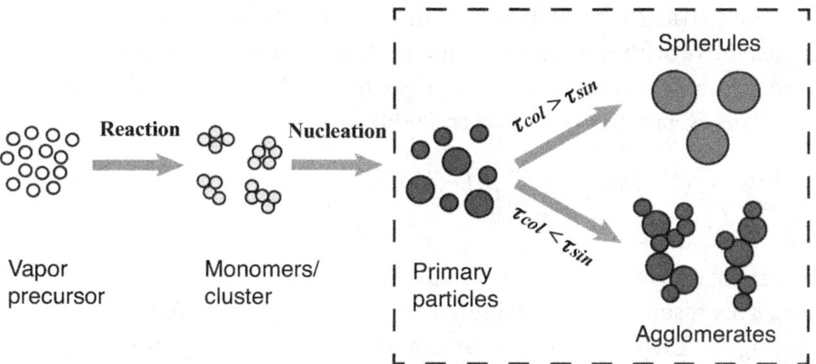

Figure 9.14. Key time scales related to the collision-coalescence mechanism of vapor-phase nanoparticle formation.

agglomerates may form. In the case where the sintering time is large compared with the collision time scale, the time-accurate DEM approach based on an appropriate adhesive contact model is well suited to describe the restructuring and break-up of particle agglomerates. If sintering time is small compared with the collision time scale, a population balance method might be the best way to model the particle transport and coagulation, in which colliding particles are immediately merged into a single quasi-particle. This section focuses on determination of the time scales for different phases of the collision-coalescence mechanism, as well as clarifying the underlying physics related to the two key processes of collision and sintering for nanoparticles.

9.3.1. Time Scale of Particle Collisions

Coagulation is a process where small particles collide with each other to form larger coalesced or aggregated particles. There are various mechanisms that can bring particles to collide with each other, including Brownian motion, gravitational settling, laminar shear, turbulent flow, and field-induced coagulation. Which mechanism dominates the process depends on physicochemical properties of the particles and the suspending fluid as well as the flow conditions. Nanoparticles have small inertia and tend to have low volume concentrations, so Brownian motion is usually the dominant collision mechanism for most nanoparticle applications.

During the Brownian coagulation process, the particles move about in a random manner due to the thermal collisions of the particles with gas molecules. As noted in our discussion of particle collision by turbulent flows in Section 6.3, the rate of collisions per unit volume $\dot{n}_{c12}$ between a group of particles with radius r_1 and a second group of particles with radius r_2 can be written in terms of the number of particles per unit volume of each group, n_1 and n_2, as

$$\dot{n}_{c12} = \alpha_{12} n_1 n_2. \tag{9.3.1}$$

The collision kernel α_{12} depends on the particle radii and relative velocities. When the collision process is dominated by Brownian motion, the expression for collision kernel depends on Knudsen number. In the free-molecular regime ($Kn \gg 1$), the probability for a gas molecule to collide with the particle is low, and the particles therefore move along straight trajectories relatively long distances between collisions. The particles move with a purely ballistic motion between collisions, and the resulting collision kernel is similar to that for a gas molecule, which is given by the kinetic theory of gases (Glassman and Yetter, 2008) as

$$\alpha_{12} = \left(\frac{3}{4\pi}\right)^{1/6} \left(\frac{6k_B T}{\rho_p}\right)^{1/2} \left(\frac{1}{V_1} + \frac{1}{V_2}\right)^{1/2} \left(V_1^{1/3} + V_2^{1/3}\right)^2 \tag{9.3.2}$$

where ρ_p is the particle density and V_1 and V_2 are the volumes of individual particles in the two groups. This result can be obtained from the Abrahamson collision kernel (6.3.13) by replacing the turbulent flow mean-square velocities $\langle v_{1m}^2 \rangle$ and $\langle v_{2m}^2 \rangle$ with the Brownian motion mean-square velocities $3k_B T/m_1$ and $3k_B T/m_2$, respectively.

In the continuum regime ($\text{Kn} \ll 1$), the gas molecules collide frequently with the particle and the particle moves in a random, meandering path. In this regime, Brownian collision can be described as a Brownian diffusion process, for which the collision kernel is given by

$$\alpha_{12} = 4\pi D_b (r_1 + r_2), \tag{9.3.3}$$

where D_b is determined by the Einstein-Stokes expression (9.1.6). Using the general form of the Brownian diffusion coefficient for unequal-size particles and substituting into (9.3.3), the Brownian collision kernel in the continuum regime can be expressed in terms of the particle volumes as

$$\alpha_{12} = \frac{2k_B T}{3\mu C_C} \left(\frac{1}{V_1^{1/3}} + \frac{1}{V_2^{1/3}}\right) \left(V_1^{1/3} + V_2^{1/3}\right). \tag{9.3.4}$$

The characteristic collision time for a mono-disperse particle system can be approximately determined using (9.3.1) as

$$\tau_{col} = \frac{2}{\alpha_{11} n_1}. \tag{9.3.5}$$

This collision time scale provides an estimate of the time interval between particle collisions. It is noted that the collision time scale is different from the contact time scale $\tau_C \cong 2.868(m^2/E^2 R v_0)^{1/5}$ given in (3.2.22), where the latter represents the characteristic time that particles remain in contact during a collision.

The effect of particle long-range interactions on collision frequency is a long-standing problem in aerosol dynamics. In general, long-range interaction refers to forces with a range of action that is of the order of the particle radius or longer. For nanoparticles, the long-range interactions mainly include van der Waals interaction, Coulomb electrostatic forces, and dipole-dipole interaction. Fuchs (1964) proposed that the effect of long-range interaction be accounted for with use of an enhancement factor in the equation for collision rate and collision kernel, giving

$$\dot{n}_C = W_{enh} \dot{n}_{C0}, \quad \alpha = W_{enh} \alpha_0, \tag{9.3.6}$$

where $\dot{n}_C$ and α are the collision rate and collision kernel including the effects of long-range interaction and $\dot{n}_{C0}$ and α_0 are the same quantities for Brownian coagulation without long-range interaction. Experimental results shown in Figure 9.15 indicate the rapid growth in value of the enhancement factor as the particle

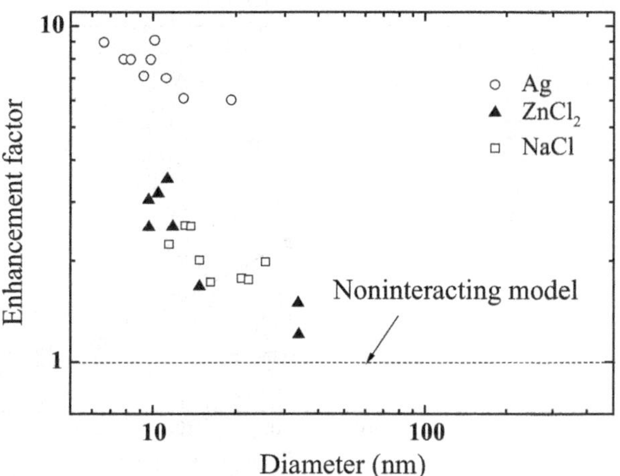

Figure 9.15. Experimental enhancement factors as a function of diameters for Ag, ZnCl$_2$, and NaCl obtained by Okuyama et al. (1984). The dashed line illustrates the case with no particle interaction, for which the enhancement factor is equal to unity. The experiment confirms the effect of particle interaction on collision rate when the particle size decreases to the nanoscale size range.

size decreases within the nanoparticle size range. A wide variety of expressions have been proposed for the enhancement factor W_{enh}, both for the continuum regime and for the free-molecule regime with both charged and neutral particles. A review of various expressions for enhancement factor in these two regimes for problems in which the particle interaction is governed by a potential field was given by Ouyang et al. (2012).

Zhang et al. (2011) applied molecular dynamics (MD) to study the role of dipole-dipole interaction on enhancing the Brownian collision of neutral TiO$_2$ nanoparticles in the free molecular regime. To quantitatively evaluate the enhancement factor, four characteristic dipole-moment directions/orientations were examined, as illustrated in Figure 9.16. The parallel-to-path, co-orientated dipoles exhibited the greatest enhancement factor values, and the parallel-to-path, counter-orientated dipoles gave the least enhancement factors. Comparatively, the perpendicular-to-path dipoles, either co-orientated or counter-orientated, straddle the two limiting cases. The dipole-dipole interaction is found to have a significant influence on the enhancement factor values, particularly at lower temperatures. Figure 9.17 illustrates the enhancement factors including both van der Waals and dipole-dipole interaction forces for temperatures ranging from 273 K to 1,273 K. As the temperature increases,

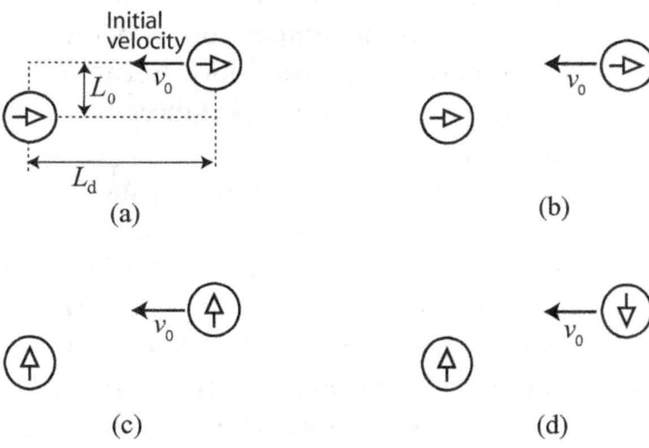

Figure 9.16. Four initial dipole directions/orientations investigated in the MD simulation: (a) parallel-to-path, co-oriented dipoles; (b) parallel-to-path, counter-oriented dipoles; (c) perpendicular-to-path, co-oriented dipoles; (d) perpendicular-to-path, counter-oriented dipoles.

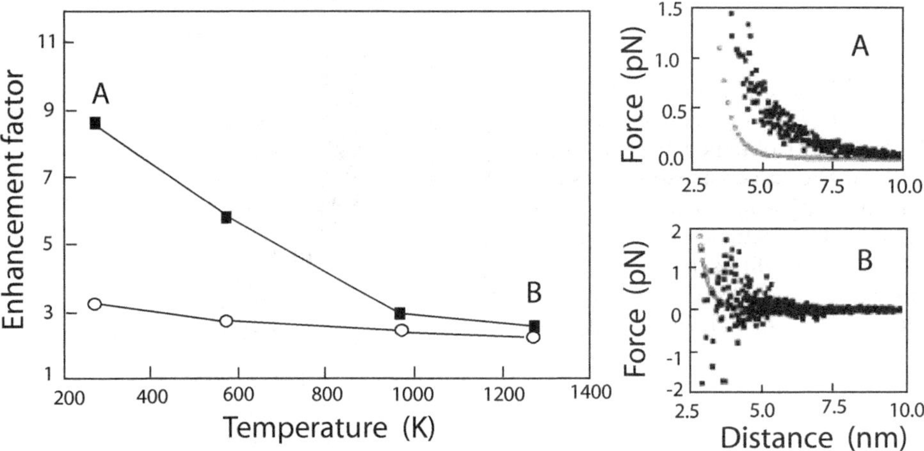

Figure 9.17. Effect of system temperature on enhancement factors of coagulation. Open circles indicate the case with only van der Waals force, and the filled square symbols indicate the case with both dipole-dipole force and van der Waals force. Subplots A and B show the forces between anatase TiO_2 nanoparticles at 273 K and 1,273 K, respectively. [Reprinted with permission from Zhang et al. (2011).]

the enhancement factor drops rapidly due to the reduction of the particle dipole moments as a result of thermal fluctuations.

9.3.2. Time Scale of Nanoparticle Sintering

Coalescence of particles via sintering is a common occurrence in nanoparticle processing, and it is primarily responsible for determining the final morphology and size of the particles. As discussed in Section 4.6, sintering is a thermal process that involves mass transport at the atomic scale. The coalescence of nanoparticles shares the same physical process as observed for microscale particles, but nanoparticle sintering also has a number of distinguishing features due to the small particle size.

9.3.2.1. Mass Transport in Sintering

The nature of mass transport during sintering processes depends on the temperature relative to the melting point of the material. If the particle temperature is above the melting point, or if the material is amorphous, the mass transport is governed by fluid advection. If the temperature is below the melting point, atomic diffusion governs the sintering mass transport. Because most nanoparticle coagulation processes of interest occur with temperatures below the melting point, the latter process is of primary interest.

Several different diffusion mechanisms have been reported to describe the transport mechanism of the atoms (German, 1996). These mechanisms include *surface diffusion* (SD), *volume diffusion* (VD), *grain-boundary diffusion* (GBD), and *evaporation-condensation processes* (E-C). An illustration of these various mechanisms is given in Figure 9.18. Surface diffusion is caused by the presence of a concentration gradient on surfaces of different curvature. Volume diffusion refers to the transport of atoms from the interior of the particle to the surface, as well as transport

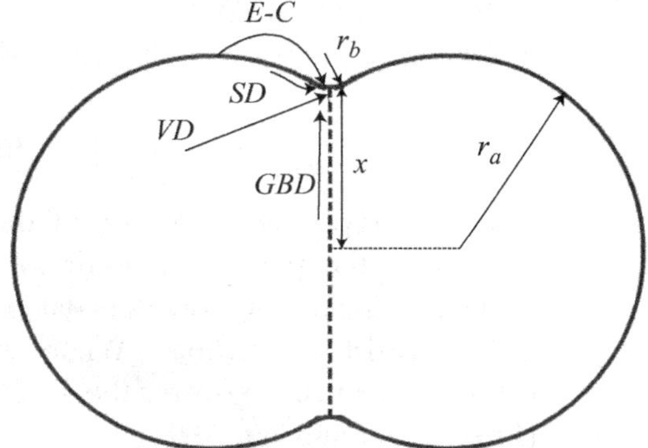

Figure 9.18. Diffusion mechanisms of mass transport in sintering. SD: surface diffusion. VD: volume diffusion. GBD: grain boundary diffusion. E-C: evaporation-condensation model.

of vacancies from the surface to the interior. Grain-boundary diffusion occurs at the grain boundary separating the two particles. The evaporation-condensation process involves phase change of the atoms under a given vapor pressure, where the fluid pressure is influenced by local surface curvature due to the effect of surface tension. All these mechanisms tend to decrease the surface area and surface energy of the system.

For nanoparticle sintering processes, surface diffusion is usually considered to be the dominant mechanism due to the small size (and high curvature) of the particles. Volume diffusion can be neglected as a consequence of the large surface-volume ratio of the particles. Grain boundary diffusion is complex for nanoscale sintering, especially when atomic reorientation or alignment occurs (Zachariah and Carrier, 1999; Ding et al., 2009). Lehtinen and Zachariah (2001, 2002) pointed out that the heat release resulting from the decrease in surface area and surface energy during sintering acts to increase the particle temperature, which in turn enhances the coalescence rate of the nanoparticles.

9.3.2.2. Characteristic Sintering Time

Coalescence between two nanoparticles in an aerosol system generally begins with Brownian collision of the particles and ends when a new (larger) particle is formed from two colliding particles. The time scale for this whole process, called the *coalescence time*, ranges from nanoseconds to milliseconds according to the size of the particles.

Koch and Friedlander (1990) developed a model for the effect of particle coalescence on the change in particle surface area as

$$\frac{dS_a}{dt} = -\frac{1}{\tau_{\text{sin}}}(S_a - S_{a,\,final}),\qquad(9.4.1)$$

where S_a is the total surface area of the colliding particles before sintering, $S_{a,\,final}$ is the particle surface area after sintering, and τ_{sin} denotes the sintering time scale (or the coalescence time). If the particles are initially perfect spheres, then $S_{a,\,final}$ will be 79% of the initial area S_a. As mentioned in the previous section, viscous flow governs the sintering mass transport when the particle temperature is above the material melting point. In this case, the sintering time is affected by surface tension and material viscosity, so that (Frenkel, 1945)

$$\tau_{\text{sin}} = \frac{\mu_p d}{\sigma_p}, \tag{9.4.2}$$

where μ_p represents viscosity of the melted particle material, σ_p is the surface tension of the particle melt-air system, and d is the particle diameter before sintering. The driving force of coalescence is surface tension, and the melt viscosity acts to resist coalescence. When the temperature is below the melting point, solid-state diffusion governs the mass transport and the sintering time scale becomes (Friedlander and Wu, 1994)

$$\tau_{\text{sin}} = \frac{3k_B T V_p}{64\pi D \sigma_p \Omega_a}, \tag{9.4.3}$$

where k_B is the Boltzmann constant, T is particle absolute temperature, D is the solid-state diffusion coefficient, σ_p is the surface tension, and V_p and Ω_a denote the particle volume and the molecular volume, respectively.

The material properties for the nanomaterial melt, such as diffusion coefficient and surface tension, are usually quite different from those of the bulk material and not known a priori. Various empirical formulas have been reported to estimate the coalescence time. For example, Xiong and Pratsinis (1993) in studies of TiO_2 nanoparticles estimated sintering time as

$$\tau_{\text{sin}} = 8.3 \times 10^{24} T d^4 \exp\left(\frac{3700}{T}\right), \; d \text{ in } m; \tag{9.4.4}$$

where d is particle diameter before sintering. Ehrman et al. (1998), also working with TiO_2 nanoparticles, proposed

$$\tau_{\text{sin}} = 1.87 \times 10^9 T d^3 \exp\left(-\frac{34372}{T}\right), \; d \text{ in } m. \tag{9.4.5}$$

Kobata et al. (2004) proposed a similar formula

$$\tau_{\text{sin}} = 1.19 \times 10^{18} T r_p^4 \exp\left[\frac{258 \, kJ/mol}{RT}\right], \; r_p \text{ in } m. \tag{9.4.6}$$

These formulas all have the basic form of some factor times a term that varies exponentially in the inverse temperature; nevertheless, the final estimate for the coalescence time can differ widely between the formulas. Some researchers attribute the difference to the overestimation of coalescence time when the particles are relatively small (Windeler et al., 1997; Wu et al., 1993). These results imply that sintering at nanoscales requires further investigation and reconsideration of model assumptions.

Figure 9.19 shows typical Brownian collision time and sintering time (calculated using Equation (9.4.5)) values as functions of particle diameter for different temperatures. The characteristic times exhibit strong dependence on particle size. When the sintering time is larger than the collision time for a given temperature and particle size, coalescence between the colliding nanoparticles cannot be completed in the time between subsequent collisions of the particle. In this case, agglomerates are formed with soft interparticle bonding and the size of the primary particle remains unchanged. On the other hand, if the collision time is larger than the sintering time scale, new particles form with approximately spherical shape.

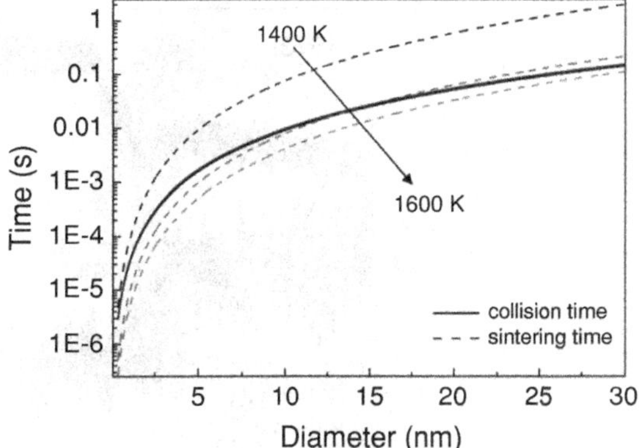

Figure 9.19. Typical Brownian collision time and sintering time scales for different temperatures.

9.3.3.3. Lattice Structure in Coalescence

The lattice structure of nanoparticles is no longer uniform during coalescence, during which it is influenced by the extremely high surface curvature that occurs near the sintering region. Molecular dynamics simulation is frequently used to study lattice structure of nanoparticles during coalescence. For example, Zhu (1996) used MD to examine grain boundary rotation during nanoparticle sintering. Other researchers have calculated particle coalescence time using MD and made comparisons with different models (Zachariah and Carrier, 1999; Buesser et al., 2011).

The lattice structure of TiO_2 nanoparticles during coalescence was recently investigated by Zhang et al. (2012b), who found the structure to be sensitive to particle size and local temperature. For a 3 nm TiO_2 (anatase) nanoparticle at a temperature below the melting point, the typical lattice structure is a crystal core surrounded by an amorphous shell. When the particle size is below a critical value at a specific temperature, a structural transition occurs from the core-shell structure to a completely amorphous structure. Figure 9.20 shows the dependence of the lattice structure regimes on particle diameter and temperature for TiO_2 particles.

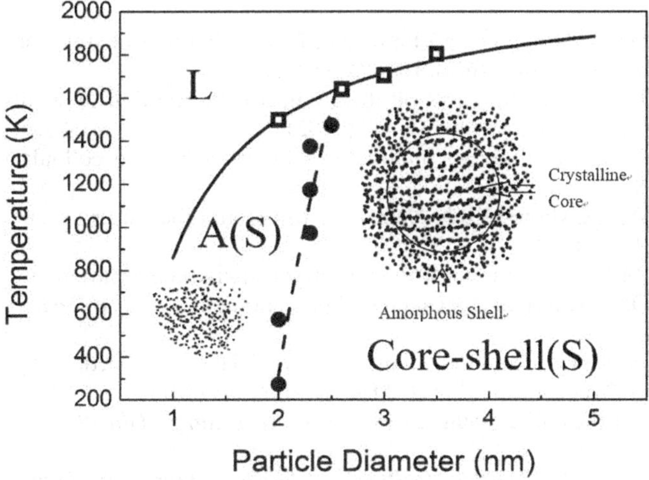

Figure 9.20. Lattice structure map of TiO_2 (anatase) as a function of temperature and particle diameter. L: liquid. A(S): solid in amorphous. Core-shell (S): solid in core-shell structure. (Based on data from Zhang et al., 2012b.)

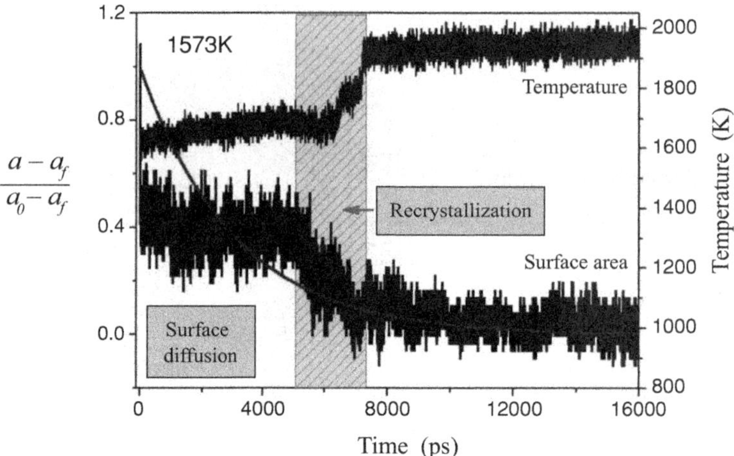

Figure 9.21. Coalescence process between two nanoparticles with initial temperature at 1,573 K.

Sintering between two crystal nanoparticles at a temperature well below the melting point is initially dominated by surface diffusion with amorphous shell fusion and neck formation, and subsequently dominated by grain-boundary reorientation within the later stages of sintering. As the temperature approaches the melting point, recrystallization of the particle structure occurs. Figure 9.21 reveals the predicted temperature increase and the surface area reduction during MD simulations of coalescence of two TiO_2 nanoparticles (anatase with core-shell structure) at an initial temperature of 1573 K. First, the amorphous shell fuses by surface diffusion, then a new crystal core appears in the center of the newly coalesced particle accompanied a sharp temperature increase due to the recrystallization process. The surface area reduction is far below the prediction of the Koch-Friedlander model (9.4.1) at the start of coalescence due to the fact that the core-shell structure of the nanoparticle violates the assumption of homogeneity made in the model.

REFERENCES

Adachi M, Kousaka Y, Okuyama K. Unipolar and bipolar diffusion charging of ultrafine aerosol particles. *Journal of Aerosol Science* **16**(2), 109–123 (1985).

Allen MD, Raabe OG. Re-evaluation of Millikan's oil drop data for the motion of small particles in air. *Journal of Aerosol Science* **13**(6), 537–547 (1982).

Becker V, Briesen H. Tangential-force model for interactions between bonded colloidal particles. *Physical Review E* **78**, 061404 (2008).

Bishop KJM, Wilmer CE, Soh SL, Grzybowski BA. Nanoscale forces and their uses in self-assembly. *Small* **5**(14), 1600–1630 (2009).

Blanton SA, Leheny RL, Hines MA, Guyot-Sionnest P. Dielectric dispersion measurements of CdSe nanocrystal colloids: Observation of a permanent dipole moment. *Physical Review Letters* **79**, 865–868 (1997).

Buesser B, Gröhn A J, Pratsinis S E. Sintering rate and mechanism of TiO_2 nanoparticles by molecular dynamics. *Journal of Physical Chemistry C* **115**, 11030–11035 (2011).

Davies CN, Definitive equations for the fluid resistance of spheres. *Proceedings of the Physical Society* **57**(4), 259–270 (1945).

Davis EJ, Schweiger G. *The Airborne Microparticle: Its Physics, Chemistry, Optics and Transport Phenomena*. Springer-Verlag, Heidelberg, pp. 755–807 (2002).

Ding L, Davidchack RL, Pan J. A molecular dynamics study of sintering between nanoparticles. *Computational Materials Science* **45**, 247–256 (2009).

Eggersdorfer ML, Kadau D, Herrmann HJ, Pratsinis SE. Fragmentation and restructuring of soft-agglomerates under shear. *Journal of Colloid and Interface Science* **342**, 261–268 (2010).

Ehrman SH, Friedlander SK, Zachariah MR. Characteristics of SiO2/TiO2 nanocomposite particles formed in a premixed flat flame. *Journal of Aerosol Science* **29**, 687–706 (1998).

Epstein PS. On the resistance experienced by spheres in their motion through gases. *Physical Review* **23**, 710–733 (1924).

Fanelli M, Feke DL, Manas-Zloczower I. Prediction of the dispersion of particle clusters in the nano-scale. I. Steady shearing responses. *Chemical Engineering Science* **61**, 473–488 (2006).

Feng XD, Sayle DC, Wang ZL, Paras MS, Santora B, Sutorik AC, Sayle TXT, Yang Y, Ding Y, Wang XD, Her YS. Converting ceria polyhedral nanoparticles into single-crystal nanospheres. *Science* **312**, 1504–1508 (2006).

Frenkel J. Viscous flow of crystalline bodies under the action of surface tension. *Journal of Physics* **9**, 385–391 (1945).

Friedlander SK. *Smoke, Dust and Haze: Fundamentals of Aerosol Dynamics*. 2nd ed., Oxford University Press, New York (2000).

Friedlander SK, Wu MK. Linear rate law for the decay of the excess surface area of a coalescing solid particle. *Physical Review B* **49**, 3622–3624 (1994).

Fuchs NA, *The Mechanics of Aerosols*. Macmillan, New York (1964).

German RM. *Sintering Theory and Practice*. John Wiley & Sons: Wiley-Interscience, New York (1996).

Glassman I, Yetter RA. *Combustion*. 4th ed., Academic Press, Burlington, Vermont (2008).

Goesmann H, Feldmann C. Nanoparticulate functional materials. *Angew. Chem. Int. Ed.* **49**, 1362–1395 (2010).

Hawa T, Zachariah MR. Molecular dynamics study of particle–particle collisions between hydrogen-passivated silicon nanoparticles. *Physical Review B* **69**, 035417 (2004).

Higashitani K, Iimura K, Sanda H. Simulation of deformation and breakup of large aggregates in flows of viscous fluids. *Chemical Engineering Science* **56**, 2927–2938 (2001).

Kobata A, Kusakabe K, Morooka S. Growth and transformation of TiO_2 crystallites in aerosol reactor. *AIChE Journal* **37**, 347–359 (2004).

Koch W, Friedlander SK. The effect of particle coalescence on the surface area of a coagulating aerosol. *Journal of Colloid and Interface Science* **140**, 419–427 (1990).

Lehtinen KE, Zachariah MR. Effect of coalescence energy release on the temporal shape evolution of nanoparticles. *Physical Review B* **63**, 205402 (2001).

Lehtinen KE, Zachariah MR. Energy accumulation in nanoparticle collision and coalescence processes. *Journal of Aerosol Science* **33**, 357–368 (2002).

Li SQ, Marshall JS. Discrete element simulation of micro-particle deposition on a cylindrical fiber in an array. *Journal of Aerosol Science* **38**, 1031–1046 (2007).

Li ZG. Gas kinetic theory and molecular dynamics simulation of nanomaterial transfer in dilute gases. Ph.D. thesis, University of Delaware, pp. 71–89 (2005).

Li ZG, Wang H. Drag force, diffusion coefficient, and electric mobility of small particles. I. Theory applicable to the free-molecule regime. *Physical Review E* **68**(6), 061206 (2003a).

Li ZG, Wang H. Drag force, diffusion coefficient, and electric mobility of small particles. II. Application. *Physical Review E* **68**(6), 061207 (2003b).

Li ZG, Wang H. Thermophoretic force and velocity of nanoparticles in the free molecule regime. *Physical Review E* **70**, 021205 (2004).

Loyalka SK. Thermophoretic force on a single particle. I. Numerical solution of the linearized Boltzmann equation. *Journal of Aerosol Science* **23**, 291–300 (1992).

Luan B, Robbins MO. The breakdown of continuum models for mechanical contacts. *Nature Letters* **435**, 929–932 (2005).

Madler L, Friedlander SK. Transport of nanoparticles in gases: Overview and recent advances. *Aerosol and Air Quality Research* **7**(3), 304–342 (2007).

Madler L, Lall AA, Friedlander SK. One-step aerosol synthesis of nanoparticle agglomerate films: Simulation of film porosity and thickness. *Nanotechnology* **17**(19), 4783–4795 (2006).

Matsui M, Akaogi M. Molecular dynamics simulation of the structural and physical properties of the four polymorphs of TiO_2. *Molecular Simulation* **6**, 239–244 (1991).

Okuyama K, Kousaka Y, Hayashi K. Change in size distribution of ultrafine aerosol particles undergoing Brownian coagulation. *Journal of Colloid and Interface Science* **101**, 98–109 (1984).

Ouyang H, Gopalakrishnan R, Hogan GJ. Nanoparticle collisions in the gas phase in the presensce of singular contact potentials. *Journal of Chemical Physics* **137**, 064316 (2012).

Pantina JP, Furst EM. Elasticity and critical bending moment of model colloidal aggregates. *Physical Review Letters* **94**, 138301 (2005).

Park SH, Kim WJ, Kim SS. Thermophoretic transport and deposition of particles in vertical tube flow with variable wall temperature and thermal radiation. *KSME International Journal* **13**(3), 253–263 (1999).

Penn RL, Banfield JF. Morphology development and crystal growth in nanocrystalline aggregates under hydrothermal conditions: Insights from titania. *Geochimica et Cosmochimica Acta* **63**(10), 1549–1557 (1999).

Pui DYH, Fruin S, McMurry PH. Unipolar diffusion charging of ultrafine aerosols. *Aerosol Science and Technology* **8**(2), 173–187 (1988).

Sahu M, Park JH, Biswas P. In situ charge characterization of TiO_2 and Cu–TiO_2 nanoparticles in a flame aerosol reactor. *Journal Nanoparticle Research* **14**, 687 (2012).

Schwarz UD, Zwörner O, Köster P, Wiesendanger R. Quantitative analysis of the frictional properties of solid materials at low loads. II. Mica and germanium sulfide. *Physical Review B* **56**(11), 6997–7000 (1997).

Shim M, Guyot-Sionnest P. Permanent dipole moment and charges in colloidal semiconductor quantum dots. *Journal of Chemical Physics* **111**(15), 6955–6964 (1999).

Sone Y, Aoki K. A similarity solution of the linearized Boltzmann equation with application to thermophoresis of a spherical particle. *J. Mecanique Theorique Appliquee* **2**(1), 3–12 (1983).

Stratmann F, Fissan H, Papperger A, Friedlander S. Suppression of particle deposition to surfaces by the thermophoretic force. *Aerosol Science and Technology* **9**(2), 115–121 (1988).

Sun WF, Zeng QH, Yu AB, Kendall K. Calculation of normal contact forces between silica nanospheres. *Langmuir* **29**, 7825–7837 (2013).

Talbot L. Thermophoresis: A review. In *Progress Astro Aero*, Vol. 74, Rarefied Gas Dynamics, SS Fisher, editor, New York, AIAA (1981).

Talbot L, Cheng R. K, Schefer RW, Willis DR. Thermophoresis of particles in a heated boundary layer. *Journal of Fluid Mechanics* **101**, 737–758 (1980).

Tang Z, Kotov N A, Giersig M. Spontaneous organization of single CdTe nanoparticles into luminescent nanowires. *Science* **297**, 237–240 (2002).

Tolmachoff ED, Abid AD, Phares DJ, Campbell CS, Wang H. Synthesis of nano-phase TiO_2 crystalline films over premixed stagnation flames. *Proc Combust Inst* **32**, 1839–1845 (2009).

Tyndall J. On dust and disease. *Proc. Royal Inst.* **6**, 1–14 (1870).

Waldmann L. Über die Kraft eines inhomogenen Gases auf kleine suspendierete Kugeln. *Z. Naturforsch A* **14a**, 589–599 (1959).

Wenning L, Müser MH. Friction laws for elastic nanoscale contacts. *Europhysics Letters* **54**(5), 693–699 (2001).

White FM. *Viscous Fluid Flow*, 3rd ed., McGraw-Hill, New York, pp. 149–151 (2006).

Windeler RS, Lehtinen KE, Friedlander SK. Production of nanometer-sized metal oxide particles by gas phase reaction in a free jet. II. Particle size and neck formation: Comparison with theory. *Aerosol Science and Technology* **27**, 191–205 (1997).

Wu MK, Windeler RS, Steiner CKR, Börs CK, Friedlander SK. Controlled synthesis of nanosized particles by aerosol processes. *Aerosol Science and Technology* **19**, 527–548 (1993).

Xing YC, Koylu UO, Rose RDE. Synthesis and restructuring of inorganic nano-particles in counterflow diffusion flames. *Combustion & Flame* **107**, 85–102 (1996).

Xiong Y, Pratsinis SE. Formation of agglomerate particles by coagulation and sintering. Part I. A two-dimensional solution of the population balance equation. *Journal of Aerosol Science* **24**, 283–300 (1993).

Yamamoto K, Ishihara Y. Thermophoresis of a spherical particle in a rarefied gas of a transition regime. *Physics of Fluids* **31**, 3618–3624 (1988).

Yan W, Li S, Zhang Y, Yao Q, Tse SD. Effects of dipole moment and temperature on the interaction dynamics of titania nanoparticles during agglomeration. *Journal Physical Chemistry C* **114**, 10755–10760 (2010).

Ye Y, Pui DYH, Liu BYH, Opiolka S, Blumhorst S, Fissan H. Thermophoretic effect of particle deposition on a free standing semiconductor wafer in a clean room. *Journal of Aerosol Science* **22**(1), 63–72(1991).

Zachariah MR, Carrier MJ. Molecular dynamics computation of gas-phase nanoparticle sintering: A comparison with phenomenological models. *Journal of Aerosol Science* **30**, 1139–1152 (1999).

Zhang YY, Li SQ, Yan W, Yao Q, Tse SD. Role of dipole–dipole interaction on enhancing Brownian coagulation of charge-neutral nanoparticles in the free molecular regime. *Journal of Chemical Physics* **134**, 084501 (2011).

Zhang YY, Li SQ, Deng SL, Yao Q, Tse SD. Direct synthesis of nanostructured TiO2 films with controlled morphologies by stagnation swirl flames. *Journal of Aerosol Science* **44**, 71–82 (2012a).

Zhang YY, Li SQ, Yan W, Tse SD. Effect of size-dependent grain structures on the dynamics of nanoparticle coalescence. *Journal of Applied Physics* **111**, 124321 (2012b).

Zhang YY, Li SQ, Yan W, Yao Q. Nanoparticle transport and deposition in boundary layer of stagnation-point premixed flames. *Powder Technology* **227**, 24–34 (2012c).

Zhu H. Sintering processes of two nanoparticles: A study by molecular dynamics simulations. *Philosophical Magazine Letters* **73**, 27–33 (1996).

 # Computer Implementation and Data Analysis

The physical models used to describe the forces and torques on particles in various types of DEM simulations are discussed in the previous chapters of this book. This chapter focuses on numerical aspects related to implementation of these models, as well as on different ways to characterize and interpret the results of the DEM simulations.

10.1. Particle Time Stepping

Because the momentum and moment of momentum equations governing particle motion are ordinary differential equations in time, the first task that must be addressed in computer implementation of DEM is the discretization of these equations in time. In order to select an appropriate time step for calculation of the particle motion, it is necessary to first understand the various time scales involved in the problem. The largest time scale is usually the *fluid convective time scale* $\tau_F = O(L/U)$, where L and U denote characteristic macroscopic length and velocity scales of the flow field, respectively. This is itself somewhat of a simplification, as fluid flows often have very different time and length scales depending on position in the flow (e.g., whether or not the particle is in a boundary layer region) and on wavenumber of the fluctuation in a turbulent flow. For present purposes, however, let us assume that a distinct fluid velocity and length scale can be defined. The next smallest scale is typically the *particle convective time scale* $\tau_{CP} = O(d/U)$, where d is the particle diameter. It is important in a DEM simulation that the time step used to move the particles be significantly less than τ_{CP} in order not to miss particle collisions during the time step. In the presence of particle collisions, we must also set the time step so as to resolve elastic response of the particle during the collision. The particle contact time based on the Hertz theory is given in (3.2.22) as $\tau_C \cong 2.868(m^2/E^2 R v_0)^{1/5}$. For an order-of-magnitude analysis, we take $m = O(\rho_p d^3)$, $E = O(E_p)$, $R = O(d)$, and $v_0 = O(U)$, so that *particle contact time scale* $\tau_C = d(\rho_p^2/E_p^2 U)^{1/5}$, where E_p is the particle elastic modulus. Adhesion forces acting between colliding particles have approximately the same time scale as the elastic repulsion force.

10.1.1. Numerical Stability

An additional time scale, called the *particle aerodynamic time scale* $\tau_{AP} = m/\mu d$, is related to resolution of the particle response to change in the fluid velocity field. One way in which this time scale arises is evident by consideration of the numerical stability of the particle moment equation when computed using an explicit scheme. To see this, we examine the simple model in which the particle's inertia is balanced by its drag force, the governing equation for which is given by (1.2.2) as

$$m\frac{d\mathbf{v}}{dt} = -3\pi\mu d(\mathbf{v} - \mathbf{u}). \tag{10.1.1}$$

Nondimensionalizing velocity by U and time by L/U and indicating dimensionless variables with a prime, (10.1.1) becomes

$$\frac{d\mathbf{v}'}{dt'} = -\frac{1}{St}(\mathbf{v}' - \mathbf{u}'), \tag{10.1.2}$$

where the Stokes number St is defined by (1.2.5). The differential equation (10.1.2) is discretized using the simple forward Euler scheme, which is typical of an explicit difference scheme, giving

$$\mathbf{v}'^{n+1} = \left[1 - \frac{\Delta t'}{St}\right]\mathbf{v}'^n - \frac{\Delta t'}{St}\mathbf{u}'^n, \tag{10.1.3}$$

where $\mathbf{v}'^n = \mathbf{v}'(t_n)$ and $\Delta t'$ is the dimensionless time step. Now suppose that the initial velocity value $\hat{\mathbf{v}}'^0$ is perturbed by an amount $\delta\mathbf{v}'$, so that the initial condition in the computation becomes

$$\mathbf{v}'^0 = \hat{\mathbf{v}}'^0 + \delta\mathbf{v}'. \tag{10.1.4}$$

Substituting (10.1.4) into (10.1.3) gives the solution at any time step n as

$$\mathbf{v}'^n = \hat{\mathbf{v}}'^n + \left[1 - \frac{\Delta t'}{St}\right]^n \delta\mathbf{v}', \tag{10.1.5}$$

where $\hat{\mathbf{v}}'^n$ is the solution at time t_n with the exact initial condition $\mathbf{v}'^0 = \hat{\mathbf{v}}'^0$. The magnitude of the error term in (10.1.5) changes in proportion to an amplification term $|\xi|^n$, where $\xi = 1 - \Delta t'/St$, which will increase as time (n) increases for $|\xi| > 1$ and decrease with time for $|\xi| < 1$. Consequently, the condition for stability of the forward Euler scheme for this equation is that $|\xi| \leq 1$, from which we see that computations are stable provided that $\Delta t' \leq 2St$. In terms of dimensional variables, the stability criterion becomes

$$\Delta t \leq \frac{2m}{3\pi\mu d}. \tag{10.1.6}$$

Although this discussion pertains only to the forward Euler scheme, other explicit schemes exhibit a similar stability condition that limits $\Delta t'$ to be less than some multiple of the Stokes number. Of course, it is possible to solve (10.1.2) implicitly using an integrating factor to write the equation as

$$\frac{d}{dt'}[\mathbf{v}'\exp(t'/St)] = \frac{1}{St}\mathbf{u}'\exp(t'/St). \tag{10.1.7}$$

Solving for particle velocity then gives

$$\mathbf{v}'^{n+1} = \mathbf{v}'^{n} \exp(-\Delta t'/\mathrm{St}) + \frac{1}{\mathrm{St}} \exp(-t'^{n+1}/\mathrm{St}) \int_{t'^{n}}^{t'^{n+1}} \mathbf{u}'(\tau) \exp(\tau/\mathrm{St}) \, d\tau. \quad (10.1.8)$$

The integral in (10.1.8) must in general be evaluated numerically. For instance, if we assume that the fluid velocity $\mathbf{u}'$ is approximately constant during the time step, (10.1.8) becomes

$$\mathbf{v}'^{n+1} = \mathbf{v}'^{n} \exp(-\Delta t'/\mathrm{St}) + \mathbf{u}'^{n}[1 - \exp(-\Delta t'/\mathrm{St})]. \quad (10.1.9)$$

The implicit scheme (10.1.9), as well as the more general form (10.1.8), is stable for all values of the time step $\Delta t'$. However, if $\Delta t'$ is larger than O(St), the exponential term in (10.1.9) quickly approaches zero, giving the result that the particle velocity will always be nearly equal to the local fluid velocity. In this case, the particle response to changes in the fluid velocity is nearly instantaneous and the inertial response of the particle is not resolved.

10.1.2. Multiscale Time-Stepping Approaches

The principal challenge in implementing DEM for large numbers of particles is the fact that the various time scales associated with particle transport and collision described earlier have very different orders of magnitude. For instance, dust transport in air is a fairly typical aerosol that is of importance for many different applications, including fouling of electronics and construction equipment. Assuming an aerosol with 10 μm diameter dust particles ($\rho_p = 2650$ kg/m^3) in air flow ($\mu = 1.8 \times 10^{-5}$ Pa·s) with fluid length and velocity scales $L = 1$ cm and $U = 1$ m/s, respectively, the Stokes number is obtained as St $= 0.08$. The fluid time scale is $\tau_F = L/U = 0.01$ s, and the particle convection and aerodynamic time scales are $\tau_{CP} = d/U = 10^{-5}$ s and $\tau_{AP} = 2L\mathrm{St}/U = 0.0016$ s. Assuming an elastic modulus for the dust particle (quartz) of $E_p \cong 70$ GPa, the contact time scale is $\tau_C \cong 10^{-8}$ s. The ratio of the fluid time scale to the contact time scale in this example is $\tau_F/\tau_C = 10^6$, and the ratio of the fluid time scale to the particle time scale is $\tau_F/\tau_{CP} = 10^3$. This large difference in the observed time scales indicates a very significant degree of numerical stiffness in computational solution of the particle transport.

A similar situation exists in a typical colloidal solution, although here the relationship between the two particle time scales τ_{CP} and τ_{AP} can often be reversed. Flow of red blood cells in a large artery is perhaps a typical example. For this estimate, let us treat red blood cells as rigid particles with effective diameter of approximately 3 μm. The fluid length and velocity scales in an artery correspond to the artery diameter and mean velocity, given approximately by $L \cong 1$ cm and $U \cong 10$ cm/s, respectively. The fluid time scale and the particle convection time scale for this problem are $\tau_F \cong 0.1$ s and $\tau_{CP} \cong 3 \times 10^{-4}$ s. The Stokes number is computed using the properties of blood plasma, with viscosity $\mu \cong 0.003$ Pa · s and density $\rho_f = 1000$ kg/m^3, as St $\cong 1.7 \times 10^{-6}$, giving the particle aerodynamic time scale as $\tau_{AP} \cong 3.3 \times 10^{-7}$ s. The effective elastic modulus for red blood cells was determined experimentally by Dulinska et al. (2006) as $E_p \cong 2.6 \times 10^4$ Pa, giving the particle contact time scale as $\tau_C \cong 1.3 \times 10^{-6}$ s. Because red blood cells are relatively soft, the contact time scale τ_C is about two orders of magnitude larger for this application than it is for the dust aerosol considered earlier. Similarly, the high viscosity of blood plasma compared with air makes the aerodynamic time scale τ_{AP} about four orders of magnitude smaller than it is for the aerosol. As a result, the particle aerodynamic time scale is the smallest time scale in this problem, with a ratio of the fluid to particle aerodynamic time scales of $\tau_F/\tau_{AP} \cong 3 \times 10^5$.

Because of the high numerical stiffness involved in particulate flow systems, it is necessary to implement various methods to try to accelerate the calculation while resolving motion at the different time scales listed earlier. In the absence of particle collisions and for systems with small Stokes number, the relative velocity is known to obey the scaling estimate $|\mathbf{v} - \mathbf{u}| = O(\mathrm{St}\, U)$, as noted in Chapter 1. In this case, the particle acceleration in the particle momentum equation can be approximated to leading order in the Stokes number by the fluid acceleration, or $d\mathbf{v}/dt \cong D\mathbf{u}/Dt$. This approximation was introduced in Section 6.2 with a more detailed derivation, in the context of measuring particle dispersion in turbulent flows. For systems with only particle inertia and particle drag force, the dimensionless particle transport equation (10.1.2) can be approximated by

$$\mathbf{v}' = \mathbf{u}' - \mathrm{St}\, \mathbf{a}' + O(\mathrm{St})^2, \qquad (10.1.10)$$

where $\mathbf{a}'$ is the dimensionless fluid acceleration $\mathbf{a}' \equiv D\mathbf{u}'/Dt'$ and D/Dt is the material time derivative with respect to fluid elements. The approximation (10.1.10) entirely removes the stiffness imposed by the particle aerodynamic stability condition (10.1.6) on the time step, thus allowing use of much larger time steps than would be the case with the full particle momentum equation. Extension of this method for cases with other fluid forces, such as lift and added mass force, was discussed by Ferry and Balachandar (2001) and Ferry et al. (2003). However, this approximation has not yet been successfully applied for problems where particles collide or adhere with each other.

A multiple-time step DEM algorithm was proposed by Marshall (2009) using three time steps – a fluid time step, a particle time step, and a contact time step. For the particle time step, either the particle collision time τ_{CP} or the particle aerodynamic time τ_{AP} is used, whichever is smaller. The algorithm performs time-consuming computational tasks at the longer fluid time step, such as identifying a list of neighboring particles, while minimizing the number of computations that need be performed at the shorter collision and particle time steps.

10.2. Flow in Complex Domains

The various models for fluid force and torque on the particle discussed in Chapter 5, as well as the models for electrostatic force discussed in Chapter 8, require that the fluid velocity and vorticity field and the electric field vector be known at each particle position at every time step. Because DEM simulations usually require a very small time increment, as discussed in the previous section, it is necessary to utilize a fast method for determining these fields if we are to enable simulations with large numbers of particles. In typical applications, the fluid flow is solved using a grid-based computational fluid dynamics approach, so that the fluid velocity and vorticity vectors will be known at the nodes of the grid used for the fluid flow computation. Many different types of grids are used in computational fluid dynamics for different problems and solution approaches, including tetrahedral unstructured grids, multi-block structured grids, and even Cartesian grids. The last type is used, for instance, in formulations where the motion of bodies or interfaces within the flow field are represented using level-set, volume-of-fluid, or related approaches in which the interface is represented by a jump in fluid density and viscosity in a continuous fluid domain, with any interfacial force distributed to neighboring grid cells (Sethian and Smereka, 2003).

In electrostatics, the electric field is often obtained using a boundary element method, as discussed in Section 8.3. However, if there are M panels used to discretize the domain boundary using a BEM approach, computation of the electric field at a point in space will require $O(M^3)$ computations at the initial time step and $O(M^2)$ computations at subsequent time steps. For large values of M and large number of particles, as is typical in BEM with domains of complex shape, it will be too time consuming to repeat this computation at each particle location at each time step. An alternative is to evaluate the electric field vector on a grid covering the flow field and then interpolate onto particle locations from this grid. Usually, the electric field values on the grid are computed using a larger time interval than that used for particle motion.

Prior to interpolation of a field from a grid onto a particle centroid location, it is necessary to identify which grid cell contains the particle. Although grid cell identification is a simple matter with a Cartesian grid using integer division, this task can be quite time consuming for other grid types. We therefore recommend that the grid used for the fluid flow computation be mapped onto a Cartesian grid prior to the particle transport computation. If the fluid grid is fixed in time, this mapping can be done once and stored for later use. With fluid velocity and vorticity fields known on a Cartesian grid, it is a simple matter to determine the grid cell containing a given particle and to interpolate these fields onto the particle location. However, because the flow boundaries no longer correspond with the grid boundaries on the Cartesian grid, it is necessary to utilize an additional function to indicate the distance of any point on the grid to the domain boundary. For this purpose we recommend use of a level-set distance function, which is a function whose absolute value is equal to the distance to the closest point on the flow boundary and whose sign changes as one travels across the boundary. In the two sections that follow, an efficient search algorithm is discussed that can be used to map the flow computation grid onto a Cartesian grid and a method is described for computation of the level-set distance function used to identify domain boundaries on the Cartesian grid.

10.2.1. Particle Search Algorithm

A particle search algorithm proposed by Allievi and Bermejo (1997) can be used to determine which grid cell contains a given space point with arbitrary grid types. For definiteness, this algorithm is illustrated in the current section for quadrilateral elements in a two-dimensional space with coordinates (x, y). Let us suppose that the computational domain $\bar{D}$ can be decomposed into a set of M nonoverlapping regions D_j, such that the union of all of the D_j is equal to $\bar{D}$. As shown in Figure 10.1, a mapping $\mathbf{F}_j$ is defined that maps the unit square $\hat{D}$ in a two-dimensional space with

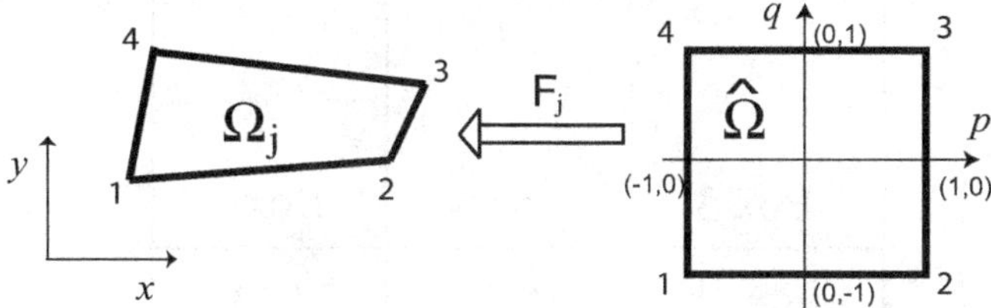

Figure 10.1. Illustration of mapping of a rectangular element into a quadrilateral element, used in the search algorithm of Allievi and Bermejo (1997).

coordinates (p, q) into the element D_j in the x–y plane. We seek to know whether a given point (X, Y) is contained within an element D_j.

For a quadrilateral element, the x- and y-components of the mapping $\mathbf{F}_j : \hat{D} \to D_j$ are given by

$$F_{1j}(p, q) = x = \sum_{i=1}^{4} x_{v,i} G_i(p, q), \quad F_{2j}(p, q) = y = \sum_{i=1}^{4} y_{v,i} G_i(p, q), \quad (10.2.1)$$

where $(x_{v,i}, y_{v,i})$, $i = 1, \ldots, 4$ denote the positions of the four vertices of D_j. The basis functions G_i are given by

$$G_1(p, q) = \tfrac{1}{4}(1 - p)(1 - q), \qquad (10.2.2a)$$

$$G_2(p, q) = \tfrac{1}{4}(1 + p)(1 - q), \qquad (10.2.2b)$$

$$G_3(p, q) = \tfrac{1}{4}(1 + p)(1 + q), \qquad (10.2.2c)$$

$$G_4(p, q) = \tfrac{1}{4}(1 - p)(1 + q). \qquad (10.2.2d)$$

The algorithm consists of two parts, referred to as "search" and "locate." In the search part, we seek to determine which grid cell contains the point (X, Y). The algorithm is initiated by selecting an initial grid cell D_j to examine. This could be done randomly, but if a simple method presents itself by which an initial cell can be selected that is close to the point (X, Y), the search algorithm will converge to the correct cell more rapidly. Once an initial cell is selected, we seek to determine whether or not the point (X, Y) is in this cell and, if not, which neighboring cell should next be examined to optimally find this point. For this purpose, we utilize the iterative expression

$$\begin{Bmatrix} p^{k+1} \\ q^{k+1} \end{Bmatrix} = \begin{Bmatrix} p^k \\ q^k \end{Bmatrix} + \frac{1}{\Delta^k} \begin{bmatrix} b_2 + b_3 p^k & -a_2 - a_3 p^k \\ -b_1 - b_3 q^k & a_1 + a_3 q^k \end{bmatrix} \begin{Bmatrix} X - x^k \\ Y - y^k \end{Bmatrix}, \quad (10.2.3)$$

where x^k and y^k are the values of x and y corresponding to p^k and q^k, as determined from (10.2.1), and

$$a_1 = \tfrac{1}{4}(x_{v,2} - x_{v,1} + x_{v,3} - x_{v,4}), \; b_1 = \tfrac{1}{4}(y_{v,2} - y_{v,1} + y_{v,3} - y_{v,4}),$$

$$a_2 = \tfrac{1}{4}(x_{v,3} - x_{v,1} + x_{v,4} - x_{v,2}), \; b_2 = \tfrac{1}{4}(y_{v,3} - y_{v,1} + y_{v,4} - y_{v,2}),$$

$$a_3 = \tfrac{1}{4}(x_{v,1} - x_{v,2} + x_{v,3} - x_{v,4}), \; b_3 = \tfrac{1}{4}(y_{v,1} - y_{v,2} + y_{v,3} - y_{v,4}),$$

$$\Delta^k = (a_1 b_2 - a_2 b_1) + (a_1 b_3 - a_3 b_1)p^k + (a_3 b_2 - a_2 b_3)q^k.$$

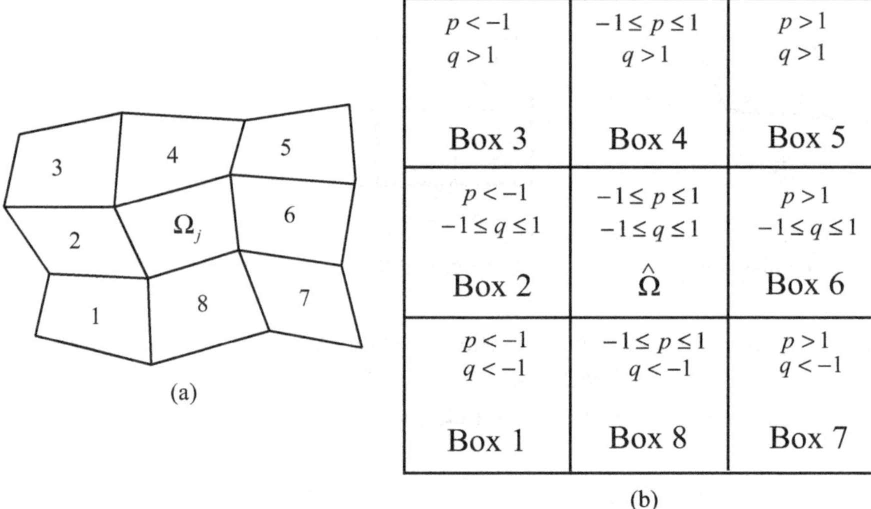

Figure 10.2. Numbering of neighbor grid cells used in the search algorithm of Allievi and Bermejo (1997): (a) in the original grid in x–y space; (b) in the transformed grid in p-q space.

This expression is simply the Newton-Raphson iteration formula for the mapping (10.2.1). In the search algorithm, the initial value of p and q is selected as $(p^0, q^0) = (0, 0)$, corresponding to the center of cell $\hat{D}$. The formula (10.2.3) is used to find the value of (p^1, q^1). The point (X, Y) is contained within the given grid cell if and only if (p^1, q^1) satisfies

$$-1 \leq p^1 \leq 1, \quad -1 \leq q^1 \leq 1. \tag{10.2.4}$$

If the condition (10.2.4) is not satisfied, a different grid cell must be selected for examination. This is done by moving to one of the neighboring cells of the original grid cell, based on the value of (p^1, q^1) as indicated in Figure 10.2b. For instance, if $p^1 < -1$ and $q^1 < -1$, then we move to the neighbor indicated by the number 1 in Figure 10.2a, and so forth. The procedure is repeated until a cell is found that satisfies (10.2.4).

Once the correct cell is found containing the point (X, Y), it is necessary to determine the value of p and q corresponding to this point, denoted as (P, Q). This can be accomplished by iterating on (p^k, q^k) using (10.2.3) until some appropriate convergence threshold is met. Once the transformed point (P, Q) is found, interpolation of any field $f(x, y)$ from the grid onto the point (X, Y) can be performed using

$$f(X, Y) = \sum_{i=1}^{4} f_{v,i} G_i(P, Q), \tag{10.2.5}$$

where G_i are the basis functions given in (10.2.2) and $f_{v,i} = f(x_{v,i}, y_{v,i})$ are the values of $f(x, y)$ on the four vertices of the grid cell.

The search procedure presented here can be used at each time step to locate each particle within the grid used to compute the fluid flow, in order that the fluid velocity and vorticity fields can be interpolated onto the particle locations. However, even with an optimized search algorithm it is usually far too time consuming to perform such a search at every time step, particularly because DEM computations generally require a large number of time steps. Alternatively, this procedure can be used to generate a mapping between the fluid computation grid and a Cartesian grid that covers the entire flow field. Assuming that both grids are fixed in time, such an approach allows the fluid velocity to be rapidly mapped onto the Cartesian grid and then interpolated from there onto the particle locations. Although this procedure involves two interpolation steps, it requires that the time-consuming search step only be performed once at the beginning of the computation.

10.2.2. Level Set Distance Function

Interpolation of the fluid velocity and vorticity fields onto a Cartesian grid results in internalization of the flow domain boundaries, with the result that the domain boundaries no longer coincide with the grid boundaries. In order to include collisions between particles and the domain boundaries in the computations, a level-set distance function $\varphi(\mathbf{x}, t)$ can be used, which is defined such that $\varphi(\mathbf{x}, t) = 0$ on the domain boundary. The absolute value of $\varphi(\mathbf{x}, t)$ is equal to the distance to the nearest domain boundary and the sign of $\varphi(\mathbf{x}, t)$ is positive inside the flow domain and negative outside the flow domain. Based on this definition, it is evident that the level-set function satisfies the Eikonal equation

$$|\nabla\varphi| = 1. \tag{10.2.6}$$

The value of the level-set function is determined on the Cartesian grid points. For cases where the flow computation grid is fixed, the level-set function can be computed once at the start of the computation and then interpolated onto the centroid location of the particles at each time step. Collision of a particle with a domain boundary (e.g., a wall) occurs when the value φ_i of the level-set function at the centroid of particle i is less than the particle radius r_i. The normal overlap between the particle and the wall is given by $\delta_N = r_i - \varphi_i$, and the particle-wall contact point is given by $\mathbf{x}_C = \mathbf{x}_i - \hat{\mathbf{n}}\varphi_i$, where $\hat{\mathbf{n}} = \nabla\varphi/|\nabla\varphi|$ is the wall unit normal (pointing into the flow domain).

The level-set distance function can be constructed by solving (10.2.6) using the fast marching method of Sethian (1996a). In this method, the level-set function is first initialized at Cartesian grid points that surround the domain boundaries (i.e., the $\varphi = 0$ surface). The value of the level-set function is then obtained for other grid points by solution of (10.2.6) in the discretized form

$$\left[\max\left(D_{ijk}^{-x}\varphi, -D_{ijk}^{+x}\varphi, 0\right)^2 + \max\left(D_{ijk}^{-y}\varphi, -D_{ijk}^{+y}\varphi, 0\right)^2 + \max\left(D_{ijk}^{-z}\varphi, -D_{ijk}^{+z}\varphi, 0\right)^2\right]^{1/2} = 1 \tag{10.2.7}$$

where D_{ijk}^{-x} and D_{ijk}^{+x} denote the backward and forward differences in x, and so forth for the y and z directions. This equation results in a quadratic equation that can be solved to obtain the level-set function at each Cartesian grid point using given values of the level-set function on neighboring grid points. The fast marching method solves (10.2.7) on the remaining Cartesian grid points using only upwind values of φ, which is achieved by propagating information from smaller values of $|\varphi|$ (i.e., from grid points lying closer to the $\varphi = 0$ curve) to larger values of $|\varphi|$. A detailed explanation

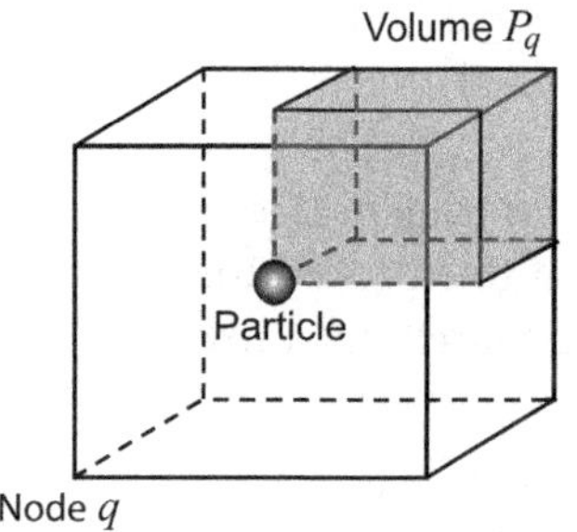

Figure 10.3. Illustration showing volume P_j (shaded) opposite to node j. [Reprinted with permission from Marshall and Sala (2013).]

of the fast marching algorithm for solving (10.2.7) is given in the monograph by Sethian (1996b, Chapter 8).

10.3. Measures of Local Concentration

It is often desirable to compute the concentration field for particulate flows. The concentration field is used as a way to illustrate variation of particle number density or segregation of particles of different volume within the flow field, and it is an important step in computing the body force imposed on the fluid flow due to the particle interfacial force. Computation of the concentration field can often be challenging, particularly when the particle diameter d is a significant fraction of the grid increment size Δx. A comparative study of five different methods for computation of concentration field was given by Marshall and Sala (2013), using two methods based on use of a Cartesian grid and three methods based on use of a radial basis function representation.

The most common approach for computing the concentration field distributes the volume of each particle to the nodes of the grid cell containing the particle centroid. The change in the concentration $\Delta\phi_{q,j}$ at the jth node of a grid cell q due to the nth particle is given by

$$\Delta\phi_{q,j} = \frac{V_{p,n}P_j}{G_{cell}^2} \tag{10.3.1}$$

where P_j, $j = 1, \ldots, 8$, is the volume of the part of grid cell q opposite to node j, as illustrated in Figure 10.3. In this equation, $V_{p,n}$ is the volume of particle n and G_{cell} is the volume of cell q. The *box-counting method* given by (10.3.1) satisfies the discretely conservative property, which requires that the numerical approximation of the integral of the concentration field over the flow volume is equal to the sum of the volume of particles in the flow, or

$$M_0 \equiv \sum_{q=1}^{Q} \phi_q G_{cell} = \sum_{n=1}^{N} V_{p,n} \tag{10.3.2}$$

where ϕ_q is the average concentration value in grid cell q. One difficulty with this method is that it produces an excessive amount of noise for cases where $d/\Delta x$ is not sufficiently small.

The *concentration blob method* for computation of the concentration field uses a radial basis function approximation (Babic, 1997; Zhu and Yu, 2002), in which the

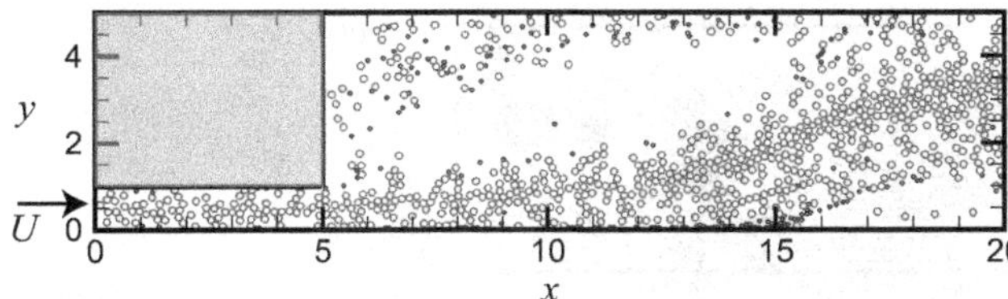

Figure 10.4 Plot showing particles of large and small sizes in a backward-facing step flow, which originates in the small channel on the left and exits the computational domain from the large channel on the right-hand side of the figure.

concentration at a position $\mathbf{x}$ is written as a sum of concentration "blobs," in which the centroid $\hat{\mathbf{x}}_n$ of blob n coincides with that of particle n. The concentration field associated with each blob is decomposed as the product of an amplitude coefficient A_n and a weighting function f, such that the total concentration field is given by

$$\phi(\mathbf{x}, t) = \sum_{n=1}^{N} A_n f(\mathbf{x} - \hat{\mathbf{x}}_n, R_n) \tag{10.3.3}$$

where R_n is a characteristic distance associated with blob n, called the blob radius. The function f is normalized so that its integral over all space equals unity. A common weighting function is the Gaussian function

$$f(\mathbf{x} - \hat{\mathbf{x}}_n, R_n) = \frac{2}{3\pi R_n^3} \exp\left[- \left|\mathbf{x} - \hat{\mathbf{x}}_n\right|^2 / R_n^2 \right]. \tag{10.3.4}$$

If the blob amplitude A_n is set equal to the particle volume $V_{p,n}$, the exact integral of (10.3.3) over the flow field gives

$$\int_{V_{flow}} \phi(\mathbf{x}, t)\, dv = \sum_{n=1}^{N} V_{p,n}. \tag{10.3.5}$$

The concentration blob method yields a smoothly varying concentration function for arbitrary grid increment size. However, the method does not satisfy the discretely conservative property (10.3.2).

A third method for concentration field calculation, proposed by Marshall and Sala (2013), is given by a discretely conservative form of the concentration blob method, in which instead of setting the blob amplitude equal to the particle volume, it is specified as

$$A_n = \frac{V_{p,n}}{G_{cell} \sum_{j=1}^{Q} f(\mathbf{x}_j - \hat{\mathbf{x}}_n, R_n)}, \tag{10.3.6}$$

where $\mathbf{x}_j$ is the location of the centroid of grid cell j and $\hat{\mathbf{x}}_n$ is the centroid of particle n. The discretely conservative property (10.3.2) is identically satisfied when the blob amplitude is set using (10.3.6).

An example illustrating different concentration computation methods is shown for the backward-facing step flow field shown in Figure 10.4. In this flow, particles are carried into the computational domain within the small channel on the left-hand side of the figure. Two particle sizes are used, with a large particle size with dimensionless

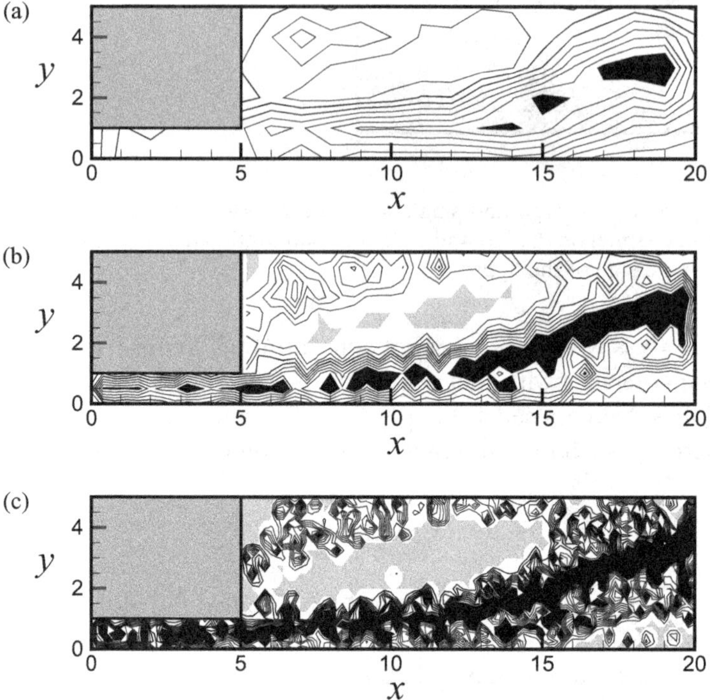

Figure 10.5. Concentration contours obtained using the box-counting method with different numbers of grid cells: (a) 25×5 grid; (b) 50×10 grid; (c) 100×20 grid. Regions with $\phi < 0.05$ are shaded gray and regions with $\phi > 0.4$ are shaded black.

diameter $d_1 = 0.16$ (nondimensionalized by the inlet channel width L) and small particles with diameter $d_2 = 0.08$. These two particle sizes are selected to mimic flow of red blood cells and platelets, respectively, in both size and number ratio in a two-dimensional inlet channel with fluid Reynolds number $\mathrm{Re}_F = 125$. In keeping with the usual particle migration in blood flows, the larger particles (RBCs) are primarily located in the central region at the inlet and the small particles (platelets) are primarily located near the sides (Aarts, 1988). The flow field separates at the step and reattaches to the top boundary $x = 16$. However, the pressure gradient downstream of the step causes a second flow separation along the bottom surface at about $x = 15$. Most large particles are swept downstream in the region of highest flow velocity. However, some large particles and a relatively high number of small particles become caught in the recirculation region near the top stagnation point and are carried backward into the recirculation region.

Concentration computations for this flow were conducted with all three methods described earlier; however, the results with the radial basis function method (10.3.3) and the discretely conservative form of this method given in (10.3.6) were visually almost identical, so we only show the former results. Figure 10.5 shows results for the concentration fields computed using the box-counting method with three different grid sizes. For the coarsest grid size, the contour lines are poorly resolved and exhibit jagged corners. The middle-size grid begins to exhibit significant noise, and the finest grid exhibits considerable noise in the concentration field. This noise arises because the grid cell size for this grid is close to the large particle diameter, so the number

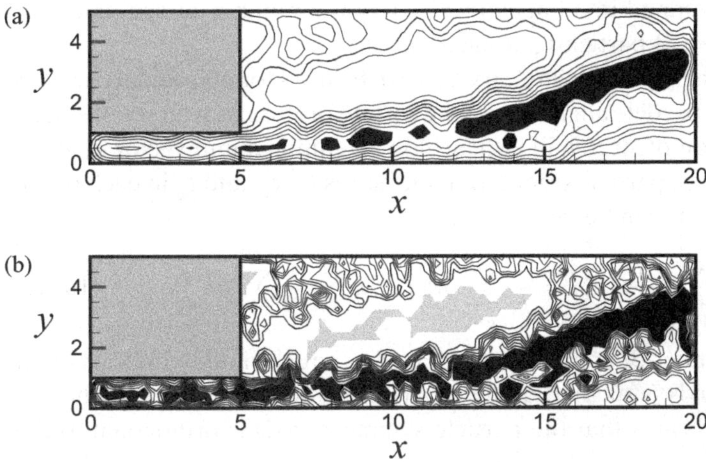

Figure 10.6. Concentration contours obtained using the concentration blob method on a 100×20 grid, with blob radius: (a) $R = 0.4$; (b) $R = 0.2$. Regions with $\phi < 0.05$ are shaded gray and regions with $\phi > 0.4$ are shaded black.

of particles in a grid cell is small. Figure 10.6 shows concentration results obtained using the concentration blob approach on the finest grid (100×20), with blob radius $R = 0.4$ in Figure 10.6a and $R = 0.2$ in Figure 10.6b. Figure 10.6a exhibits a smooth concentration field that appears to be consistent with the particle locations in Figure 10.4. The results with $R = 0.2$, on the other hand, exhibit significant noise because the blob radius has been reduced to significantly less than the distance between neighboring particles. In general, both the box-counting and concentration blob methods are capable of generating accurate concentration fields provided that blob radius and grid increment used in the computations are carefully selected, although in general the results of the concentration blob computations are smoother than those for the box counting method.

10.4. Measures of Particle Agglomerates

The structure of an agglomerate is related to the number and mass of particles making up the agglomerate, the volume of the region occupied by the agglomerate, the relative locations of contact points of touching particles, and (for nonspherical particles) the relative orientations of these particles. A variety of different measures are utilized for analysis of these different features of agglomerate structure. The following discussion presents some of these measures for systems composed of spheroidal particles, which are used because although spheroids still offer a relatively simple description of the particle geometry, they have the additional feature of a preferential orientation that is not present in spherical particles.

10.4.1. Particle Count and Orientation Measures

A number of simple agglomerate measures can be developed by tracking the *agglomerate size distribution*, which gives the percentage of agglomerating particles contained in agglomerates consisting of some number n particles. Dividing the total number of particles contained in all agglomerates by the number of agglomerates gives the average number of particles per agglomerate.

The orientation of the particle symmetry axis can be used for ellipsoidal particles to measure tendency for particle alignment along certain preferential directions, such as the maximum direction of stretch in a shear flow. Defining a unit vector **m** along the particle symmetry axis, particle orientation measures ζ_x, ζ_y, and ζ_z in each of the three coordinate directions can be introduced as

$$\zeta_x = \frac{1}{N}\sum_{i=1}^{N}|m_x|, \quad \zeta_y = \frac{1}{N}\sum_{i=1}^{N}|m_y|, \quad \zeta_z = \frac{1}{N}\sum_{i=1}^{N}|m_z|, \tag{10.4.1}$$

where N is the total number of particles. A value of ζ_x equal to unity indicates that the particle symmetry axis is aligned along the x-direction (streamwise), whereas a value of ζ_x of zero indicates that the particle symmetry axis is orthogonal to the streamwise direction.

10.4.2. Agglomerate Orientation Measures

Chesnutt and Marshall (2010) introduced a number of orientational measures for agglomerates formed of spheroidal particles. A *symmetry-axis-angle orientation measure* O_I is defined based on the angle θ_{ij} between the symmetry axes of two touching spheroidal particles, labeled i and j. This measure provides information about the relative orientation of particles within an agglomerate. This orientation measure is defined by

$$O_I = \frac{1}{2N_T}\sum_{i=1}^{N}\sum_{j=1}^{N}a_{ij}|\cos\theta_{ij}|, \tag{10.4.2}$$

where N_T is the number of touching particle pairs and the indicator coefficient a_{ij} equals unity if particles i and j are touching each other and zero otherwise. When $O_I = 1$ the symmetry axes of all particles are parallel, and when $O_I = 0$ the symmetry axes of all touching particle pairs are perpendicular.

Information about the location of the particle contact points is given by the *contact-point orientation measure* O_{II}. This measure is defined in terms of the angle ϑ_{ij} between the symmetry axis of particle i and the line joining the center of particle i and its contact point with another particle j. This measure is defined by

$$O_{II} = \frac{1}{2N_T}\sum_{i=1}^{N}\sum_{j=1}^{N}a_{ij}|\cos\vartheta_{ij}|. \tag{10.4.3}$$

When $O_{II,ij} \equiv |\cos\vartheta_{ij}| = 1$, the contact point of particle i with particle j is located at one end of the symmetry axis (a "pole") of particle i. For $O_{II,ij} = 0$, the contact point of particle i with particle j is located on the "equator" of particle i.

10.4.3. Equivalent Agglomerate Ellipse

The orientation measures introduced in the previous section deal with the relative orientation of different particles contained in an agglomerate. In the current section, we describe a method to instead examine the orientation of the agglomerate itself.

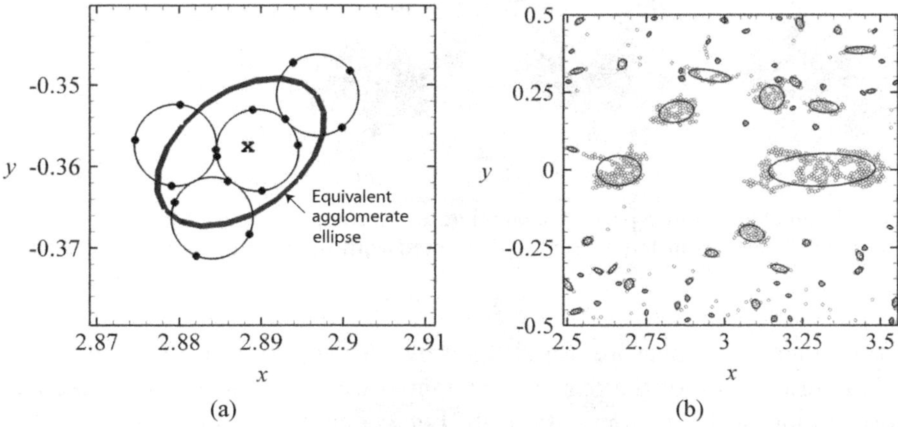

Figure 10.7. (a) Illustration showing the extrema points of a set of (two-dimensional) elliptical particles, along with the equivalent agglomerate ellipse shown as a heavy solid line, where the ellipse center is marked by an x. (b) Equivalent aggregate ellipses (bold) fit to aggregates in a close-up of a 2D channel flow.

Specifically, the size, shape, and orientation of an agglomerate is characterized by fitting an ellipse to the projection of the agglomerate along a plane formed of two coordinates, such as the x–y or x-z planes. The first step prior to fitting this ellipse is to identify a set of *extrema points* corresponding with the intersection of the particle principal axes with the particle surface. For spheroidal particles projected into the y-z plane, the extrema points along the particle equator are chosen to be the farthest points in the y- and z-directions, respectively, as shown in Figure 10.7a. The extrema points for all particles in the agglomerate are projected onto the projection plane.

An ellipse can be fit to the set of extrema particles on the projection plane using the direct least squares method described by Fitzgibbon et al. (1999), which was subsequently modified by Halir and Flusser (1998) to improve the numerical stability. Using this method, an ellipse can be expressed as a quadratic form as

$$F(x, y) = ax^2 + bxy + cy^2 + dx + ey + f = 0, \qquad (10.4.4)$$

where the requirement that the general conic be of elliptical shape is satisfied subject to the inequality constraint

$$b^2 - 4ac < 0. \qquad (10.4.5)$$

The least-square fit of the elliptic curve to the given set of extrema points results in a constrained minimization problem, the numerical solution of which is simplified if we scale the ellipse coefficients so that the right-hand side of (10.4.5) has a specific value. For instance, if this value is set equal to -1, then the constraint becomes

$$4ac - b^2 < 1. \qquad (10.4.6)$$

The coefficients in (10.4.4) can be obtained using a direct least-squares method with the constraint (10.4.6) imposed by the Lagrange multiplier method, as described by Halir and Flusser (1998). An example showing the equivalent agglomerate ellipses (in bold) computed using this method for elliptical particles in a two-dimensional shear flow is shown in Figure 10.7b.

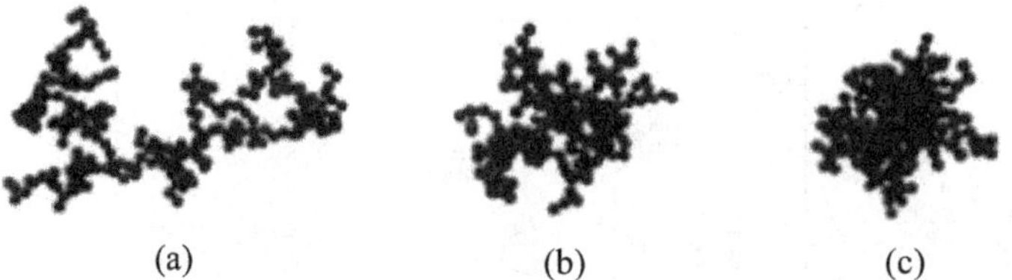

(a)   (b)   (c)

Figure 10.8. Examples of three particle agglomerates of complex shape, with agglomerate compactness increasing from left to right. [Reprinted with permission from Eggersdorfer et al. (2011).]

A third orientation measure was defined by Chesnutt and Marshall (2010) in terms of the orientation of the particles contained within an agglomerate relative to the orientation of the agglomerate itself. Let J_k denote the set of M_k particles contained in an agglomerate k, and let α_{ik} be the angle between the symmetry axis of particle i and the major axis of the equivalent ellipse of agglomerate k. A *particles-within-agglomerate orientation measure* $O_{III,k}$ is defined by

$$O_{III,k} = \frac{1}{|M_k|} \sum_{i \in J_k} |\sin \alpha_{ik}|. \tag{10.4.7}$$

When $|\sin \alpha_{ik}| = 1$ the symmetry axis of particle i is orthogonal to the major axis of the equivalent agglomerate ellipse, and when $|\sin \alpha_{ik}| = 0$ the symmetry axis of particle i is parallel to the major axis of the equivalent ellipse. The average of $O_{III,k}$ over all agglomerates is

$$O_{III} = \frac{1}{N_A} \sum_{k=1}^{N_A} O_{III,k}, \tag{10.4.8}$$

where N_A is the number of agglomerates.

10.4.4. Agglomerate Fractal Dimension

One of the distinguishing features of particle agglomerates is that they exhibit complex geometry as the number of particles in the agglomerate increases. Experimental results from Eggersdorfer et al. (2011) showing complex particle agglomerates with varying degrees of compactness are shown in Figure 10.8. This geometry can be characterized by measures such as the number of particles in an agglomerate, the area shaded by the agglomerate when projected onto a plane, or some other measure of the size of an agglomerate. A common size measure is the *radius of gyration R_G*, which is defined by

$$R_G^2 = \sum_{i=1}^{N} |\mathbf{x}_i - \bar{\mathbf{x}}|^2, \tag{10.4.9}$$

where N is the number of particles in the agglomerate, $\mathbf{x}_i$ is the centroid position for the ith particle, and $\bar{\mathbf{x}}$ is the centroid of all of the particles in the agglomerate. The radius of gyration provides a measure of the effective radius of the agglomerate.

As the number of particles within an agglomerate grows large, certain relationships are observed to form between the number of particles in the agglomerate and

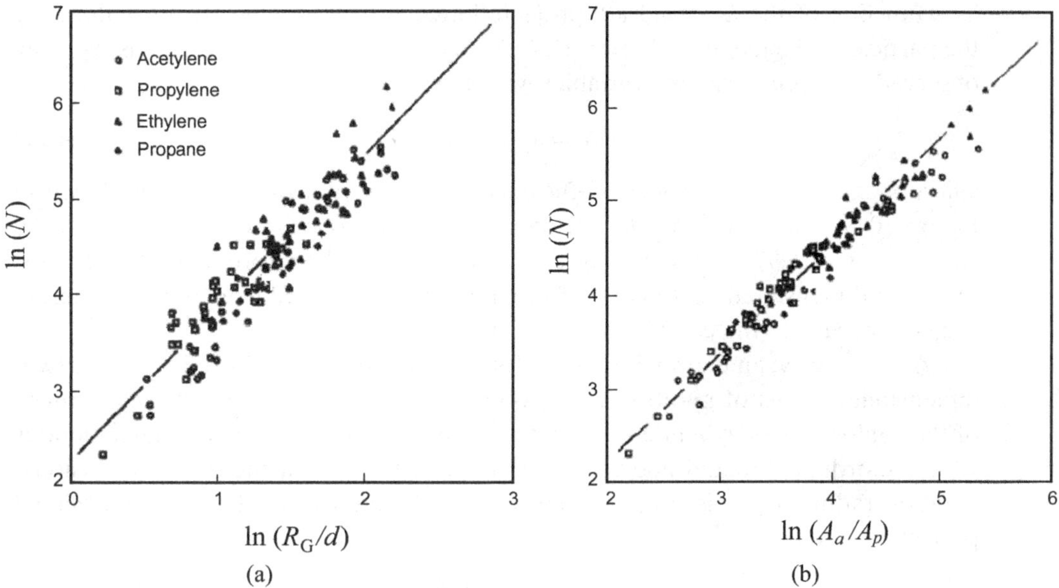

Figure 10.9. Plots showing power-law scaling for agglomerates formed of soot particles from different combustible gases obtained by plotting the logarithm of the number of particles versus (a) the logarithm of the radius of gyration R_G normalized by the particle diameter d and (b) the logarithm of the agglomerate projected area A_a normalized by the projected particle area $A_p = \pi d^2/4$. [Reprinted with permission from Köylü et al. (1995).]

the various measures of agglomerate size. These relationships usually take the form of a power law with noninteger exponent, which is characteristic of geometric objects called *fractals* (e.g., Mandlebrot, 1982). An important power-law relationship occurs between the number of particles in the agglomerate and the radius of gyration R_G, which has the form

$$N = k_G(R_G/r_p)^{d_f}. \qquad (10.4.10)$$

In this expression, it is assumed that the N particles making up the agglomerate have a uniform radius r_p. The coefficient k_G is called the *fractal pre-factor* and d_f is the *fractal dimension* of the agglomerate. An example showing experimental data for N and R_G for agglomerates of soot particles obtained from flames generated with different combustible gases is shown in Figure 10.9a, from Köylü et al. (1995). Although there is some natural variability in the agglomerates, the data agree reasonably well with the power-law expression (10.4.10) over two orders of magnitude of variation in agglomerate size. Similar results are reported by a number of other investigators, including agglomerates generated using numerical simulations by Brasil et al. (1999). Based on a review of many studies reported in the literature, Brasil et al. (1999) report typical values of the coefficient k_G to be in the range 1.2 to 3.5. Reported values of the fractal dimension d_f vary between about 1.5 and 2.3 depending on the agglomerate formation process, although the most common value is about 1.8.

It is often very difficult to experimentally determine R_G for particle agglomerates formed in the laboratory, whereas it is a simple matter to estimate agglomerate size using the projected area of the agglomerate onto a planar surface. For this reason, many experimental studies instead plot the number of particles in an agglomerate as a function of the agglomerate projected area, which is typically normalized by the particle projected area $A_p = \pi d^2/4$. A second power-law expression is typically observed between these two variables, given by

$$N = k_a (A_a/A_p)^\alpha, \tag{10.4.11}$$

where k_a and α are unknown coefficients. A typical plot of this type is shown in Figure 10.9b, for which case $k_a = 1.05$ and $\alpha = 1.09$. A procedure is described by Brasil et al. (1999) for recovering the coefficients in the expression (10.4.10) for the three-dimensional agglomerate from quantities measured from the agglomerate projection, such as the coefficients in (10.4.11).

Alternatively, an expression of the form (10.4.10) can be plotted using the two-dimensional radius of gyration $R_{G,2D}$ obtained directly from the projected image of the agglomerates. Köylü et al. (1995) examined simulated particle agglomerates with two orders of magnitude variation in N and found that the three-dimensional gyration radius $R_{G,3D}$ is proportional to that obtained from the two-dimensional projection, such that

$$R_{G,3D}/R_{G,2D} = 1.24 \pm 0.01. \tag{10.4.12}$$

As a consequence, the fractal dimension d_f in (10.4.10) will be approximately the same whether the expression is written in terms of $R_{G,3D}$ or $R_{G,2D}$.

10.4.5. Particle Packing Measures

Some of the traditional measures used for characterization of packed beds are also useful for particle agglomerates. One such measure is called the *coordination number*, which is defined as the average number of spheres in contact with any given sphere. The coordination number is a simple topological measure that is commonly used in computational studies to assess the degree of particle contact within a flow (Yang et al., 2000). Experimental studies typically do not have sufficient information on the positions and contact of all spheres, so this measure is not as frequently reported in the experimental literature.

Another measure of the particle spacing is obtained by computing the average number of neighboring particles, $N(r)$, whose centroids are located within a radial distance r from a given particle centroid. The pair distribution function $g(r)$ is defined as

$$g(r) = \frac{1}{4\pi \rho_0 r^2} \frac{dN}{dr}, \tag{10.4.13}$$

where ρ_0 is the average particle number per unit volume, which is related to the particle concentration ϕ by $\rho_0 = 6\phi/\pi$. The function $g(r)$, which is also known as the *radial distribution function*, can be interpreted as the probability distribution of finding the center of a particle at a distance r from a reference particle center. The radial distribution function is widely used for characterizing packing chain structures (Aste et al., 2005; Zou et al., 2009). A plot of the radial distribution function as a function of distance r is shown in Figure 10.10, based on computational data from Aste et al. (2005) for a large packed bed of particles. A large peak in the radial distribution function curve is observed at $r \cong d$, which corresponds to neighboring

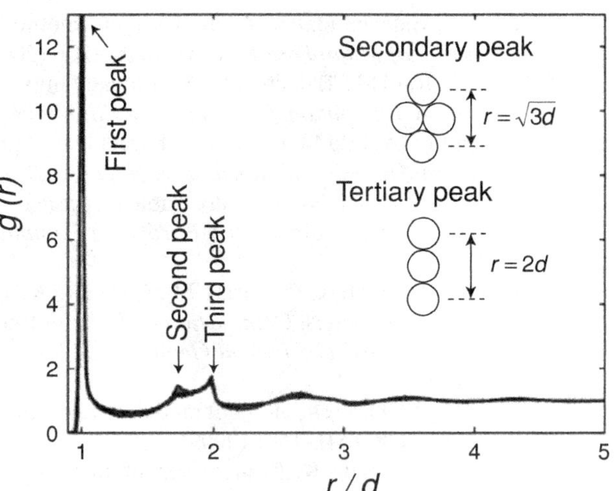

Figure 10.10. Radial distribution function of dense packing of particle chains, showing structures associated with secondary and tertiary peaks (based on computational data of Aste et al., 2005, for packed particle beds).

particles that are in contact with the given particle (first-neighbors). A series of additional peaks are observed for higher values of r, with the second peak occurring at $r = \sqrt{3}d$ and the third peak at $r = 2d$. As shown by Zou et al. (2009), the second peak is associated with particles that form a planar rhomboid structure between the given particle and two first-neighbor particles that are in contact with the given particle as well as with each other (see inset of Figure 10.10). The second peak is caused by a fourth particle that is touching the two first-neighbor particles, thus completing the rhombus shape. The third peak corresponds to structures in which three particles are located along a line, which include the given particle, a first neighbor that is in contact with the given particle, and a second neighbor that is in contact with the first neighbor. These peaks occur even though the particle packing shows no signs of a crystalline structure. For very large r, the value of $g(r)$ approaches unity.

REFERENCES

Aarts PAMM, van den Broek SAT, Prins GW, Kuiken GDC, Sixma JJ, Heethaar RM. Blood platelets are concentrated near the wall and red blood cells, in the center in flowing blood. *Arterioscler* **8**, 819–824 (1988).

Allievi A, Bermejo R. A generalized particle search-locate algorithm for arbitrary grids. *Journal of Computational Physics* **132**, 157–166 (1997).

Aste T, Saadatfar M, Senden TJ. Geometrical structure of disordered sphere packings. *Physical Review E* **71**, 061302 (2005)

Babic M. Average balance equations for granular materials. *International Journal of Engineering Science* **35**, 523–548 (1997)

Brasil AM, Farias TL, Carvalho MG. A recipe for image characterization of fractal-like aggregates. *Journal of Aerosol Science* **30**(10), 1379–1389 (1999).

Chesnutt JKW, Marshall JS. Structural analysis of red blood cell aggregates under shear flow. *Annals of Biomedical Engineering* **38**(3), 714–728 (2010).

Dulinska I, Targosz M, Strojny W, Lekka M, Czuba P, Balwierz W, Szymonski M. Stiffness of normal and pathological erythrocytes studied by means of atomic force microscopy. *Journal of Biochemical and Biophysical Methods* **66**, 1–10 (2006).

Eggersdorfer ML, Kadau D, Herrmann HJ, Pratsinis SE. Multiparticle sintering dynamics: From fractal-like aggregates to compact structures. *Langmuir* **27**, 6358–6367 (2011).

Ferry J, Balachandar S. A fast Eulerian method for disperse two-phase flow. *International Journal of Multiphase Flow* **27**, 1099–1226 (2001).

Ferry J, Rani SL, Balachandar S. A locally implicit improvement of the equilibrium Eulerian method. *International Journal of Multiphase Flow* **29**, 869–891 (2003).

Fitzgibbon A, Pilu M, Fisher RB. Direct least square fitting of ellipses. *IEEE Transactions on Pattern Analysis and Machine Intelligence* **21**, 476–480 (1999).

Halir R, J Flusser. Numerically stable direct least squares fitting of ellipses. *The Sixth International Conference in Central Europe on Computer Graphics and Visualization '98* **1**, 125–132 (1998).

Köylü ÜÖ, Faeth GM, Farias TL, Carvalho MG. Fractal and projected structure properties of soot aggregates. *Combustion and Flame* **100**, 621–633 (1995).

Mandlebrot BB. *The Fractal Geometry of Nature*. W.H. Freeman and Company, New York (1982).

Marshall JS. Discrete-element modeling of particulate aerosol flows. *Journal of Computational Physics* **228**, 1541–1561 (2009).

Marshall JS, Sala K. Comparison of methods for computing the concentration field of a particulate flow. *International Journal of Multiphase Flow* **56**, 4–14 (2013).

Mousel J, Marshall JS. Aggregate growth and breakup in particulate suspension flow through a micro-nozzle. *Microfluidics and Nanofluidics* **8**(2), 171–186 (2010).

Sethian JA. A fast marching level set method for monotonically advancing fronts. *Proceedings National Academy of Sciences* **93**, 1591–1595 (1996a).

Sethian JA. *Level Set Methods and Fast Marching Methods*. Cambridge University Press, Cambridge UK (1996b).

Sethian JA, Smereka P. Level set methods for fluid interfaces. *Annual Review of Fluid Mechanics* **35**, 341–372 (2003).

Yang RY, Zou RP, Yu AB. Computer simulation of the packing of fine particles. *Physical Review E* **62**, 3900–3907 (2000).

Zhu HP, Yu AB. Averaging method of granular materials. *Physical Review E* **66**, 021302 (2002).

Zou LN, Cheng X, Rivers ML, Jaeger HM, Nagel SR. The packing of granular polymer chains. *Science* **326**, 408–410 (2009).

11 Applications

Adhesive particle flows are involved in a wide variety of applications, including industrial, energy, environmental, biological, and geological processes of many types. Rather than trying to treat a large number of such processes here, we restrict attention to four specific types of particle flows that are of particular interest. This interest arises in part because the particles in these systems exhibit interesting behavior that has important consequences in many different applied systems. Although this chapter is by no means a comprehensive coverage of particle flow phenomena, it will perhaps serve as an introduction to some of the interesting and unusual behaviors that particulate systems can exhibit.

11.1. Particle Migration in Tube and Channel Flows

The problem of a suspension of particles carried within a tube or channel by a fluid flow arises in a large number of problems. For example, this flow is found in pneumatic intake systems used to feed fuel particles (e.g., coal or biomass pellets) into a combustion unit, in the flow of blood through the cardiovascular system (where the red blood cells and other types of blood cells are treated as particles), in flow of digestive fluids through the intestines, in tar-sand excavation and oil extraction, and in many other process in which a particulate slurry is transported along a channel or pipeline. For the sake of simplicity in the current discussion let us refer to flow in a tube, but similar phenomena occur also with channel flows. It is also convenient to simplify the problem by neglecting gravity, or else considering a vertically oriented flow in which gravity points in the direction of flow. We simplify the problem further to consider laminar flow in a straight tube, so that the fluid streamlines are all aligned in the direction of the tube axis and the particles are simply carried downstream by the fluid flow. However, as the particles are transported along the tube axis by the fluid flow, a very interesting thing happens – they begin to drift laterally in the tube. This phenomenon is known as *particle migration*. As a result of particle migration, the concentration profile of the particles within the tube becomes nonuniform, where the particle concentration will become higher either at the tube center or at some other finite radial location and it will be come lower in other regions, typically near the tube walls, even if the concentration is uniform at the tube entrance.

Particle migration is most often observed in low Reynolds number laminar flows, as fluid turbulence tends to rapidly mix the particles so as to eliminate concentration nonuniformities. In flows where significant particle migration occurs, it can have an important influence on the flow mechanics. One important example of this occurs in blood flow, for which the migration of red blood cells (RBCs) toward the low-shear region at the center of the tube (i.e., the blood vessel) leads to the phenomena of marginalization and plasma skimming, discussed in Section 1.3.4. The first of these phenomena causes the different constituents of blood (e.g., the red and white blood cells and the platelets) to have highly nonuniform concentration distributions within the blood vessel. These nonuniformities influence the overall blood flow resistance as well as the reaction rates between the white blood cells and platelets with the endothelial cells along the vessel walls. The plasma skimming phenomenon occurs when an upstream channel with strong particle migration enters into a bifurcation with nonuniform flow rates. The outlet channel with low flow rate tends to primarily entrain fluid with low RBC concentration from near the blood vessel walls, whereas the outlet channel with high flow rate will tend to entrain more of the RBC-rich fluid from within the center of the channel. As a result, the average RBC concentration within the outlet channels can be very different from each other.

It is known that there can be no lateral particle transport in flows with vanishing particle Reynolds number and no particle collisions (Bretherton, 1962). The effects of particle inertia and of collisions on particle migration are examined in the following sections, followed by a discussion of the enhancement of particle migration rate that can occur due to waviness of the tube wall.

11.1.1. Inertial Particle Migration in Straight Tubes

We consider a dilute particulate suspension undergoing a fully developed laminar flow in a tube, in which the flow is oriented vertically upward so that gravity is oriented opposite to the direction of flow. If the particles are heavier than the fluid, the downward drift of the particles will induce a negative relative slip velocity of the particles, $v_s < 0$, and the particles will lag the fluid flow. Both the Saffman and Magnus lift forces discussed in Section 5.2 are oriented such that particles will be drawn inward toward the tube axis when $v_s < 0$. If the particles are lighter than the fluid, the opposite occurs and the Saffman lift force drives the particles toward the tube wall. A well-known experimental study performed by Segré and Silberburg (1962) for neutrally buoyant particles in tube flow shows that for very dilute suspensions and finite particle Reynolds numbers, particles collect at a radial position of approximately 60% of the tube radius. This phenomenon is called the *tubular pinch effect*, and is characterized by the particles collecting at some radial position midway between the wall and the tube axis. An example illustrating this effect is shown in Figure 11.1, based on the experimental data of Matas et al. (2004).

The lateral force on a neutrally buoyant particle due to weakly nonlinear inertia for two-dimensional laminar shear flow was examined by Ho and Leal (1974) using a perturbation analysis. This paper shows that both shear rate variation and the lateral boundaries at the channel walls lead to an inward radial force for particles close to the wall. This inward force acting on particles close to the wall combined with the outward lateral force on particles near the channel center is consistent with

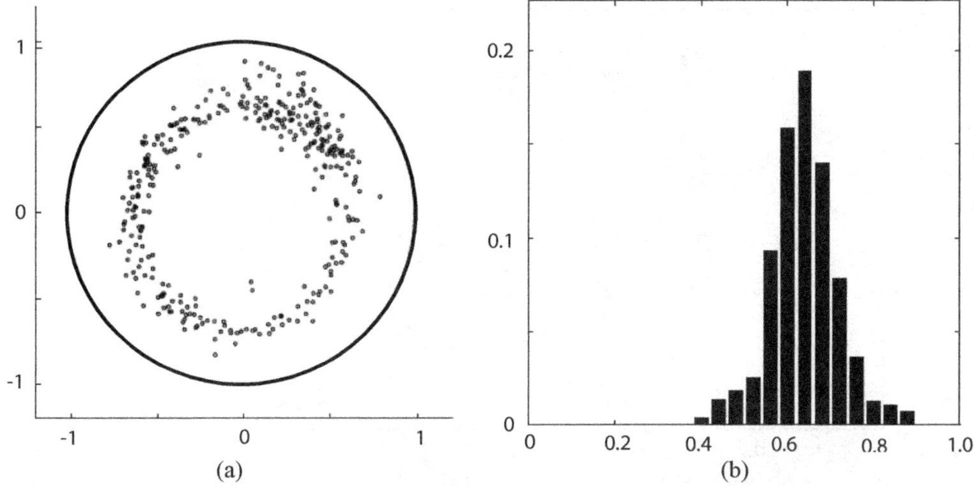

Figure 11.1. Experimental results for (a) particle distribution and (b) histogram of radial locations for particles of diameter d flowing with mean velocity U in a tube of diameter D with tube Reynolds number $\mathrm{Re}_F = UD/\nu = 67$ and $D/d = 9$. [Reprinted with permission from Matas et al., (2004).]

the experimental observations of Segré and Silberburg (1962). Extensions to this asymptotic theory have been developed for Poiseuille flow in a tube (Schonberg and Hinch, 1989), for particles that are not neutrally buoyant (Hogg, 1994), and for higher Reynolds numbers (Asmolov, 1999). An experimental study by Matas et al. (2004) showed that the radial position at which the particles collect moves outward toward the wall as the Reynolds number is increased. These authors also observe appearance of a second inner ring on which particles collect, located closer to the center than the primary ring.

11.1.2. Collision-Induced Particle Migration

The effect of finite particle concentration on lateral migration in a tube was examined by Han et al. (1999) and Hampton et al. (1997). Experimental results from Han et al. (1999) are plotted in Figure 11.2 for a nominal particle Reynolds number of 0.2 with four different values of the average concentration. For small average concentration values, the particles collect preferentially at a finite radial position. As the average concentration increases, the particles increasingly drift to the low-shear region near the tube center. For moderate average concentration values there are local concentration maxima at both a finite radial location and at the tube center. At large values of the average concentration, the only concentration maximum occurs at the tube center, which exhibits a local cusp in concentration value.

　　The tendency for particles to preferentially drift toward the center of the tube as the average concentration increases is caused by the phenomenon of *shear-induced migration*. This phenomenon was initially identified by Leighton and Acrivos (1987), who also developed a scaling theory to estimate the particle diffusive flux. The diffusion associated with this phenomenon comes from irreversible particle interactions (e.g., collisions), the rates of which vary as a function of the shear rate, the

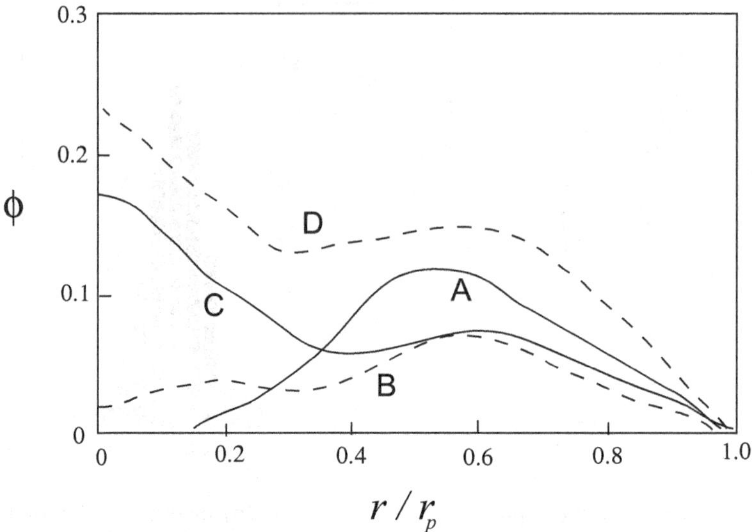

Figure 11.2. Plot of concentration profiles in tube flow with nominal particle Reynolds number of 0.2, for bulk concentrations of $\phi_0 = 0.06$ (Line A, solid), 0.1 (Line B, dashed), 0.2 (Line C, solid), and 0.28 (Line D, dashed). Based on experimental data of Han et al. (1999).

particle concentration, and the effective viscosity of the multiphase flow. Specifically, Leighton and Acrivos argue that a particle placed in a flow field with a nonzero gradient in the particle concentration ϕ or shear rate $\dot{\gamma}$ will experience a higher rate of collisions with other particles on the side of the particle with higher concentration or higher shear rate, where the excess rate of interactions on this side of the particle is proportional to $-r_p\nabla(\dot{\gamma}\phi)$. Assuming that the displacement of the given particle after each collision is proportional to the particle radius r_p, then the diffusive flux associated with the concentration or shear rate gradient should be proportional to $-r_p^2\phi\nabla(\dot{\gamma}\phi)$.

The governing equation for particle concentration ϕ in the continuum particulate flow theory has the form

$$\frac{D\phi}{Dt} + \nabla \cdot \mathbf{N} = 0, \tag{11.1.1}$$

where $\mathbf{N}$ is the particle diffusive flux. Phillips et al. (1992) developed a constitutive equation for $\mathbf{N}$ that has three terms, where

$$\mathbf{N} = \mathbf{N}_b + \mathbf{N}_c + \mathbf{N}_\mu. \tag{11.1.2}$$

In this equation, $\mathbf{N}_b$ is the diffusive flux from Brownian motion, given by

$$\mathbf{N}_b = -D_b\nabla\phi, \tag{11.1.3}$$

where D_b is the Brownian diffusion coefficient defined in (5.4.11). The second term in (11.1.2) is the shear-induced diffusion related to concentration or shear rate gradient. Using the Leighton-Acrivos scaling, we can write

$$\mathbf{N}_c = -K_c r_p^2(\phi^2\nabla\dot{\gamma} + \phi\dot{\gamma}\nabla\phi), \tag{11.1.4}$$

where K_c is a constant. The third term in (11.1.2) is a second type of shear-induced diffusion that is related to the fact that a gradient in particle concentration induces

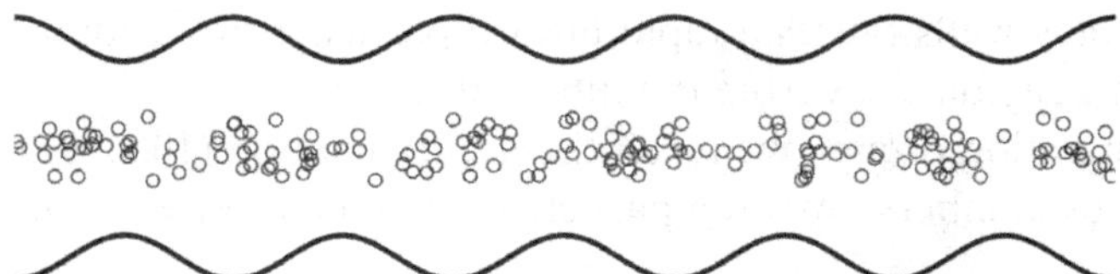

Figure 11.3. Result of a DEM simulation of particle transport in a tube with a wavy wall and periodic boundary conditions. This figure shows only a slice of the particles within a section passing through the channel centerline. The particles initially fill the tube with uniform concentration, but over time the particles drift toward the central part of the tube. [Reprinted with permission from Hewitt and Marshall (2010).]

an associated gradient in the effective viscosity of the two-phase mixture. Assuming that the shear-induced collision frequency is proportional to $\varphi\dot{\gamma}$ and that the net drift associated with viscosity gradient is proportional to the change in effective viscosity over the particle radius divided by the effective viscosity, or $r_p(\nabla\mu_{eff})/\mu_{eff}$, the diffusive flux associated with viscosity variation can be written as

$$\mathbf{N}_\mu = -K_\mu\phi^2\dot{\gamma}\frac{r_p^2}{\mu_{eff}}\frac{d\mu_{eff}}{d\phi}\nabla\phi, \tag{11.1.5}$$

where K_μ is a constant. A typical empirical formula relating effective viscosity with concentration is that proposed by Krieger (1972):

$$\mu_{eff} = \mu(1 - \phi/\phi_m)^{-1.82}, \tag{11.1.6}$$

where ϕ_m is the maximum packing concentration, equal to approximately 0.68 for spheres.

Comparison of the Phillips et al. (1992) theory with experimental data was reported by Hampton et al. (1997) for pressure-driven flow in a circular tube and by Koh et al. (1994) for flow in a rectangular channel. The predictions for concentration profile are in reasonable agreement with the data, except in the immediate vicinity of the channel center. The Phillips et al. theory predicts a sharp peak in concentration at the tube or channel center, leading to the result that the centerline concentration will always be equal to the maximum packing concentration ϕ_m no matter what the value of the average concentration within the tube. This prediction at the centerline disagrees with the experiments, particularly for lower values of the average concentration. An extension to this theory that does not require that concentration approach ϕ_m at the tube centerline was proposed by Nott and Brady (1994), who also presented a validation of their theory using Stokesian dynamics simulations. This theory was extended for Brownian suspensions by Frank et al. (2003) and to account for streamline curvature effects by Kim et al. (2008).

11.1.3. Particle Migration in the Presence of Wavy Tube Walls

It was reported by Hewitt and Marshall (2010) that the inward migration of particles in a pressure-driven tube flow is dramatically accelerated if the tube has wavy walls. An example is shown in Figure 11.3, showing results of a DEM simulation with periodic end conditions, where the particle concentration was initially uniform throughout the tube. The interaction of the suspended particles with the flow oscillations induced by the wavy walls causes a rapid inward particle drift, leading the particles to collect within a region away from the tube walls.

To see why this rapid migration occurs, we recall the approximation (10.1.10), which for small particle Stokes numbers gives the particle drift velocity $\mathbf{v}_d = \mathbf{v} - \mathbf{u}$ in dimensionless form as

$$\mathbf{v}'_d = -\mathrm{St}\,\mathbf{a}', \tag{11.1.7}$$

where velocity is nondimensionalized by the mean fluid velocity U, acceleration is nondimensionalized by U^2/D, and D is the tube mean diameter. The furrow-averaged radial location of particle n is defined as

$$\bar{r}_n(z_n) \equiv \frac{1}{\lambda_F} \int_{z_n-\lambda_F}^{z_n} r_n(z)\,dz, \tag{11.1.8}$$

where (r_n, z_n) denotes the instantaneous radial and axial coordinates of particle n and λ_F is the wavelength of the furrow. Assuming small wave amplitude, $\eta \equiv A/D \ll 1$, the fluid acceleration at the particle centroid can be expanded in a Taylor series about the furrow-averaged position $\bar{r}_n(t)$ to write

$$\mathbf{a}'(r'_n) = \mathbf{a}'(\bar{r}_n) + (r'_n - \bar{r}_n)\frac{\partial \mathbf{a}'}{\partial r'}\bigg|_{r=\bar{r}_n} + \cdots, \tag{11.1.9}$$

where the omitted terms are higher order in the small parameter η. Two functions, $\mathbf{v}'_{d1}$ and $\mathbf{v}'_{d2}$, called the first and second particle drift velocities, are defined by

$$\mathbf{v}'_{d1} \equiv -\mathrm{St}\,\overline{\mathbf{a}'(\bar{r}')}, \quad \mathbf{v}'_{d2} \equiv -\mathrm{St}\,\overline{[(r' - \bar{r}')\partial \mathbf{a}'/\partial r']}, \tag{11.1.10}$$

where the total drift velocity is the sum of $\mathbf{v}_{d1}$ and $\mathbf{v}_{d2}$. The first drift velocity is the average of the fluid acceleration along the tube at the furrow-averaged radial position of the particle. The second drift velocity is related to the correlation between the radial oscillation of the particle about the furrow-averaged location and the variation of the radial acceleration gradient.

In a steady flow, the definition of a streamline can be integrated to write the radial perturbation from the furrow-averaged position as

$$r' - \bar{r}' = \int (u'/w')\,dz', \tag{11.1.11}$$

where u and w are the radial and axial components of the fluid velocity, respectively. Using the lubrication approximation for small flow Reynolds numbers and $\eta \ll 1$, the fluid velocity components can be approximated as

$$u' = -\frac{2r'\alpha\eta}{h'^3}[1 - (r'/h')^2]\sin(\alpha z'), \quad w' = \frac{2}{h'^2}[1 - (r'/h')^2], \tag{11.1.12}$$

where $\alpha \equiv kD$, $k = 2\pi/\lambda_F$ is the wave number of the wall corrugation, and $h(z)$ is the tube diameter at axial position z. Substituting (11.1.12) into (11.1.9), (11.1.10), and (11.1.11) gives the radial component of the two parts of the drift velocity to leading order in η as

$$v'_{d1} = -2\mathrm{St}\,(\alpha\eta)^2\bar{r}'(1 - \bar{r}'^2)(3 - 7\bar{r}'^2), \tag{11.1.13a}$$

$$v'_{d2} = 2\,\mathrm{St}\,(\alpha\eta)^2\bar{r}'(1 - \bar{r}'^2)(1 - 5\bar{r}'^2). \tag{11.1.13b}$$

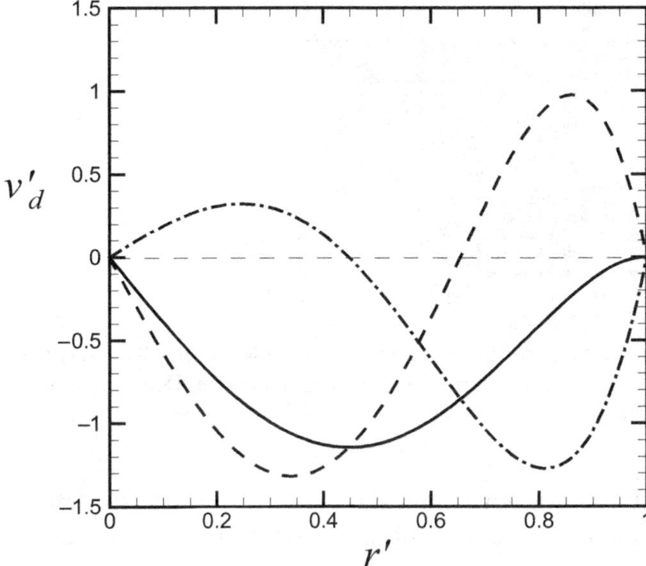

Figure 11.4. Plot showing first particle drift velocity v'_{d1} (dashed line), second particle drift velocity v'_{d2} (dashed-dotted line), and net particle drift velocity $v'_d = v'_{d1} + v'_{d2}$ (solid line) in the radial direction for low Stokes number particle transport in a tube given by the lubrication theory. All velocities are normalized by St $(\alpha\eta)^2$. [Reprinted with permission from Hewitt and Marshall (2010).]

The particle drift velocities, normalized by St $(Ak)^2$, are plotted in Figure 11.4 as functions of radius. The magnitude of the first drift velocity is significantly greater than that of the second drift velocity within the central part of the tube. The drift velocities each change sign within the outer part of the tube, and in this region both drift velocities have similar magnitude. The net drift velocity is observed to have a negative sign throughout the tube, implying that all particles will drift toward the tube center at a radial velocity proportional to St $(Ak)^2$. Extension of this lubrication theory result to finite flow Reynolds numbers was examined computationally using DEM by Hewitt and Marshall (2010), who found that with increase of the Reynolds number the particles drift to a region with nonzero radius within the tube, rather than drifting to the tube axis.

11.2. Particle Filtration

Collision of particles on surfaces and their subsequent adhesion is an important area of research for applications ranging from air or water filtration to fouling of microfluidic circuits (Perry and Kandlikar, 2008). Problems of this type involve particles suspended in a fluid flow past a blunt body, such as a fiber or the fin at the end of a microchannel. The fluid drag force attempts to move the particles around the body, but drift due to particle inertia allows the particles to move closer to the surface of the bluff body at conditions of finite Stokes number than they would if the particles followed the fluid flow streamlines. Collisions of the particles with the surface occur due to a combination of the particle drift within the region of high

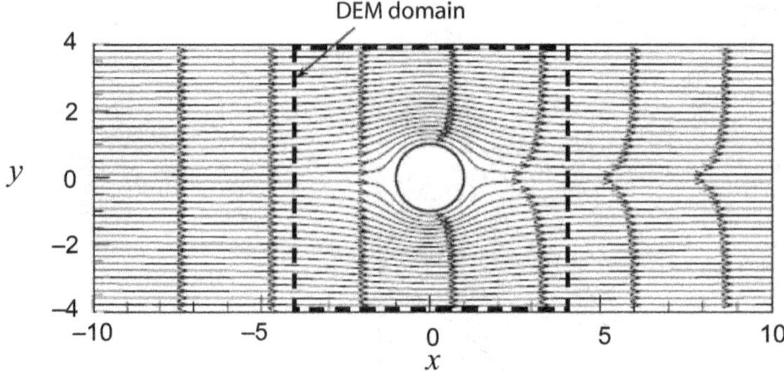

Figure 11.5. Fluid flow past a 10 μm diameter fiber at a fiber Reynolds number $\mathrm{Re}_F = 0.178$. The inner region used for DEM simulations is indicated by a dashed line. [Reprinted with permission by Li and Marshall (2007).]

streamline curvature in the stagnation-point flow and the finite particle size effect within the high shear regions along the particle sides, as discussed in Section 1.3.1.

11.2.1. Fiber Filtration

Fiber filtration, with either a single fiber or a fiber array, has been studied by a number of investigators as being representative of particle interaction with bluff bodies immersed in the flow field. The initial stages of particle collection on the fiber can be adequately handled using the one-way coupling approximation, because the flow field is not very much affected by the particles. As the particle deposits on the fiber build up over time, it is necessary to utilize some method to account for modification of the fluid flow near the fiber by the particle deposits. A typical flow field used to study particle capture on a fiber is plotted in Figure 11.5. Because the time interval and number of particles that can be practically simulated with DEM is limited, it is common in filtration computations to use DEM only in a small region around the fiber, indicated by dashed lines in Figure 11.5. It is also common to accelerate the process of particle collection by significantly increasing the particle concentration at the inlet to the computational domain compared to what is observed in practice.

A sketch of the fluid streamlines in flow past a cylindrical fiber at low Reynolds number is shown in Figure 11.6, along with the motion of a select particle. The fiber surface is divided into three regions in this sketch. At the front of the fiber (Region A) is a region in which particles that disperse rapidly from the fluid flow, that is, particles with intermediate or large Stokes numbers, will collide with the

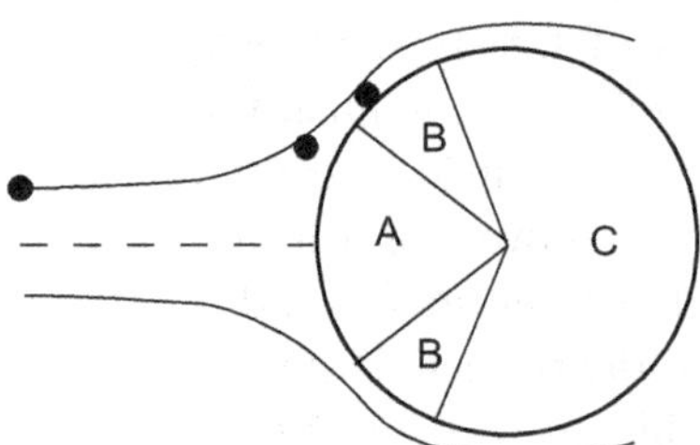

Figure 11.6. Sketch showing fluid streamlines in the flow past a cylindrical fiber as well as motion of a select particle. The fiber surface is divided into three regions: (A) rebound region, (B) collision region, and (C) no-impaction region.

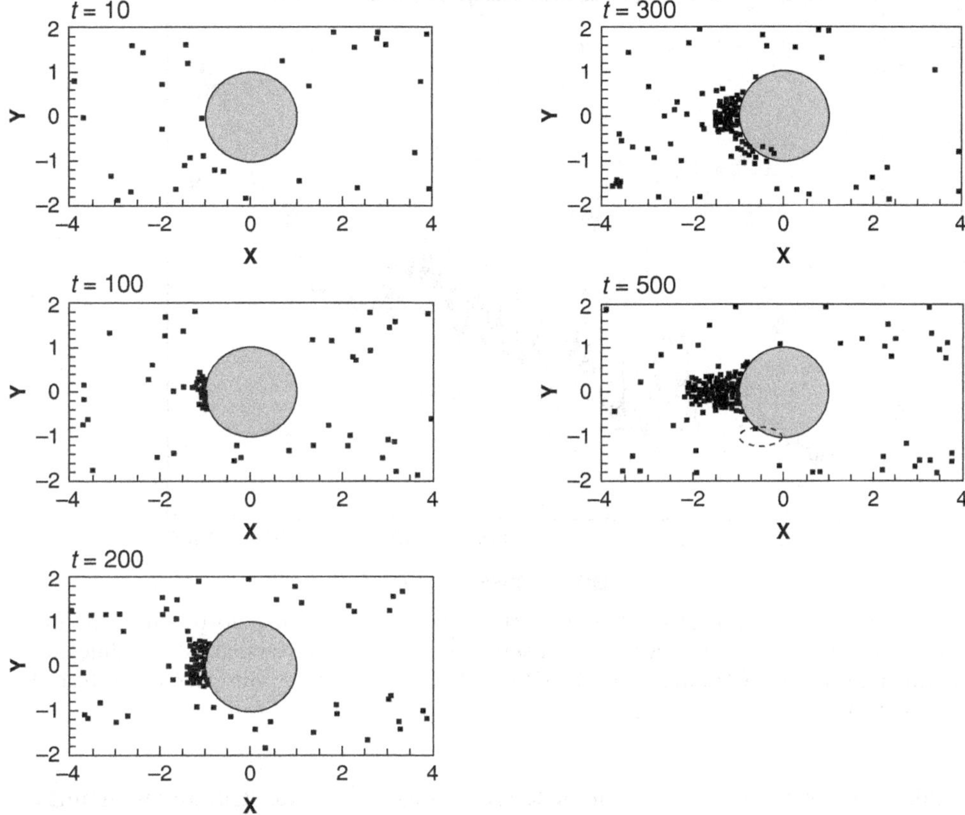

Figure 11.7. Time series showing results of a DEM simulation of particles collecting on the front surface of a cylindrical fiber. The agglomerate observed in the dashed ellipse at $t = 300$ has been stripped off and carried downstream by $t = 500$. [Reprinted with permission from Li and Marshall (2007).]

fiber. These particles will strike the fiber surface at a large angle and with relatively high velocity, so they will be most likely to bounce from the fiber surface. Smaller particles with smaller Stokes numbers will tend to more closely follow the fluid flow. Even if these particles are aimed directly toward the fiber far upstream, they will tend to be swept aside by the fluid flow near the fiber front surface and collide with the fiber closer to the shoulders of the cylinder, as indicated by Region B in Figure 11.6. Because these particles are smaller than those colliding in Region A, they have lower inertia and are less inclined to bounce following collision with the fiber surface. For noncharged fibers it is rare to find particles adhering to the rear of the fiber, indicated by Region C in Figure 11.6, as the particles instead tend to be swept downstream with the fluid flow and do not often enter this region.

A time series showing results of a DEM simulation by Li and Marshall (2007) of particle collision and adhesion on an array of cylindrical fibers is shown in Figure 11.7. For the case examined, the adhesive surface energy and the normal dissipation coefficient of the particles were sufficiently high that even particles colliding near the cylinder leading edge did not bounce significantly from the surface. The particles are observed to continue collecting on the front face of the fiber, leading to development of a particulate structure that projects a length $L(t)$ upstream of the

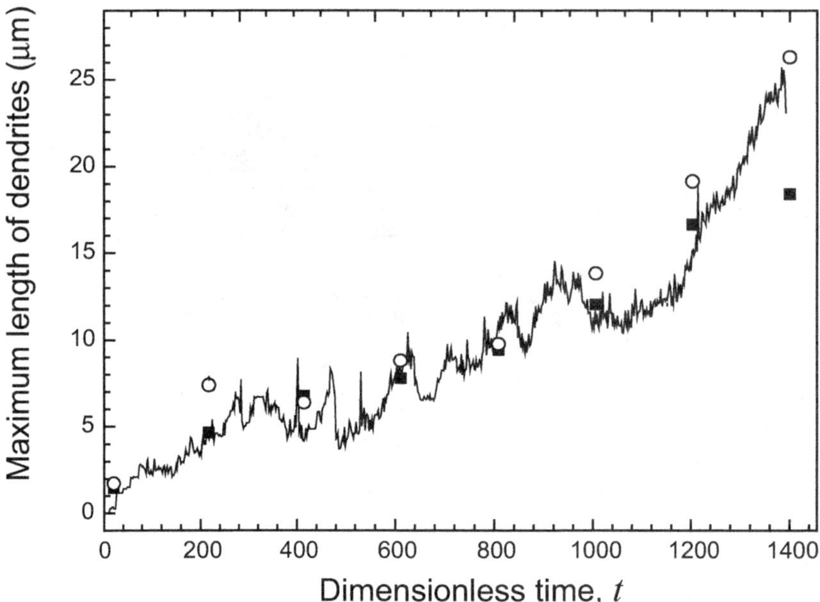

Figure 11.8. The variation of the length L of the dendritic structure upstream of the fiber with dimensionless time, comparing DEM simulations of Li and Marshall (2007) (line) with experimental results of Huang et al. (2006) (symbols). [Reprinted with permission from Li and Marshall (2007).]

cylinder leading edge. Variation of the length of this particulate structure with time as computed by the DEM simulations is compared with experimental results obtained by Huang et al. (2006) in Figure 11.8, showing reasonable agreement between the computational results and experimental data. The soft-sphere DEM method used by Li and Marshall allows for particle agglomerates to break off of the fiber if they become too long to sustain the fluid drag force exerted on them by the flow field. This break-off can be seen in Figure 11.7, for instance, by noting the agglomerate present within the dashed elliptical region at dimensionless time $t = 300$ that is no longer present at time $t = 500$, but instead has broken off of the fiber due to the high shear stress in this region and has been carried downstream out of the field of view. The agglomerate break-off process allows the particles adhering to the fiber to approach a state of statistical equilibrium after long time. This attainment of a quasi-equilibrium state associated with intermittent break-off of particle agglomerates is a key observation of the experimental study of Huang et al. (2006), but such a state is not allowed by many previous models of particle filtration processes, which assume that once a particle collides with the surface it is determined to have adhered to the surface and the adhesive bond is "frozen" in place (Kanaoka et al., 1983; Ramarao et al., 1994; Konstandopoulos, 2000).

An important measure of fiber filtration processes is the fiber efficiency η, defined as the fraction of incident particles that collide with the surface minus the fraction of incident particles that rebound from the surface. A plot comparing predictions for single-fiber efficiency from the soft-sphere DEM simulation by Yang et al. (2013) with experimental data from Rembor et al. (1999) is given in Figure 11.9. The fiber efficiency is plotted as a function of Stokes number over an interval

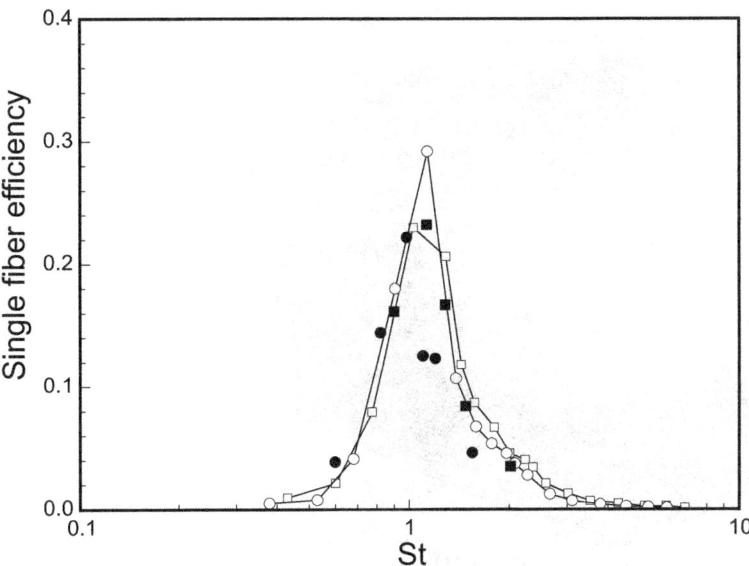

Figure 11.9. Comparison of soft-sphere DEM simulation predictions for single-fiber capture efficiency (lines with open symbols) with experimental data of Rembor et al. (1999) (filled symbols), for different particle sizes as a function of Stokes number. Square symbols are for particle diameter 2.83 μm and fiber diameter 30 μm, and circular symbols are for particle diameter 1.77 μm and fiber diameter 20 μm. [Reprinted with permission from Yang et al. (2013).]

$0.35 < \mathrm{St} < 7$. The DEM results are again found to agree closely with the corresponding experimental data. In addition to the Stokes number, the fiber efficiency depends on the adhesive surface energy of the particles. In order to model the tendency of particles to rebound from the fiber surface upon collision, it is convenient to nondimensionalize the particle surface energy by its inertia, which gives the *adhesion parameter* Ad, defined by

$$\mathrm{Ad} = \frac{2\gamma}{\rho_p U^2 d},\tag{11.2.1}$$

where d is the particle diameter. A plot of DEM predictions for the single fiber efficiency as a function of Stokes number and adhesion parameter is given in Figure 11.10, from Yang et al. (2013). The fiber efficiency is observed to be large within a region with large Stokes number (between 1 and 10) and large adhesion parameter (between 1 and 100). A Stokes number greater than unity is necessary in order for the incident particles to have a sufficiently large drift velocity relative to the fluid to collide with the fiber surface. A sufficiently large adhesion parameter is necessary so that these colliding particles will stick to the surface.

An important component in determining the fiber efficiency is the probability that a colliding particle will stick to the fiber surface, called the *sticking probability h*. Several empirical fits for sticking probability have been proposed, including those of Hiller and Löffler (1980) and Ptak and Jaroszcyk (1990), in which h is expressed as a function of Stokes number, flow Reynolds number, and particle-to-fiber size ratio. While these correlations fit the data from which they were developed, they miss

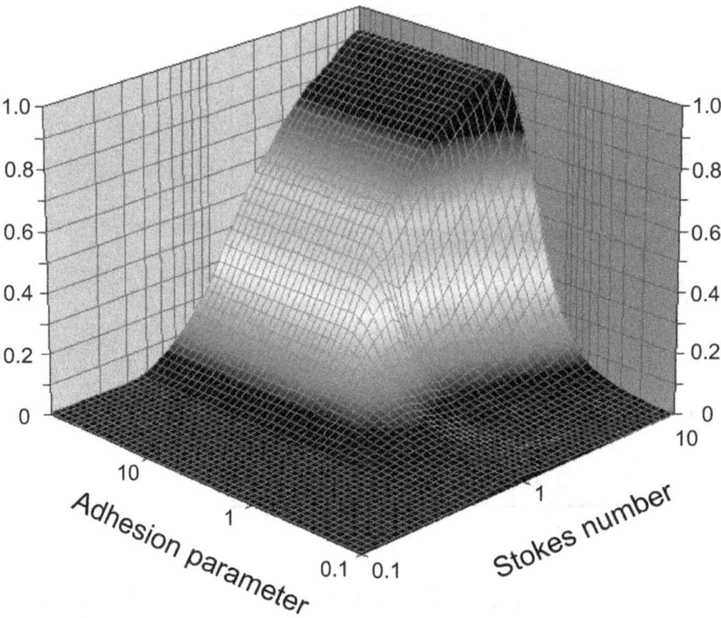

Figure 11.10. Plot showing variation of the single-fiber efficiency η as a function of Stokes number and adhesion parameter, obtained from soft-sphere DEM simulations. [Reprinted with permission from Yang et al. (2013).]

the important roles played by the particle adhesive force and the particle inertia. Based on a set of simulations using a soft-sphere DEM approach, Yang et al. (2013) found that the computational results for sticking probability when $\mathrm{Re}_F < 16$ can be collapsed onto a power-law expression as a function of the adhesion parameter Ad, given by

$$h = \begin{cases} 0.0558\,\mathrm{Ad}^{5/3} & \text{when Ad} < 5.65 \\ 1 & \text{when Ad} > 5.65 \end{cases}. \qquad (11.2.2)$$

A plot showing the collapse of the computational results is given in Figure 11.11. Data are plotted both with a fixed particle size and different Stokes numbers, as well as with a fixed Stokes number and different particle sizes. Higher flow Reynolds numbers change the location of particle collision, resulting in significantly higher sticking probabilities than indicated by this correlation. It is noted that these computational results were obtained using a fixed value of the particle restitution coefficient. As shown in Figures 3.6 and 3.8, the restitution coefficient is a function of particle impact velocity, and hence of the Stokes number, although in the viscoelastic regime the dependence is often quite weak.

11.2.2. Enhancement of Filtration Rate by Particle Mixtures

A number of approaches have been proposed to enhance capture rate of particles by fibrous filters. In a study of capture of fly ash particles resulting from coal combustion, Xu et al. (2010) examined the role of simultaneous capture of small and large particles on the capture efficiency. This study observed formation of fly ash particles in both a coarse mode (with diameter of approximately 1 μm or larger) and a submicron

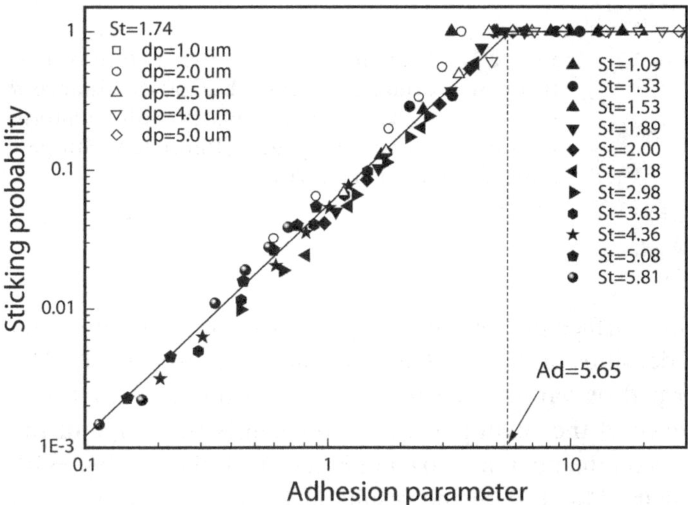

Figure 11.11. Plot showing collapse of computational results for sticking probability as a function of adhesion parameter. Solid symbols represent data from 2.6 μm diameter particles with different Stokes numbers, and open symbols represent different size particles with St = 1.74. [Based on data from Yang et al. (2013).]

fine mode. Because the fine particles in this combustion process have a relatively higher surface energy than the coarse particles, due to the alkali-rich composition of the small particles, they act as a type of adhesive between the coarse particles and the collecting cylinder (i.e., the fiber). A demonstration of this adhesive effect was given by Li et al. (2011), who reported results of DEM simulations for capture of particles by a fiber with a mixture of small and large particle sizes. As shown in Figure 11.12a, the number of large particles deposited onto the fiber at a given time

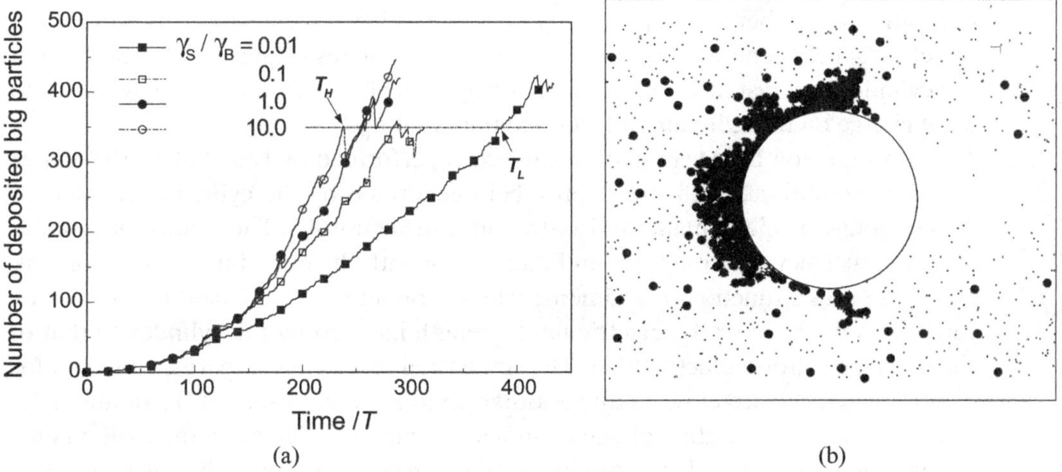

Figure 11.12. Plots showing effect of a set of small particles with surface energy density γ_S on deposition of a set of big particles with surface energy γ_B onto a fiber. (a) Number of deposited big particles as a function of γ_S/γ_B obtained from DEM simulations with all other parameters held fixed. (b) Plot showing big and small particles deposited onto fiber front surface. [Reprinted with permission from Li et al. (2011).]

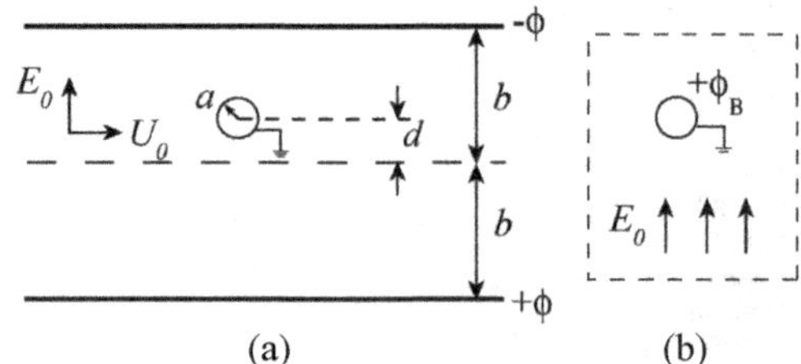

Figure 11.13. Configuration for the capture of particles by a cylinder positioned at offset distance d between two charged infinite plates, with a uniform fluid flow between the plates. [Reprinted with permission from Liu et al. (2010).]

increases significantly as the surface energy ratio γ_S/γ_B is increased, where γ_S and γ_B are the surface energy densities of the small and big particles, respectively. The computations were performed by varying γ_S with a constant value of γ_B. At high values of γ_S/γ_B, the presence of the small particles also changes the shape of the region of deposited particles on the fiber, as shown in Figure 11.12b for $\gamma_S/\gamma_B = 10$. Specifically, comparing Figure 11.12b with the results in Figure 11.7 with only a single particle size, it is evident that the region of deposited particles at the front of the cylinder covers a larger region on the cylinder surface in Figure 11.12b than in the case without the small particles present. A similar phenomenon was observed experimentally by Xu et al. (2010) and Kupka et al. (2009).

11.2.3. Enhancement of Filtration Rate by Electric Fields

Precharging of particles before they come into contact with the fiber may initially enhance particle deposition on the fiber surface, but after a certain time the repulsive interaction between deposited and incident particles will limit further deposition. On the other hand, approaches such as electrically charging the fiber itself or exposing the entire filter to an external electric field have been shown to be highly effective methods for enhancing the efficiency of fibrous filters (Wang, 2001). A well-known experimental study illustrating the effect of a background DC electric field on particle deposition on a fiber was reported by Oak et al. (1985) for a single fiber placed in the electric field generated by two parallel charged plates (Figure 11.13). Shifting of the cylinder by a distance d from the halfway plane between the two plates leads to a net charge on the cylinder with surface potential $E_0 d$.

A companion DEM simulation study was performed by Liu et al. (2010) under the same conditions, with the electric field generated by the cylinder represented using a boundary element method as described in Section 8.3. The results for cylinder capture efficiency are compared in Figure 11.14 with the experimental data of Oak et al. (1985) as a function of a dimensionless parameter $W_p = d/a \ln(4b/\pi a)$, which represents the ratio of the electric field strength induced by the cylinder to that of the external uniform electric field. The simulation results and experimental data for capture efficiency are close to one another up to a critical value of W_p of about 12, beyond which the cylinder voltage is sufficiently high that the air in the experiments becomes ionized. This effect is not included in the computations. Photographs taken by an electronic microscope during the experiments exhibited different deposition patterns at high and low voltage values, which supports the observations of particle deposition in the simulations shown in Figure 11.15. For the low-voltage case, most particles are captured on the side of the cylinder with y > 0, which is attributed to

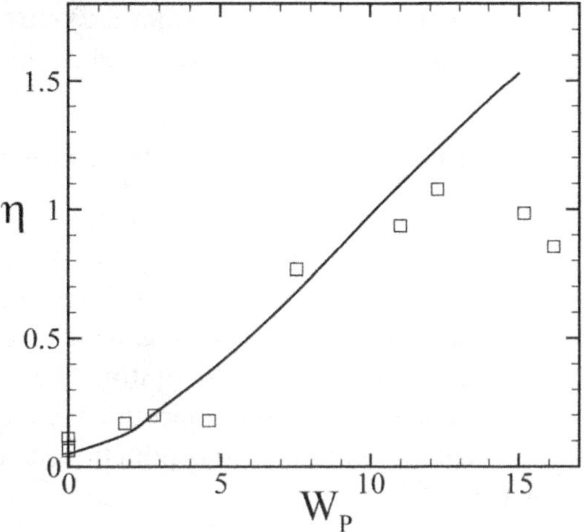

Figure 11.14. Variation of the capture efficiency η with the dimensionless parameter $W_p = d/a \ln(4b/\pi a)$, showing the DEM simulation predictions (solid line) and the experimental data (symbols) of Oak et al. (1985). The point $W_p \cong 12$ where the two results deviate corresponds to the point where air starts to become ionized by the electric field, which was not included in the simulations. [Reprinted with permission from Liu et al. (2010).]

the fact that the electric field is relatively higher on the upper half of the cylinder. For the high-voltage case, the electric field is sufficiently strong that particles deposit on both sides of the cylinder. For both cases, particles agglomerate in the form of particle chains pointing radially outward into the fluid, due to the presence of strong

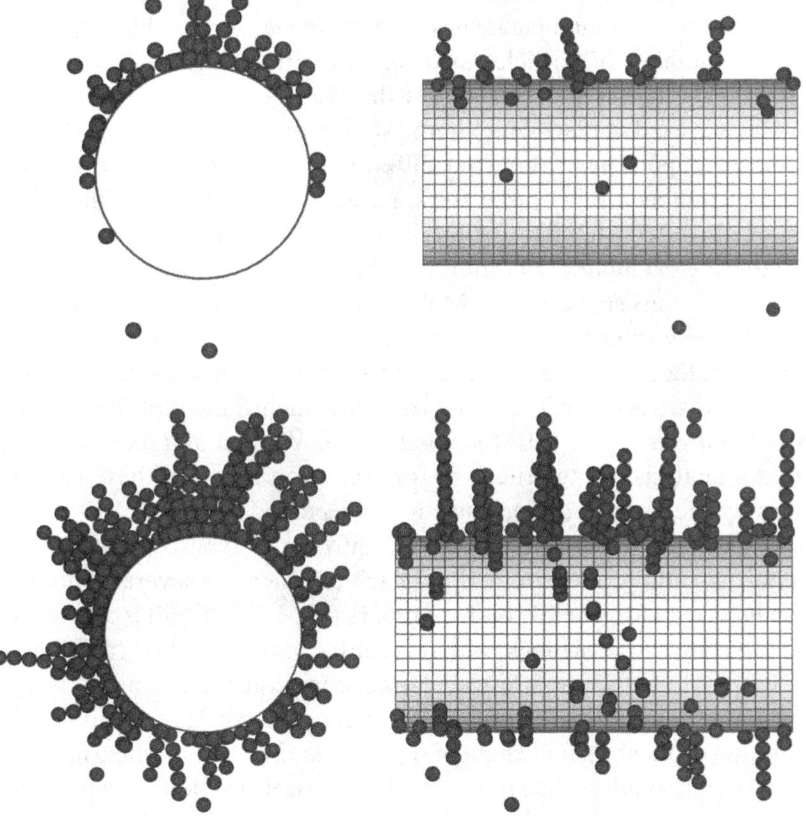

Figure 11.15. Snapshots from a DEM simulation of particle deposition on a cylinder exposed to a low voltage electrostatic field (top) and a high voltage electrostatic field (bottom). [Reprinted with permission from Liu et al. (2010).]

DEP force on the (uncharged) particles. These chains can grow to a length scale on the order of the cylinder radius.

11.3. Rotating Drum Mixing Processes

Mixing and separation processes play a central role in many applications involving particulate flow. Mixing processes are necessary both to combine different ingredients into a homogeneous mixture and to maintain a well-mixed suspension in the presence of gravity or mean particle drift. A common device used to mix particles in granular flows is the rotating drum (or rotating kiln), which is used in applications such as mixing pharmaceutical components, washing and drying operations, mixing components in food production, cement processing, and particle coating operations.

11.3.1. Flow Regimes

Let us consider a horizontal drum of radius R and length L partially filled with particles of diameter d, in which the drum axis is orthogonal to the direction of gravitational acceleration. The drum rotates at a rate Ω_D, where for the present we take Ω_D to be a constant. The primary dimensionless parameter characterizing the flow is the Froude number $\mathrm{Fr} \equiv \Omega_D R^2/g$. A second important dimensionless parameter is the filling degree J, which is equal to the percentage of the drum volume filled with particles at the maximum packing concentration ϕ_m, or $J = V_{drum}\phi_m/NV_p$ where N is the total number of particles in the drum with particle volume V_p. A third parameter that influences flow regimes is the particle-wall dynamic friction coefficient μ_W, which is used in place of μ_f in the sliding resistance equation (3.5.7).

Several different papers have reported different regimes of particle motion within the rotating drum as a function of these parameters, based both on experiments (Henein et al., 1983; Mellmann, 2001) and on computations (Yang et al., 2008). Although there are a number of differences between the various regime classifications, the classifications suggested by Mellmann (2001) serve well to illustrate the different particle flow transitions as the rotation rate of the system is changed. Schematic diagrams of these different regimes and a listing of the parameter ranges for which they are valid are given in Table 11.1 from Mellmann, and plots illustrating several of these regimes using the DEM simulations of Yang et al. (2008) is given in Figure 11.16. An analysis giving criterions for transition between these various regimes was given by Mellmann (2001), which is not repeated here.

Mellmann categorized the particle behavior into three types of motion (or regimes) – slipping, cascading, and cataracting – each of which has several subtypes. The *slipping regime* occurs for small Froude numbers ($\mathrm{Fr} < 10^{-4}$), and is characterized by motion of the particles as a block, with no relative particle motion (and hence no mixing). For cases where the particle-wall dynamic friction coefficient μ_W is less than a critical value $\mu_{W,C}$, a *sliding* motion occurs where the particles continuously slide over the rotating drum at a small angle of repose β and with no particle motion. If μ_W is greater than $\mu_{W,C}$ but less than the coefficient of static friction μ_S, a periodic *surging* motion is observed in which the particle mass rotates with the drum up to a critical angle $\beta_{\max}$, beyond which the particles slide back down as a single mass to a lower angle or repose $\beta_{\min}$, and the cycle repeats. Even though a periodic particle

Table 11.1. *Listing of different flow regimes for a rotating drum, along with values of key parameters. [Reprinted with permission from Mellmann (2001)]*

Basic form subtype	Slipping regime		Cascading regime			Cataracting regime	
	Sliding	Surging	Slumping	Rolling	Cascading	Cataracting	Centrifuging
Schematic							
Physical process	Slipping		Mixing			Crushing	Centrifuging
Froude number, Fr	$Fr < 10^{-4}$		$10^{-5} < Fr < 10^{-3}$	$10^{-4} < Fr < 10^{-2}$	$10^{-3} < Fr < 0.1$	$0.1 < Fr < 1$	$Fr > 1$
Filling degree, J	$J < 0.1$	$J > 0.1$	$J < 0.1$	$J > 0.1$		$J > 0.2$	
Wall friction coeff, μ_W	$\mu_W < \mu_{W,C}$	$\mu_W > \mu_{W,C}$	$\mu_W > \mu_{W,C}$			$\mu_W > \mu_{W,C}$	
Application	No use		Rotary kilns and reactors; rotary dryers and coolers; mixing drum			Ball mills	No use

motion is observed in this latter motion relative to a fixed frame, the particles do not move relative to each other. Because the primary purpose of a rotating drum is to mix the particles either with each other or with a surrounding fluid, the slipping regime is not useful for practical applications.

The *cascading regime* occurs over a broad range of Froude numbers in the interval $10^{-5} < Fr < 0.1$, for cases with sufficient wall friction. This regime is characterized by either continuous or periodic tumbling of particles along the top of the bed from the high side to the low side of the particle bed. Mellmann identified three subtypes of motion within this regime, referred to as slumping, rolling, and cascading. In the *slumping* motion, the particle tumbling occurs periodically as a sequence of "avalanches." The particle bed is raised with the rotating drum, and then a top layer of particles avalanche downward from the top of the raised bed to the low portion of

Figure 11.16. Computed particle flow patterns at different rotation speeds showing different flow regimes: (a) slumping, (b) slumping-rolling transition, (c) rolling, (d) cascading, (e) cataracting, and (f) centrifuging. Shading represents particle velocity. The sliding and surging regimes were not examined in the computations. [Reprinted with permission from Yang et al. (2008).]

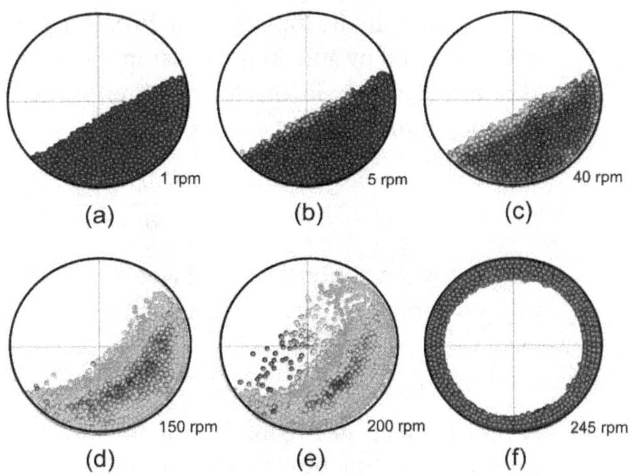

the bed. This avalanche causes the angle of repose of the bed to be lower, allowing the cycle to repeat. In the *rolling* motion, the particles roll continuously from the upper part to the lower part of the particle bed, with the bed surface approximately flat. The particle flow exhibits two distinct regions – a "fluid-like" cascading region of rapidly moving particles along the top of the bed and a "solid-like" plug-flow region of slowly transported particles within the rest of the bed, which are slowly raised upward by the drum rotation. At higher rotation rates a *cascading* motion is observed, which is similar to the rolling motion except that the top of the particle bed is no longer flat, but instead arches upward within the raised part of the bed. The rolling and cascading motions generate strong mixing, and as a consequence most practical applications of rotating drums operate in the cascading regime.

At very high rotation rates, corresponding to Fr > 0.1, the particles exhibit a *cataracting* regime, in which the friction from the rotating drum is sufficiently large to carry the particles nearly up to the top of the drum. Depending on the rotation rate, they will either fall back downward in a *cataracting* motion onto the lower part of the bed, or they will remain on the drum and spin back down the other side in a *centrifuging* motion. Relatively little mixing occurs in a very rapid centrifuging motion, since the particles circulate with the rotating drum in nearly a solid-body motion.

11.3.2. Mixing and Segregation

Within the cascading regime, the DEM simulations of Yang et al. (2008) showed that the particle velocity magnitude v normalized by the mean wall speed, given by $v^* = v/\Omega_D R$, can be collapsed for different rotation rates onto a single log-normal distribution of the form

$$P(v^*) = a \exp\{-b[\ln(v^*/v_0^*)]^2\}, \tag{11.3.1}$$

where $a = 2.622$, $b = 1.510$, and $v_0^* = 0.227$. This distribution is compared with various DEM computational data in Figure 11.17.

A study of longitudinal and transverse mixing within a rotating drum was reported by Finnie et al. (2005) based on DEM simulations, building on previous work by Ding et al. (2001) and Schutyser et al. (2001, 2002). Because there is no mean motion in the longitudinal direction for a horizontal drum, the diffusion is well approximated by a Fickian diffusion approximation. Consequently, mixing of nearly identical particles in the longitudinal direction is governed by a one-dimensional diffusion equation of the form

$$\frac{\partial \phi}{\partial t} = \frac{\partial}{\partial x}\left[D\frac{\partial \phi}{\partial x}\right], \tag{11.3.2}$$

for the concentration $\phi(x, t)$ of a certain particle type and D is a diffusion coefficient with units m²/s. A dimensionless diffusion coefficient D' can be defined as

$$D' = 2\pi D/\Omega_D R^2. \tag{11.3.3}$$

A plot is given in Figure 11.18a showing estimated values of D' from DEM simulations as a function of Froude number for different filling degree values. The

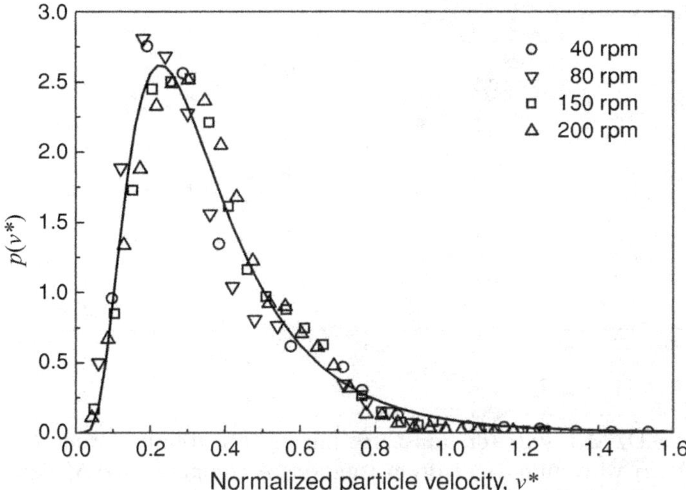

Figure 11.17. Plot showing collapse of the probability distribution for the normalized particle velocity v^* onto a log-normal curve within the cascading regime of rotating drum flow. [Reprinted with permission from Yang et al. (2008).]

simulations indicate that D is relatively insensitive to Froude number and filling angle, having a value between about 1.4–1.6 over a broad range of Fr and J values. A comparison of DEM results and predictions of (11.3.3) is given in Figure 11.18b, which shows excellent agreement for average particle concentration profiles across the longitudinal direction along the drum as a function of time.

While a diffusion-like equation can also be used to quantify transverse mixing, the diffusion coefficient is spatially-variable and dependent on the particle velocities (Heydenrych, 2001). An alternative "entropic" approach was proposed by Schutyser

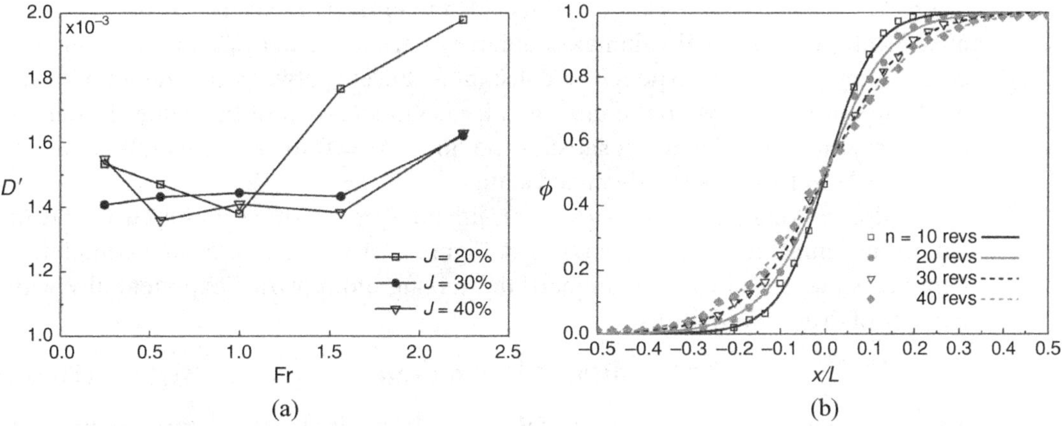

Figure 11.18. Results of DEM simulation for longitudinal mixing in a rotating drum: (a) dimensionless particle diffusion coefficient as a function of Froude number for different filling degree values and (b) concentration profile along longitudinal direction as predicted by DEM simulations (symbols) after n drum revolutions and as given by Equation (11.3.2) (solid lines). [Reprinted with permission from Finnie et al. (2005).]

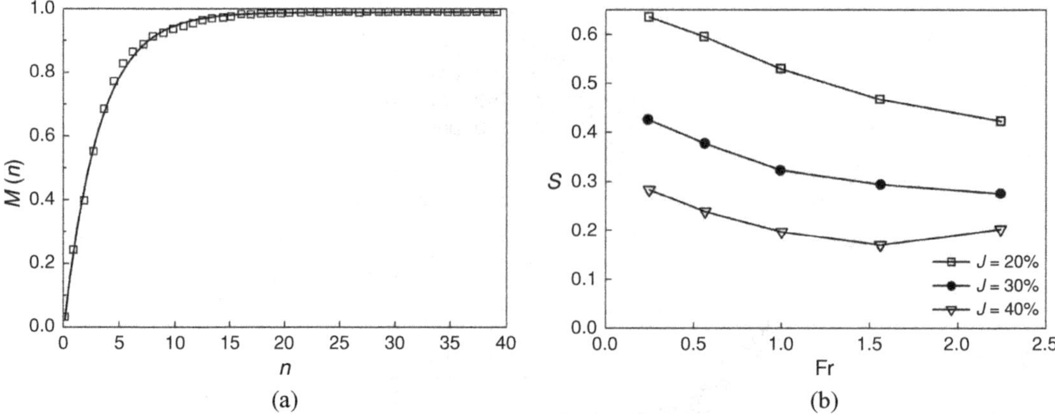

Figure 11.19. Plots illustrating DEM results for transverse mixing in a rotating drum: (a) variation of mixing index $M(n)$ with number of drum rotations n comparing DEM data (symbols) with exponential fit (11.3.6) (line); (b) dependence of computed transverse mixing rate S on filling degree and Froude number. [Reprinted with permission from Finnie et al. (2005).]

et al. (2001, 2002). This approach places a Cartesian grid over the domain of interest, such that for each grid cell (i, j), a quasi-entropy quantity S_{ij} is defined by

$$S_{ij} = -kV_{ij}[\alpha_{1,ij} \ln \alpha_{1,ij} + \alpha_{2,ij} \ln \alpha_{2,ij}]. \tag{11.3.4}$$

Here, V_{ij} is the volume occupied by particles in grid cell (i, j), $\alpha_{1,ij}$ is the fraction of V_{ij} occupied by particles of type 1, $\alpha_{2,ij}(= 1 - \alpha_{1,ij})$ is the fraction of V_{ij} occupied by particles of type 2, and k is a constant. Using this entropy measure, a *mixing index* $M(n)$ is defined as a function of the number n of drum rotations as

$$M(n) = \sum_{i,j} \frac{S_{ij}(n)}{S_\infty}, \tag{11.3.5}$$

where $S_\infty = -kV[\alpha_{1,0} \ln \alpha_{1,0} + \alpha_{2,0} \ln \alpha_{2,0}]$ is the entropy measure for the perfectly mixed state, V is the total volume occupied by particles, and $\alpha_{i,0}$ is the fraction of V occupied by particles of type i. The constant k cancels between the numerator and denominator in (11.3.5), so the mixing index is independent of the value chosen for k. The mixing index definition satisfies the limits $M = 0$ for a completely unmixed state and $M = 1$ for a perfectly mixed state.

A plot showing the variation of $M(n)$ with number of drum rotations using DEM data from Finnie et al. (2005) is given in Figure 11.19a, starting from a completely unmixed state. In all cases examined, the mixing index varies exponentially with number of drum rotations as

$$M(n) \cong 1 - \exp(-Sn), \tag{11.3.6}$$

where the coefficient S is a measure of the rate of mixing (defined in terms of number of drum rotations, rather than time). A plot of S as a function of Froude number for cases with three different values of the filling degree is given in Figure 11.19b. These results indicate that the particles within the rotating drum mix with fewer rotations for smaller values of the Froude number (i.e., lower rotation rate or smaller drum

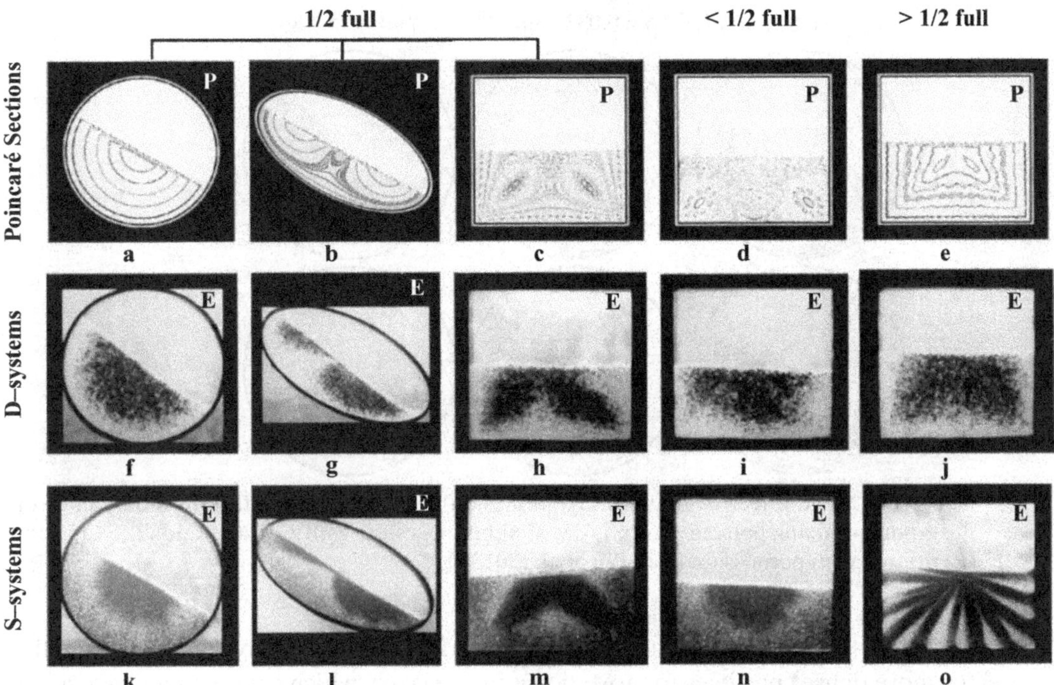

Figure 11.20. Results for two-dimensional rotating tumblers of different shapes, comparing Poincaré sections (top row) with experimental results (middle and bottom rows) for systems containing particles with different size (S-systems) or density (D-systems). [Copyright (1999) National Academy of Sciences. Reprinted with permission from Hill et al. (1999).]

radius) and for smaller filling degree. One exception occurs for the highest filling degree examined, for which S increases with Froude number for the highest value of Fr examined.

The analysis presented here is for mixing of two sets of similar particles, such as might occur with two sets of otherwise identical particles that are painted different colors. When there are differences in physical characteristics of the particles, such as size or density, analysis of mixing in the rotating drum becomes much more complicated because both mixing and segregation can occur at the same time. A number of studies of this segregation phenomenon have been reported in the literature, principally from Ottino, Lueptow, and their students at Northwestern University. An early study of rotating drum segregation phenomena was performed by Hill et al. (1999), who examined radial band formation in two-dimensional tumblers of different shapes, including circles, ellipses, and squares. The systems examined all consist of particles of two different properties, either different sizes (S-systems) or different densities (D-systems), and were operated in the rolling regime. Results comparing Poincaré sections of the granular flow field with experimental results for these systems are shown in Figure 11.20. The mixtures were formed of binary D-systems (with 2 mm glass and steel spheres) and of ternary S-systems (0.8 mm, 1.2 mm, and 2.0 mm glass spheres), shown as cases (f) to (j), and of ternary S-systems (0.8 mm, 1.2 mm, and 2.0 mm glass spheres), shown as cases (k) to (o). In particular, Figures 11.20f and 11.20k illustrate the classic radially segregated structure in circular

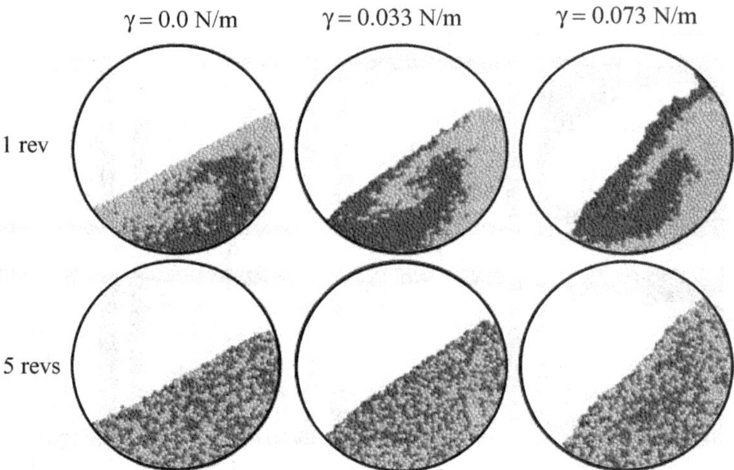

Figure 11.21 The effect of liquid-bridge adhesion with different surface tension values on the granular mixing patterns, from DEM simulation results with $J = 40\%$ and $\Omega_D = 15$ rpm. [Reprinted with permission from Liu et al. (2013).]

mixers. In these two cases, the dynamics in the flowing surface layer cause the smaller (or more dense) particles to move down in the layer, which causes them have a low probability of reaching the end of the layer, leading to a segregated core. A circular core is time invariant, because it coincides with the streamlines. The segregation structures obtained in noncircular mixers are much complicated. For instance, this is quite evident in the case of a half-filled square mixer, shown in Figures 11.20h and m. A region with a high concentration of small (dense) particles is located away from the center of rotation and is nearly separated into two regions spanning the corners of the square and the center of rotation. The corresponding segregation structure for the ellipse also shows that the smaller (denser) particles are pulled away from the center of rotation. The final case shown (Case o) never settled down to an equilibrium flow but instead persisted as a changing pattern of streaks. This study was extended to three-dimensional systems by Gilchrist and Ottino (2003) and to systems with combined size and density differences by Jain et al. (2005). Fiedor and Ottino (2005) examined the effect of unsteady rotation of a circular drum, in which the rotation rate has a nonzero mean value plus a time-periodic part. An extensive study of the relationship between mixing and segregation processes in rotating drums and aspects of dynamical systems theory was given by Meier et al. (2007).

11.3.3. Cohesive Mixing and Segregation

The study of the mixing and segregation of cohesive particles is of considerable interest for numerous industrial and agricultural processes. By incorporating liquid bridging force into the Hertz contact force model, Liu et al. (2013) conducted DEM simulations of the mixing of cohesive particles in rotating drums. Sample results are shown in Figure 11.21 for a case with $J = 40\%$, $\Omega_D = 15$ rpm, and $d = 1.3$ mm. The presence of capillary force caused by a small amount of moisture mixed with the particles is observed to significantly inhibit mixing between two sets of otherwise identical colored particles. Chaudhuri et al. (2006) had earlier used both DEM

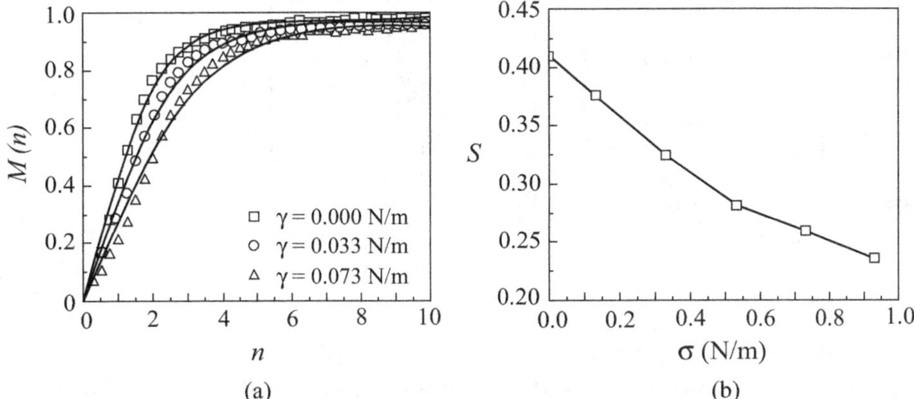

Figure 11.22. Plots illustrating DEM results for cohesion on mixing in a rotating drum: (a) effect of cohesion on mixing index $M(n)$; (b) dependence of mixing rate S on surface tension of liquid bridges. [Reprinted with permission from Liu et al. (2013).]

simulations and experiments to demonstrate that small capillary forces enhance mixing whereas high values of the capillary force inhibit mixing. Liu et al. (2013) interpreted the DEM results in terms of the mixing index $M(n)$ and mixing rate S defined in (11.3.4) and (11.3.5), showing the variation of $M(n)$ and S as a function of the liquid-gas surface tension (Figure 11.22). As the value of surface tension increases from 0 to 0.073 N/m (the value for an air-water interface), the mixing index continues to be well fit by the exponential function shown in (11.3.6), but the mixing rate S decreases from 0.41 to 0.24 as surface tension increases.

For systems of particles of different sizes (S-systems) or different densities (D-systems), the processes of segregation and mixing occur simultaneously. Specifically, capillary forces between particles of a similar type (cohesion) tend to enhance segregation of particles, whereas capillary forces between particles of different types (adhesion) tend to inhibit (or mitigate) segregation. Li and McCarthy (2003) developed a phase diagram of the particle-scale mixing/segregation in the asymptotic (long-time) state by determining which conditions favor adhesive versus cohesive behavior for binary particle mixtures. They restricted their study to rotation rates that are sufficiently small that the shearing force is negligible compared to the capillary force $\mathbf{F}_{cap}$ and the gravitational force $\mathbf{F}_g$ acting on the particles. The ratio of the magnitude of capillary and gravitational forces yields the *granular Bond number*, Bo_g, which using (4.5.14) can be written as

$$\mathrm{Bo}_g = \frac{F_{cap}}{F_g} = \frac{3\sigma R \cos\theta}{g r_{p,i}^3 \rho_p} G_f, \qquad (11.3.7)$$

where R is the effective particle radius, $r_{p,i}$ is the radius of particles of type i (where $i = 1$ or 2), and the coefficient G_f, defined in (4.5.15), approaches unity as two particles approach each other. The values of Bo_g for each potential particle pair in a binary mixture are denoted by $\mathrm{Bo}_{g,1-1}$, $\mathrm{Bo}_{g,1-2}$, and $\mathrm{Bo}_{g,2-2}$, where particle type 2 is defined as the larger particles. The phase boundaries are identified by finding where in the parameter space of the particle size ratio $r_{p,1}/r_{p,2}$, density ratio $\rho_{p,1}/\rho_{p,2}$, and wetting angle ratio $\cos\theta_1/\cos\theta_2$ different regimes of mixing/segregation are observed. When

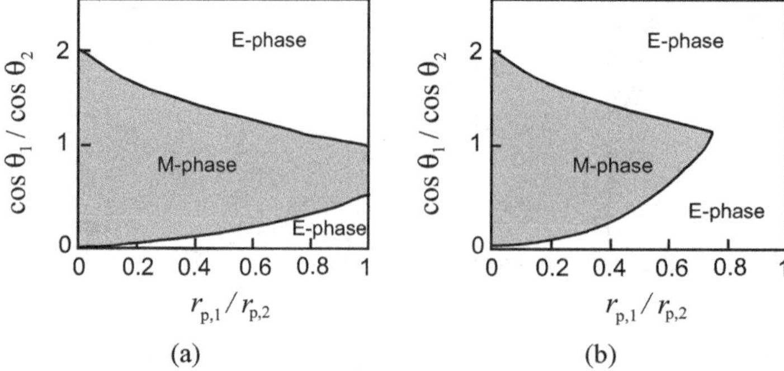

Figure 11.23 Phase diagram of cohesive mixing/segregation in a binary system with density ratio (a) $\rho_{p,1}/\rho_{p,2} = 0.56$ and (b) the inverse case with $\rho_{p,1}/\rho_{p,2} = 1.78$. Based on data from Li and McCarthy (2003).

the granular material is adhesion-dominated, such that $\mathrm{Bo}_{g,1-2} > \{\mathrm{Bo}_{g,1-1}, \mathrm{Bo}_{g,2-2}\}$, it is expected that a small amount of segregation will occur because capillary bonds joining particles of different types dominate other forces. When the material is cohesion-dominated, such that $\mathrm{Bo}_{g,1-2} < \mathrm{Bo}_{g,1-1}$ or $\mathrm{Bo}_{g,1-2} < \mathrm{Bo}_{g,2-2}$, capillary forces between particles within a given type will dominate, acting to enhance segregation of particle types compared to the dry granular case.

Figure 11.23 shows regime diagrams indicating regions of mitigated (M-phase) and enhanced (E-phase) segregation of particles in a rotating drum for a case with density ratio $\rho_{p,1}/\rho_{p,2} = 0.56$ (Figure 11.23a) and the inverse case with density ratio $\rho_{p,1}/\rho_{p,2} = 1.78$ (Figure 11.23b), based on the data of Li and McCarthy (2003). The white region corresponds to the E-phase and the gray region to the M-phase. The size of the M-phase region grows when the smaller particle is less dense than the larger particle and shrinks when it is more dense. These regime diagrams indicate only whether addition of moisture will enhance or mitigate the degree of segregation relative to the dry state; they do not indicate anything about the absolute amounts of mixing/segregation.

11.4. Dust Removal Processes

Removal of dust from surfaces is an essential part of numerous cleaning processes. It is of particular importance in manufacture of electronic components and in exploration of dusty planetary bodies, such as the Moon and Mars. There are many approaches that can be used for particle removal, several of which require contact with the surface, such as brushing dust off the surface or removal from capillary force by a liquid film. In this section, we examine two noncontact methods for particle removal from a surface, which illustrate several of the interaction phenomena discussed in this book.

11.4.1. Hydrodynamic Dust Mitigation

A traditional method for particle removal from a surface is to aim a jet flow at the surface, thereby blowing the particles along the surface and entraining the particles into

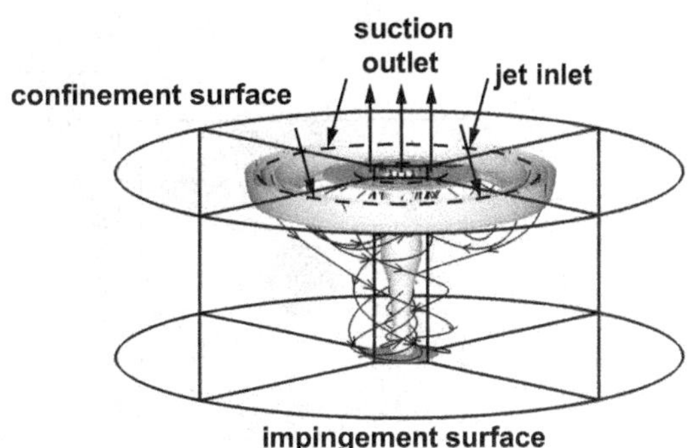

Figure 11.24. Plot of a bounded vortex flow, showing the annular jet inlet at the top and the wall-normal vortex at the center. The shaded surface is an iso-vorticity magnitude surface and the lines are streamlines originating from the jet inlet. [Reprinted with permission from Maynard and Marshall (2011).]

the jet flow. Fluid dynamics issues associated with jet impingement onto a flat plate surface have been examined by a number of investigators, including two-dimensional slot jets (Chiriac and Ortega, 2002; Sakakibara et al., 2001), axisymmetric jets (Chatterjee, 2008), and annular slot jets (Chattopadhyay, 2004). Particle transport due to impingement of a circular jet on a surface was examined experimentally by Masuda et al. (1994), Anderson and Longmire (1995), and Zhang et al. (2002) for constant jet flow rates and by Otani et al. (1995) and Ziskind et al. (2002) for pulsed jets. Both pulsed and steady impinging jet flows generate ring-like coherent vortices that surround the jet and increase shear stress magnitude on the impingement plate. Impinging jet methods are highly effective at removing particles, but they tend to scatter the particles on the surface.

A combination of a swirling impinging jet flow within an annular slot and a central suction flow was used by Maynard and Marshall (2011) and Huang and Marshall (2011) as an effective approach for hydrodynamic particle removal from a surface. This combination of a swirling jet and a suction flow forms a wall-normal vortex structure immediately below the suction intake that is bounded by the axisymmetric slot jet, forming what is called a *bounded vortex flow*, as illustrated in the flow field computation shown in Figure 11.24. The bounded vortex flow is generated by a nozzle with both inlet and outlet ports, which is held at a distance H above the impingement surface and oriented normal to this surface. The injection jets on the nozzle are located in a circular pattern at a radial distance R from the nozzle center, with inlet flow oriented both downward toward the impingement surface and tangentially around the impingement point. The outlet suction port is located at the center of the nozzle, which is connected to a vacuum line. The bounded vortex flow is characterized by two key dimensionless parameters – the ratio h/R of separation distance to nozzle injection radius and the ratio $\eta = Q_{out}/Q_{in}$ of the outlet suction flow rate to the total of the inlet jet flow rates. Depending on the value of η, the jet may be entrained into the suction port with little influence on the near-surface flow or it may spread radially outward scattering particles along the impingement surface. However, the flow is most effective when the flow rate ratio is set approximately equal to unity, for which the shear stress streamlines on the impingement surface swirl inward just underneath the nozzle, carrying the particles in toward the suction outlet. The wall-normal vortex acts to direct fluid streamlines originating from the jet inlet downward into the boundary layer along the impingement plate and then upward along the vortex core into the suction outlet. The wall-normal vortex flow

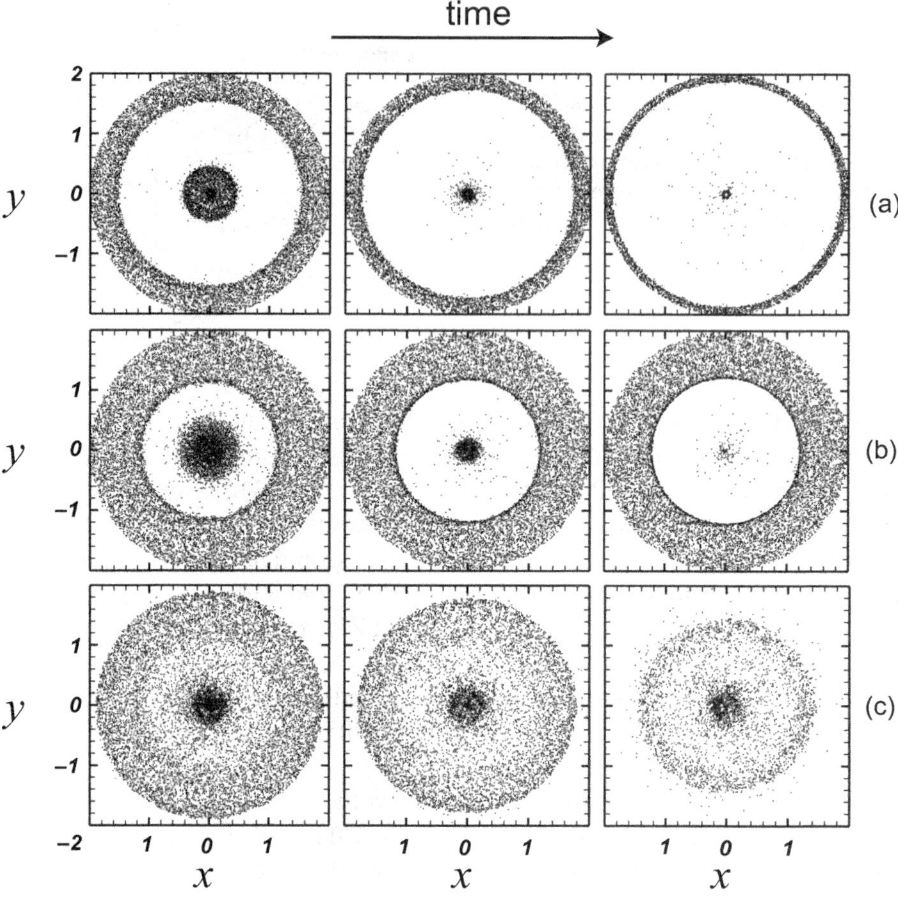

Figure 11.25. Time series (from left to right) showing DEM results for particle locations on the impingement surface for a bounded vortex with flow rate ratio (a) $\eta = Q_{out}/Q_{in} = 0.5$; (b) $\eta = 1$; and (c) $\eta = 2$. [Reprinted with permission from Maynard and Marshall (2011).]

generates high shear stress levels along the wall (Phillips, 1985; Phillips and Khoo, 1987; Parras and Fernandez-Feria, 2007), which aids in removal of the particles from the surface.

A series of numerical simulations of particle removal underneath a bounded vortex flow were reported by Maynard and Marshall (2011), who used a soft-sphere DEM approach with near-wall corrections to the fluid-induced forces on the particles similar to those discussed in Section 5.6. The particles are initially spread evenly on the surface, and they are followed in time by the simulation until an equilibrium state is reached. Results for particle positions are shown in Figure 11.25 at three different values of the flow rate ratio η, with three times shown for each value of η. For jet-dominated flows with $\eta = 0.5$, a particle front forms that gradually moves outward and an inner island of particles is entrained into the suction outlet. For the suction-dominated flows with $\eta = 2$, particles from across the impingement plane are drawn inward toward the suction outlet, with the results that no cleaned regions on the impingement surface are formed. In the case of balanced inflow and outflow ($\eta = 1$), regions of inward- and outward-flowing streamlines near the impingement surface are separated by the circular separatrix, so that a circle with constant radius is formed within which particles are removed and outside which the particles remain on the surface. Extensions of the bounded vortex method for particle removal have been reported by Vachon and Hitt (2010, 2011) for turbulent vortex flows and by Furhmann et al. (2013), who showed that addition of acoustic disturbances enhances efficiency of the process by helping to break the particle bonds with the impingement surface.

11.4.2. Electric Curtain Mitigation for Charged Particles

When the particles on the surface are charged, many standard methods for particle removal do not work well. An exception is the electric curtain device discussed in Section 1.3.2 and shown in Figure 1.10. The electric curtain is formed of a series of parallel electrodes embedded into a dielectric material, which are coated by a thin dielectric film. The electrodes have oscillating potential, creating an n-phase wave pattern in the electric field that propagates across the dielectric surface. Early work on electric curtains was reported by Masuda et al. (1972) for confinement of an aerosol cloud in a ring-type electric curtain device. Electric curtain technology was later used as a charged toner conveyer device in xerographic copy machines (Melcher et al., 1989; Schmidlin, 1991). The electric curtain technology is of particular interest for designs for handling dust adhesion during extraterrestrial exploration of dusty planets, as proposed by Calle et al. (2008, 2009) and Atten et al. (2009), as dust particles in places such as the Moon and Mars carry significant electrical charge. Other proposed applications of electric curtains include liquid droplet transport (Kawamoto and Hayashi, 2006), particle size classification (Kawamoto and Hasegawa, 2004), and particle separation and sorting (Weiss and Thibodeaux, 1984; Masuda et al., 1987; Machowski and Balachandran, 1997; Kawamoto, 2008).

Schmidlin (1995) described three different modes of motion for individual particles on the electric curtain, referred to as surfing mode, hopping mode, and curtain mode. These different modes are illustrated in Figure 11.26, along with plots of time variation of computed particle horizontal velocity and particle height for each mode. In the *surfing mode*, particles roll or slide on the dielectric surface at an average velocity equal to the wave speed. Consequently, a given particle always experiences the same phase of the electric curtain traveling wave. In the *hopping mode*, particles are transported by successive hops, and between each hop particles remain fixed to the dielectric surface. Particle hop distance can vary from hop to hop, and in some cases particles perform several small hops, or even a backwards hop, before resuming large forward hops. Average forward speed of a particle in hopping mode motion can vary from the wave speed to much less than the wave speed. In the *curtain mode*, particles are suspended above the dielectric surface and travel in a spiraling path with a mean drift velocity that is much smaller than the electric curtain wave speed.

The motion of an individual particle on an electric curtain is more complex and irregular than represented by these three modes of motion due to the effects of collisions and electrostatic interactions between particles, as well as the effects of adhesion between the particles and the dielectric surface. The amount of data that can be obtained from experimental studies of multiparticle systems is limited because it is difficult to determine the charge of each particle. Masuda and Kamimura (1975)

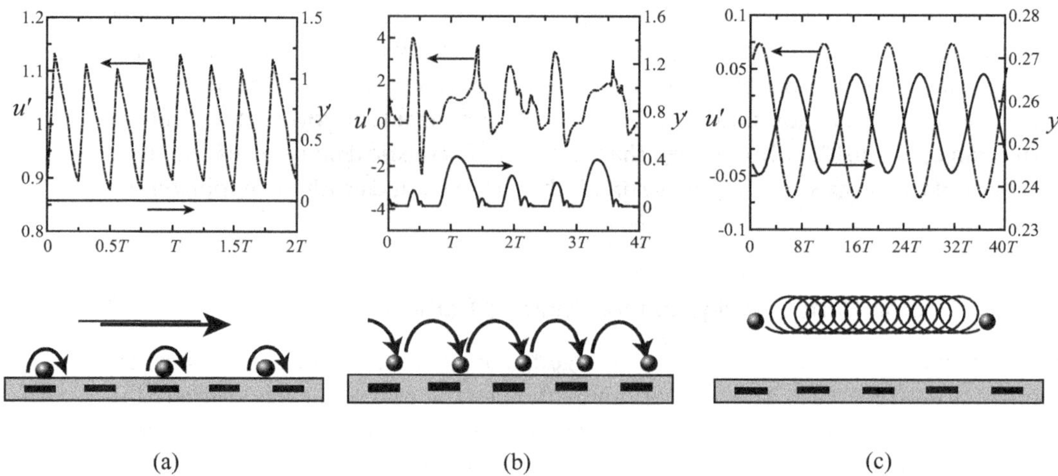

(a)   (b)   (c)

Figure 11.26. Schematic diagram of modes of particle motion on an electric curtain, showing (a) particles rolling or sliding along the curtain in surfing mode; (b) particles intermittently jumping along the curtain in hopping mode; and (c) particles spiraling above the electric curtain in curtain mode. Plots show computed time variation of tangential velocity component (left) and height (right) of particles in each mode. [Reprinted with permission from Liu and Marshall (2010a).]

examined different approximate methods for modeling the electric field induced by a traveling electric wave above the electric curtain, which are often used in computational studies. Computations of nonadhesive particle motion on an electric curtain were reported by Gartstein and Shaw (1999) and Kawamoto et al. (2006) using the hard-sphere discrete-element method.

Computations using a soft-sphere discrete-element method with adhesive particles were performed by Liu and Marshall (2010a,b) for both traveling waves and standing waves on the electric curtain. The computations compared particle transport for traveling waves both with and without particle-particle collisions. Results are shown in Figure 11.27 for bins representing the fraction of particles that propagate forward, propagate backward, and have essentially no average forward motion. At low driving frequencies the particles primarily propagate in the forward direction, whereas at high frequencies particles primarily propagate in the backward direction or have essentially no forward velocity. Inclusion of particle-particle collisions and electrostatic interactions increases the prevalence of the majority motion type, so that when particle-particle interactions are included at low frequencies a higher fraction of particles moves forward and at high frequencies a higher fraction of particles moves backward. The fraction of nontransported particles is always decreased by inclusion of particle-particle interactions except within a band of intermediate frequencies, where collision of essentially equal numbers of forward- and backward-moving particles increases the fraction of particles with essential zero average velocity.

11.5. Final Comments

The discrete element method has been used for a wide variety of problems involving both adhesive and nonadhesive particles. The applications described in this

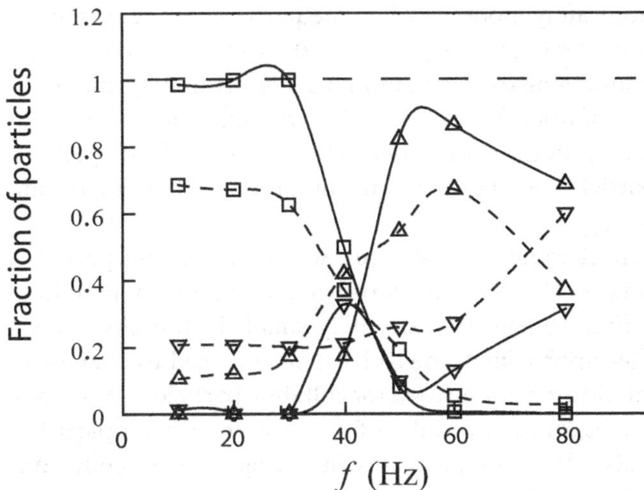

Figure 11.27. Variation of fractions of forward-transported (square symbols), backward-transported (triangle symbols), and nontransported (delta symbols) particles versus frequency for particle transport on an electric curtain, both with particle-particle interaction (solid lines) and without particle-particle interaction (dashed lines). [Reprinted with permission from Liu and Marshall (2010a).]

chapter provide just a few examples of problems in which DEM has been employed to provide insight into mesoscale particle transport. DEM allows investigators to perform studies with sufficient numbers of particles to study particle collective behavior, influenced by collision and adhesion of particles with each other, while still modeling the particle interactions at a fundamental level. The various models used in DEM for particle interaction and for forces induced by fluid, electric, and acoustic fields have developed over the past several decades to greatly extend the range of problems for which this method can be reliably employed.

At the same time, there are certain aspects of DEM that still require significant development. Primary among these is the problem of accurately modeling hydrodynamic interaction between particles without use of excessive computational resources. Computational methods used for microscale simulation methods, which compute the flow field around each particle, require too much time to allow for simulations with large numbers of particles while still accurately resolving the particle interaction. The Stokesian dynamics method is one approach for handling this problem, but this method only works for problems with small values of the flow Reynolds number.

For turbulent flows, stochastic Lagrangian methods can be used to accurately model time variation of particle fluctuations in the turbulence. These methods also yield good predictions for large-scale particle dispersion in the turbulence. Further work must be performed to improve modeling of the turbulent flows to yield improved predictions of particle collision and adhesion rate. Stochastic Lagrangian models are not accurate for this purpose because they lack spatial correlation in the random forcing term.

Most DEM simulations have been performed for spherical particles. However, rolling motion of the particles is highly sensitive to variation in particle shape, and such processes might not be accurately modeled when the particles are represented by spheres. Models have been presented for ellipsoidal particles that can be used to partially surmount issues with modeling of rolling motions. Unfortunately, methods for computation of the collision of nonspherical particles are much more time consuming than those for spherical particles. Faster methods need to be developed for handling arbitrarily shaped particles, as these are encountered in many important applications.

A challenge with the discrete element method occurs when transporting particles with very small Stokes numbers. In this case the aerodynamic time step limitation (10.1.6) requires the particle time step to be extremely small. In the absence of particle collisions, the fast Euler approximation (10.1.10) can be used to overcome this problem, but this approximation cannot be used for colliding particles. This time-stepping limitation particularly becomes a problem for transport of nanoparticles in liquids. Time scale issues also arise for problems involving particle collective phenomena that take a long time to develop, such as thrombosis development in blood flows. Typical time scales covered by a DEM simulation with adhesive particles are on the order of several seconds. Bridging between DEM simulations with results of both larger-scale models (such as the population-balance model and the discrete parcel model) and smaller-scale models (including different types of microscale simulation models) remains of central importance for many problems.

REFERENCES

Anderson SL, Longmire EK. Particle motion in the stagnation zone of an impinging air jet. *Journal of Fluid Mechanics* **299**, 333–366 (1995).

Asmolov ES. The inertial lift on a spherical particle in a plane Poiseuille flow at large channel Reynolds number. *Journal of Fluid Mechanics* **381**, 63–87 (1999).

Atten P, Pang HL, Reboud JL. Study of dust removal by standing wave electric curtain for application to solar cells on Mars. *IEEE Transactions on Industry Applications* **45**, 75–86 (2009).

Bretherton FP. The motion of rigid particles in a shear flow at low Reynolds number. *Journal of Fluid Mechanics* **14**, 284–304 (1962).

Calle CI, Mazumder MK, Immer CD, Buhler CR, Clements JS, Lundeen P, Chen A, Mantovani JG. Controlled particle removal from surfaces by electrodynamic methods for terrestrial, lunar and Martian environmental conditions. *Journal of Physics: Conference Series* **142**, 012073 (2008).

Calle CI, Buhler CR, McFall JL, Snyder SJ. Particle removal by electrostatic and dielectrophoretic forces for dust control during lunar exploration missions. *Journal of Electrostatics* **67**, 89–92 (2009).

Chatterjee AJ. Multiple vortex formation in steady laminar axisymmetric impinging flow. *Computer and Fluids* **37**, 1061–1076 (2008).

Chattopadhyay H. Numerical investigations of heat transfer from impinging annular jet. *International Journal of Heat and Mass Transfer* **47**, 3197–3201 (2004).

Chaudhuri B, Mehrotra A, Muzzio FJ, Tomassone MS. Cohesive effects in powder mixing in a tumbling blender. *Powder Technology* **165**, 105–114 (2006).

Chiriac VA, Ortega A. A numerical study of the unsteady flow and heat transfer in a transitional confined slot jet impinging on an isothermal surface. *International Journal of Heat and Mass Transfer* **45**, 1237–1248 (2002).

Ding YL, Forster RN, Seville JPK, Parker DJ. Scaling relationships for rotating drums. *Chemical Engineering Science* **56**, 3737–3750 (2001).

Fiedor SJ, Ottino JM. Mixing and segregation of granular matter: Multi-lobe formation in time-periodic flows. *Journal of Fluid Mechanics* **533**, 223–236 (533).

Finnie GJ, Kruyt NP, Ye M, Zeilstra C, Kuipers JAM. Longitudinal and transverse mixing in rotary kilns: a discrete element method approach. *Chemical Engineering Science* **60**, 4083–4091 (2005).

Frank M, Anderson D, Weeks ER, Morris JF. Particle migration in pressure-driven flow of a Brownian suspension. *Journal of Fluid Mechanics* **493**, 363–378 (2003).

Furhmann A, Marshall JS, Wu J-R. Effect of acoustic levitation force on aerodynamic particle removal from a surface. *Applied Acoustics* **74**, 535–543 (2013).

Gartstein YN, Shaw JG. Many-particle effects in travelling electrostatic wave transport, *Journal of Physics D: Applied Physics* **32**, 2176–2180 (1999).

Gilchrist JF, Ottino JM. Competition between chaos and order: Mixing and segregation in a spherical tumbler. *Physical Review E* **68**, 061303 (2003).

Hampton RE, Mammoli AA, Graham AL, Tetlow N, Altobelli SA. Migration of particles undergoing pressure-driven flow in a circular conduit. *Journal of Rheology* **41**(3), 621–640 (1997).

Han M, Kim C, Kim M, Lee S. Particle migration in tube flow of suspensions. *Journal of Rheology* **43**, 1157–1174 (1999).

Henein H, Brimacombe JK, Watkinson AP. Experimental study of transverse bed motion in rotary kilns. *Metallurgical Transactions B* **14B**, 191–205 (1983).

Heydenrych MD. Modelling of rotary kilns. Ph.D. Dissertation, Department of Chemical Engineering, University of Twente, The Netherlands (2001).

Hewitt GF, Marshall JS. Particle focusing in suspension flow through a corrugated tube. *Journal of Fluid Mechanics* **660**, 258–281 (2010).

Hill KM, Khakhar DV, Gilchrist JF, McCarthy JJ, Ottino JM. Segregation-driven organization in chaotic granular flows. *Proceedings of the National Academy of Sciences* **96**(21), 11701–11706 (1999).

Hiller M, Löffler M. Influence of particle impact and adhesion on the collection efficiency of fiber filters. *German Chemical Engineering* **3**, 327–332 (1980).

Ho BP, Leal LG. Inertial migration of rigid spheres in two-dimensional unidirectional flows. *Journal of Fluid Mechanics* **65**, 365–400 (1974).

Hogg AJ. The inertial migration of non-neutrally buoyant spherical particles in two-dimensional shear flows. *Journal of Fluid Mechanics* **272**, 285–318 (1994).

Huang B, Yao Q, Li SQ, Zhao HL, Song Q, You CF. Experimental investigation on the particle capture by a single fiber using microscopic image technique. *Powder Technology* **163**, 125–133 (2006).

Huang Y, Marshall JS. Experiments on bounded vortex flows and related particle transport. *Journal of Fluids Engineering* **133**(7), 071204 (2011).

Jain N, Ottino JM, Lueptow RM. Regimes of segregation and mixing in combined size and density granular systems: an experimental study. *Granular Matter* **7**, 69–81 (2005).

Kanaoka C, Emi H, Tanthapanichakoon W. Convective diffusional deposition and collection efficiency of aerosol on a dust-loaded fiber. *AIChE Journal*, **29**(6), 895–902 (1983).

Kawamoto H. Some techniques on electrostatic separation of particle size utilizing electrostatic traveling-wave field. *Journal of Electrostatics* **66**, 220–228 (2008).

Kawamoto H, Hasegawa N. Traveling wave transport of particles and particle size classification. *Journal of Imaging Science and Technology* **48**, 404–411 (2004).

Kawamoto H, Hayashi S. Fundamental investigation on electrostatic travelling-wave transport of a liquid drop. *Journal of Physics D: Applied Physics* **39**, 418–423 (2006).

Kawamoto H, Seki K, Kuromiya N. Mechanism of traveling-wave transport of particles. *Journal of Physics D: Applied Physics* **39**, 1249–1256 (2006).

Kim JM, Lee SG, Kim C. Numerical simulations of particle migration in suspension flows: Frame-invariant formulation of curvature-induced migration. *Journal of Non-Newtonian Fluid Mechanics* **150**, 162–176 (2008).

Koh CJ, Hookham P, Leal LG. An experimental investigation of concentrated suspension flows in a rectangular channel. *Journal of Fluid Mechanics* **256**, 1–32 (1994).

Konstandopoulos AG. Deposit growth dynamics: Particle sticking and scattering phenomena. *Powder Technology* **109**, 262–277 (2000).

Krieger IM, Rheology of monodisperse lattices. *Advances in Colloid and Interface Science* **3**(2), 111–136 (1972).

Kupka T, Zajac K, Weber R. Effect of fuel type and deposition surface temperature on the growth and structure of an ash deposit collected during co-firing of coal with sewage sludge and sawdust. *Energy & Fuels* **23**, 3429–3436 (2009).

Leighton D, Acrivos A. The shear-induced migration of particles in concentrated suspensions. *Journal of Fluid Mechanics* **181**, 415–439 (1987).

Li HM, McCarthy JJ. Controlling cohesive particle mixing and segregation. *Physical Review Letters* **90**(18), 184301 (2003).

Li SQ, Marshall JS. Discrete element simulation of micro-particle deposition on a cylindrical fiber in an array. *Journal of Aerosol Science* **38**, 1031–1046 (2007).

Li SQ, Marshall JS, Liu G, Yao Q. Adhesive particulate flow: The discrete element method and its application in energy and environmental engineering. *Progress in Energy and Combustion Science* **37**(6), 633–668 (2011).

Liu G, Marshall JS. Effect of particle adhesion and interactions on motion by traveling waves on an electric curtain. *Journal of Electrostatics* **68**, 179–189 (2010a).

Liu G, Marshall JS. Particle transport by standing waves on an electric curtain. *Journal of Electrostatics* **68**, 289–298 (2010b).

Liu G, Marshall JS, Li SQ, Yao Q. Discrete-element method for particle capture by a body in an electrostatic field. *International Journal for Numerical Methods in Engineering* **84**(13), 1589–1612 (2010).

Liu PY, Yang RY, Yu AB. DEM study of the transverse mixing of wet particles in rotating drums. *Chemical Engineering Science* **86**, 99–107 (2013).

Machowski W, Balachandran W. Electrodynamic control and separation of charged particles using traveling wave field technique. *Journal of Electrostatics* **40**–41, 325–330 (1997).

Masuda S, Kamimura T. Approximate methods for calculating a non-uniform traveling field. *Journal of Electrostatics* **1**, 351–370 (1975).

Masuda S, Fujibayashi K, Ishida K, Inaba H. Confinement and transportation of charged aerosol clouds via electric curtain. *Electrical Engineering in Japan* **92**, 43–52 (1972).

Masuda S, Washizu M, Iwadare M. Separation of small particles suspended in liquid by non-uniform traveling field. *IEEE Transactions on Industry Applications* **IA-23**, 478–480 (1987).

Masuda H, Gotoh K, Fukada H, Banba Y. The removal of particles from flat surfaces using a high-speed air jet. *Advanced Powder Technology* **5**, 205–217 (1994).

Matas JP, Morris JF, Guazzelli E. Inertial migration of rigid spherical particles in Poiseuille flow. *Journal of Fluid Mechanics* **515**, 171–195 (2004).

Maynard AB, Marshall JS. Particle removal from a surface by a bounded vortex flow. *International Journal of Heat and Fluid Flow* **32**(5), 901–914 (2011).

Meier SW, Lueptow RM, Ottino JM. A dynamical systems approach to mixing and segregation of granular materials in tumblers. *Advances in Physics* **56**(5), 757–827 (2007).

Melcher JR, Warren EP, Kotwal RH. Travelling wave delivery of single component developer. *IEEE Transactions on Industry Applications* **25**, 956–961 (1989).

Mellmann J. The transverse motion of solids in rotating cylinders: Forms of motion and transition behavior. *Powder Technology* **118**, 251–270 (2001).

Nott PB, Brady JF. Pressure-driven flow of suspensions: Simulation and theory. *Journal of Fluid Mechanics* **275**, 157–199 (1994).

Oak MJ, Saville DA, Lamb GER. Particle capture on fibers in strong electric fields. I. Experiment studies of the effects of fiber charge, fiber configuration and dendrite structure. *Journal of Colloid and Interface Science* **106**, 490–501 (1985).

Otani Y, Namiki N, Emi H. Removal of fine particles from smooth flat surfaces by consecutive pulse air jets. *Aerosol Science and Technology* **23**, 665–673 (1995).

Parras L, Fernandez-Feria R. Interaction of an unconfined vortex with a solid surface. *Physics of Fluids* **19**, 067104 (2007).

Perry JL, Kandlikar SG. Fouling and its mitigation in silicon microchannels used for IC chip cooling. *Microfluidics and Nanofluidics* **5**, 357–371 (2008).

Phillips RJ, Armstrong RC, Brown RA, Graham AL, Abbott JR. A constitutive equation for concentrated suspensions that accounts for shear-induced particle migration. *Physics of Fluids A* **4**(1), 30–40 (1992).

Phillips WRC. On vortex boundary layers. *Proceedings of the Royal Society A* **400**, 253–261 (1985).

Phillips WRC, Khoo BC. The boundary layer beneath a Rankine-like vortex. *Proceedings of the Royal Society A* **411**, 177–192 (1987).

Ptak T, Jaroszcyk T. Theoretical-experimental aerosol filtration model for fibrous filters at intermediate Reynolds numbers. In *Proceedings of the 5th World Filtration Congress*, Nice (1990).

Ramarao BV, Tien C, Mohan S. Calculation of single-fiber efficiencies for interception and impaction with superposed Brownian motion. *Journal of Aerosol Science* **25**(2), 295–313 (1994).

Rembor HJ, Maus R, Umhauer H. Measurements of single fiber efficiencies at critical values of the Stokes number. *Particle and Particle Systems Characterization* **16**, 54–59 (1999).

Rubinow SI, Keller JB. The transverse force on a spinning sphere moving in a viscous fluid. *Journal of Fluid Mechanics* **11**, 447–459 (1961).

Saffman PG. The lift on a small sphere in a slow shear flow. *Journal of Fluid Mechanics* **22**, 385–400 (1965).

Sakakibara J, Hishida K, Phillips WRC. On the vortical structure in a plane impinging jet. *Journal of Fluid Mechanics* **434**, 273–300 (2001).

Schmidlin FW. A new nonlevitated mode of traveling wave toner transport. *IEEE Transactions on Industrial Applications* **27**, 480–487 (1991).

Schmidlin FW. Modes of traveling wave particle transport and their applications. *Journal of Electrostatics* **34**, 225–244 (1995).

Schonberg JA, Hinch EJ. Inertial migration of a sphere in Poiseuille flow. *Journal of Fluid Mechanics* **203**, 517–524 (1989).

Schutyser MAI, Padding JT, Weber FJ, Briels WJ, Rinzema A, Boom R. Discrete particle simulations predicting mixing behaviour of solid substrate in a rotating drum fermenter. *Biotechnology and Bioengineering* **75**, 666–675 (2001).

Schutyser MAI, Weber FJ, Briels WJ, Boom RM, Rinzema A. Three-dimensional simulation of grain mixing in three different rotating drum designs for solid-state fermentation. *Biotechnology and Bioengineering* **79**, 284–294 (2002).

Segré G, Silberberg A. Behaviour of macroscopic rigid spheres in Poiseuille flow. Part 2. Experimental results and interpretation. *Journal of Fluid Mechanics* **14**, 136–157 (1962).

Vachon N, Hitt DL. A bound-vortex surface impingement method for adhered dust particle removal. *Proc. 40th AIAA Fluid Dynamics Conference*, AIAA paper 2010–4297 (2010).

Vachon N, Hitt DL. An experimental investigation of a bound-vortex surface impingement method for dust removal. *Proc. 41st AIAA Fluid Dynamics Conference*, AIAA paper 2011–3893 (2011).

Wang CS. Electrostatic forces in fibrous filters: A review. *Powder Technology* **118**(1–2), 166–170 (2001).

Weiss LC, Thibodeaux DP. Separation of seed by-products by an AC electric field. *Journal of the American Oil Chemists' Society* **61**(5), 886–890 (1984).

Xu XG, Li SQ, Li GD, Yao Q. Effect of co-firing straw with two coals on the ash deposition behavior in a down-fired pulverized coal combustor. *Energy & Fuels* **24**, 241–249 (2010).

Yang M, Li SQ, Yao Q. Mechanistic studies of initial deposition of fine adhesive particles on a fiber using discrete-element methods. *Powder Technology* (in press, 2013).

Yang RY, Yu AB, McElroy L, Bao J. Numerical simulation of particle dynamics in different flow regimes in a rotating drum. *Powder Technology* **188**, 170–177 (2008).

Zhang XW, Yao ZH, Hao PF, Xu HQ. Study on particle removal efficiency of an impinging jet by an image-processing method. *Experiments in Fluids* **32**(3), 376–380 (2002).

Ziskind G, Yarin LP, Peles S, Gutfinger C. Experimental investigation of particle removal from surfaces by pulsed air jets. *Aerosol Science and Technology* **36**, 652–659 (2002).

Index